INTRODUCTION TO THE THEORY OF

ALGEBRAIC NUMBERS AND FUNCTIONS

PURE AND APPLIED MATHEMATICS

A Series of Monographs and Textbooks

Edited by

PAUL A. SMITH and SAMUEL EILENBERG

Columbia University, New York

1: ARNOLD SOMMERFELD. Partial Differential Equations in Physics. 1949 (Lectures on Theoretical Physics, Volume VI)

2: REINHOLD BAER. Linear Algebra and Projective Geometry. 1952

3: HERBERT BUSEMANN AND PAUL KELLY. Projective Geometry and Projective Metrics. 1953

4: STEFAN BERGMAN AND M. SCHIFFER. Kernel Functions and Elliptic Differential Equations in Mathematical Physics. 1953

5: RALPH PHILIP BOAS, JR. Entire Functions. 1954

6: HERBERT BUSEMANN. The Geometry of Geodesics. 1955

7: CLAUDE CHEVALLEY. Fundamental Concepts of Algebra. 1956

8: SZE-TSEN HU. Homotopy Theory. 1959

9: A. M. OSTROWSKI. Solution of Equations and Systems of Equations. Second Edition. 1966

10: J. DIEUDONNÉ. Foundations of Modern Analysis. 1960

11: S. I. GOLDBERG. Curvature and Homology. 1962

12: SIGURDUR HELGASON. Differential Geometry and Symmetric Spaces. 1962

13: T. H. HILDEBRANDT. Introduction to the Theory of Integration. 1963

14: SHREERAM ABHYANKAR. Local Analytic Geometry. 1964

15: RICHARD L. BISHOP AND RICHARD J. CRITTENDEN. Geometry of Manifolds. 1964

16: STEVEN A. GAAL. Point Set Topology. 1964

17: BARRY MITCHELL. Theory of Categories. 1965

18: ANTHONY P. MORSE. A Theory of Sets. 1965

19: GUSTAVE CHOQUET. Topology. 1966

20: Z. I. BOREVICH AND I. R. SHAFAREVICH. Number Theory. 1966

21: JOSÉ LUIS MASSERA AND JUAN JORGE SCHÄFFER. Linear Differential Equations and Function Spaces. 1966

22: RICHARD D. SCHAFER. An Introduction to Nonassociative Algebras. 1966

23: MARTIN EICHLER. Introduction to the Theory of Algebraic Numbers and Functions. 1966

24: SHREERAM ABHYANKAR. Resolution of Singularities of Embedded Algebraic Surfaces. 1966

In preparation:

FRANÇOIS TREVES. Topological Vector Spaces, Distributions, and Kernels.

OYSTEIN ORE. The Four Color Problem.

PETER D. LAX and RALPH S. PHILLIPS. Scattering Theory.

INTRODUCTION TO THE THEORY OF ALGEBRAIC NUMBERS AND FUNCTIONS

Martin Eichler
DEPARTMENT OF MATHEMATICS
UNIVERSITY OF BASEL
BASEL, SWITZERLAND

Translated by George Striker
GÖTTINGEN, GERMANY

1966

ACADEMIC PRESS New York and London

First published in the German language under the title
Einführung in die Theorie der algerbraischen Zahlen und Funktionen
and copyrighted in 1963 by Birkhäuser Verlag,
Basel, Switzerland

ACADEMIC PRESS INC.
111 Fifth Avenue, New York, New York 10003

United Kingdon Edition published by
ACADEMIC PRESS INC. (LONDON) LTD.
Berkeley Square House, London W.1.

LIBRARY OF CONGRESS CATALOG CARD NUMBER: 65-26396

PRINTED IN THE UNITED STATES OF AMERICA

Preface to the English Edition

Several errors and ambiguities which had been indicated in an Addenda et Errata sheet to the German edition have been eliminated. Three important changes were made: (a) Chapter III, §1 has been shortened considerably by rearranging the material. (b) Section 4 of Chapter III, §3 has been rewritten as sketched in the Addenda et Errata. (c) Fields of genus $g \leqq 1$ had been implicitly excluded from representations of correspondences by hth degree differentials in the case $h > 1$. This assumption, in Chapter V, §2 and in the preparations in Chapter III, §5, has been avoided.

The author extends his most sincere thanks to Academic Press for publishing an English translation of this book. Indeed, the attempt to keep up with the times while not adhering to the representation of the subject matter which has today attained a virtual monopoly certainly does place certain demands upon the reader.

He also thanks the editors of the Series in Pure and Applied Mathematics, Professors Samuel Eilenberg and Paul Smith, for choosing this book for their esteemed series. He particularly thanks Mr. George Striker for his clear translation into a pleasant English as well as his agreement to special wishes during the translation work.

Basel, Switzerland
May 1966

M. EICHLER

Preface to the German Edition

The purpose of this book is twofold. First it serves to introduce the general notions, the concepts, and the methods which underlie the theories of algebraic numbers and algebraic functions, primarily in one variable. A certain mastery of algebra and function theory, as might be gained from a modest one-volume text on each, is presupposed.

It seems most important to me to illuminate this subject from various points of view, projecting a multidimensional image. In choosing the topics and methods to be included in this book it became particularly clear to me that mathematics obeys the law of the biological theory of evolution: all individuals follow the general development of the species. It may well be that this law holds for other sciences as well, but the immense formative force of mathematics on the human mind assures particular adherence to this law.

After some basic ideas are presented in the first chapter, the development of the actual subject is commenced in the second chapter, with the divisibility theory of algebraic number and function fields. At this point no distinction is necessary between these two types of fields, for within the bounds of divisibility theory they obey the same laws. In an appendix to the second chapter the theory of algebraic number fields is pursued only a small step beyond this interesting point. From that point on our sole interest will be the study of algebraic fields of functions.

In the third chapter the foundations of the classical theory of algebraic function fields are laid down, using the Puiseux power series. The reader can thus omit Chapters I and II if he so desires. We find, however, that the power series themselves are not indispensable, and that this classical approach eventually merges with another method even applicable to fields of arbitrary characteristic, provided the field of constants is algebraically closed.

The second object of this book is to introduce the theory of elliptic modular functions, with its wonderful and deep applications in analytic number theory. This theory has provided an unquenchable effervescent spring of new discoveries for more than one hundred years. Some of these discoveries have crystallized in the theory of correspondences of general algebraic function fields, which is dealt with in Chapter V.

Of course, brevity of a book such as this one is possible only at the cost of omitting many interesting topics. It might have been particularly attractive to compile the knowledge of theta and zeta functions of algebraic number fields available today. This would have led to a reunion of the algebraic number and function theories at a higher plane.

Those familiar with the subject might still be interested in the following remarks. In occasional lecture courses in the past I had always used the methods of valuation theory. During the preparation of this book it became clear to me that the older development of divisibility theory, that of L. Kronecker is by far the most elegant. It leads directly to principal ideal rings, the divisors of which correspond to the ideals. It was first presented in this form by H. Weber, and seems all the more attractive today, a time when the basic properties of principal ideal rings are taken up in elementary algebra. It is not surprising that Kronecker did not realize this possibility himself, for the notion of a principal ideal ring would have seemed too abstract in his day. Instead, he proved the lemma that the contents of the product of two polynomials is equal to the product of their contents. It was the difficulty of this proof that lead to the general adherence to E. Noether's development of ideal theory. The valuation–theoretical foundations of divisibility theory, supported chiefly by H. Hasse, also avoid this lemma. But these methods each have their own disadvantages.

It would be wrong, though, to interpret this to mean that the entire theory should be developed along Kronecker's lines. On the contrary, it is R. Dedekind's notion of ideals which permits the valuable methods of linear algebra to be introduced. Similarly, the local considerations introduced by Hasse are quite indispensable. I have avoided the term *valuation*, operating instead with the corresponding valuation rings. In particular, I have avoided using the perfect closures of fields with respect to valuations, for these not only are unpleasant from an esthetic point of view but, moreover, represent an unnecessary detour.

The references at the end of later sections of this book are by no means intended to be complete.

It is a pleasure to acknowledge the stimulation provided by Dr. H. Kappus who read and edited the drafts. His thesis also contributed decisively to the final form of Sections III,5 and V,2. The support of Mr. A. Heeb who read the proofs to the German edition was most valuable. In particular, I want to express my appreciation to the publisher who made this book possible and gave it an exemplary appearance.

Basel, Switzerland
March 1963

M. EICHLER

Contents

* *The sections marked with an asterisk might be omitted in the first reading.*

INTRODUCTION TO THE THEORY OF

ALGEBRAIC NUMBERS AND FUNCTIONS

Introduction

1. The Subject

A finite algebraic number field K is a finite extension of the rational number field $k = Q$; a finite algebraic function field K is a finite extension of the rational function field $k = k_0(x_1, ..., x_m)$ in the m variables x_i over some field of constants k_0. We will, however, simply talk of number and function fields, thus taking for granted the prefix "finite algebraic," but sometimes distinguishing between "algebraic" and "rational."

In a function field, all elements of K which satisfy an algebraic equation with coefficients in k_0 form a subfield k_1, called the exact constant field. K can always be considered as a function field over the exact constant field, and in general this is convenient.

There is a remarkable similarity between algebraic number and function fields, which at the start permits a common treatment of both, and is first manifested in the formal analogy between the integral domains $\mathfrak{i} = Z$ of rational integers and $\mathfrak{i} = k_0[x]$ of rational functions. Both are Euclidean domains. In such domains the theorem of unique prime factorization holds. This property is also shared by the polynomial ring $\mathfrak{i} = k_0[x_1, ..., x_m]$ in view of the theorem: *If an integral domain $\mathfrak{O}$ has unique prime factorization, then so does the ring $\mathfrak{O}[x]$.*

In this book only marginal treatment will be given function fields with $m > 1$ variables, as long as the arguments do not require a distinction for different numbers of variables. The special problems and difficulties of these fields will not be considered. The few points of a general theory that can be included, however, will finally prove to be useful for the theory of functions of one variable.

Among the first problems is the development of a theory of divisibility for algebraic number and function fields. This is the subject matter of Chapter II. With the extension of k to K one arrives at an integral domain $\mathfrak{J}$ in K which can, in the corresponding sense, be considered an extension of $\mathfrak{i}$. In $\mathfrak{J}$, however, the theorem of unique prime factorization no longer holds. The simplest counterexample is found in the quadratic number field $K = Q(\sqrt{-6})$, where $6 = -(\sqrt{-6})^2 = 2 \cdot 3$ are two factorizations of an integer into integral irreducible factors. The uniqueness of prime factorization is restored by the introduction of ideal elements.

Chapter I gives several theorems of linear algebra, which arose in the course of the continual reformulation of the theories of algebraic numbers and functions. They are attractive enough to stand apart from their applications. Their study can be postponed, however, until they are needed.

2. The Method

To explain an important method which will be among those used, we consider the example of the rational function field $C(x)$ of one variable over the field C of complex numbers. We will often refer to this example simply as "classical function theory." One usually first expands a function in a power series (Laurent series) at some point or place (we use the term "place" immediately, anticipating the usual algebraic terminology), and asks for the first nonzero coefficient. The places $x = \xi$ are interpreted as being in the complex number plane, which is closed in the number sphere by introduction of the place symbolized $x = \infty$. The expansion of the function at $x = \infty$ is simply the expansion in a power series in x^{-1}. This preparatory local theory of functions then leads to the global theory, which makes statements about the behavior of a function in an entire domain—for the present case, in the entire number sphere. An example is given by the theorem: *A function holomorphic everywhere except for poles is a rational function.*

The power series in $x - \xi$ which contain no negative powers form an integral domain. The rational functions with power series of this property also form an integral domain, which we always designate $\mathfrak{i}_\xi$. These functions can be characterized without reference to power series; in the decomposition

$$f(x) = \gamma \prod_{\varrho} (x - \xi_\varrho)^{\nu(\xi_\varrho)}$$

to first degree factors the term $x - \xi$ appears with a positive power $\nu(\xi)$, if at all. The rational functions which have no negative powers when developed in powers of x^{-1} form an integral domain $\mathfrak{i}_\infty$. They are characterized by their degrees (Grad $= G$)

$$G(f(x)) = \sum_{\varrho} \nu(\xi_\varrho)$$

being nonpositive.

The integral domain $\mathfrak{i} = C[x]$ is clearly the intersection

$$\mathfrak{i} = \bigcap_{i \neq \infty} \mathfrak{i}_\xi .$$

This equation shows that a function holomorphic at every finite place is a polynomial.

The method of algebraic function theory consists, first of all, of finding not only a corresponding extension $\mathfrak{J}$ of $\mathfrak{i}$, but also an extension $\mathfrak{J}_\xi$ of $\mathfrak{i}_\xi$. (There will turn out to be more than one.) Functions are then examined with respect to their relation to $\mathfrak{J}_\xi$; the results lead to the "local" properties. The goal is, of course, to find "global" properties of functions, i.e., such that hold for all $\mathfrak{J}_\xi$. The separation into a local and a global theory is the natural graduation from the easier to the more difficult. It is perhaps amazing that this distinction between local and global properties, which stems from classical function theory of one variable, can be carried over profitably to algebraic numbers and functions in several variables.

From what has been said it becomes plausible that in the detailed investigation one can use power series expansions, but that they are not indispensible. Certain advantages in the study of functions over an abstract field k_0 are gained by remaining free of power series; we do so in Chapter II. On the other hand the method of power series is so standard, from classical function theory, that it would be too biased not to base some of our representation on it. Thus we reformulate function theory in Chapter III with its help.

Table of Several Abbreviations and Symbols

(a) General

g.c.d.	greatest common divisor,
$G(f(x))$	degree (Grad) of a polynomial or rational function $f(x)$,
$g(\mathfrak{a})$	degree of a divisor $\mathfrak{a}$ (II, §5,4),
$n_{K/k}(\cdots)$	norm with respect to a field extension K/k,
$s_{K/k}(\cdots)$	trace with respect to K/k,
$\sigma_{K/k}(\cdots)$	pseudotrace with respect to K/k (I, §4,4),
$\mathfrak{a}^*_{K/k}$	pseudocomplement or complement of a module $\mathfrak{a}$ with respect to K/k (II, §4,1)
$\mathfrak{d}_{K/k}$	pseudodifferent or different with respect to K/k (II, §4,2),
$\mathfrak{o}/\mathfrak{a}$	ring of residue classes of $\mathfrak{o}$ modulo an ideal $\mathfrak{a}$,
$\mathfrak{a}/a$	ring of residue classes of $\mathfrak{a}$ modulo an $a \in \mathfrak{a}$.

(b) Matrices

$E = (\delta_{ij})$	unit matrix,
A^t	transposed matrix of the matrix A,
$A^* = (A^t)^{-1}$	contragredient to a matrix A. (In Chapter V the hermitian adjoint $\bar{A}^t$ is also denoted A^*.)

(c) Fields

Q	field of rational numbers,
R	field of real numbers,
C	field of complex numbers,
k_0	any field, in particular the exact constant field of a function field K,
$k = Q$ or $k_0(x)$,	depending upon whether one wishes to consider algebraic numbers or functions. In Chapter II, however, we often designate the quotient field of an integral domain $\mathfrak{o}$ by k,
$k^\times$, $K^\times$	the multiplicative groups of the fields, k, K.

(d) Integral Domains

Z	domain of rational integers,
$\mathfrak{i} = Z$ or $k_0[x]$,	depending upon whether algebraic numbers or functions are to be considered,
$\mathfrak{J}$	principal order of a finite extension K/k with respect to $\mathfrak{i}$ (II, §1,1),
$\mathfrak{o}$	integral domain, such that it is a principal ideal domain and has quotient field k,
$\mathfrak{O}$	principal order of K with respect to $\mathfrak{o}$ (II, §1,1),
$\mathfrak{i}_p$	integral domain of all integral elements of $k = Q$ or $k_0(x)$ at a place p.

CHAPTER I

Linear Algebra

§1. Modules in Principal Ideal Domains

1. Finite Modules

We first consider *principal ideal domains*, that is, integral domains $\mathfrak{o}$ in which every ideal $\mathfrak{a} \subseteq \mathfrak{o}$ is a *principal ideal* $\mathfrak{a} = \mathfrak{o}a$.

Given two elements $a, b \in \mathfrak{o}$, there always exists a greatest common divisor (g.c.d.) d, defined by the fact that every common divisor of a and b divides d. The *theorem of the greatest common divisor* states that there exist two elements $u, v \in \mathfrak{o}$, such that $au + bv = d$.

In principal ideal domains the *theorem of unique decomposition into prime elements* holds: if

$$a = ep_1^{\alpha_1} \cdots p_r^{\alpha_r} = uq_1^{\beta_1} \cdots q_s^{\beta_s}$$

are two decompositions of a into prime elements p_ϱ, q_δ and units e, u, then $r = s$ and suitable indexing gives $\beta_\varrho = \alpha_\varrho$, $q_\varrho = u_\varrho p_\varrho$, with units u_ϱ.

The residue classes of a principal ideal domain with respect to a prime element form a field, as we know from the *theorem of residue class fields with respect to prime elements.*

These three theorems will be assumed known from elementary algebra. In many cases it is difficult to verify the assumption that not only finitely generated $\mathfrak{o}$-ideals, but even infinitely generated ones are principal ideals. A different definition becomes handier: *A principal ideal domain is one in which every finitely generated ideal is principal, and each of whose elements can be decomposed into prime elements in at least one manner.* If $\mathfrak{o}$ is a principal ideal domain in this sense, we can first prove the uniqueness of prime decomposition and then the basis theorem, the finiteness criterion, and the ascending chain theorem, as is done below. The last step shows that infinitely generated ideals can be finitely generated, assuring us that they, too, are principal ideals.

Now let $\mathfrak{o}$ be a principal ideal domain and k its quotient field. It is known that a linear vector space of dimension n over k is isomorphic to the space k^n of vectors $\alpha = (a_1, \ldots, a_n)$ with coefficients in k. An $\mathfrak{o}$-*module* $\mathfrak{a}$ in k^n is an additive group of vectors of k^n such that whenever it contains the vector α of k^n it also contains all $a\alpha$ with $a \in \mathfrak{o}$. The maximal number of vectors in $\mathfrak{a}$ linearly indepenedent with respect to $\mathfrak{o}$ (or equivalently, with respect to k) is called the *rank* or *dimension* of $\mathfrak{a}$.

An $\mathfrak{o}$-module $\mathfrak{a}$ is called *finite* if there exists a finite number of vectors $\beta_1, \ldots, \beta_m \in \mathfrak{a}$ such that every vector $\alpha \in \mathfrak{a}$ has at least one representation as a sum

$$\alpha = \sum_{i=1}^{m} a_i \beta_i \tag{1}$$

with $a_i \in \mathfrak{o}$. The vectors β_i are called a *system of generators* of $\mathfrak{a}$.

A one-dimensional $\mathfrak{o}$-module $\mathfrak{a}$ with elements in the quotient field k is called an $\mathfrak{o}$-ideal of k. If $\mathfrak{a} \subseteq \mathfrak{o}$ we call $\mathfrak{a}$ an *integral* $\mathfrak{o}$*-ideal.* Up to now we had used the term ideal to designate an integral $\mathfrak{o}$-ideal in this sense, and this is often done. In this book we will distinguish the more general $\mathfrak{o}$-ideal from the integral $\mathfrak{o}$-ideal.

Basis Theorem. *If $\mathfrak{a}$ is a finite $\mathfrak{o}$-module of rank r, then there exist r vectors α_i in $\mathfrak{a}$ such that every $\alpha \in \mathfrak{a}$ can be represented in a unique manner as $\alpha = \sum a_i \alpha_i$ with $a_i \in \mathfrak{o}$.*

The α_i are called a *basis* of $\mathfrak{a}$.

Proof. First, let $\beta_i = (b_{i1}, \ldots, b_{in})$ $(i = 1, \ldots, m)$ be a system of generators of $\mathfrak{a}$, and let $b_{ij} = c_{ij}/h$, with $c_{ij}, h \in \mathfrak{o}$. Such a "common denominator" h must exist, as k is the quotient field of $\mathfrak{o}$. Due to the existence of the representation (1), the components of any vector $\alpha = (a_1, \ldots, a_n) \in \mathfrak{a}$ satisfy $ha_j \in \mathfrak{o}$. Thus the components of all the vectors in $\mathfrak{a}$ have the common denominator h.

The first components a_1 of all $\alpha \in \mathfrak{a}$ form an $\mathfrak{o}$-ideal $\mathfrak{a}_1$ (and it is no loss of generality to assume that they are not all 0). For, if $\alpha = (a_1, \ldots)$ and $\alpha' = (a_1', \ldots)$ are two vectors, and $a, a' \in \mathfrak{o}$, then $a\alpha + a'\alpha' = (aa_1 + a'a_1', \ldots) \in \mathfrak{a}$, and thus $aa_1 + a'a_1' \in \mathfrak{a}_1$. We remember that $h\mathfrak{a}_1 \subseteq \mathfrak{o}$, and that therefore $h\mathfrak{a}_1$ is an integral $\mathfrak{o}$-ideal, and thus by our hypothesis a principal ideal, $h\mathfrak{a}_1 = c_1\mathfrak{o}$ with $c_1 \in \mathfrak{o}$. We have thus shown that $\mathfrak{a}$ contains a vector α_1 whose first component is $a_1 = c_1/h$, while the first components of all $\alpha \in \mathfrak{a}$ are of the form ac_1/h with $a \in \mathfrak{o}$.

This proves the contention for the case $r = 1$. By assuming it correct for modules of rank $r - 1$, we can continue as follows. For every vector $\alpha \in \mathfrak{a}$ consider the difference $\alpha' = \alpha - a\alpha_1$ whose first component is 0. These α' form a module of dimension $r - 1$, which has a basis $\alpha_r, \ldots, \alpha_r$ by inductive hypothesis. Thus, $\alpha_1, \ldots, \alpha_r$ form a basis of $\mathfrak{a}$. ▼†

Up to now we have not used prime decomposition. We will need it to prove the following:

† The symbol ▼ signifies "end of proof."

Criterion for Finiteness. *A submodule* $\mathfrak{b}$ *of a finite* $\mathfrak{o}$*-module* $\mathfrak{a}$ *is finite.*

Proof. Let α_i be a system of generators of $\mathfrak{a}$. Any $\beta \in \mathfrak{b}$ may be written as $\beta = \sum b_i \alpha_i$ with $b_i \in \mathfrak{o}$. With a sequence $\beta, \beta', \beta'', \ldots \in \mathfrak{b}$ we associate a sequence of $\mathfrak{o}$-ideals $\mathfrak{o}b_1, (\mathfrak{o}b_1, \mathfrak{o}b_1'), (\mathfrak{o}b_1, \mathfrak{o}b_1', \mathfrak{o}b_1''), \ldots$, each of which is contained in all the following. All are principal ideals, generated by the elements $b_{11} = b_1, b_{12}, b_{13}, \ldots$, each divisible by all the following, all $b_{1i} \in \mathfrak{o}$. As an element of $\mathfrak{o}$ can only have a finite number of divisors, this sequence must break off, so that the ideals stop increasing after a finite number of steps. Thus, there exists a β_1 whose first component b_1 divides the first components of all the other β. Now, choosing suitable $b \in \mathfrak{o}$, the first components of $\beta - b\beta_1$ are 0, and the proof can be completed as above by induction on the rank of $\mathfrak{a}$. ▼

The criterion leads immediately to the following:

Ascending Chain Theorem for $\mathfrak{o}$-Modules. *An ascending chain* $\mathfrak{a}_1 \subset \mathfrak{a}_2 \subset \cdots$ *of submodules of a finite module ends after a finite number of steps.*

Proof. From hypothesis, the union $\mathfrak{a}$ of all the $\mathfrak{a}_i$ is a submodule of a finite module, and therefore itself a finite module with, say, generators α_i. Each of these α_i is contained in some $\mathfrak{a}_{n_i}$, and then in all those following it. Thus, there exists an index n such that $\mathfrak{a}_n$ contains all the α_i, which means that $\mathfrak{a}_n = \mathfrak{a}$, that is, the chain ceases to ascend after $\mathfrak{a}_n$. ▼

2. The Theorem of Elementary Divisors

If $\mathfrak{a}, \mathfrak{b}$ *are finite* $\mathfrak{o}$*-modules and* $\mathfrak{b} \subseteq \mathfrak{a}$, *then there exists a basis* α_i *of* $\mathfrak{a}$ *and elements* $t_i \in \mathfrak{o}$ *such that* $\beta_i = t_i \alpha_i$ *is a basis of* $\mathfrak{b}$. *Furthermore, any nonzero* t_i *divides* t_{i+1}.

Several remarks will precede the proof. First, let α_i, β_i be any bases of $\mathfrak{a}, \mathfrak{b}$. As $\mathfrak{b} \subseteq \mathfrak{a}$, a system of equations

$$\beta_i = \sum_j a_{ij} \alpha_j, \qquad a_{ij} \in \mathfrak{o} \tag{2}$$

holds. If α_i', β_i' are other bases we further have

$$\alpha_i' = \sum_j u_{ij} \alpha_j, \qquad \beta_i' = \sum_j v_{ij} \beta_j, \qquad u_{ij}, v_{ij} \in \mathfrak{o} \tag{3}$$

and

$$\alpha_i = \sum_j u_{ij}' \alpha_j', \qquad \beta_i = \sum_j v_{ij}' \beta_j', \qquad u_{ij}', v_{ij}' \in \mathfrak{o}.$$

The matrices (u_{ij}'), (v_{ij}') are inverses to the matrices (u_{ij}), (v_{ij}). Now, matrices with coefficients in $\mathfrak{o}$ having inverses whose coefficients also lie in $\mathfrak{o}$ are called *unimodular matrices*; they form a group.

Equations (2) and (3) yield

$$\beta_i' = \sum_j a_{ij}'\alpha_j' \qquad \text{with} \quad (a_{ij}') = (v_{ij})(a_{ij})(u_{ij})^{-1}.$$

The theorem of elementary divisors asserts the existence of unimodular matrices (u_{ij}), (v_{ij}) such that

$$(a_{ij}') = \begin{pmatrix} t_1 & \cdots & 0 & \cdots & 0 \\ \cdot & \cdots & \cdot & \cdots & \cdot \\ \cdot & \cdots & \cdot & \cdots & \cdot \\ 0 & \cdots & t_r & \cdots & 0 \end{pmatrix}, \qquad t_1/t_2, \qquad \text{etc.} \tag{4}$$

is in diagonal form, and t_1 divides t_2 and so on. Thus, we can state our theorem as a proposition concerning matrices.

The t_i will concur, up to multiplication by units, with the *elementary divisors* commonly defined for the matrices (a_{ij}) and (a_{ij}'). The hth elementary divisor ($h = 1, 2, \ldots$) is defined as the quotient D_h/D_{h-1}, where D_h is the g.c.d. of all subdeterminants of order h, and $D_0 = 1$. It is clear by (4) that this is so for (a_{ij}'). We must assure ourselves that (a_{ij}) and (a_{ij}') have the same elementary divisors. Along with the theorem we prove the following:

Corollary. *The group of unimodular matrices with $n > 2$ columns is generated by the special matrices formed from the n-column unit matrix by replacing the elements at the places $\mu\mu$, $\mu\nu$, $\nu\mu$, $\nu\nu$ by the elements a, b, c, d of a two-column unimodular matrix. For example,*

$$V_{12}\begin{pmatrix} a & b \\ c & d \end{pmatrix} = \begin{pmatrix} a & b & & & \\ c & d & & & \\ & & 1 & & \\ & & & \ddots & \\ & & & & 1 \end{pmatrix},$$

(*with zeros in the empty spaces*). *For $\mathfrak{o} = Z$ it suffices to let*

$$\begin{pmatrix} a & b \\ c & d \end{pmatrix} = \begin{pmatrix} 1 & b \\ 0 & 1 \end{pmatrix}, \quad \begin{pmatrix} 0 & 1 \\ 1 & 0 \end{pmatrix}.$$

Left multiplication of a matrix A with row vectors $\zeta_1, \ldots, \zeta_n$ by $V_{\mu\nu}\left(\begin{smallmatrix} a & b \\ c & d \end{smallmatrix}\right)$ is equivalent to replacing the μth and νth rows by $a\zeta_\mu + b\zeta_\nu$ and $c\zeta_\mu + d\zeta_\nu$, respectively. Multiplication from the right does the same to the columns.

We will now demonstrate the invariance of the determinant divisor D_h of A under multiplication by $V_{\mu\nu}\left(\begin{smallmatrix} a & b \\ c & d \end{smallmatrix}\right)$ from the left. Certainly for those subdeterminants not containing any elements of the rows μ or ν, or containing both, there is no change in the g.c.d. Let $\varrho_1, \ldots, \varrho_{h-1}$ be a set of indices different from each other and from μ and ν. Using self-explanatory symbols, the subdeterminants of $V_{\mu\nu}\begin{pmatrix} a & b \\ c & d \end{pmatrix}A = A'$ arising from elements of rows $\varrho_1, \ldots, \varrho_{h-1}, \mu$ are

$$D'_{\varrho_1,\ldots,\varrho_{h-1},\mu} = D(\zeta_{\varrho_1}, \ldots, \zeta_{\varrho_{h-1}}, a\zeta_\mu + b\zeta_\nu)$$
$$= aD_{\varrho_1,\ldots,\varrho_{h-1},\mu} + bD_{\varrho_1,\ldots,\varrho_{h-1},\nu}$$

and similarly for those from rows $\varrho_1, \ldots, \varrho_{h-1}, \nu$.

Thus, the hth determinant divisor of A' is divisible by that of A. But the converse is also true, for the matrix inverse to $V_{\mu\nu}\begin{pmatrix} a & b \\ c & d \end{pmatrix}$ is also unimodular. The same argument holds for multiplication from the right, by replacing the rows by columns.

Proof of the Elementary Divisor Theorem. We prove the second form of our assertion, remarked upon above, by induction on the number of elements of (a_{ij}). The assertion is trivial for one-row matrices.

In the general case, a suitable permutation of rows and columns yields $a_{11} \neq 0$; this permutation is accomplished by left and right multiplication with certain $V_{ij}\begin{pmatrix} 0 & 1 \\ 1 & 0 \end{pmatrix}$. The matrix (a_{ij}) is then multiplied from the left and right by suitable unimodular $V_{1i}\begin{pmatrix} a & b \\ c & d \end{pmatrix}$ in such a way that the first coefficient a'_{11} of the product matrix becomes a proper divisor of the first coefficient a_{11} of the former. This process breaks off when this is no longer possible, and it is the case after a finite number of such multiplications, since a_{11} has but a finite number of divisors. We maintain that the final a'_{11} is even a divisor of all the a'_{ij}.

Our indirect proof of this is simplified by again writing a_{ij} for a'_{ij}. Say a_{11} does not divide a_{21}, and set $a'_{11} = (a_{11}, a_{21})$. There then exist $a, b \in \mathfrak{o}$ such that

$$a'_{11} = aa_{11} + ba_{21},$$

where $(a, b) = 1$. This means that there exist $c, d \in \mathfrak{o}$ with $ad - bc = 1$. The matrix $V_{12}\begin{pmatrix} a & b \\ c & d \end{pmatrix}$ is thus unimodular, and applying it as left multiplier for (a_{ij}) replaces a_{11} by the proper divisor a'_{11}, contrary to our assumption.

The same method shows that a_{11} divides all the a_{i1} and, using right multiplication, all the a_{1i}. This shown, the matrix (a_{ij}) can be put into a first normal form in which $a_{1i} = a_{i1} = 0$ for $i \neq 1$, using repeated left and right multiplication by suitable $V_{1i}\begin{pmatrix} 1 & 0 \\ t & 1 \end{pmatrix}$ and $V_{1i}\begin{pmatrix} 1 & t \\ 0 & 1 \end{pmatrix}$, respectively.

We finally show that all the a_{ij} are divisible by a_{11}. If, say, a_{ij} with $i, j \neq 1$ were not divisible by a_{11}, we could set $V_{1j}\begin{pmatrix} 1 & 1 \\ 0 & 1 \end{pmatrix}(a_{ij}) = (a'_{ij})$ and would have $a'_{11} = a_{11}$, but $a'_{1j} = a_{ij}$ is not divisible by a'_{11}, which was shown to be impossible.

All further transformations using left and right unimodular multiplication can be carried out with matrices whose first row and column are $1, 0, \ldots, 0$.

The first row and column remain unchanged, while the rest of the matrix is transformed as desired. Use of the inductive hypothesis thus permits the matrix to be brought into the diagonal form asserted, while a_{11} remains as divisor of all the elements occurring. The theorem is proved. ▼

The same argument applied, in particular, to a unimodular matrix (a_{ij}) yields the corollary to the elementary divisor theorem.

3. Dual Spaces and Complementary Modules

In addition to k^n we consider another vector space over k of the same dimension n, and denote it k^{n*}. We agree to write scalar multiplication of vectors $\alpha^* \in k^{n*}$ by elements $a \in k$ in inverse order: $\alpha^* a$.

We further assume that to every pair of vectors $\alpha \in k^n$, $\alpha^* \in k^{n*}$ a product with the following properties is defined:

(1) $\alpha\alpha^* \in k$;
(2) $(a_1\alpha_1 + a_2\alpha_2)\beta^* = a_1\alpha_1\beta^* + a_2\alpha_2\beta^*$,
$\alpha(\beta_1{}^*b_1 + \beta_2{}^*b_2) = \alpha\beta_1{}^*b_1 + \alpha\beta_2{}^*b_2$;
(3) if $\alpha \in k^n$ is such that $\alpha\alpha^* = 0$ for all $\alpha^* \in k^{n*}$, then $\alpha = 0$;
if $\alpha^* \in k^{n*}$ is such that $\alpha\alpha^* = 0$ for all $\alpha \in k^n$, then $\alpha^* = 0$.

Let β_i, $\beta_i{}^*$ be bases of k^n and k^{n*}. Multiplication of $\alpha = \sum a_i\beta_i$ by $\beta^* = \sum \beta_i{}^* b_i$ can be written

$$\alpha\beta^* = \sum a_i B_{ij} b_j, \qquad B_{ij} = \beta_i\beta_j{}^*.$$

The third postulate assures the nonsingularity of the matrix (B_{ij}). Conversely, starting with such a matrix, one can define a product of vectors $\alpha = (a_i)$ by $\beta^* = (b_i)$; it will satisfy the above requirements.

Under these assumptions one calls the two vector spaces k^n and k^{n*} *dual*.†

With every $\mathfrak{o}$-module $\mathfrak{a} \subseteq k^n$ we associate its *complementary* $\mathfrak{o}$-module $\mathfrak{a}^* \subseteq k^{n*}$ as follows: $\mathfrak{a}^*$ consists of all $\alpha^* \in k^{n*}$ such that

$$\alpha\alpha^* \in \mathfrak{o} \qquad \text{for all} \quad \alpha \in \mathfrak{a}; \tag{5}$$

$\mathfrak{a}^*$ is also called the *complement* of $\mathfrak{o}$. The following is evident:

$$\mathfrak{a}_1 \subseteq \mathfrak{a}_2 \qquad \text{implies} \quad \mathfrak{a}_1{}^* \supseteq \mathfrak{a}_2{}^*.$$

Let $\mathfrak{a}$ be a finite $\mathfrak{o}$-module of rank n and β_i, $\beta_i{}^*$, α_i bases of k^n, k^{n*}, and $\mathfrak{a}$; further let

$$\alpha_i = \sum_j m_{ij}\beta_j.$$

† By the first part of the second postulate one can consider k^{n*} to be the module of all linear functions on k^n with values in k. The rest of the postulates must then be proved.

We formally define elements

$$\alpha_i^* = \sum_j \beta_j^* m_{ji}^*$$

in k^{n*} with indeterminate coefficients m_{ij}^* and require

$$\alpha_i \alpha_j^* = \delta_{ij} = \begin{cases} 1 & \text{for} \quad i = j \\ 0 & \text{for} \quad i \neq j \end{cases} \tag{6}$$

to hold; this can be written in the matrix form

$$(m_{ij})(\beta_i \beta_j^*)(m_{ij}^*) = (m_{ij})(B_{ij})(m_{ij}^*) = (\delta_{ij}).$$

For $\mathfrak{a}$ of rank n there is a unique solution (m_{ij}^*) for this equation.

The α_i^* are what is called the *complementary basis* to the α_i; they are a basis of the complement $\mathfrak{a}^*$. For clearly $|m_{ij}^*| \neq 0$ and thus every $\alpha^* \in \mathfrak{a}^*$ may be written as $\alpha^* = \sum \alpha_i^* a_i$ with $a_i \in k$. As $\alpha_i \alpha^* = a_i$ we have $\alpha_i \alpha^* \in \mathfrak{o}$ for all i, or equivalently $\alpha\alpha^* \in \mathfrak{o}$ for all $\alpha \in \mathfrak{a}$ as long as all $a_i \in \mathfrak{o}$. This gives the contention.

The symmetry of the complementary basis gives

$$\mathfrak{a}^{**} = \mathfrak{a}, \tag{7}$$

so that we know complementation is an involution.

4*. Noetherian Rings

Certain applications of module theory require only a finite system of generators of a module in place of a finite basis. The term *noetherian ring* denotes a commutative ring $\mathfrak{o}$ with the property that every integral $\mathfrak{o}$-ideal $\mathfrak{a}$ has a finite system of generators. Manipulation within noetherian rings is, essentially, similar to that within principal ideal rings. Of course the assumptions and results are weaker.

Two theorems will give examples of noetherian rings which in general are not principal ideal rings.

If $\mathfrak{o}$ is a noetherian ring and $\mathfrak{a}$ an $\mathfrak{o}$-ideal, then the residue class ring $\mathfrak{o}/\mathfrak{a}$ is also noetherian.

Proof. Residue class formation $\mathfrak{o} \to \bar{\mathfrak{o}} = \mathfrak{o}/\mathfrak{a}$ maps an $\mathfrak{o}$-ideal $\mathfrak{b}$ onto an $\bar{\mathfrak{o}}$-ideal $\bar{\mathfrak{b}}$. Conversely, every $\bar{\mathfrak{o}}$-ideal $\bar{\mathfrak{b}}$ corresponds to the set containing all elements of all residue classes of $\bar{\mathfrak{b}}$, an $\mathfrak{o}$-ideal $\mathfrak{b}$. Thus, a finite system of generators of $\mathfrak{b}$ yields the same for $\bar{\mathfrak{b}}$. ▼

If $\mathfrak{o}$ is a noetherian ring, so is $\mathfrak{o}[x_1, \ldots, x_m]$.

Proof. Of course we need only consider the case $m = 1$, for the variables may be adjoined one after the other. Let $\mathfrak{a}$ be an $\mathfrak{o}[x]$-ideal. The leading coefficients c_0 of all polynomials $a(x) = c_0x^n + c_1x^{n-1} + \cdots + c_n$ of $\mathfrak{a}$, of some fixed (formal) degree n, including 0, give an $\mathfrak{o}$-ideal $\mathfrak{c}_n$. For if $a'(x) = c_0'x^n + \cdots$ is another polynomial in $\mathfrak{a}$, and d some element of $\mathfrak{o}$, then $a(x) - a'(x)$ and $da(x)$ are elements of $\mathfrak{a}$ and have nth coefficients $c_0 - c_0'$ and dc_0, respectively.

The same argument shows that the leading coefficients of all polynomials $a(x) \in \mathfrak{a}$, along with 0, form an $\mathfrak{o}$-ideal $\mathfrak{c}$. For, now setting $a'(x) = c_0'x^{n'} + \cdots$ with, say, $n' \leqq n$, we find that $a(x) - x^{n-n'}a'(x)$ has leading coefficient $c_0 - c_0'$.

Let $c_{01}, c_{02}, \ldots$ be a finite system of generators of $\mathfrak{c}$, and $a_1(x) = c_{01}x^{n_1} + \cdots$, $a_2(x) = c_{02}x^{n_2} + \cdots, \ldots$ be polynomials in $\mathfrak{a}$ with these generators as leading coefficients. Say that n is the highest degree occurring among them. We start forming the desired system of generators of $\mathfrak{a}$ by choosing the polynomials $b_i(x) = a_i(x)x^{n-n_i}$, certainly contained in $\mathfrak{a}$.

Now, if $d(x)$ is a polynomial in $\mathfrak{a}$ of some degree $m \geqq n$, our construction of the $b_i(x)$ permits us to find $e_i \in \mathfrak{o}$ such that $d'(x) = d(x) - x^{m-n} \sum e_i b_i(x)$ is of degree $m' < m$, and continuation of the process yields a polynomial of degree less than n. We need now only extend our $b_i(x)$ by a finite system of polynomials $b(x)$, with leading coefficients that generate each of the ideals $\mathfrak{c}_m$, $m = 0, 1, \ldots, n-1$. This gives a finite system of generators of $\mathfrak{a}$. ▼

In noetherian rings we have the following:

Ascending Chain Theorem for $\mathfrak{o}$-Ideals. *Every ascending chain $\mathfrak{a}_1 \subset \mathfrak{a}_2 \subset \cdots$ of $\mathfrak{o}$-ideals breaks off after finitely many steps.*

The *proof* is identical to that for $\mathfrak{o}$-modules of §1,1.

Let $\mathfrak{o}$ be a *noetherian domain* (integral domain and noetherian ring), and let k be the quotient field. We then have the following:

Let ω_i be some basis of k^n, and write the vectors $\alpha = \sum a_i\omega_i$ as component vectors (a_i). An $\mathfrak{o}$-module $\mathfrak{a}$ in k^n is finite if and only if there exists a nonzero $h \in \mathfrak{o}$ such that for all $\alpha \in \mathfrak{a}$ we have $ha_i \in \mathfrak{o}$ for every i. (We think of h as a common denominator for the a_i.)

As in §1,1, this leads to the following:

Finiteness Criterion. *A submodule of a finite module is finite.*

Proof. As $\mathfrak{a}$ may be replaced by $h\mathfrak{a}$, it is no loss of generality to assume $h = 1$. For $n = 1$ the contention is identical to that of the ascending chain theorem. Assume the theorem proved for $n - 1$, and consider the $\mathfrak{o}$-ideal formed by the first vector components a_1. Choose some finite system of vectors $\alpha_1, \alpha_2, \ldots \in \mathfrak{a}$ whose first components generate this ideal. For any $\alpha \in \mathfrak{a}$ our construction permits us to choose $b_i \in \mathfrak{o}$ such that $\alpha' = \alpha - b_1\alpha_1 - b_2\alpha_2 - \cdots$ has zero for its first component. But the inductive hypothesis assures the finite generation of the ideal, in $\mathfrak{a}$, of vectors with vanishing first components. The assertion is proved. ▼

5*. A Further Basis Theorem

As before, let $\mathfrak{o}$ be a principal ideal domain with quotient field k. The ring $\mathfrak{o}[x]$ of polynomials in one variable x with coefficients in $\mathfrak{o}$ is no longer a principal ideal domain unless $\mathfrak{o} = k$.

A finite $\mathfrak{o}[x]$-module $\mathfrak{a}$ of rank n, contained in a linear space L over $k(x)$, has a basis consisting of n elements if and only if it satisfies the following condition: if a and b are mutually prime elements of $\mathfrak{o}[x]$ and $\alpha \in L$ such that $a\alpha \in \mathfrak{a}$, $b\alpha \in \mathfrak{a}$, then $\alpha \in \mathfrak{a}$.

The necessity of the condition is obvious.

To prove the sufficiency, consider the principal ideal domain $\mathfrak{o}' = k[x]$ and the $\mathfrak{o}'$-module $\mathfrak{a}'$ of all finite sums $\sum a_i\alpha_i$ with $a_i \in \mathfrak{o}'$, $\alpha_i \in \mathfrak{a}$. Now, $\mathfrak{a}'$ is a finite $\mathfrak{o}'$-module of rank n and thus, by §1,1, it has a basis α_i'. Since k is the quotient field of $\mathfrak{o}$, there exists an $a \in \mathfrak{o}$ such that $a\alpha_i' \in \mathfrak{a}$. But this is also a basis of $\mathfrak{a}'$, so that we lose no generality in assuming that $\alpha_i' \in \mathfrak{a}$.

If the α_i' are not yet a basis of $\mathfrak{a}$, then there exists a vector

$$\alpha = \frac{1}{p}\sum f_i\alpha_i', \qquad p \in \mathfrak{o}, \quad f_i \in \mathfrak{o}[x], \tag{8}$$

in $\mathfrak{a}$, with the f_i not all divisible by p. There is even such an α for which p is a prime element in $\mathfrak{o}$.

We consider the polynomials f_i with respect to the module p, that is as polynomials over the residue class field $\mathfrak{o}/p$. As such they have a g.c.d., say d. Then, $f_i = dg_i + pv_i$, with polynomials v_i and g_i, the latter of which are mutually prime modulo p. Substitution into (8) yields

$$\alpha = d\frac{1}{p}\sum g_i\alpha_i' + \sum v_i\alpha_i' = d\alpha' + \sum v_i\alpha_i'.$$

The vectors $d\alpha'$ and $p\alpha'$ are in $\mathfrak{a}$, and since d and p are mutually prime, we see that $\alpha' = 1/p \sum g_i\alpha_i'$ is in $\mathfrak{a}$, by our assumption.

This shows that we may assume, henceforth, that the f_i in (8) are mutually prime mod p. We can therefore find polynomials g_i, v in $\mathfrak{o}[x]$ such that

$$\sum g_i f_i = 1 + pv. \tag{9}$$

We must study the solution of the diophantine equation (9) in detail. We again calculate within the polynomial ring $(\mathfrak{o}/p)[x] = \mathfrak{o}[x]/p$. Transform the one-row matrix composed of the f_i by subtracting multiples of one element from another repeatedly (Euclidean algorithm) until the row $1, 0, 0, \ldots$ remains. These operations correspond to the right multiplication by a unimodular matrix $(g_{ij}) = G$ in $(\mathfrak{o}/p)[x]$:

$$(f_1, \ldots, f_n)G = (1, 0, \ldots, 0).$$

And the first column $g_1, \ldots, g_n$ solves (9) as a congruence mod p. By the corollary in §1,2, G is a product of the special matrices

$$\begin{pmatrix} 0 & 1 & & & & \\ 1 & 0 & & & & \\ & & 1 & & & \\ & & & \cdot & & \\ & & & & \cdot & \\ & & & & & \cdot \end{pmatrix}, \quad \begin{pmatrix} 1 & m & & & & \\ 0 & 1 & & & & \\ & & 1 & & & \\ & & & \cdot & & \\ & & & & \cdot & \\ & & & & & \cdot \end{pmatrix}, \quad \text{etc.}$$

We then construct a matrix H with coefficients in $\mathfrak{o}[x]$ by substituting, in G, the 0 and 1 of k for the 0 and 1 of $\mathfrak{o}/p$, and any representative of its residue class in $\mathfrak{o}[x]$ for a nonzero element of $(\mathfrak{o}/p)[x]$. The matrix H is unimodular in $\mathfrak{o}[x]$, and its residue class modulo p is $G = (g_{ij})$. Designate the elements of the first row of H as g_i; they satisfy (9).

The equation $\alpha_i' = \sum h_{ij}\alpha_j''$ can now be used to introduce a new basis α_i'' for the module spanned by the α_i'. With (9) we can write the elements (8) in the form

$$\alpha = \frac{1}{p}\sum_{i,j} f_i h_{ij}\alpha_j'' = \frac{1}{p}\left((1 + pv)\alpha_1'' + \sum_{j=2}^{n} f_j'\alpha_j''\right) = v\alpha_1'' + \frac{1}{p}\left(\alpha_1'' + \sum_{j=2}^{n} f_j'\alpha_j''\right)$$

with $f_i' \in \mathfrak{o}[x]$. The vector

$$\alpha'' = \frac{1}{p}\left(\alpha_1'' + \sum_{j=2}^{n} f_j'\alpha_j''\right)$$

also lies in $\mathfrak{a}$. The elements $\alpha'', \alpha_2'', \ldots, \alpha_n''$ form a basis of a larger module contained in $\mathfrak{a}$. As we presupposed $\mathfrak{a}$ to be a finite module, a finite number of these steps must suffice to arrive at a basis of $\mathfrak{a}$.

Exercise (nontrivial). Let x and y be independent variables, p a prime element in $\mathfrak{o}$, and μ_1, μ_2 two linearly independent vectors. The $\mathfrak{o}[x, y]$-module $\mathfrak{a}$ generated by the vectors μ_1, μ_2 and $1/p(x\mu_1 + y\mu_2)$ satisfies the conditions of the above theorem for $\mathfrak{o}[x, y]$ instead of $\mathfrak{o}[x]$. But it has no basis consisting of only two elements.

§2. Systems of Linear Inequalities

1. Minkowski's Point Lattice Theorem

As always let R^n be the n-dimensional (affine) vector space over the field R of real numbers. Z^n will signify the module or, using geometric language, the lattice of vectors $\alpha = (a_1, \ldots, a_n)$ with components $a_i \in Z$.

Minkowski's Point Lattice Theorem. *Let $\mathfrak{R}$ be an open convex region in*

R^n having the origin as centerpoint (i.e., if the vector $\alpha \in \mathfrak{R}$ then also $-\alpha \in \mathfrak{R}$), which has the volume V in the sense of Riemann.† *Then including $\alpha = 0$, there are at least $2^{-n}V$ vectors in $Z^n \cap \mathfrak{R}$.*

Proof. Let N be a natural number and $Z_N{}^n$ the lattice of all vectors $\alpha = (a_i)$ such that $Na_i \in Z$; this point lattice consists of cubes of edge length N^{-1}. We take a small fixed positive constant ε and form two new regions similar to $\mathfrak{R}$ but shrunk in the ratios $1 : \frac{1}{2}$ and $1 : (\frac{1}{2} + \varepsilon)$ and denote them $\frac{1}{2}\mathfrak{R}$ and $(\frac{1}{2} + \varepsilon)\mathfrak{R}$. The numbers $b_N(\frac{1}{2})$ and $b_N(\frac{1}{2} + \varepsilon)$ of cubes of the lattice $Z_N{}^n$ lying completely within these two regions, respectively, multiplied by the volume N^{-n} of each cube yield, essentially, the volumes of these regions:

$$N^{-n}b_N(\tfrac{1}{2}) = (\tfrac{1}{2})^nV - \eta_N, \qquad N^{-n}b_N(\tfrac{1}{2} + \varepsilon) = (\tfrac{1}{2} + \varepsilon)^nV - \eta_N',$$

where η_N, η_N' approach zero as N increases.

These numbers of *cubes* are essentially the same as the numbers of *points* of the lattice $Z_N{}^n$ in the regions, for one can associate each cube with that corner point which has the smallest coordinates. Counting thus is correct, except in the boundary zone. The number of points a_N in $Z_N{}^n \cap \frac{1}{2}\mathfrak{R}$ can certainly be approximated, for sufficiently large N, as follows:

$$b_N(\tfrac{1}{2}) \leqq a_N \leqq b_N(\tfrac{1}{2} + \varepsilon).$$

Thus, for any given small positive ε_1 and a sufficiently large N we have

$$2^{-n}V - \varepsilon_1 \leqq N^{-n}a_N \leqq 2^{-n}V + \varepsilon_1.$$

On the other hand, $Z_N{}^n$ and Z^n are additive groups, the latter being a subgroup of the former with index N^n. The a_N points in $Z_N{}^n \cap \frac{1}{2}\mathfrak{R}$ are distributed among the N^n residue classes of $Z_N{}^n$ mod Z^n, and thus clearly some residue class must contain at least $N^{-n}a_N$ points: denote them α_i, $i = 1, 2, \ldots$. As $\mathfrak{R}$ is symmetric, $-\alpha_1 \in \frac{1}{2}\mathfrak{R}$, and as $\mathfrak{R}$ is convex, even $\frac{1}{2}(\alpha_i - \alpha_1) \in \frac{1}{2}\mathfrak{R}$ for all i. Thus $\alpha_i - \alpha_1 \in \mathfrak{R}$. Since all α_i lie in the same residue class mod Z^n, we have also $\alpha_i - \alpha_1 \in Z^n$. The number of these vectors is $\geqq N^{-n}a_N \geqq 2^{-n}V - \varepsilon_1$ for every $\varepsilon_1 > 0$, which proves the theorem. ▼

We shall need the following application:

Given the system

$$\left|\sum_{i=1}^{n} x_i m_{ij}\right| < c_j \qquad (j = 1, \ldots, n) \tag{1}$$

of inequalities, where the m_{ij} are complex, and the c_j positive real numbers such that if the m_{ij_1} of a column are not real, the complex conjugate column $m_{ij_2} = \overline{m}_{ij_1}$ occurs, with $c_{j_2} = c_{j_1}$. Further, assume $(\pi/2)^{r_2} \prod_j c_j > |\det(m_{ij})|$

† The last assumption always holds, in fact, for convex regions. We need not take the trouble, however, to prove this.

(absolute value of the determinant), with $2r_2$ being the number of nonreal columns of (m_{ij}). Then (1) *has a nontrivial solution in rational integers x_i.*

Proof. An open convex symmetric region is given by (1) in the space R^n with coordinates x_i. Let the nonreal columns of (m_{ij}) be

$$m_{ij_1} = m'_{ij_1} + im''_{ij_1}, \qquad m_{ij_2} = m'_{ij_1} - im''_{ij_1}.$$

Consider the real n-column matrix formed by the real m_{ij} and m'_{ij_1}, m''_{ij_1}; the absolute value of its determinant is $2^{-r_2}|\det(m_{ij})|$. A linear transformation with this matrix puts the inequalities (1) in the forms

$$|x_j'| < c_j \quad \text{and} \quad x'^2_{j_1} + x'^2_{j_2} < c^2_{j_1}.$$

This defines a region of volume $2^{n-2r_2}\pi^{r_2}\prod_j c_j$. Thus, the region defined by (1) has the volume

$$V = 2^n(\pi/2)^{r_2}\prod_j c_j|\det(m_{ij})|^{-1}.$$

By our hypothesis we have $2^{-n}V > 1$. By the lattice point theorem there is more than one solution to (1), and thus at least one nontrivial solution. ▼

2*. Siegel's Proof

A second proof of Minkowski's theorem, which is in fact a strengthening of the part essential in application, is due to Siegel†:

Let $\mathfrak{R}$ be a convex region centered at the origin, which contains no further lattice points. Then

$$2^n = V + \frac{1}{V}\sum_{\mu\neq 0}\left|\int_{\mathfrak{R}}\cdots\int e^{-\pi i\xi\mu}\,dx_1\cdots dx_n\right|^2 \qquad \left(\xi\mu = \sum_i x_i m_i\right) \tag{2}$$

where V is the volume of $\mathfrak{R}$, and the sum on the right is taken over all lattice points $\mu \neq 0 \in Z^n$.

The consequence is, then, that such a region cannot exist with $V \geqq 2^n$.

For the proof one defines a function $\varphi(\xi)$ in R^n by setting $\varphi(\xi) = 1$ whenever the point ξ with coordinates x_i lies within $\mathfrak{R}$, and $\varphi(\xi) = 0$ otherwise. Further, set

$$f(\xi) = \sum_\mu \varphi(2\xi + 2\mu),$$

summed over all $\mu \in Z^n$. Now, $f(\xi)$ is periodic in all variables with period 1 and can thus be expanded in the (convergent, at least in the mean) Fourier series

$$f(\xi) = \sum_\mu c_\mu e^{2\pi i\xi\mu}.$$

---

† *Über Gitterpunkte in konvexen Körpern und ein damit zusammenhängendes Extremalproblem,* Acta Math. **65**, 307–323 (1935).

The coefficients are

$$c_\mu = \int_0^1 \cdots \int f(\xi) e^{-2\pi i \xi \mu} \, dx_1 \cdots dx_n, \tag{3}$$

and we have the relation of completeness

$$\int_0^1 \cdots \int f(\xi)^2 \, dx_1 \cdots dx_n = \sum_\mu |c_\mu|^2, \tag{4}$$

which, properly analyzed, will yield the theorem.

First let $g(\xi)$ be some integrable function with period 1 in all x_i. Then,

$$\begin{aligned} &\int_0^1 \cdots \int f(\xi) g(\xi) \, dx_1 \cdots dx_n \\ &\qquad = 2^{-n} \sum_\mu \int_0^2 \cdots \int \varphi(\xi + 2\mu) g(\tfrac{1}{2}\xi) \, dx_1 \cdots dx_n. \end{aligned} \tag{5}$$

Let $\mathfrak{R}_\mu$ be the intersection of the cube $0 \leqq x_i \leqq 2$ with the region $\mathfrak{R}$ translated by -2μ. $\mathfrak{R}_\mu$ and $\mathfrak{R}_\nu$ are disjoint for $\mu \neq \nu$, for if ξ were in both, then $\xi + 2\mu$ and $\xi + 2\nu$ would be in $\mathfrak{R}$. By the symmetry and convexity of $\mathfrak{R}$ we would have $\frac{1}{2}(\xi + 2\mu - \xi - 2\nu) = \mu - \nu \in \mathfrak{R}$, contrary to hypothesis. The integrals to the right in (5) are actually taken over the regions $\mathfrak{R}_\mu$ with the integrand $g(\frac{1}{2}\xi)$. The substitution $\xi \to \xi - 2\mu$ is applied in the integrals, leaving $g(\frac{1}{2}\xi)$ invariant because of the assumed periodicity. The translated regions $\mathfrak{R}_\mu$ form, together, the region $\mathfrak{R}$, and one has

$$\int_0^1 \cdots \int f(\xi) g(\xi) \, dx_1 \cdots dx_n = 2^{-n} \int_{\mathfrak{R}} \cdots \int g(\tfrac{1}{2}\xi) \, dx_1 \cdots dx_n.$$

This formula is applied on the one hand to (3),

$$c_\mu = 2^{-n} \int_{\mathfrak{R}} \cdots \int e^{-\pi i \xi \mu} \, dx_1 \cdots dx_n, \qquad \text{in particular} \quad c_0 = 2^{-n} V,$$

while on the other

$$\begin{aligned} \int_0^1 \cdots \int f(\xi)^2 \, dx_1 \cdots dx_n &= 2^{-n} \int_{\mathfrak{R}} \cdots \int f(\tfrac{1}{2}\xi) \, dx_1 \cdots dx_n \\ &= 2^{-n} \sum_\mu \int_{\mathfrak{R}} \cdots \int \varphi(\xi + 2\mu) \, dx_1 \cdots dx_n. \end{aligned}$$

For $\mu = 0$ the integrand to the right has the constant value 1, while for $\mu \neq 0$ it is $= 0$. For, $\varphi(\xi + 2\mu) \neq 0$ and $\xi \in \mathfrak{R}$ would, with the symmetry and convexity of $\mathfrak{R}$, imply that $\frac{1}{2}(\xi + 2\mu - \xi) = \mu \in \mathfrak{R}$, while $\mathfrak{R}$ contains no such lattice point. Thus,

$$\int_0^1 \cdots \int f(\xi)^2 \, dx_1 \cdots dx_n = 2^{-n} V.$$

This result, inserted into (4), gives (2).

3. Generalization to Function Fields

The theorem of §2,1 has its counterpart in function theory, which strengthens the resemblance, mentioned in the introduction, between numbers and functions. Let $k = k_0(x)$ be the field of rational functions in one indeterminate over the constant field k_0, and $\bar{k}$ the field of formal power series in x^{-1}:

$$f(x) = \sum_{\nu=\nu_0}^{\infty} a_\nu x^{-\nu}, \qquad a_{\nu_0} \neq 0, \quad \nu_0 \gtreqless 0$$

with coefficients in k_0. The negative of first exponent $-\nu_0$ of such a power series is called the *degree* $G(f(x))$. We have

$$\begin{aligned} G(f(x) + g(x)) &\leqq \max\bigl(G(f(x)), G(g(x))\bigr), \\ G(f(x)\, g(x)) &= G(f(x)) + G(g(x)). \end{aligned} \tag{6}$$

This concept of degree corresponds to the usual one for rational functions, when expanded in powers of x^{-1}.

In this connection the following fields and integral domains correspond to each other:

$$Q, \quad R, \quad Z \quad \text{and} \quad k = k_0(x), \bar{k}, k_0[x].$$

In place of the absolute value in R we use the degree of a function in $\bar{k}$. The problem analogous to (1) and its solution are as follows.

Let (m_{ij}) be an n-rowed nonsingular matrix with coefficients in $\bar{k}$ and (m_{ij}^*) be its contragradient. Further, let there be given n rational integers d_j. The solutions of

$$G\left(\sum_{i=1}^{n} x_i m_{ij}\right) \leqq d_j \qquad (j = 1, \ldots, n) \tag{7}$$

$$G\left(\sum_{i=1}^{n} x_i^* m_{ij}^*\right) \leqq d_j^* = -2 - d_j \tag{8}$$

in vectors x_i, x_i^* with components in $k_0[x]$ each form a linear space of finite dimension over k_0. The equation

$$l - l^* = \sum_{j=1}^{n} d_j - G(|m_{ij}|) + n \tag{9}$$

holds between the dimensions l and l^* of these spaces. For each pair of solutions x_i, x_i^* of (7) *and* (8) *we have*

$$\sum_{i=1}^{n} x_i x_i^* = 0. \tag{10}$$

This theorem supplies information concerning the number of solutions to (7) similar to that which the theorem in §2,1 supplies for (1), as (9) gives an approximation from below because $l^* \geqq 0$. The link between the problems

(7) and (8) and the exact formula (9) are, on the other hand, characteristic of function theory and have no direct analogies in number theory.

Formula (10) is, incidentally, a simple consequence of (7) and (8). For, as the matrices (m_{ij}) and (m_{ij}^*) are assumed contragredient, we have

$$\sum_i x_i x_i^* = \sum_{\mu,\nu,j} x_\mu m_{\mu j} x_\nu^* m_{\nu j}^*,$$

while with (6) through (8) the degrees satisfy

$$G\left(\sum_i x_i x_i^*\right) \leqq \max(d_j - 2 - d_j) = -2.$$

Nevertheless, the term on the left in (10) is a polynomial which cannot have negative degree and must therefore be 0.

In proving (9) we first assume the m_{ij} all to lie in $k = k_0(x)$. This proof, which specially covers the case of terminating power series for the m_{ij}, will be given in the paragraph after next, where concurrently a new interpretation of the theorem will evolve.† At this point we will only derive the general case from this special one, although we will make no use of it below.

To this end we set

$$\sum_i x_i^* m_{ij}^* = y_j^*, \qquad \text{that is,} \quad x_i^* = \sum_j m_{ij} y_j^*$$

and, instead of (8), consider the equivalent system

$$G(y_j^*) \leqq -2 - d_j, \qquad \sum_j m_{ij} y_j^* \in k_0[x]. \tag{11}$$

The theorem is assumed to hold for terminating power series, and the general m_{ij} are approximated by their partial sums $m_{ij}^{(\nu)}$:

$$m_{ij} = m_{ij}^{(\nu)} + r_{ij}^{(\nu)},$$

the remainders $r_{ij}^{(\nu)}$ starting with the power $x^{-\nu}$ at the earliest. The equation

$$(m_{ij}^{(\nu)*}) = ((m_{ij}^{(\nu)})^t)^{-1}$$

is introduced. The matrix identity $M^{-1} - N^{-1} = M^{-1}(N - M)N^{-1}$ applied to (m_{ij}) and $(m_{ij}^{(\nu)})$ yields

$$(m_{ij}^*) = (m_{ij}^{(\nu)*}) - (m_{ij}^{(\nu)*})(r_{ij}^{(\nu)})^t(m_{ij}^*) = (m_{ij}^{(\nu)*}) + (r_{ij}^{(\nu)*}),$$

with $r_{ij}^{(\nu)*}$ as an abbreviating symbol. The powers of x^{-1} with which the sequences $r_{ij}^{(\nu)*}$ start increase unbounded as ν increases; thus the $m_{ij}^{(\nu)*}$ also approximate the m_{ij}^*.

---

† Another proof is given in M. Eichler, *Ein Satz über Linearformen in Polynombereichen,* Arch. Math. **10,** 81–84 (1959).

Let $l^{(v)}, l^{(v)*}$ be the number of linearly independent solutions $x_i^{(v)}, x_i^{(v)*}$ of (7) and (11) with $(m_{ij}^{(v)})$ in place of (m_{ij}). From some certain v onwards the degree $G(|m_{ij}^{(v)}|)$ ceases to change. Further, from $\sum m_{ij}^{(v)} y_j^{(v)*} = x_i^{(v)*}$ we have the uniform boundedness of the degrees of the polynomials $x_i^{(v)*}$, the same argument holding for the $x_i^{(v)}$. Hence the degrees of the power series $\sum x_i^{(v)} r_{ij}^{(v)}, \sum x_i^{(v)*} r_{ij}^{(v)*}$ remain under a certain bound, which goes to $-\infty$ as v increases. As a consequence the $x_i^{(v)}, x_i^{(v)*}$ are solutions of (7) and (11) for some sufficiently large v and then $l^{(v)} = l, l^{(v)*} = l^*$. As $l^{(v)}, l^{(v)*}$ satisfy Eq. (9) the theorem is proved. ▼

§3. Linear Divisors

1. Basic Concepts

We now come to a first and very important application of the methodical principle sketched in the introduction. First, let $k = k_0(x)$ be the field of rational functions in one variable over some constant field k_0. Every rational function $f(x) \in k_0(x)$ can be decomposed in a unique manner into prime polynomials $p = p(x)$:

$$f(x) = \gamma \prod_p p^{v(p)}, \qquad \gamma \in k_0.$$

The p which occur here are, in case k_0 is algebraically closed, all of the form $p = x - \xi$ with $\xi \in k_0$, while in general, prime polymonials of higher degrees occur. In place of the $\mathfrak{i}_\xi$ mentioned in the introduction we have here the integral domains $\mathfrak{i}_p$ of those rational functions $f(x)$ in whose prime decomposition p occurs with a power $v(p) \geqq 0$. In addition to the $\mathfrak{i}_p$ we have the integral domain $\mathfrak{i}_\infty$ of those $f(x)$ with degrees $G(f(x)) \leqq 0$.

The $\mathfrak{i}_p$ can be defined not only for $k_0(x)$, but even in general for all such fields k in which a unique prime decomposition is possible. These include, in particular, the fields $k = Q$ and $k = k_0(x_1, \ldots, x_m)$ in $m > 1$ variables. It is easy to see that the $\mathfrak{i}_p$ are always principal ideal domains.

In the last case one introduces $\mathfrak{i}_\infty$ as the integral domain of functions having a nonpositive degree (the degree of the quotient of polynomials being the difference between the degree of the numerator and the degree of the denominator). One could also introduce several integral domains $\mathfrak{i}_{\infty i}$, which would then consist of those functions having nonpositive degrees as functions of some x_i. The proof that even here $\mathfrak{i}_\infty$ or $\mathfrak{i}_{\infty i}$ are principal ideal domains may be left as an exercise for the reader, as these fields will play no part in what follows. He must convince himself that every function can be expressed in the form $f(x_i) = e(x_i) y^k$, where $e(x_i)$ is a unit of $\mathfrak{i}_\infty$ or $\mathfrak{i}_{\infty i}$, respectively, and y is any function of degree one.

Although we are concerned, in the end, only with the field $k = k_0(x)$, we do not exclude the possibilities $k = Q$ and $k = k_0(x_1, ..., x_m)$ for the time being in order to accentuate the generality of the method.

We take the opportunity to introduce a terminology which has become customary. As we saw in the classical case, the prime polynomials $p(x)$ and their associated integral domains $\mathfrak{i}_p$ correspond to the points or places $x = \xi$, including $\xi = \infty$ of the Riemann number sphere. One now associates *places* with the $\mathfrak{i}_p$, in a formal manner, without necessarily bringing a geometric concept into play. We will never properly speak of the places as such, but rather in connections of the following variety: an element is integral "at the place" p if p occurs in its prime decomposition with a degree $\nu(p) \geqq 0$. Basically the places are but elements of an index set. The correct use of the expression and its practicality will appear by itself in what follows.

For $\mathfrak{i} = Z$ or $\mathfrak{i} = k_0[x_1, ..., x_m]$ we have the important and self-evident equation

$$\mathfrak{i} = \bigcap_{p \neq \infty} \mathfrak{i}_p \tag{1}$$

to which we remark that, for the case $k = Q$, i.e., $\mathfrak{i} = Z$, we have not defined $\mathfrak{i}_\infty$ and that therefore the restriction $p \neq \infty$ is superfluous.

Let k^n be the n-dimensional vector space over k and ω_i a basis of k^n which will be fixed from now on. A *linear divisor* $\mathfrak{a}$ of k^n will be understood to be a system of $\mathfrak{i}_p$-modules $\mathfrak{a}_p$ with the following properties:

(a) With every place p (including $p = \infty$ if it is defined) there is associated a finite $\mathfrak{i}_p$-module $\mathfrak{a}_p$ in k^n of rank n. These $\mathfrak{a}_p$ will be called the *p-components* of $\mathfrak{a}$.

(b) For almost all p (that is, for all but a finite number of the p) the ω_i form a basis of $\mathfrak{a}_p$ with respect to $\mathfrak{i}_p$.

The basis ω_i in the definition can be replaced by any arbitrary basis ω_i'. For then there exists a system of linear equations $\omega_i = \sum c_{ij}\omega_j'$ in k. The denominators of the c_{ij} contain but a finite number of primes. If p is not a divisor of the denominator of some c_{ij} or of the numerator of the determinant $|c_{ij}|$, then ω_i and ω_i' are bases of the same $\mathfrak{i}_p$-module. This is almost always the case.

The intersection

$$\mathfrak{a}_\mathfrak{i} = \bigcap_{p \neq \infty} \mathfrak{a}_p \tag{2}$$

is a finite $\mathfrak{i}$*-module of rank n.*

Proof. For almost all p, $\omega_i \in \mathfrak{a}_p$. For the finitely many exceptions, let α_{pi} $(i = 1, 2, ...)$ be generating systems for $\mathfrak{a}_p$. All these finitely many α_{pi}, together with the ω_i, generate, on the one hand, finite $\mathfrak{i}_p$-modules $\mathfrak{a}_p' \subseteq \mathfrak{a}_p$, and on the other a finite $\mathfrak{i}$-module $\mathfrak{a}_\mathfrak{i}'$. We have

$$\mathfrak{a}_\mathfrak{i}' = \bigcap_{p \neq \infty} \mathfrak{a}_p' \supseteq \mathfrak{a}_\mathfrak{i}.$$

By the finiteness criterion of §1,4 we thus know that $\mathfrak{a}_{\mathfrak{i}}$ is a finite $\mathfrak{i}$-module. It remains to show that it is of rank n. As all the $\mathfrak{a}_p$ are of rank n there exists, for each place p, an exponent h such that $p^h\omega_i \in \mathfrak{a}_p$. Because of property (*b*) we can take $h = 0$ almost everywhere. Letting a be the product of all the p^h, we have $a\omega_i \in \mathfrak{a}_{\mathfrak{i}}$ for all i, completing the proof. ▼

In this connection we consider the dual space k^{n*}.

The complementary modules $\mathfrak{a}_p{}^*$ to the $\mathfrak{a}_p$ form the components of a linear divisor $\mathfrak{a}^*$ of k^{n*}; $\mathfrak{a}$ and $\mathfrak{a}^*$ are said to be complementary.

Proof. From §1,3 we immediately have property (*a*). Letting $\omega_i{}^*$ be the basis complementary to ω_i, the $\omega_i{}^*$ always form a basis of $\mathfrak{a}_p{}^*$ with respect to $\mathfrak{i}_p$ when the ω_i form a basis of $\mathfrak{a}_p$, and this is almost always true. ▼

Let

$$\alpha_{p,i} = \sum_j a_{p,ij}\omega_j \tag{3}$$

be bases of $\mathfrak{a}_p$ ($p = \infty$ and $p \neq \infty$). Further, let $A = (a_{ij})$ be some nonsingular matrix in k. Then

$$\beta_{pi} = \sum_{j,l} a_{p,ij}a_{jl}\omega_l \tag{4}$$

are bases of the p-components of a linear divisor $\mathfrak{b}$. This divisor is uniquely determined by A and the divisor $\mathfrak{a}$, for if other bases of the $\mathfrak{a}_p$ are used, the matrices $(a_{p,ij}) = A_p$ and A_pA appearing in (3) and (4) are simply multiplied from the left by the matrices giving rise to the basis transformations. We write

$$\mathfrak{b} = \mathfrak{a}A \tag{5}$$

and call $\mathfrak{a}$ and $\mathfrak{b}$ *equivalent*. The following is clear:

$$(\mathfrak{a}A)B = \mathfrak{a}(AB) \tag{6}$$

for two matrices A and B. This shows that our equivalence relationship has the familiar property of transitivity; reflexivity and symmetry are obvious. Therefore equivalent divisors may be grouped into *divisor classes*.

The divisor complementary to (5) is

$$\mathfrak{b}^* = \mathfrak{a}^*A^*, \qquad A^* = (A^t)^{-1}. \tag{7}$$

For, if $\omega_i{}^*$ is the basis complementary to ω_i, and one sets

$$\alpha_{pi}^* = \sum a_{p,ij}^*\omega_j{}^*$$

using the matrix $A_p{}^* = (a_{p,ij}^*) = (A_p{}^t)^{-1}$ contragredient to $A_p = (a_{p,ij})$, one finds that

$$\alpha_{pi}\alpha_{pj}^* = \omega_i\omega_j{}^* = \delta_{ij}.$$

Thus the α_{pi}^* form a basis of $\mathfrak{a}_p{}^*$, so that Eq. (7) follows from the identity $(A_pA)^* = A_p{}^*A^*$.

2. NORM AND DEGREE OF A LINEAR DIVISOR

The matrices $(a_{p,ij})$ in (3) are determined up to a unimodular factor on the left, thus their determinants are fixed up to a unit factor. Let

$$|a_{p,ij}| = \begin{cases} u_p p^{h_p} & \text{for} \quad p \neq \infty \\ u_\infty (1/y)^{h_\infty} & \text{for} \quad p = \infty \end{cases} \tag{8}$$

with a function y of degree 1, and with units $u_p \in \mathfrak{i}_p$, $u_\infty \in \mathfrak{i}_\infty$. The exponents h_p are invariants of $\mathfrak{a}_p$. They are $h_p = 0$, because of property (*b*), for almost all p. Now associate to $\mathfrak{a}$ the linear divisor $\mathfrak{a}^1$, defined by $\mathfrak{a}_p{}^1 = p^{h_p}\mathfrak{i}_p$, $\mathfrak{a}_\infty{}^1 = (1/y)^{h_\infty}\mathfrak{i}_\infty$. It is called the *linear norm* of $\mathfrak{a}$, symbolically $\mathfrak{a}^1 = N(\mathfrak{a})$.

This definition immediately yields the equation

$$N(\mathfrak{a}A) = N(\mathfrak{a})|A|. \tag{9}$$

In the case of a function field one also defines the *linear degree* of the linear divisor $\mathfrak{a}$ by

$$G(\mathfrak{a}) = \sum_{p \neq \infty} h_p \, G(p) + h_\infty, \tag{10}$$

where $G(p)$ is the degree of the prime polynomial p.

Equivalent divisors have the same linear degree. This. implies, incidentally, that the degree remains invariant under a change of basis, while this does not hold for the norm.

Proof. Using the notation of (5), let $|A|$ have the prime decomposition

$$|A| = c \prod_{p \neq \infty} p^{v_p}$$

with a constant c. The degree of $|A|$ is then $G(|A|) = \sum_{p \neq \infty} v_p G(p)$. Further,

$$|A|\mathfrak{i}_p = \begin{cases} p^{v_p}\mathfrak{i}_p & \text{for} \quad p \neq \infty \\ (1/y)^{-G(|A|)}\mathfrak{i}_\infty & \text{for} \quad p = \infty \end{cases}$$

and thus

$$G(\mathfrak{a}A) = \sum_{p \neq \infty} (h_p + v_p) \, G(p) + h_\infty - G(|A|) = G(\mathfrak{a}). \quad \blacktriangledown$$

3. THE DIMENSION OF A LINEAR DIVISOR

From now on throughout we will assume k to be a function field. The intersection of all the $\mathfrak{a}_p$ (including $\mathfrak{a}_\infty$) is a linear space over the field of constants k_0. The dimension of this intersection is called the *dimension* of $\mathfrak{a}$ and is denoted $l(\mathfrak{a})$. A vector contained in all $\mathfrak{a}_p$ is called a *multiple* of $\mathfrak{a}$.

Equivalent divisors have the same dimension.

Proof. Let $\alpha = \sum a_i \omega_i$ be contained in all $\mathfrak{a}_p$. Then, because of (3), the system of linear equations

$$\sum_i x_{p,i} a_{p,ij} = a_j$$

is solvable for all p with $x_{p,i} \in \mathfrak{i}_p$ and, conversely, the solvability implies $\alpha \in \mathfrak{a}_p$. Passing now to the divisor $\mathfrak{b} = \mathfrak{a}A$ we see that

$$\sum_i x_{p,i} \sum_j a_{p,ij} a_{jl} = \sum_j a_j a_{jl} = b_l$$

is simultaneously solvable. This means that the vector $\beta = \sum b_l \omega_l$ is contained in every $\mathfrak{b}_p$. The correspondence between α and β is unique in both directions and linear. This yields the assertion. ▼

The dimension of a linear divisor is finite.

Proof. All multiples are contained in the intersection $\mathfrak{a}_i$ defined by (2), which is a finite i-module. Were $l(\mathfrak{a}) = \infty$, then there would exist multiples of arbitrarily high degree in a representation with a basis of $\mathfrak{a}_i$. But, as these multiples also are contained in $\mathfrak{a}_\infty$ and $\mathfrak{a}_\infty$ is a finite module, this is impossible. ▼

Riemann-Roch Theorem for Linear Divisors. *Let $k = k_0(x)$. Associate, with the linear divisor $\mathfrak{a}$ in k^n, the divisor $\mathfrak{a}'$ in k^{n*} by setting*

$$\mathfrak{a}_p' = \begin{cases} \mathfrak{a}_p^* & \text{for } p \neq \infty, \\ x^{-2}\mathfrak{a}_\infty^* & \text{for } p = \infty. \end{cases} \tag{11}$$

Then,

$$l(\mathfrak{a}) - l(\mathfrak{a}') = -G(\mathfrak{a}) + n \tag{12}$$

holds. Furthermore, $\alpha\alpha' = 0$ for multiples α, α' of $\mathfrak{a}$, $\mathfrak{a}'$.

The fact that (11) defines a divisor follows from §3,1. By §1,(7) we have $(\mathfrak{a}_p^*)^* = \mathfrak{a}_p$ and $(x^{-2}\mathfrak{a}_\infty^*)^* = x^2\mathfrak{a}_\infty$. Thus, the divisor associated with $\mathfrak{a}'$ is $\mathfrak{a}'' = \mathfrak{a}$. Thus (12) implies

$$G(\mathfrak{a}) + G(\mathfrak{a}') = 2n,$$

which could also be proved directly by setting $y = x$ in (8).

An immediate consequence of (7) is

$$(\mathfrak{a}A)' = \mathfrak{a}'A^*, \qquad A^* = (A^t)^{-1}. \tag{13}$$

Thus the correspondence $\mathfrak{a} \leftrightarrow \mathfrak{a}'$ between the linear divisors is also an involution of the divisor classes. Correspondingly, Eq. (12) contains only invariants of divisor classes.

The exceptional status of $p = \infty$ in (11) is doubtlessly an esthetic flaw, but it will dissolve of itself in the applications of the theorem below. Furthermore, we can carry out the following argument here. By taking a projective transformation

$$x' = (ax + b)/(cx + d)$$

of the variables, the places p are caused to permute; in particular, if $c \neq 0$ the place $p = cx + d$ passes to $p' = \infty$. It is therefore plausible to replace $p = \infty$ in (11) by any other place $p = cx + d\,(c \neq 0)$ and set $\mathfrak{a}_p' = p^2\mathfrak{a}_p$ there. In doing this the correspondence between divisors is in fact altered, but not that between the classes, and it is in them that we are really interested.

The *proof* of the Riemann–Roch theorem dates back to Dedekind and Weber.† As $\mathfrak{a}_\infty$ is a finite $\mathfrak{i}_\infty$-module, there exists to every $\alpha \in k^n$ a smallest exponent $\nu(\alpha)$ such that $x^{-\nu(\alpha)}\alpha \in \mathfrak{a}_\infty$ holds. As the basis theorem of §1,1 assures us that $\mathfrak{a}_\mathfrak{i}$ has a basis, we can construct a special basis α_i of $\mathfrak{a}_\mathfrak{i}$ as follows. Among all the possible bases we choose one such that $\nu(\alpha_1)$ has the least possible value. If there is more than one such basis choose, from among them, one such that $\nu(\alpha_2)$ is as small as possible, and so forth. One then has, of course,

$$\nu(\alpha_1) \leqq \nu(\alpha_2) \leqq \cdots \leqq \nu(\alpha_n). \tag{14}$$

Now we claim that the vectors $x^{-\nu(\alpha_i)}\alpha_i$ form a basis of $\mathfrak{a}_\infty$ with respect to $\mathfrak{i}_\infty$. For, if this were not so, there would exist elements $f_i \in \mathfrak{i}_\infty$ not all divisible by $1/x$ such that the vector

$$\beta = x \sum f_i x^{-\nu(\alpha_i)}\alpha_i \tag{15}$$

would lie in $\mathfrak{a}_\infty$. One could then set

$$f_i = f_i(0) + (1/x)g_i, \qquad f_i(0) \in k_0, \quad g_i \in \mathfrak{i}_\infty .$$

Then

$$\beta' = x \sum f_i(0) x^{-\nu(\alpha_i)}\alpha_i$$

is also in $\mathfrak{a}_\infty$ and, in our assumption, not all of the $f_i(0)$ are 0. Thus there is no loss of generality in assuming that $f_i = f_i(0) \in k_0$ in (15). Letting r be the largest index for which $f_i \neq 0$, we see by (14) that $\alpha_r' = x^{\nu(\alpha_r)-1}\beta$ is a vector in $\mathfrak{a}_\mathfrak{i}$. $\alpha_1, \ldots, \alpha_{r-1}, \alpha_r'$ would generate the same submodule of $\mathfrak{a}_\mathfrak{i}$ as $\alpha_1, \ldots, \alpha_r$. Thus one could replace α_r by α_r'. But, since $\nu(\alpha_r') \leqq \nu(\alpha_r) - 1$, this contradicts our construction of the α_i.

If one takes the α_i for the ω_i then, in the notation of (8), $(a_{p,ij})$ is the unit matrix for all $p \neq \infty$ and $(a_{\infty,ij})$ the diagonal matrix with entries $x^{-\nu(\alpha_i)}$. By §3.2 this yields

$$G(\mathfrak{a}) = \sum \nu(\alpha_i). \tag{16}$$

Now, a multiple α of $\mathfrak{a}$ is represented in two ways:

$$\alpha = \sum a_i\alpha_i = \sum b_i x^{-\nu(\alpha_i)}\alpha_i , \qquad a_i \in \mathfrak{i}, \quad b_i \in \mathfrak{i}_\infty .$$

This gives $a_i = b_i x^{-\nu(\alpha_i)}$. The conditions $a_i \in \mathfrak{i}, b_i \in \mathfrak{i}_\infty$ cannot be satisfied

†*Theorie der algebraischen Funktionen einer Veränderlichen*, J. Reine Angew. Math. **92**, 181–290 (1882).

when $\nu(\alpha_i) > 0$; when $\nu(\alpha_i) \leqq 0$ they are satisfied by all polynomials $a_i = a_i(x)$ of degree $\leqq -\nu(\alpha_i)$. Hence there exist

$$l(\mathfrak{a}) = \sum_{\nu(\alpha_i) \leq 0} (1 - \nu(\alpha_i)) = \sum_{\nu(\alpha_i) < 1} (1 - \nu(\alpha_i)) \tag{17}$$

linearly independent multiples with respect to k_0.

The basis α_i^* complementary to α_i is a basis of $\mathfrak{a}_i^* = \mathfrak{a}_i'$, and the basis $x^{\nu(\alpha_i)}\alpha_i^*$ complementary to $x^{-\nu(\alpha_i)}\alpha_i$ is a basis of $\mathfrak{a}_\infty^* = x^2\mathfrak{a}_\infty'$. Hence a multiple α' of $\mathfrak{a}$ is represented doubly as

$$\alpha' = \sum a_i^*\alpha_i^* = \sum b_i^* x^{\nu(\alpha_i)-2}\alpha_i^*, \qquad a_i^* \in \mathfrak{i}, \quad b_i^* \in \mathfrak{i}_\infty.$$

Using the same argument, among these there are

$$l(\mathfrak{a}') = \sum_{-\nu(\alpha_i)+2 \leq 1} (1 - (2 - \nu(\alpha_i))) = -\sum_{\nu(\alpha_i) > 1} (1 - \nu(\alpha_i)) \tag{18}$$

that are linearly independent. Equations (16)–(18) yield the asserted equation (12). ▼

4*. The Riemann–Roch Theorem and the Minkowski Linear Form Theorem

Let α_i be an arbitrary basis of $\mathfrak{a}_i$ and $\alpha_{\infty i}$ a basis of $\mathfrak{a}_\infty$. As the basis ω_i we use $\omega_i = \alpha_i$. Let

$$\alpha_i = \sum m_{ij}\alpha_{\infty j}. \tag{19}$$

As above, where we proved (16), this yields

$$G(\mathfrak{a}) = G(|m_{ij}|). \tag{20}$$

The multiples $\alpha = \sum x_i\alpha_i$ of $\mathfrak{a}$ can now be determined by the conditions $x_i \in \mathfrak{i}$ and $\sum x_i m_{ij} \in \mathfrak{i}_\infty$. The latter condition can, due to the meaning of $\mathfrak{i}_\infty$, be formulated as $G(\sum x_i m_{ij}) \leqq d_j = 0$. Thus $l(\mathfrak{a})$ turns out to be the number of linearly independent solutions of the special system of inequalities (7) in §2.

Also $l(\mathfrak{a}')$ can be similarly interpreted. Let α_i^* be the basis complementary to α_i, and thus a basis of the module $\mathfrak{a}_i^* = \mathfrak{a}_i'$, and $\alpha_{\infty i}^*$ the basis complementary to $\alpha_{\infty i}$, i.e., a basis of $x^2\mathfrak{a}_\infty'$. Then, using the argument of §3,1 (proof of (7)), similarly to (19)

$$\alpha_i^* = \sum m_{ij}^*\alpha_{\infty j}^*, \tag{21}$$

where (m_{ij}^*) is the contragredient matrix to (m_{ij}). Here the multiples of $\mathfrak{a}'$ are $\alpha' = \sum x_i^*\alpha_i^*$ with $x_i \in \mathfrak{i}$, $\sum x_i^* m_{ij}^* \in x^{-2}\mathfrak{i}_\infty$. As above, the latter conditions can be replaced by $G(\sum x_i^* m_{ij}^*) \leqq -2$. Thus, $l(\mathfrak{a}')$ is the number of solutions of §2,(8), and the Riemann–Roch theorem coincides with the linear form theorem of §2,3.

Conversely, let (m_{ij}) be an arbitrary nonsingular matrix. Choose n arbitrary linearly independent vectors $\alpha_{\infty i}$ in k^n and define the α_i by (19). Call the $\mathfrak{i}_\infty$-modules and $\mathfrak{i}$-modules, generated by $\alpha_{\infty i}$ and α_i, $\mathfrak{a}_\infty$ and $\mathfrak{a}_\mathfrak{i}$, respectively. Here the latter is the intersection (2) of the $\mathfrak{i}_p$-modules $\mathfrak{a}_p$ with the same basis α_i. For the complementary bases (21) holds, and the number of solution, l, l^* of §2.3 (with $d_j = 0$), coincides with $l(\mathfrak{a})$ and $l(\mathfrak{a}')$.

It is seen, incidentally, that the assumption $d_j = 0$ in §2,3 does not inhibit the generality, for one can replace m_{ij} by $m_{ij}x^{-d_j}$.

§4. Traces, Norms, and Discriminants

1. Representations by Matrices

Let K be an extension of the field k of finite degree $[K:k] = n$. Whenever we wish so accentuate this fact, we write K/k (k over K). Confusion with residue class formation is impossible. Using a basis $\omega_1, \ldots, \omega_n$ of K/k one arrives at a *representation* of K by n-row matrices in k. For every $\alpha \in K$ one has

$$\alpha\omega_j = \sum_{i=1}^{n} \omega_i a_{ij}, \qquad a_{ij} \in k.$$

Abbreviating the one-row matrices with elements ω_i and $\alpha\omega_i$ as ω and $\alpha\omega$, and setting $(a_{ij}) = A$, this system of equations is written as

$$\alpha\omega = \omega A.$$

The matrix A corresponding to α is the zero matrix if and only if $\alpha = 0$.

For another element $\beta \in K$ with $\beta\omega = \omega B$,

$$(\alpha \dotplus \beta)\omega = \omega(A \dotplus B)$$

holds (where the symbol indicates addition or multiplication). The mapping $\alpha \to A$ is thus an isomorphic mapping of K onto (or a representation by) a ring of n-row matrices in k.

With a change of basis $\omega \to \omega' = \omega C$, C being a nonsingular matrix over k, it is well known that A goes into $A' = C^{-1}AC$.

Letting E denote the unit matrix and x a variable, the determinant

$$|xE - A| = x^n - s(A)x^{n-1} + - \cdots + (-1)^n n(A)$$

is known as the *characteristic polynomial* of A or also of α. It does not change when the basis changes. The coefficients

$$s(\alpha) = s(A) = \sum_{i=1}^{n} a_{ii}, \qquad n(\alpha) = n(A) = |a_{ij}|$$

are called the *trace and norm* of α. It is easy to verify the *sum* and *product formulas*

$$s(\alpha+\beta)=s(\alpha)+s(\beta), \qquad n(\alpha\beta)=n(\alpha)n(\beta). \tag{1}$$

If $\alpha\in k$ the corresponding matrix is $A=\alpha E$, and thus

$$s(a)=na, \qquad n(a)=a^n, \qquad s(a\alpha)=as(\alpha) \qquad \text{for} \quad a\in k. \tag{2}$$

In particular let the minimal polynomial

$$x^n-a_1x^{n-1}+-\cdots\pm a_n=\prod_{i=1}^{n}(x-\alpha_i) \qquad (\alpha_1=\alpha)$$

of $\alpha\in K$ be of degree n. Then $1, \alpha, \ldots, \alpha^{n-1}$ form a basis of K/k, and the corresponding matrix has the form

$$A=\begin{pmatrix} 0 & 0 & \cdots & 0 & \mp a_n \\ 1 & 0 & \cdots & 0 & \pm a_{n-1} \\ \cdot & \cdot & \cdots & \cdot & \cdot \\ 0 & 0 & \cdots & 1 & a_1 \end{pmatrix},$$

which implies

$$s(\alpha)=a_1=\sum_{i=1}^{n}\alpha_i, \qquad n(\alpha)=a_n=\prod_{i=1}^{n}\alpha_i.$$

2. The Transitivity Formulas

Let an intermediate field L be given: $k\subset L\subset K$. One can take traces and norms both with respect to L and to k, the difference being clearly expressed in an obvious terminology. We will prove the *transitivity formulas*

$$s_{K/k}(\alpha)=s_{L/k}(s_{K/L}(\alpha)), \qquad n_{K/k}(\alpha)=n_{L/k}(n_{K/L}(\alpha)). \tag{3}$$

Let $\omega_1, \ldots, \omega_n$ and $\eta_1, \ldots, \eta_m$ be bases of K/L and L/k. Then, $\omega_1\eta_1, \ldots, \omega_n\eta_1; \ldots; \omega_1\eta_m, \ldots, \omega_n\eta_m$ is a basis of K/k. We again express these bases as one-column matrices ω, η and $\omega\times\eta$, where $\omega\times\eta$ consists of the products $\omega_i\eta_j$ in lexicographical order. Now, first

$$\alpha\omega=\omega A' \tag{4}$$

holds with an n-row matrix A' over L, and then, for every coefficient α_{ij} of A', one has

$$\alpha_{ij}\eta=\eta A''_{ij} \tag{5}$$

with an m-row matrix A''_{ij} over k. Both equations can be combined to

$$\alpha(\omega\times\eta)=(\omega\times\eta)A,$$

where A is the mn-row matrix

$$A = \begin{pmatrix} A''_{11} & \cdots & A''_{1n} \\ \cdot & \cdots & \cdot \\ A''_{n1} & \cdots & A''_{nn} \end{pmatrix}.$$

Its trace is

$$s_{K/k}(\alpha) = s(A) = \sum_{i=1}^{n} s(A''_{ii}).$$

By (5), using the sum formula (1), this comes out to

$$s_{K/k}(\alpha) = \sum_{i=1}^{n} s_{L/k}(\alpha_{ii}) = s_{L/k}\left(\sum_{i=1}^{n} \alpha_{ii}\right). \tag{6}$$

The sum on the right, $\sum \alpha_{ii}$, is equal to $s_{K/L}(\alpha)$, according to (4). Thus the first of the formulas (3) is proved.

By interchanging columns and adding multiples of one column to another one can, as is well known, put the matrix $A' = (\alpha_{ij})$ into triangular form with zeros above the principal diagonal. This leaves the determinant invariant:

$$B' = \begin{pmatrix} \beta_{11} & & & \\ & \cdot & & \\ & & \cdot & \\ * & & & \cdot \\ & & & \beta_{nn} \end{pmatrix}, \qquad n_{K/L}(\alpha) = |A'| = |B'| = \prod_{i=1}^{n} \beta_{ii}.$$

Now, the m-row matrices A''_{ik} are representations of the α_{ij}. Transformations of the matrix $A = (A_{ij})$ corresponding to those of $A' = (\alpha_{ij})$ bring it into the form

$$B = \begin{pmatrix} B''_{11} & & & \\ & \cdot & & \\ & & \cdot & \\ * & & & \cdot \\ & & & B''_{nn} \end{pmatrix} \qquad \text{with} \quad \beta_{ii}\eta = \eta B''_{ii}$$

and then

$$n_{K/k}(\alpha) = |A| = |B| = \prod_{i=1}^{n} |B''_{ii}| = \prod_{i=1}^{n} n_{L/k}(\beta_{ii})$$

$$= n_{L/k}\left(\prod_{i=1}^{n} \beta_{ii}\right) = n_{L/k}(n_{K/L}(\alpha)) . \ \blacktriangledown$$

3. The Discriminant

With any n elements $\omega_i \in K$ one associates the determinant

$$D(\omega_i) = D(\omega) = |s(\omega_i \omega_j)|$$

and calls it their *discriminant*. Under the change of basis $\omega \to \omega' = \omega C$ with a matrix C over k we have, for the matrix of this determinant,

$$(s(\omega_i'\omega_j')) = C^t(s(\omega_i\omega_j))C,$$

so that

$$D(\omega') = D(\omega)|C|^2 \qquad (\omega' = \omega C).$$

Thus it is clear that $D(\omega) = 0$ whenever the ω_i are linearly dependent, although the converse does not always hold.

Now let the configuration of §4,2 prevail. We want to prove the *transitivity formula*

$$D_{K/k}(\omega \times \eta) = n_{L/k}(D_{K/L}(\omega))D_{L/k}(\eta)^{[K:L]}. \tag{7}$$

To avoid confusion we use subscrpits to denote the extension which is meant.

Equations (2) and (3) yield

$$D_{K/k}(\omega \times \eta) = |s_{K/k}(\omega_i\eta_{i'}\omega_j\eta_{j'})| = |s_{L/k}(\eta_{i'}s_{K/L}(\omega_i\omega_j)\eta_{j'})|,$$

where we are dealing with an mn-row determinant whose columns and rows are each given by double subscripts ii' and jj', respectively. Into this formula we substitute the representation of $s_{K/L}(\omega_i\omega_j)$ by the basis, i.e.,

$$s_{K/L}(\omega_i\omega_j)\eta = \eta S_{ij}.$$

For any fixed pair of subscripts i, j we have the equation

$$(s_{L/k}(\eta_{i'}s_{K/L}(\omega_i\omega_j)\eta_{j'})) = (s_{L/k}(\eta_{i'}\eta_{j'}))S_{ij} = DS_{ij}$$

with the m-row matrices D and S_{ij}. These matrices, ordered into an n-row scheme, yield the matrix of $D_{K/k}(\omega \times \eta)$. Thus, now

$$D_{K/k}(\omega \times \eta) = |S_{ij}|\ |D|^n.$$

The S_{ij} are representations of the elements $\sigma_{ij} = S_{K/L}(\omega_i\omega_j)$ of L. Elementary transformations of the n-row matrix (σ_{ij}) over L and the corresponding transformations of the mn-row matrix (S_{ij}) over k lead, as in §4,3, to

$$|S_{ij}| = n_{L/k}(|\sigma_{ij}|) = n_{L/k}(D_{K/L}(\omega)).$$

Thus (7) is proven. ▼

4. Separable and Inseparable Extensions

If K/k is separable there exists, as is known, an (primitive) element ϑ that generates K/k. Let $\vartheta = \vartheta_1, \ldots, \vartheta_n$ be the elements conjugate to ϑ. Then

$$D(1, \vartheta, \cdots, \vartheta^{n-1}) = \left|\sum_{\nu=1}^{n} \vartheta_\nu^{i+j}\right| = \begin{vmatrix} 1 & \cdots & 1 \\ \vartheta_1 & \cdots & \vartheta_n \\ \cdot & \cdots & \cdot \\ \vartheta_1^{n-1} & \cdots & \vartheta_n^{n-1} \end{vmatrix} \begin{vmatrix} 1 & \vartheta_1 & \cdots & \vartheta_1^{n-1} \\ \cdot & \cdot & \cdots & \cdot \\ \cdot & \cdot & \cdots & \cdot \\ 1 & \vartheta_n & \cdots & \vartheta_n^{n-1} \end{vmatrix}.$$

As the value of the Vandermonde determinant on the right is known, we now have

$$D(1, \vartheta, \ldots, \vartheta^{n-1}) = \prod_{i<j} (\vartheta_i - \vartheta_j)^2.$$

The discriminant of a particular and thus, of course, of every basis of K/k is $\neq 0$ if K/k is separable.

In the case of a simple inseparable extension $K = k(\sqrt[p]{a})$ of a field of characteristic p all traces are 0, and therefore so is the discriminant of any basis. By the transitivity formula the same holds true for arbitrary inseparable extensions.

This state of affairs renders the trace and the discriminant useless for the study of inseparable extensions. As a replacement we introduce a *pseudotrace.*† What we need is a function $\sigma(\alpha)$ of the elements $\alpha \in K$ with values in k with the properties

$$\sigma(\alpha + \beta) = \sigma(\alpha) + \sigma(\beta), \qquad \sigma(a\alpha) = a\sigma(\alpha) \qquad \text{for} \quad a \in k, \tag{8}$$

$$\Delta(\omega) = |\sigma(\omega_i \omega_j)| \neq 0. \qquad \text{for a basis } \omega_i \text{ of } K/k. \tag{9}$$

The last determinant is called the *pseudodiscriminant.*

Given three fields $k \subset L \subset K$ with pseudotraces for K/L and L/k, then

$$\sigma_{K/k}(\alpha) = \sigma_{L/k}(\sigma_{K/L}(\alpha)) \tag{10}$$

is a pseudotrace for K/k. Equations (8) obviously hold. The proof of the transitivity formula (7) of §4,3 carries over word for word, yielding the transitivity formula

$$\Delta_{K/k}(\omega \times \eta) = n_{L/k}(\Delta_{K/L}(\omega))\Delta_{L/k}(\eta)^{[K:L]} \tag{11}$$

for pseudotraces. Thus (9) also holds for the pseudotrace defined by (10).

For separable extensions the trace is a pseudotrace. For a purely inseparable extension $K = k(\sqrt[p]{a})$ every $\alpha \in K$ can be represented in the form

$$\alpha = s_0 + s_1\sqrt[p]{a} + \cdots + s_{p-1}(\sqrt[p]{a})^{p-1}.$$

The last coefficient is a pseudotrace:

$$\sigma(\alpha) = s_{p-1}.$$

Equation (8) is immediately evident, and the pseudodiscriminant becomes

$$\Delta(1, \sqrt[p]{a}, \ldots) = (-1)^{p(p-1)/2}.$$

---

† The concept is due to E. Lamprecht, *Über s-Differenten und Differentiale algebraischer Funktionenkörper einer Veränderlichen*, Arch. Math. **4**, 412–424 (1953). The term pseudotrace (Pseudospur) is suggested here for the first time.

Using (10) one can now prove the existence of a pseudotrace for an arbitrary extension K/k.

If σ and σ' are two pseudotraces, then there exists a uniquely determined element $\delta \neq 0$ in K such that

$$\sigma'(\alpha) = \sigma(\delta\alpha) \tag{12}$$

holds for all $\alpha \in K$. Conversely, if $\delta \neq 0$, then $\sigma(\delta\alpha)$ is always a pseudotrace.

Proof. With a basis ω_i of K/k, set

$$\alpha = \sum \omega_i a_i, \qquad \delta = \sum \omega_i d_i$$

with as yet undetermined coefficients d_i. Also put

$$\sigma(\delta\alpha) = \sum_{i,j} d_i\sigma(\omega_i\omega_j)a_j = \sum_j \sigma'(\omega_j)a_j = \sigma'(\alpha).$$

This is correct for all α if and only if

$$\sum_i d_i\sigma(\omega_i\omega_j) = \sigma'(\omega_j) \qquad (j = 1, 2, \ldots)$$

holds. By our assumptions the determinant of this linear system of equations is not zero, and thus such a δ exists. If conversely such a δ is given, then define σ' with (12) so that (8) holds. Further,

$$\delta\omega = \omega D$$

implies

$$|\sigma'(\omega_i\omega_j)| = |\sigma(\delta\omega_i\omega_j)| = |\sigma(\omega_i\omega_j)|\ |D| = \Delta(\omega)n(\delta),$$

which is $\neq 0$. ▼

Appendix to Chapter I

The Theta Function

This function is among the most important tools of analytic number theory. Its theory is built primarily on completely elementary considerations. This justifies its insertion at this point.

§1*. The Symplectic Group

1. The Basic Properties

Let A, B, C, D be n-row matrices with coefficients in R, E the n-row unit matrix, and 0 the zero matrix. Finally let

$$J = \begin{pmatrix} 0 & E \\ -E & 0 \end{pmatrix}.$$

The *symplectic group* Σ^n of the nth degree over the field of real numbers consists of all $2n$-row matrices $M = \begin{pmatrix} A & B \\ C & D \end{pmatrix}$ which satisfy the equation

$$M^t J M = \begin{pmatrix} A^t & C^t \\ B^t & D^t \end{pmatrix} \begin{pmatrix} 0 & E \\ -E & 0 \end{pmatrix} \begin{pmatrix} A & B \\ C & D \end{pmatrix} = J$$

or, written otherwise,

$$M^{-1} = J^{-1} M^t J = \begin{pmatrix} D^t & -B^t \\ -C^t & A^t \end{pmatrix}. \tag{2}$$

Computation yields the three equivalent equations

$$A^t C - C^t A = 0, \qquad B^t D - D^t B = 0, \qquad A^t D - C^t B = E. \tag{3}$$

The group Σ^1 consists of all real two-row matrices with determinant 1.

The group property of Σ^n follows from (2), and since $J \in \Sigma^n$, M^t is in Σ^n whenever M is.

Special symplectic matrices are

$$U_V = \begin{pmatrix} V & 0 \\ 0 & V^* \end{pmatrix} \quad \text{with} \quad V^* = (V^t)^{-1}, \qquad J, \qquad T_S = \begin{pmatrix} E & S \\ 0 & E \end{pmatrix} \quad \text{with} \quad S^t = S. \tag{4}$$

These matrices generate the symplectic group Σ^n.

Proof. The given symplectic matrix $\begin{pmatrix} A & B \\ C & D \end{pmatrix}$ is first multiplied from the left by a U_V and from the right by a $U_{V'}$. This transforms A to VAV'. Suitably chosen V, V' put VAV' into the diagonal form with only ones and zeros as entries: $\begin{pmatrix} E_1 & 0 \\ 0 & 0 \end{pmatrix}$. Thus it is no loss of generality to assume A has this form from the start. Subdividing C into submatrices of the same dimensions, the first of Eqs. (3) yields

$$C = \begin{pmatrix} C_{11} & 0 \\ C_{21} & C_{22} \end{pmatrix}, \qquad C_{11}^t = C_{11}.$$

Here $|C_{22}| \neq 0$, for otherwise the first n columns of $\begin{pmatrix} A & B \\ C & D \end{pmatrix}$ would be linearly dependent, which would make the determinant 0, incompatible with (2). Now multiply $\begin{pmatrix} A & B \\ C & D \end{pmatrix}$ from the left by $T_{\lambda E}$, λ being an as yet undetermined parameter. A transforms into

$$A' = A + \lambda C = \begin{pmatrix} E_1 + \lambda C_{11} & 0 \\ \lambda C_{21} & \lambda C_{22} \end{pmatrix}.$$

Choice of a suitable λ assures that the determinant $|A'| \neq 0$. Thus we may assume in full generality that $|A| \neq 0$, and thus even arrive at $A - E$ with our first argument. For $A = E$, C must be symmetric because of the first of Eqs. (3). Multiply from the left with $J^{-1} T_C J = \begin{pmatrix} E & 0 \\ -C & E \end{pmatrix}$ so that A and C go into E and 0, and only the matrix $\begin{pmatrix} E & B \\ 0 & D \end{pmatrix}$ remains to be treated. Now the two last equations (3) yield $D = E$ and $B^t = B$, and the matrix is equal to T_B.

This completes the proof. ▼ As an incidental result we see that all the matrices (4) have determinant 1, and thus:

The determinant of a symplectic matric is 1.

2. Symplectic Geometry

In the following sections we will consider the space $\mathfrak{H}^n$ of n-row complex symmetric matrices

$$\mathrm{T} = X + iY,$$

whose imaginary components $Y = (y_{ij})$ are matrices of positive definite quadratic forms in n variables ξ_i. This means that

$$f(\xi) = \sum_{i,j=1}^{n} y_{ij}\xi_i\xi_j > 0$$

for all real values of the ξ_i not identically 0. We will call the matrix Y itself *positive definite* for short.

Under linear fractional transformations

$$\mathrm{T} \to \mathrm{T}' = (A\mathrm{T} + B)(C\mathrm{T} + D)^{-1} = M(\mathrm{T}) \tag{5}$$

with symplectic matrices $M = \left(\begin{smallmatrix} A & B \\ C & D \end{smallmatrix}\right)$, *the space* $\mathfrak{H}^n$ *goes into itself.*

Proof. If

$$\begin{pmatrix} A & B \\ C & D \end{pmatrix}\begin{pmatrix} A' & B' \\ C' & D' \end{pmatrix} = \begin{pmatrix} A'' & B'' \\ C'' & D'' \end{pmatrix} \quad \text{or} \quad MM' = M'',$$

then

$$M(M'(\mathrm{T})) = M''(\mathrm{T}).$$

Hence it suffices to prove the assertion for the special matrices (4) which, as we know from the last section, generate the entire group Σ^n. For these we have

$$\mathrm{T}' = V\mathrm{T}V^t, \qquad \mathrm{T}' = -\mathrm{T}^{-1}, \qquad \mathrm{T}' = \mathrm{T} + S \quad (S^t = S),$$

respectively. In all three cases T′ is again a symmetric matrix, and in the first and third it is easy to see that the imaginary component is also positive definite. As one can, with real transformations, make a positive definite quadratic form the sum of the squares of the variables, there always exists a real matrix V and symmetric matrix Z such that

$$\mathrm{T} = X + iY = V(Z + iE)V^t = U_V(Z + iE).$$

It remains to show that $-(Z + iE)^{-1}$ has a positive definite imaginary component. $E + Z^2 = E + ZZ^t$ is positive definite and thus has an inverse. Verifying

$$-(Z + iE)^{-1} = -Z(E + Z^2)^{-1} + i(E + Z^2)^{-1}$$

gives the assertion. ▼

The space $\mathfrak{H}^n$ is convex and thus simply connected.

Proof. Let $T_1 = X_1 + iY_1$ and $T_2 = X_2 + iY_2$ be two matrices in $\mathfrak{H}^n$. For all s between 0 and 1 it is seen that $sY_1 + (1-s)Y_2$ is positive definite, and thus $sT_1 + (1-s)T_2 \in \mathfrak{H}^n$. ▼

3. The Hyperbolic Plane and Hyperbolic Space

For $n = 1$, $T = \tau = x + iy$ is a complex number, and $\mathfrak{H}^1$ is the complex upper half plane. The transformations (5) are

$$\tau \to \tau' = (a\tau + b)/(c\tau + d).$$

As Poincaré showed, $\mathfrak{H}^1$ is a model of the hyperbolic plane, and the fractional linear substitutions represent hyperbolic motions. The lines and circles perpendicular to the real axis play the part of the hyperbolic lines. Angles are given faithfully by the model. The hyperbolic surface element is $y^{-2}\,dx\,dy$.

Three-dimensional hyperbolic space can be generated by a specialization of symplectic geometry. As is well known, complex numbers $\alpha = a_0 + ia_1$ can be represented by two-row matrices

$$G(\alpha) = \begin{pmatrix} a_0 & a_1 \\ -a_1 & a_0 \end{pmatrix}.$$

Set

$$H = \begin{pmatrix} 1 & 0 \\ 0 & -1 \end{pmatrix},$$

so that

$$HG(\alpha)H = G(\alpha)^t = G(\bar{\alpha})$$

is a representation of the conjugate complex number. Let $\alpha, \beta, \gamma, \delta$ be four given complex numbers with determinant $\alpha\delta - \beta\gamma = 1$. It is easy to verify that the four-row matrix

$$\begin{pmatrix} A & B \\ C & D \end{pmatrix} = \begin{pmatrix} H & 0 \\ 0 & E \end{pmatrix} \begin{pmatrix} G(\alpha) & G(\beta) \\ G(\gamma) & G(\delta) \end{pmatrix} \begin{pmatrix} H & 0 \\ 0 & E \end{pmatrix} = \begin{pmatrix} HG(\alpha)H & HG(\beta) \\ G(\gamma)H & G(\delta) \end{pmatrix} \tag{6}$$

is symplectic.

Fractional linear transformations (5) *belonging to symplectic matrices of the form* (6) *applied to the variable matrix*

$$T = \begin{pmatrix} x_0 + ix_2 & x_1 \\ x_1 & -x_0 + ix_2 \end{pmatrix} \tag{7}$$

with real x_0, x_1, x_2 lead back to matrices of the special form (7). *Moreover, the half plane $x_2 > 0$ goes into itself.*

It is easily verified that the matrices (6) form a group isomorphic to the complex two-row matrices $\begin{pmatrix} \alpha & \beta \\ \gamma & \delta \end{pmatrix}$ of determinant 1. The latter group is generated by the matrices $\begin{pmatrix} \alpha & 0 \\ 0 & \alpha^{-1} \end{pmatrix}$, $\begin{pmatrix} 0 & 1 \\ -1 & 0 \end{pmatrix}$, $\begin{pmatrix} 1 & 1 \\ 0 & 1 \end{pmatrix}$. Thus, to prove the asserted theorem, one need but consider the transformations corresponding to these special matrices. In these cases

$$\mathrm{T}' = HG(\alpha)H\mathrm{T}G(\alpha^{-1})^{-1} = G(\bar{\alpha})\mathrm{T}G(\alpha), \qquad \mathrm{T}' = -H\mathrm{T}^{-1}H, \qquad \mathrm{T}' = \mathrm{T} + H,$$

and the assertion follows from a simple calculation each time. ▼

The matrices (6) form a subgroup of Σ^2, and the space of real dimension 3 spanned by the matrices (7) with $x_2 > 0$ is a subspace of $\mathfrak{H}^2$, which is of real dimension 6. *It is the three-dimensional hyperbolic space, the transformations* (5) *with matrices* (6) *being the hyperbolic motions.*†

To prove this, one first verifies that the hemispheres

$$(x_0 - c_0)^2 + (x_1 - c_1)^2 + x_2{}^2 = r^2, \qquad x_2 > 0 \tag{8}$$

are invariant under the three special transformations. The planes $x_0 c_0 + x_1 c_1 = c_2$ must also be included as a special case. Together they play the part of hyperbolic planes. Each plane can be transformed into one fixed plane, for example, the plane (8) under $\begin{pmatrix} 0 & E \\ -E & 0 \end{pmatrix}\begin{pmatrix} E & HG(c_0 - r + ic_1) \\ 0 & E \end{pmatrix}$ into $x_0 = -(2r)^{-1}$, or the plane $x_0 c_0 + x_1 c_1 = c_2$ under $\begin{pmatrix} E & HG(-ic_2) \\ 0 & E \end{pmatrix}\begin{pmatrix} HG(\alpha)H & 0 \\ 0 & G(\alpha^{-1}) \end{pmatrix}$ with $\alpha^2(c_0 - ic_1) = -i$ into $x_1 = 0$.

The plane $x_1 = 0$ is transformed into itself under the substitutions (6) with real $\alpha, \beta, \gamma, \delta$, these having the form

$$x_0' + ix_2' = \frac{\alpha(x_0 + ix_2) + \beta}{\gamma(x_0 + ix_2) + \delta}.$$

The geometry induced here is, thus, hyperbolic. This confirms the assertion. ▼

4. The Symplectic Modular Group

The symplectic modular group is the subset $\Gamma^n \subset \Sigma^n$ whose matrices have rational integral coefficients. The group property becomes obvious when one calculates the inverse element using (2).

If $M = \begin{pmatrix} A & B \\ C & D \end{pmatrix}$ is symplectic, then by (3) the matrices $C^t A$ and $B^t D$ are symmetric. Consider the subset $\Theta^n \subset \Gamma^n$ for which the diagonal elements of $C^t A$, $B^t D$ are even numbers; we will show that it also is a group. A quick

---

† Hyperbolic geometry of n dimensions and its group of motions are investigated in K. Th. Vahlen, *Über Bewegungen und komplexe Zahlen,* Math. Ann. **55**, 585–593 (1902), using Clifford algebras.

calculation becomes necessary. Let $M_1, M_2 \in \Gamma^n$ be given. For their product matrix $M = M_1 M_2$ we have

$$C^t A = A_2{}^t C_1{}^t A_1 A_2 + C_2{}^t D_1{}^t B_1 C_2 + C_2{}^t D_1{}^t A_1 A_2 + A_2{}^t C_1{}^t B_1 C_2,$$
$$B^t D = B_2{}^t A_1{}^t C_1 B_2 + D_2{}^t B_1{}^t D_1 D_2 + D_2{}^t B_1{}^t C_1 B_2 + B_2{}^t A_1{}^t D_1 D_2.$$

Applying (3), this yields

$$C^t A = A_2{}^t C_1{}^t A_1 A_2 + C_2{}^t B_1{}^t D_1 C_2 + C_2{}^t D_1{}^t A_1 A_2 + A_2{}^t A_1{}^t D_1 C_2 - C_2{}^t A_2,$$
$$B^t D = B_2{}^t C_1{}^t A_1 B_2 + D_2{}^t B_1{}^t D_1 D_2 + B_2{}^t A_1{}^t D_1 D_2 + D_2{}^t D_1{}^t A_1 B_2 - B_2{}^t D_2.$$

In the present connection only the diagonals of these matrices are of interest, and in particular only their residue classes mod 2. The third and fourth summands are mirror images and may be omitted. Further, we call to mind the fact that a symmetric matrix $M = (m_{ij})$ with integral coefficients and even diagonal entries m_{ii} has the property that, for integral values of the variables, its corresponding quadratic form $\sum m_{ij} x_i x_j$ gives only even values, and that this property in a quadratic form already implies that all the m_{ij} are integral and the m_{ii} are even. Applying an integral linear substitution with matrix N, the transformed form will again yield only even values. Thus, the diagonal entries of the matrix $N^t M N$ are again even. For this reason one can replace $C_1{}^t A_1$ and $B_1{}^t D_1$ in the above equations by the diagonal matrix formed with their diagonal elements. Using natural abbreviations this yields

$$(C^t A)_j \equiv \sum_i (C_1{}^t A_1)_i (A_2)^2_{ij} + \sum_i (B_1{}^t D_1)_i (C_2)^2_{ij} + (C_2{}^t A_2)_j \bmod 2$$

and another congruence of a similar sort. As $a^2 \equiv a \bmod 2$ holds for every rational integer, one can rewrite these congruences by making column vectors ξ'_M, η'_M out of the diagonal elements of $C^t A$ and $B^t D$, the components of which will only be considered to be residue classes mod 2, however. These n-term column vectors are further combined to $2n$-term column vectors $\zeta'_M = (\xi'_M \; \eta'_M)$, and yield the equation

$$\xi'_M = \zeta'_{M_1 M_2} = \zeta'_{M_1} M_2 + \zeta'_{M_2}.$$

As Θ^n is characterized by $\xi'_M = 0$, we can immediately see that Θ^n is a semigroup. However, we shall hereafter produce a system of generators of Θ^n whose inverse elements also belong to Θ^n. Together this gives: Θ^n *is a group*.

Taking this result for granted, (2) or $M^t = J M^{-1} J^{-1}$ imply that M^t belongs to Θ^n along with M. We now set $\xi_M = \xi'_{M^t}, \eta'_M = \eta'_{M^t}$; these are column vectors whose components are the residue classes of the diagonal entrics of BA^t and CD^t mod 2. Further, set $\zeta_M = (\xi_M \eta_M)$. With this, the last equation can be expressed as

$$\zeta_M = \zeta_{M_1 M_2} = \zeta_{M_2} M_1{}^t + \zeta_{M_1}, \tag{9}$$

in which form we shall apply it later.

The group Γ^n is generated by the matrices U_V, J, T_S of (4), *with unimodular V and integral symmetric S. The group Θ^n is generated by these U_V, J, and T_S, where the diagonal elements of S are even, in addition to the elements*

$$J_\nu = \begin{pmatrix} E_\nu & E - E_\nu \\ E_\nu - E & E_\nu \end{pmatrix}, \quad E_\nu = \begin{pmatrix} 1 & & & & & & \\ & \cdot & & & & & \\ & & \cdot & & & & \\ & & & 1 & & & \\ & & & & 0 & & \\ & & & & & \cdot & \\ & & & & & & \cdot \end{pmatrix}, \quad \nu = 1, 2, \ldots, n-1.$$

E_ν has ν ones in the diagonal, and zeros otherwise. $J_\nu^4 = E$.

The *proof* dates back to Witt.† Say that a given element $\left(\begin{smallmatrix} A & B \\ C & D \end{smallmatrix}\right)$ of Γ^n or Θ^n has been brought into such a form by left multiplication with permitted U_V, J, and T_S, that the product of the g.c.d. of the first columns of A and C is as small as possible while the first is not smaller than the second:

$$\text{g.c.d.}(a_{11}, \ldots, a_{n1}) \cdot \text{g.c.d.}(c_{11}, \ldots, c_{n1}) = \min,$$

$$\text{g.c.d.}(a_{11}, \ldots, a_{n1}) \geqslant \text{g.c.d.}(c_{11}, \ldots, c_{n1}).$$

Then multiply from the left with a suitable unimodular U_V, so that the first column of C contains only zeros, with the possible exception of c_{11}.

If $c_{11} \neq 0$, continue to multiply from the left, with a T_S. If $S = (s_{ij})$, then a_{i1} goes into $a_{i1} + c_{11}s_{i1}$. Suitable choice of the s_{i1} will give $|a_{i1}| < |c_{11}|$ for all i and then also g.c.d. $(a_{i1}) < |c_{11}|$, which, however, contradicts the assumption made. When dealing with Θ^n, one is limited to an even number for s_{11}. Nevertheless the same argument holds with one exception: $a_{21} = \cdots = a_{n1} = 0$, while a_{11} is divisible by c_{11} but not by $2c_{11}$. But then the determinant is divisible by c_{11}, which must therefore be ± 1. Further, C^tA has the element $a_{11}c_{11}$ as entry with subscript 11, and it is odd. But this cannot occur for an element of Θ^n.

Consequently, under the assumptions made the first column of C consists of zeros only. By multiplying from the left with a suitable U_V, one can bring the first column of A to be 1, 0, ..., 0.

Applications of the first and last of Eqs. (3) now shows that the first row

† *Eine Identität zwischen Modulformen zweiten Grades*, Abh. Math. Sem. Univ. Hamburg **14**, 323–337 (1941).

of C is $0, \ldots, 0$ and the first row of D is $1, 0, \ldots, 0$. Left multiplication by U_V with

$$V = \begin{pmatrix} 1 & d_{21} & \cdots & d_{n1} \\ 0 & 1 & \cdots & 0 \\ \cdot & \cdot & \cdots & \cdot \\ 0 & 0 & \cdots & 1 \end{pmatrix}$$

brings the first column of D to $1, 0, \ldots, 0$, while the already fixed rows and columns remain invariant.

Finally multiplication from the left with a suitable T_S makes the first column of B into $0, \ldots, 0$, leaving all of C, D and the first column of A unchanged. This works not only for Γ^n, but also for Θ^n. For in the latter case, $\left(\begin{smallmatrix} A & B \\ C & D \end{smallmatrix}\right) \in \Theta^n$ together with the normed form of D implies that b_{11} is even; b_{11} is made into 0 be setting $s_{11} = -b_{11}$. Applying the second of Eqs. (3) shows that the first row of B consists only of zeros. And by the third of these equations the first row of A is $1, 0, \ldots, 0$.

If n was 1, then the remaining element M is the unit matrix, and the proof is finished. Assume the assertion to be correct for $n - 1$ in place of n, and cross out the first and $(n+1)$th columns and rows of M. The remaining matrix $M^{(n-1)} \in \Gamma^{n-1}$ or Θ^{n-1}, respectively. It can be represented by the respective $U^{(n-1)}, J^{(n-1)}, T^{(n-1)}$, and $J_\nu^{(n-1)}$. By completing these matrices to $2n$-row symplectic matrices by annexation of two further columns and rows of the forms $1, \ldots, 0, 0, \ldots, 0$, and $0, \ldots, 0, 1, \ldots, 0$, one arrives at a representation of M as desired, at least in the case of the group Θ^n.

In the case of the group Γ^n one must still demonstrate that the matrix J_1, arising from completion of $J^{(n-1)}$, can be represented by the given generating elements. This is done by verifying the simple identity

$$J_\nu = -(JT_{-E_\nu})^3 \tag{10}$$

E_ν as in (9) for $\nu = 1$. ▼ (We shall need (10) later for $\nu > 1$.)

5. The Fundamental Domain

Consider the case $n = 1$, in particular. The modular group Γ^1 is generated by the three elements

$$U_{-1} = \begin{pmatrix} -1 & 0 \\ 0 & -1 \end{pmatrix}, \quad J = \begin{pmatrix} 0 & 1 \\ -1 & 0 \end{pmatrix}, \quad T = \begin{pmatrix} 1 & 1 \\ 0 & 1 \end{pmatrix}.$$

Under the corresponding fractional linear substitutions of the complex upper half plane, U_{-1} is the identical mapping.

By a *fundamental domain* $\mathfrak{F}$ of Γ^1 in $\mathfrak{H}^1$, a point set with the following two properties is meant: (1) to every $\tau \in \mathfrak{H}^1$ there exists exactly one point $\sigma \in \mathfrak{F}$

and one $M \in \Gamma^1$ with $M(\tau) = \sigma$; (2) in every neighborhood of every $\sigma \in \mathfrak{F}$ there are inner points of $\mathfrak{F}$.

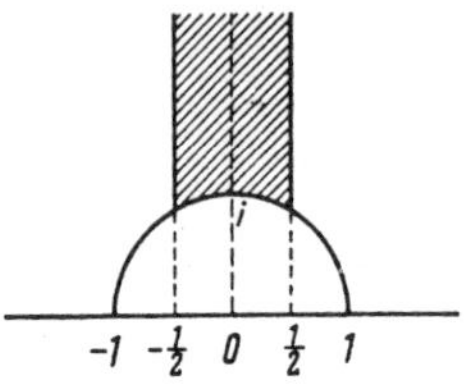

FIG. 1

Figure 1 depicts a fundamental domain of Γ^1 in $\mathfrak{H}^1$, the so-called *modular triangle*. Here the left half of the boundary, up to and including the point $\sigma = i$, is to be included in $\mathfrak{F}$, but not the other boundary points.

Proof. Using T, T^{-1}, and J, further "equivalent" triangles are attached to $\mathfrak{F}$ on its three sides. Further, the substitutions TJ, $(TJ)^2$, JT, and $(JT)^2$ represent rotations of angles $2\pi/3$ and $4\pi/3$ about the points $1/2 + i\sqrt{3}/2$ and $-1/2 + i\sqrt{3}/2$ of $\mathfrak{F}$. The images $\mathfrak{F}, T^{-1}\mathfrak{F}, JT\mathfrak{F}, JTT^{-1}\mathfrak{F}, (JT)^2\mathfrak{F}, (JT^2)T^{-1}\mathfrak{F}$ fill out the entire neighborhood of the left corner. The neighborhood of the right corner can be covered similarly.

The images $M\mathfrak{F}$ for all $M = T^{\alpha_1}JT^{\alpha_2} \cdots T^{\alpha_n}$ combine smoothly and without gaps to a subregion $\overline{\mathfrak{H}}^1$ of $\mathfrak{H}^1$. For if $M_i\mathfrak{F}$ are the neighbors of $\mathfrak{F}$ and M is arbitrary in Γ^1, then $MM_i\mathfrak{F}$ are the neighbors of $M\mathfrak{F}$.

We must now show that $\overline{\mathfrak{H}}^1 = \mathfrak{H}^1$. Were $\overline{\mathfrak{H}}^1 \subset \mathfrak{H}^1$, then $\overline{\mathfrak{H}}^1$ would have a boundary point τ_1. Let τ be an inner point of $\overline{\mathfrak{H}}^1$ near τ_1. There exists an $M = T^{\alpha_1}JT^{\alpha_2} \cdots$ such that $\sigma = M(\tau)$ lies in $\mathfrak{F}$. As M is a hyperbolic motion, the hyperbolic distance between the points $\sigma_1 = M(\tau_1)$ and σ is the same as between the points τ_1 and τ. This would imply the existence of a point σ of $\mathfrak{F}$ and a neighboring point σ_1 belonging neither to $\mathfrak{F}$ nor to a neighboring region. But this is impossible. The consequence is that every point of $\mathfrak{H}^1$ can be transformed to a point of $\mathfrak{F}$ by some $M = T^{\alpha_1}JT^{\alpha_2} \cdots$.

The $M\mathfrak{F}$ with $M = T^{\alpha_1}JT^{\alpha_2} \cdots$ cover $\mathfrak{H}$ simply. Indeed each $M\mathfrak{F}$ is surrounded by the three neighbors $MT^{-1}\mathfrak{F}$, $MT\mathfrak{F}$, $MJ\mathfrak{F}$ which, along with $M\mathfrak{F}$, cover the neighborhood of $M\mathfrak{F}$ simply. Since $\mathfrak{H}$ is simply connected the covering can also be but simple in the large. As a consequence, no point of $\mathfrak{F}$ can be transformed to another point of $\mathfrak{F}$ by such an M. It is only possible that a point of $\mathfrak{F}$ remains fixed under such a transformation, and this is the case for the points $\frac{1}{2}(-1 + \sqrt{-3})$ and i under the substitutions JT, $(JT)^2$, and J. This implies that $\mathfrak{F}$ is a fundamental domain for the group generated by these substitutions.

The last section showed that this group coincides with the entire modular group Γ^1. The proof is complete. ▼

The discussion also yields the fact that those substitutions M_i which transform the fundamental domain into a neighboring domain generate the group Γ^1. This clearly holds true for all groups which have fundamental domains. On the other hand, a system of generators of a group need, by no means, correspond to a fundamental domain in this sense.

The existence of a fundamental domain for Γ^n is proved by Siegel.† It is difficult to describe because of the behavior of the subgroup of the U_V. There is, however, even an arithmetically distinguished system of only four generators for Γ^n.‡

6. The Theta Function

We introduce a new notation before defining this function. $\mathrm{T} = (\tau_{ij})$ will always denote an n-row symmetric matrix with complex coefficients whose imaginary part is positive definite in the sense of the beginning of this section. Also, m, x, and y will be one-column matrices with the entries m_i, x_i, and y_i $(i = 1, \dots, n)$, respectively. The m_i are always rational integers, the x_i and y_i arbitrary complex numbers. We use the abbreviation

$$\mathrm{T}[m] = m^t\mathrm{T}m = \sum_{i,j=1}^{n} \tau_{ij} m_i m_j, \qquad x^t y = y^t x = \sum_{i=1}^{n} x_i y_i.$$

The function announced in the title is

$$\vartheta(\mathrm{T}, x, y) = \sum_{m} \exp(\pi i \mathrm{T}[m - y] + 2\pi i m^t x - \pi i x^t y), \tag{11}$$

where the summation is over all integral m. The series, along with all its partial derivatives with respect to all variables, converges absolutely and uniformly in every compact domain in which the imaginary part of T is positive definite.

In the next section we shall prove the *reciprocity formula*

$$\vartheta(\mathrm{T}, x, y) = |-i\mathrm{T}|^{-1/2} \vartheta(-\mathrm{T}^{-1}, y, -x). \tag{12}$$

The square root in (12) is to be taken positive when T is purely imaginary. In general the sign is found by analytic continuation of the function $|i\mathrm{T}|^{-1/2}$ in $\mathfrak{H}^n$. No ambiguity can arise, as $\mathfrak{H}^n$ was shown to be simply connected.

A simple calculation directly from the definition yields the functional equation for a unimodular V

$$\vartheta(\mathrm{T}, x, y) = \vartheta(V\mathrm{T}V^t, Vx, V^*y), \qquad V^* = (V^t)^{-1}. \tag{13}$$

Finally, for an integral symmetric matrix S,

$$\vartheta(\mathrm{T}, x, y) = \exp[-\pi i \xi(S)^t y] \vartheta(\mathrm{T} + S, x + Sy - \xi(S), y) \tag{14}$$

---

† *Einführung in die Theorie der Modulfunktionen ersten Grades*, Math. Ann. **116**, 617–657 (1939); also *Symplectic geometry*, Amer. J. Math. **65**, 1–86 (1943).

‡ L. K. Hua und I. Reiner, *On the generators of the symplectic modular group*, Trans. Amer. Math. Soc. **65**, 415–426 (1949); also H. Klingen, *Über die Erzeugenden gewisser Modulgruppen*, Nachr. Akad. Wiss. Göttingen Math.-Phys. Kl. **IIa**, 173–185 (1956).

holds, where $\xi(S)$ is the one-column matrix with entries $\frac{1}{2}s_{ii}$ $(S=(s_{ij}))$. To prove (14) we must, because of (11), verify

$$\tfrac{1}{2}\mathrm{T}[m-y]+(m-\tfrac{1}{2}y)^t x \equiv \tfrac{1}{2}(\mathrm{T}+S)[m-y] + (m-\tfrac{1}{2}y)^t(x+Sy-\xi(S)) - \tfrac{1}{2}\xi(S)^t y \bmod 1.$$

A quick calculation reduces this to

$$\tfrac{1}{2}m^t Sm - \xi(S)^t m \equiv 0 \bmod 1.$$

This congruence arises from

$$\tfrac{1}{2}m^t Sm = \tfrac{1}{2}\sum_i m_i^2 s_{ii} \equiv \tfrac{1}{2}\sum_i m_i s_{ii} = \xi(S)^t m \bmod 1.$$

The functional equations (12) to (14) can be combined to one relation:

$$\vartheta(\mathrm{T},x,y)=|C\mathrm{T}+D|^{-1/2} \times \exp[\pi i(\lambda_M - \tfrac{1}{2}\zeta_M JMz)]\vartheta(M(\mathrm{T}), Ax+By-\tfrac{1}{2}\xi_M^t, Cx+Dy-\tfrac{1}{2}\eta_M^t) \quad (15)$$

with $\lambda_M=0$; here ξ_M, η_M, and $\zeta_M=(\xi_M\eta_M)$ are integral row vectors and z is the column vector $\binom{x}{y}$. In the cases $M=J$ and $M=U_V$, ζ_M is even the zero vector, while for $M=T_S$, η_M is also zero and $\xi_M=2\xi(S)$ lies in the residue class mod 2 of the vector ξ_{T_S} introduced in §1,4.

The functional equation (15) *holds for all* $M\in\Gamma^n$. *Here* λ_M *is a rational number with denominator dividing* 4. ξ_M, η_M, and $\zeta_M=(\xi_M\eta_M)$ *are integral row vectors, representing the residue classes introduced under these notations in* §1,4; *when* ζ_M *is changed for another representative,* λ_M *must be changed accordingly.*

Proof. One first verifies that the right side of (15) goes into itself under an arbitrary variation of ζ_M mod 2 and a suitably chosen variation of λ_M. It suffices to do this for the unit matrix M.

As Γ^n is generated by J, U_V, and T_S and (15) holds for these elements, one must verify the following three equations (where $M_1M_2=M$):

$$(C_1M_2(\mathrm{T})+D_1)(C_2\mathrm{T}+D_2)=C\mathrm{T}+D, \quad (16)$$

$$\zeta_{M_2}JM_2z+\zeta_{M_1}JM_1M_2z=\zeta_M JMz,$$

$$\zeta_{M_2}M_1^t+\zeta_{M_1}=\zeta_M. \quad (17)$$

Equation (16) is easy to check. The second equation follows from (17). Equation (17) is identical to (9); however, first view (17) as a recursion formula for ζ_M as a vector with integral coefficients, and then remember that ζ_M in (15) can be arbitrarily changed up to mod 2. Finally, one can also compute λ_M by a corresponding recursion formula; it is uninteresting in this connection though. ▼

In the case $M \in \Theta^n$ one can set $\xi_M = \eta_M = \zeta_M = 0$ in (15) *arriving at the equation*

$$\vartheta(\mathrm{T}, x, y) = \chi(M)|C\mathrm{T} + D|^{-1/2}\vartheta(M(\mathrm{T}), Ax + By, Cx + Dy), \tag{18}$$

where $\chi(M)$ is an eighth root of unity.

The determination of $\chi(M)$ is difficult, because it requires the determination of the square root $|C\mathrm{T} + D|^{-1/2}$, not only for $M = \begin{pmatrix} A & B \\ C & D \end{pmatrix}$ but also for all other matrices which occur while M is traced back to the generators of the group Θ^n.

The function $\vartheta(\mathrm{T}) = \vartheta(\mathrm{T},0,0)$ as any funciton with certain regularity properties satisfying the functional equation

$$\vartheta(\mathrm{T}) = \chi(M)|C\mathrm{T} + D|^{-1/2}\vartheta(M(\mathrm{T})), \qquad M \in \Theta^n$$

is called *a symplectic modular form of the dimension $-\frac{1}{2}$ with the multiplier system $\chi(M)$*.

The solutions of these functional equations with the general factor $|C\mathrm{T} + D|^{-k}$ on the right are modular forms of dimension $-k$. For $k = 0$ one speaks of *symplectic modular functions.* Those modular functions of the entire group Γ^n with multiplier system $\chi\begin{pmatrix} A & B \\ C & D \end{pmatrix} = 1$, which are meromorphic analytic functions for all finite values of the variables, form a finite algebraic function field in $\frac{1}{2}n(n + 1)$ variables, if $n > 1$. † In the case $n = 1$ an additional assumption on the behavior of the functions as $\mathrm{T} = \tau$ approaches the real axis is needed; see Chapter IV, §4. For $n > 1$ these modular functions do not fall within the scope of this book.

7. Proof of the Reciprocity Formula

The proof of (12) is carried out under the additional assumption that T is purely imaginary. As an analytic function in the complex variables τ_{ij}, x_i, y_i is being dealt with, the functional equation continues to hold under analytic continuation. In this manner the sign of the square root in (12) is simultaneously fixed.

The function

$$\begin{aligned}\vartheta(\mathrm{T}, x, y)\exp(-\pi i y' x) &= \sum_m \exp(i\pi \mathrm{T}[m - y] - 2\pi i(y - m)' x) \\ &= \sum_m \exp(i\pi \mathrm{T}[y - m - \mathrm{T}^{-1}x] - \pi i \mathrm{T}^{-1}[x])\end{aligned} \tag{19}$$

is periodic of period 1 in all the y_i, and can therefore be expanded in the Fourier series

$$\vartheta(\mathrm{T}, x, y)\exp(-\pi i y' x) = \sum_{m'} c_{m'} \exp(2\pi i m''y) \tag{20}$$

with coefficients

† C. L. Siegel, *Über die Abhängikeit von Modulfunktionen n-ten Grades*, Nachr. Akad. Wiss. Göttingen Math.-Phys. Kl., 257–272 (1960).

$$c_{m'} = \int_0^1 \cdots \int \sum_m \exp(\pi i \mathrm{T}[y - m - \mathrm{T}^{-1}x] - \pi i \mathrm{T}^{-1}[x] - 2\pi i m''y)\, dy_1 \cdots dy_n. \tag{21}$$

The proof consists of the evaluation of this integral.

First, summation and integration may be interchanged. Then, the substitution $y - m - \mathrm{T}^{-1}x = y' + \mathrm{T}^{-1}m'$ is carried out in each summand. The argument of the exponential function in the integrand then becomes

$$\pi i \mathrm{T}[y'] - \pi i \mathrm{T}^{-1}[m' + x] - 2\pi i m^t m',$$

and the last summand can be dropped off. Now summation and integration can be combined, so that one finally integrates over the entire y'-space. Setting

$$\gamma = \int_{-\infty}^{+\infty} \cdots \int \exp(\pi i \mathrm{T}[y'])\, dy_1' \cdots dy_n',$$

we have, by (20),

$$\begin{aligned} \vartheta(\mathrm{T}, x, y) &= \gamma \sum_{m'} \exp(-\pi i \mathrm{T}^{-1}[m' + x] + 2\pi i m''y + \pi i x^t y) \\ &= \gamma\, \vartheta(-\mathrm{T}^{-1}, y, -x). \end{aligned}$$

The positive definite quadratic form $-i\mathrm{T}[y']$ becomes a sum of squares under a linear substitution with determinant $|-i\mathrm{T}|^{-1/2}$. This implies

$$\begin{aligned} \gamma = |-i\mathrm{T}|^{-1/2} \int \cdots \int \exp[-\pi(y_1{}^2 + \cdots + y_n{}^2)]\, dy_1 \cdots dy_n \\ = |-i\mathrm{T}|^{-1/2}\left(\int_{-\infty}^{+\infty} \exp(-\pi y^2)\, dy\right)^n = |-i\mathrm{T}|^{-1/2} \end{aligned}$$

completing the proof. ▼

§2*. Theta Functions for Quadratic Forms

1. Simple Gaussian Sums

In the following we use the abbreviating notation

$$e(x) = e^{2\pi i x}. \tag{1}$$

By a *simple Gaussian sum* we mean the expression

$$G(a, b) = \sum_{r \bmod b} e(ar^2/b) \tag{2}$$

where a and b are relatively prime rational integers, and r runs through a complete residue system mod b.

For odd b

$$G(a, b) = (a/b)\, G(1, b), \tag{3}$$

where (a/b) is the Legendre symbol and

$$G(1,b) = \pm\sqrt{(-1)^{(b-1)/2}\, b}. \tag{4}$$

More precisely,

$$G(1, b) = \begin{cases} |b|^{1/2} & \text{for} \quad b \equiv 1 \bmod 4, \\ i|b|^{1/2} & \text{for} \quad b \equiv -1 \bmod 4. \end{cases} \tag{5}$$

The *proof* starts with a reduction. Let $b = b_1 b_2$ with relatively prime b_1, b_2, so that integers c_1, c_2 exist with

$$b_1 c_1 + b_2 c_2 = 1. \tag{6}$$

Set

$$r = b_1 c_1 r + b_2 c_2 r = b_1 r_1 + b_2 r_2 .$$

If r_1 and r_2 run through a complete residue system mod b_2 and b_1, respectively, then r runs through a complete residue system mod b, and conversely. Thus,

$$G(a, b) = \sum_{r \bmod b} e\big(a(b_1c_1 + b_2c_2)(b_1r_1 + b_2r_2)^2/b\big) = G(ac_1, b_2)\, G(ac_2, b_1).$$

Assuming (3) to be proved for both factors on the right, one has

$$G(a, b) = (a/b_1)(a/b_2)(c_1/b_2)(c_2/b_1)\, G(1, b_1)G(1, b_2).$$

By (6) this gives

$$G(a, b) = (a/b)(b_1/b_2)(b_2/b_1)\, G(1, b_1)\, G(1, b_2).$$

Further, if one assumes (5) to hold for both Gaussian sums on the right, then (5) is verified for the left side using the quadratic reciprocity law.

Hence one need but prove (3) and (5) for powers of primes. Let p be an odd prime and $b = p^h$, $h > 1$. Set

$$r = r_1 + p^{h-1} r_2$$

and let r_1, r_2 traverse a residue system mod p^{h-1} and p, respectively. If $r_1 \not\equiv 0 \bmod p$, then $2r_1r_2$ runs through a complete residue system mod p along with r_2 so that one has

$$G(a, p^h) = p \sum_{\substack{r_1 \equiv 0 \bmod p \\ r_1 \bmod p^{h-1}}} e(ap^{-h}{r_1}^2) + \sum_{\substack{r_1 \not\equiv 0 \bmod p \\ r_1 \bmod p^{h-1}}} e(ap^{-h}{r_1}^2) \sum_{r_2 \bmod p} e(ap^{-1}r_2).$$

The second term is clearly 0. This leaves

$$G(a, p^h) = p\, G(a, p^{h-2}).$$

If (3) through (5) hold for the Gaussian sums on the right, then they also hold for the left side.

It thus suffices to prove the equations asserted for a prime $b = p$. If c is any quadratic residue mod p, then clearly $G(ac, p) = G(a, p)$. If, however, c is not a quadratic residue, one has

$$G(a, p) + G(ac, p) = 2 \sum_{r \bmod p} e(r/p) = 0.$$

Thus for a c relatively prime to p,

$$G(ac, p) = (c/p)\, G(a,p)$$

holds, and this implies (3).

Because of (3)

$$\begin{aligned}(-1/p)\, G(1, p)^2 = G(1, p)\, G(-1, p) &= \sum_{r_1, r_2 \bmod p} e(p^{-1}(r_1{}^2 - r_2{}^2)) \\ &= \sum_{r_1, r_2 \bmod p} e(p^{-1}(r_1 - r_2)(r_1 + r_2)).\end{aligned}$$

As p is odd, when r_1, r_2 run through complete residue systems independently so do $r_1 + r_2$ and $r_1 - r_2$. Hence

$$(-1/p)\, G(1, p)^2 = \sum_{r_1, r_2 \bmod p} e(p^{-1} r_1 r_2) = p.$$

By inserting the value of the Legendre symbol on the left from the first complementation theorem to the quadratic reciprocity law, one has (4).

It is more difficult to determine the sign of (4); this will be done in the next section. In anticipation we can remark that our calculations, up to now, remain valid when in place of (1) we set $e(x) = e^{2\pi i d x}$ with an arbitrary number d relatively prime to b. This can be interpreted as a rearrangement of the bth roots of unity in the sums, and the algebraic equations dealt with thus far are invariant under automorphisms of the field of roots of unity. The problem of determining the sign in (4) does not appear until one bases the convention (1) on more than purely algebraic information.

2. The Quadratic Reciprocity Law and the Sign of Gaussian Sums

Let a and b be relatively prime odd numbers and τ a complex variable with positive imaginary component. Taking $n = 1$, $x = y = 0$ in (11) of §1* gives the function

$$\vartheta(\tau) = \sum_{m=-\infty}^{+\infty} \exp(\pi i \tau m^2). \tag{7}$$

We will substitute $\tau = 2a/b + i\lambda$ and let λ approach 0.

The index of summation is written $m = r + bm_1$, and r is permitted to traverse a complete residue system mod b, while m_1 takes on all integral values. This yields

$$\begin{aligned}\vartheta(\tau) &= \sum_{r \bmod b} e\left(\frac{a}{b} r^2\right) \exp(-\pi\lambda r^2) \sum_{m_1=-\infty}^{+\infty} \exp(-\pi\lambda b^2 m_1{}^2 - 2\pi\lambda b r m_1) \\ &= \sum_{r \bmod b} e\left(\frac{a}{b} r^2\right) \exp(-\pi\lambda r^2)\, \vartheta(i\lambda b^2, i\lambda b r, 0).\end{aligned}$$

The reciprocity formula (12) of §1* is applied to the theta function on the right, and then λ is allowed to approach 0. The result is

$$\lim_{\lambda\to 0} |\sqrt{\lambda}|\vartheta\left(2\frac{a}{b}+i\lambda\right) = |b|^{-1}G(a, b). \tag{8}$$

The same limit is then computed in another manner. Formula (12) of §1* is applied immediately to (7):

$$\vartheta\left(2\frac{a}{b}+i\lambda\right) = \left(\lambda - i2\frac{a}{b}\right)^{-1/2}\vartheta\left(-\frac{b}{2a}+i\lambda_1\right),$$

$$\lambda_1 = \frac{\lambda}{\dfrac{2a}{b}\left(\dfrac{2a}{b}+i\lambda\right)}. \tag{9}$$

The square root on the right is to be taken positve for $a = 0$ and general values found by analytic continuation with $2a/b + i\lambda$ remaining in the upper half-plane. This has the consequence

$$\lim_{\lambda\to 0}\left(\lambda - i2\frac{a}{b}\right)^{-1/2} = \exp[\pi i \operatorname{sign}(ab)/4]\left|\frac{2a}{b}\right|^{-1/2}. \tag{10}$$

The second factor of (9) is changed in form, as was (7), in particular by setting $m = r + 2am_1$:

$$\vartheta\left(-\frac{b}{2a}+i\lambda_1\right) = \sum_{r \bmod 2a} e\left(\frac{-b}{4a}r^2\right)\exp(-\pi\lambda_1 r^2)$$
$$\times \sum_{m_1}\exp(-4\pi\lambda_1 a^2 m_1{}^2 - 4\pi\lambda_1 a r m_1);$$

(8) is again applied to the theta function on the right. Noting the connection between λ and λ_1 in (9) one finds

$$\lim_{\lambda\to 0}|\sqrt{\lambda}|\vartheta\left(-\frac{b}{2a}+i\lambda_1\right) = |b|^{-1}\sum_{r \bmod 2a} e\left(\frac{-b}{4a}r^2\right). \tag{11}$$

Finally the sum on the right can be operated upon as in the beginning of this section by setting

$$4c_1 + ac_2 = 1$$

with integral c_1, c_2, This yields

$$\sum_{r \bmod 2a} e\left(\frac{-b}{4a}r^2\right) = \sum_{r \bmod 2} e\left(-\frac{bc_2}{4}r^2\right)G(-bc_1, a).$$

Now, c_1 is a quadratic residue mod a and $ac_2 \equiv 1 \bmod 4$. Thus,

$$\sum_{r \bmod 2a} e\left(\frac{-b}{4a}r^2\right) = \sum_{r \bmod 2} e\left(\frac{-ab}{4}r^2\right)G(-b, a).$$

Combining (8) through (11) we have

$$G(a, b) = \left|\frac{b}{2a}\right|^{1/2} \exp[\tfrac{1}{4}\pi i \operatorname{sign}(ab)] \sum_{r \bmod 2} e\left(\frac{-ab}{4} r^2\right) G(-b, a).$$

The sum still remaining on the right is easy to compute, and this leads to the *reciprocity law for Gaussian sums*:

$$G(a, b) = \varepsilon |b/a|^{1/2} G(-b, a) \tag{12}$$

with

$$\varepsilon = \varepsilon_1 \exp[\tfrac{1}{4}\pi i \operatorname{sign}(ab)], \quad \varepsilon_1 = \begin{cases} \exp[-\tfrac{1}{4}\pi i] & \text{for } ab \equiv 1 \bmod 4, \\ \exp[\;\;\tfrac{1}{4}\pi i] & \text{for } ab \equiv -1 \bmod 4. \end{cases} \tag{13}$$

By specializing $a = 1$, $b > 0$, it is seen that $G(-b, a) = 1$, so that as a first application we have the sign determination (5) for Gaussian sums.

Secondly one has a proof of the quadratic reciprocity law by putting two different odd primes in (12) as a and b. It was shown in the first section that formula (3) can be used. This leads to

$$\begin{aligned}(a/b) &= (-b/a) \exp[\tfrac{1}{4}\pi i(1 - (\nu_a - 1) + (\nu_b - 1) - \nu_{ab})] \\ &= (-b/a) \exp[\tfrac{1}{4}\pi i(1 - \nu_a + \nu_b - \nu_{ab})],\end{aligned}$$

where ν_a, ν_b, ν_{ab}, take the values 1 or -1, depending upon whether a, b, and ab, respectively, are congruent 1 or -1 mod 4. This last formula becomes the reciprocity law if one applies the first complementation theorem. The generalized form of the reciprocity law for arbitrary integers a, b is a simple consequence of the law for primes.

3. The Theta Function for a Definite Quadratic Form

A definite quadratic form in $n = 2k$ variables m_i,

$$F[m] = m^t F m = \sum_{i,j=1}^{2k} f_{ij} m_i m_j \qquad (f_{ij} = f_{ji}),$$

is given. Let the matrix $F = (f_{ij})$ have rational integral coefficients, and let the f_{ii} be even. For integral values of the m_i, $F[m]$ can take on only even values. The smallest natural number N, such that the matrix $F' = NF^{-1}$ also has integral coefficients and, moreover, even diagonal coefficients, is called the level of F. Because NF^{-1} is integral, so is $|NF^{-1}| = N^{2k}|F|^{-1}$ so that $|F|$ is divisor of a power of N. It is easy to verify that $(-1)^k|F|$ is always congruent 0 or 1 mod 4.

Now let τ be a complex variable with positive imaginary component. Then, $\mathrm{T} = \tau F$ is a matrix of the sort on which our considerations in §1*,6 were based. The theta function for F arises from the following specialization of (11) of the last section:

$$\vartheta(\tau, x, y) = \sum_m \exp(\pi i \tau F[m - y] + 2\pi i m^t x - \pi i x^t y). \tag{14}$$

For every matrix $\left(\begin{smallmatrix} a & b \\ c & d \end{smallmatrix}\right) \in \Gamma^1$ *with the property that* $c \equiv 0 \bmod N$, *the functional equation*

$$\vartheta\left(\frac{a\tau + b}{c\tau + d}, ax + bFy, cF^{-1}x + dy\right)(c\tau + d)^{-k}$$
$$= (\operatorname{sign} d)^k \left(\frac{(-1)^k |F|}{|d|}\right) \vartheta(\tau, x, y), \tag{15}$$

with the generalized Legendre symbol† *on the right holds; the symobl* $|F|$ *here is the determinant.*

The substitutions $\left(\begin{smallmatrix} a & b \\ c & d \end{smallmatrix}\right)$ with $c \equiv 0 \bmod N$ obviously form a subgroup of finite index in the modular group Γ^1.

Proof. Using the assumption on $\left(\begin{smallmatrix} a & b \\ c & d \end{smallmatrix}\right)$ it is easy to prove that

$$M = \begin{pmatrix} A & B \\ C & D \end{pmatrix} = \begin{pmatrix} aE & bF \\ cF^{-1} & dE \end{pmatrix} \tag{16}$$

is an element of the subgroup $\Theta^{2k} \subset \Gamma^{2k}$ described in §1*,4. Thus, by §1*,(18) and

$$(A\tau F + B)(C\tau F + D)^{-1} = \frac{a\tau + b}{c\tau + d} F,$$

Eq. (15) holds as maintained, with some eighth root of unity $\chi\left(\begin{smallmatrix} a & b \\ c & d \end{smallmatrix}\right)$ depending upon that matrix on the right side. It only remains to prove that

$$\chi\begin{pmatrix} a & b \\ c & d \end{pmatrix} = (\operatorname{sign} d)^k \left(\frac{(-1)^k |F|}{|d|}\right). \tag{17}$$

An easy consequence of the fact that (15) holds with $\chi\left(\begin{smallmatrix} a & b \\ c & d \end{smallmatrix}\right)$ as a factor of $\vartheta(\tau, x, y)$ on the right is the equation

$$\chi\left(\begin{pmatrix} a_1 & b_1 \\ c_1 & d_1 \end{pmatrix}\begin{pmatrix} a_2 & b_2 \\ c_2 & d_2 \end{pmatrix}\right) = \chi\begin{pmatrix} a_1 & b_1 \\ c_1 & d_1 \end{pmatrix}\chi\begin{pmatrix} a_2 & b_2 \\ c_2 & d_2 \end{pmatrix} \tag{18}$$

for two transformations belonging to our group.

Multiplication of a, b, c, d with -1 leaves the theta function on the left of (15) invariant because of §1*,(13) with $V = -E$, so that both sides are multiplied by $(-1)^k$. Thus it is permissible, from now on, to assume $d > 0$ in the proof of (15) or (17).

---

† The generalized Legendre symbol (a/b) is also defined for even b, when $b > 0$ and $a \equiv 1 \bmod 4$. Then, $(a/b) = (b/|a|)$ holds.

In determining $\chi\left(\begin{smallmatrix} a & b \\ c & d \end{smallmatrix}\right)$ one may assume $x = y = 0$. The procedure is similar to that of §2,2. By the reciprocity formula (12) of §1*, we have, for $\tau = i\lambda$, λ real,

$$\lim_{\lambda \to 0} \lambda^k \,\vartheta(i\lambda, 0, 0) = |F|^{-1/2} \lim_{\lambda \to 0} \vartheta^*(i\lambda^{-1}, 0, 0),$$

where ϑ^* is the theta series which is to be formed analogously with F^{-1} in place of F. The limit on the right is reduced to the summand with $m = 0$. Thus

$$\lim_{\lambda \to 0} \lambda^k \,\vartheta(i\lambda, 0, 0) = |F|^{-1/2}. \tag{19}$$

The same limit is now determined for the left side of (15). Writing

$$(ai\lambda + b)/(ci\lambda + d) = b/d + i\lambda_1, \qquad \lambda_1 = \lambda/d(ci\lambda + d) \tag{20}$$

and decomposing the summation index of the theta series to $m = r + dm_1$ so that r traverses a residue system of integral vectors mod d and m_1 runs through all integral vectors, the exponent of the general summand is

$$\pi i(b/d + i\lambda_1)\, F[r + dm_1] = \pi i(b/d\, F[r] - \pi\lambda_1\, F[r]$$
$$+ \pi ibd\, F[m_1] + 2\pi ibr^t\, Fm_1 - \pi\lambda_1 d^2\, F[m_1] - 2\pi\lambda_1 dm_1{}^t\, Fr.$$

The third and fourth summands on the right are integral multiples of $2\pi i$ and can be omitted. The last two summands lead to the theta function

$$\vartheta(i\lambda_1 d^2 F, id\lambda_1 Fr, 0) = \lambda_1^{-k} d^{-2k} |F|^{-1/2} \vartheta((i/\lambda_1 d^2)F^{-1}, 0, id\lambda_1 Fr).$$

Taking (20) into account, this gives

$$\lim_{\lambda \to 0} \lambda^k \,\vartheta\left(\frac{ai\lambda + b}{ci\lambda + d}, 0, 0\right)(ci\lambda + d)^{-k} = |F|^{-1/2} \sum_{r \bmod d} \exp(\pi ibd^{-1}F[r]),$$

on the left-hand side of (15). The same limit is formed for the right-hand side as well, while (19) is observed. Thus we find that (15) is true with the constant factor

$$\chi\begin{pmatrix} a & b \\ c & d \end{pmatrix} = d^{-k} \sum_{r \bmod d} \exp(\pi ibd^{-1}F[r]). \tag{21}$$

Eventually we have to transform (21) into (17). For this we first reduce the task to the case where d is odd.

The *proof* of (17) is first reduced to the case that d is odd. Let d, therefore, be even. Then c is odd and therefore so is N. But if $|F|$ is a divisor of a power of N, then $|F|$ also must be odd. Now, it can be read directly from (21) that $\chi\left(\begin{smallmatrix} a & b \\ c & d \end{smallmatrix}\right) = 1$ for $d = 1$. Set

$$\begin{pmatrix} a & b \\ c & d \end{pmatrix}\begin{pmatrix} 1 + Ns & s \\ N & 1 \end{pmatrix} = \begin{pmatrix} a' & b' \\ c' & d' \end{pmatrix}$$

with integral s. These are three substitutions of the sort permitted. Take, in particular, $s = t|F|$ with t odd, then $d' = d + ct|F| \equiv 1 \bmod 2$, and by the quadratic reciprocity law (see footnote on page 49),

$$\left(\frac{(-1)^k|F|}{d'}\right) = \left(\frac{(-1)^k|F|}{d}\right).$$

Because of (18) and the fact that $\chi\begin{pmatrix} 1+Ns & s \\ N & 1 \end{pmatrix} = 1$ we have $\chi\begin{pmatrix} a' & b' \\ c' & d' \end{pmatrix} = \chi\begin{pmatrix} a & b \\ c & d \end{pmatrix}$. Thus it suffices to prove (17) with a', b', c', d' instead of a, b, c, d. As d' is odd, we may, without losing generality, assume d to be odd.

For the case of an odd d, the proof of (17) is as follows: The quadratic form $F[r]$ can, in the residue class ring mod d, be transformed to

$$F[r] \equiv 2\sum_{\nu=1}^{2k} f_\nu s_\nu^2 \bmod d, \qquad s_\nu \equiv \sum_{\mu=1}^{2k} c_{\nu\mu} r_\mu \bmod d, \tag{22}$$

where the determinant $|c_{\nu\mu}|$ is relatively prime to d. As r traverses a residue system mod d, so does the vector $s = (s_\nu)$. Thus (21) becomes

$$\chi\begin{pmatrix} a & b \\ c & d \end{pmatrix} = d^{-k} \sum_{s \bmod d} \exp(2\pi i b d^{-1} \sum f_\nu s_\nu^2) = d^{-k} \prod_{\nu=1}^{2k} \sum_{s_\nu=1}^{d} \exp(2\pi i b d^{-1} f_\nu s_\nu^2).$$

By (2) through (4) and the first complementation theorem this is equivalent to

$$\chi\begin{pmatrix} a & b \\ c & d \end{pmatrix} = \left(\frac{-1}{d}\right)^k \left(\frac{\prod f_\nu}{d}\right).$$

But because of (22), $2^{2k} \prod f_\nu \equiv |F|\,|c_{\nu\mu}|^2 \bmod d$, and we have finished. ▼

Finally we give another important property of the function ϑ. *There is an integral power series* f_0 *with*

$$\vartheta(\tau, 0, 0) = (-1/\tau)^k f_0(q_0), \qquad q_0 = \exp(-2\pi i/N \cdot 1/\tau). \tag{23}$$

For a reduced fraction $s = a/c \neq 0$, *determine rational* b, d, *so that* $ad - bc = 1$. *Then there is a power series* f_s *with*

$$\vartheta(\tau, 0, 0) = \left(\frac{-1}{-c\tau + a}\right)^k f_s(q_s), \qquad q_s = \exp[-2\pi i/N \cdot (d\tau - b)(-c\tau + a)^{-1}]. \tag{24}$$

The first series expansion is gotten directly from the reciprocity formula §1*,(12) under the consideration that $NF^{-1}[m]$ always represents even integers. To get the second expansion, set $\tau = \tau' + s$ and

$$\frac{d\tau - b}{-c\tau + a} = -\frac{d}{c} - \frac{1}{c(c\tau - a)} = -\frac{d}{c} - \frac{1}{c^2\tau'}. \tag{25}$$

As above, in the sum over m for $\vartheta(\tau F, 0, 0)$, set $n = r + cm_1$ with r traversing a residue system mod c and m_1 all integral vectors. Then

$$\begin{aligned}\vartheta(\tau, 0, 0) &= \sum_{r \bmod c} \exp(\pi i(\tau' + s)F[r])\, \vartheta(c^2\tau' F, \tau' cFr, 0) \\ &= \left(\frac{i}{c^2\tau'}\right)^k |F|^{-1/2} \sum_{r \bmod c} \exp(\pi i(\tau' + s)F[r]) \\ &\quad \times \vartheta\left(-\frac{1}{c^2\tau'} F^{-1}, 0, -\tau' cFr\right) \\ &= \left(\frac{i}{c^2\tau'}\right)^k |F|^{-1/2} \sum_{r \bmod c} \exp(\pi i s F[r]) \\ &\quad \times \sum_{m_1} \exp(-\pi i/N \cdot 1/c^2\tau' \cdot NF^{-1}[m_1] + 2\pi i/c \cdot m_1{}^t r).\end{aligned}$$

Up to the factor $(-ic^2\tau')^{-k}|F|^{-1/2}$, the right side is an integral power series in $\exp(-2\pi i/N \cdot 1/c^2\tau')$; using (25) this can also be written as an integral power series in q_s. ▼

The function $\vartheta(\tau, 0, 0)$ can be written as

$$\vartheta(\tau, 0, 0) = \sum_{\mu=0}^{\infty} c(\mu, F)\, e^{2\pi i \mu \tau}, \tag{26}$$

where $c(\mu, F)$ is the number of representations of μ in the form $\mu = \frac{1}{2}F[m]$. As is easily observed, this number and, in fact, the function $\vartheta(\tau, 0, 0)$ are invariant under unimodular transformation of the variables of the quadratic form $F[m]$. They are *class invariants.*

The functional equation (15) and the expansions (23) and (24) are the basis of the analytic number theory of quadratic forms. They demonstrate that quotients of functions $\vartheta(\tau, 0, 0)$ of forms F with the same determinant and level are, in principle, algebraic functions of one variable. The algebraic function fields thus generated are among the most remarkable examples of our theory since they serve, to a great measure, in the illustration of the more advanced concepts. In the course of history they have often inspired the formation of extensive general concepts. We will make a particular study of these functions in IV, §4 and V, §3.

CHAPTER II

Ideals and Divisors

§1. Ideals

1. Integral Dependence

Let an arbitrary field k, and a noetherian or, more specifically, a principal ideal domain $\mathfrak{o}$ with quotient field k, be given. Further, let K/k be a finite extension. An $\alpha \in K$ is called *integrally dependent* upon $\mathfrak{o}$ if all its powers, $1, \alpha, \alpha^2, \ldots$, belong to a finite $\mathfrak{o}$-module.

If and only if α satisfies an equation

$$\alpha^n + c_1\alpha^{n-1} + \cdots + c_n = 0 \tag{1}$$

in k with coefficients in $\mathfrak{o}$, is α integrally dependent upon $\mathfrak{o}$.

Proof. If (1) holds with $c_i \in \mathfrak{o}$, then all powers of α belong to the $\mathfrak{o}$-module formed by $1, \alpha, \ldots, \alpha^{n-1}$. Conversely, assume all powers of α belong to a finite $\mathfrak{o}$-module. By the finiteness criterion of I, §1,1 or I, §1,4, respectively, they themselves generate a finite $\mathfrak{o}$-module, $\mathfrak{a}$. Let $\alpha_1, \ldots, \alpha_m$ be a system of generators of $\mathfrak{a}$ and $\alpha_i = \sum a_{ij}\alpha^j$ with $a_{ij} \in \mathfrak{o}$. Let $n-1$ be the highest power of α appearing here. Then $\mathfrak{a}$ is generated by $1, \alpha, \ldots, \alpha^{n-1}$, and we can represent $\alpha^n = -c_n - \cdots - c_1\alpha^{n-1}$, with $c_i \in \mathfrak{o}$. Hence, α satisfies an equation $f(x) = 0$ with coefficients in $\mathfrak{o}$. ▼

If α satisfies an equation (1) *with $c_i \in \mathfrak{o}$, it also satisfies an irreducible equation of this kind.* Indeed, due to the theorem of Gauss, if the left hand side of (1) is factorized into polynomials in $k[\alpha]$ with leading coefficients 1 these have coefficients in $\mathfrak{o}$.

From the theorem just proven it is seen that the trace $\mathrm{tr}_{k(\alpha)/k}(\alpha) = -c_1$ of an element α integrally dependent upon $\mathfrak{o}$ is in $\mathfrak{o}$. Applying the transitivity formula I, §4,(3) to $L = k(\alpha)$ yields: *If $\alpha \in K$ is integrally dependent upon $\mathfrak{o}$ then the trace $\mathrm{tr}_{K/k}(\alpha) \in \mathfrak{o}$.*

The set of elements of a finite extension field K/k integrally dependent upon $\mathfrak{o}$ form an integral domain $\mathfrak{O}$, with K as its quotient field. It is called the *principal order* of K with respect to $\mathfrak{o}$.

Proof. If α, β are integrally dependent upon $\mathfrak{o}$ then the $\mathfrak{o}$-modules generated by all powers of α and β, respectively, have finite generating systems, α_i, β_i.

The powers of $\alpha \pm \beta$ and $\alpha\beta$ are contained in the $\mathfrak{o}$-module generated by all α_i, β_i and all products $\alpha_i\beta_j$. This is again a finite module. Thus, the elements α integrally dependent upon $\mathfrak{o}$ form an intergal domain $\mathfrak{O} \subset K$.

Now let α be arbitrary in K and (1) be the normalized irreducible equation in k which α satisfies. As k is the quotient field of $\mathfrak{o}$ there exists an $a \in \mathfrak{o}$, such that all the $a^i c_i$, $i = 1, \ldots, n$, are in $\mathfrak{o}$. Now, $\beta = a\alpha$ satisfies the irreducible equation $\beta^n + ac_1\beta^{n-1} + \cdots + a^n c_n = 0$. Thus, β is integrally dependent upon $\mathfrak{o}$, and $\alpha = \beta a^{-1}$ is the quotient of two elements of $\mathfrak{O}$. ▼

In the same vein we still show that the linear rank of $\mathfrak{O}$ with respect to k is equal to $[K:k]$. In fact, starting with $[K:k]$ linearly independent elements $\alpha_i \in K$, there exists an $a \in \mathfrak{o}$ so that all the $a\alpha_i \in \mathfrak{O}$.

If $\mathfrak{O}_K$ and $\mathfrak{O}_L$ denote the principal orders of K and an intermediary field L with respect to $\mathfrak{o}$, then $\mathfrak{O}_K$ is also the principal order of K with respect to $\mathfrak{O}_L$. For, if the α^h are in a finite $\mathfrak{o}$-module, then because of $\mathfrak{O}_L \supseteq \mathfrak{o}$ they certainly lie in a finite $\mathfrak{O}_L$-module.

Exercise. If $\mathfrak{O}_L$ is a finite $\mathfrak{o}$-module, then the principal order of K with respect to $\mathfrak{O}_L$ is the same as the principal order of K with respect to $\mathfrak{o}$.

2. The Finiteness of the Principal Order

If $\mathfrak{o}$ is a principal ideal ring, then the principal order $\mathfrak{O}$ with respect to $\mathfrak{o}$ of a finite separable extension K/k is a finite $\mathfrak{o}$-module of rank $[K:k]$. The theorem holds even if $\mathfrak{o}$ is only a Noetherian ring, as can be seen from the proof.

Proof. Let ω_i be a field basis of K/k. We have already seen that an $a \in \mathfrak{o}$ exists such that all $a\omega_i \in \mathfrak{O}$. Thus there is no loss of generality in assuming the $\omega_i \in \mathfrak{O}$. Now, let

$$\xi = \sum x_i \omega_i$$

be some element of $\mathfrak{O}$. The $\xi\omega_j \in \mathfrak{O}$, so that their traces are in $\mathfrak{o}$. The discriminant $D(\omega_i) = |s_{K/k}(\omega_i\omega_j)|$ is $\neq 0$. Hence the x_i can be determined uniquely from the linear system of equations

$$s_{K/k}(\xi\omega_j) = \sum_i x_i s_{K/k}(\omega_i\omega_j).$$

Therefore, they are all contained in the module $\mathfrak{o}D(\omega_i)^{-1}$. We thus see that

$$\mathfrak{O} \subseteq D(\omega_i)^{-1}(\mathfrak{o}\omega_1, \ldots, \mathfrak{o}\omega_n);$$

i.e., $\mathfrak{O}$ is contained in a finite $\mathfrak{o}$-module. By the finiteness criterion of I,§1,1, $\mathfrak{O}$ itself is a finite $\mathfrak{o}$-module. ▼

This theorem does not hold unconditionally for an inseparable extension K/k. We will prove here only as much as we need for our applications. *Let the field k be of the form*

$$k = k_0(x_1, \ldots, x_m), \qquad k_0 = k_{00}(w_1, \ldots, w_r; w_{r+1}, \ldots, w_{r+s}),$$

where k_{00} is a perfect field and the $w_1, \ldots, w_r$ are independent indeterminates while $w_{r+1}, \ldots$ are algebraically dependent upon the first w_ρ. Further, let

$$\mathfrak{o} = k_0[x_1, \ldots, x_m].$$

Then, the principal order $\mathfrak{O}$ of an arbitrary finite extension K/k with respect to $\mathfrak{o}$ is a finite $\mathfrak{o}$-module of rank $[K:k]$.

Proof. For $m > 1$, $\mathfrak{o}$ is no longer a principal ideal ring, but it is a noetherian ring and we must apply the previous theorem with that hypothesis.

Let K_1 be the largest subfield of K separable over k. K is then contained in the field $K_2 = K_1{}^{p^{-e}}$, created by adjoining all p^eth roots of elements of K_1, where p is the characteristic and e is a sufficiently large number. The principal order $\mathfrak{O}_1$ of K_1 is a finite $\mathfrak{o}$-module. Letting ω_i denote a system of generators of $\mathfrak{O}_1$, we maintain that the elements

$$\Omega_{\varrho_1, \ldots; \mu_1, \ldots; i} = w_1^{\varrho_1 p^{-e}} \cdots x_1^{\mu_1 p^{-e}} \cdots \omega_i^{p^{-e}}$$

with $0 \leqq \varrho_j, \mu_j < p^e$ form a system of generators of the principal order $\mathfrak{O}_2$ of K_2 as an $\mathfrak{o}$-module. In fact, let $\alpha \in \mathfrak{O}_2$. The p^eth power of α lies in $\mathfrak{O}_1$, and can thus be represented as

$$\alpha^{p^e} = \sum_i R_i(w_1, \ldots; x_1, \ldots)\omega_i,$$

where the R_i are rational functions in the w_ϱ and polynomials in the x_μ with coefficients in k_{00}. They can be written as

$$R_i(w_1, \ldots; x_1, \ldots) = [N(w_1^{p^e}, \ldots)]^{-1} \sum S_{\varrho_1,\ldots;\mu_1,\ldots;i}(w_1^{p^e}, \ldots; x_1^{p^e}, \ldots)w_1^{\varrho_1} \cdots x_1^{\mu_1} \cdots$$

where N and the $S_{\varrho_1,\ldots;\mu_1,\ldots;i}$ are polynomials in all variables. The p^eth roots of the coefficients of these polynomials are again contained in k_{00}. They are used to form new polynomials, which are plausibly denoted $N^{p^{-e}}$, etc. Thus,

$$\alpha = [N^{p^{-e}}(w_1, \ldots)]^{-1} \sum S^{p^{-e}}_{\varrho_1,\ldots;\mu_1,\ldots;i}(w_1, \ldots; x_1, \ldots)\Omega_{\varrho_1,\ldots;\mu_1,\ldots;i}$$

holds, which verifies our assertion.

By the finiteness criterion of I,§1,4, $\mathfrak{O}$, being a submodule of the finite module $\mathfrak{O}_2$, is itself finite as an $\mathfrak{o}$-module. ▼

From the last theorem it is easy to derive another, which we shall also need. *Let k be as before, and $p(x_i)$ a prime polynomial in $k_0[x_1, \ldots, x_m]$. The integral domain of all rational functions in the x_i whose denominators are not divisible by $p(x_i)$ is denoted $\mathfrak{o}_p$. Then, the principal order $\mathfrak{O}_p$ of a finite extension K/k with respect to $\mathfrak{o}_p$ is a finite $\mathfrak{o}_p$-module.*

Proof. Let $\alpha \in \mathfrak{O}_p$ satisfy the irreducible equation (1) with $c_i \in \mathfrak{o}_p$. A common denominator a of the c_i is relatively prime to $p(x_i)$, and thus a unit in $\mathfrak{o}_p$. The element $\beta = a\alpha$ is in the principal order $\mathfrak{O}$ with respect to $\mathfrak{o} = k_0[x_i, \cdots]$, as the same argument used in §1,1 shows. By the previous theorem there exists a finite system of generators ω_i for $\mathfrak{O}$. Therefore

$$\alpha = (1/a) \sum b_i\omega_i = \sum c_i\omega_i$$

with $b_i \in \mathfrak{o}$ or $c_i \in \mathfrak{o}_p$. Hence the ω_i also form a system of generators for $\mathfrak{O}_p$ with respect to $\mathfrak{o}_p$. ▼

3. Kronecker Divisors

We make the assumption that $\mathfrak{o}$ is a principal ideal domain, and stay with it until §1,5. An indeterminate x is now adjoined to k. A polynomial in x is called primitive when the greatest common divisor of its coefficients is 1. By Gauss's theorem the product of primitive polynomials is primitive.

$\mathfrak{o}(x)$ will denote the integral domain of quotients of polynomials with coefficients in $\mathfrak{o}$, the denominator being primitive. $\mathfrak{o}(x)$ *is again a principal ideal domain.*

Proof. Every $f(x) \in \mathfrak{o}(x)$ can be represented in the form $f(x) = ag(x)$, where a is the g.c.d. of the numerator of $f(x)$, and $g(x)$ is a quotient of primitive polynomials. Let $f'(x)$ be another such function, $f'(x) = a'g'(x)$, and let d be the g.c.d. of a and a'. With suitable u, u' of $\mathfrak{o}$ we have $au + a'u' = d$, or

$$f(x)ug'(x) + f'(x)u'g(x) = dg(x)g'(x),$$

and the right side is a divisor of both $f(x)$ and $f'(x)$. It follows that every finitely generated ideal is principal.

Prime decomposition of elements follows from the same property in $\mathfrak{o}$. ▼

Similarly we adjoin infinitely many indeterminates $x_1, x_2, \ldots$ to k and form the integral domain $\mathfrak{o}(x) = \mathfrak{o}(x_1)(x_2) \ldots$. It again is a principal ideal domain.

In addition, let K/k be a finite extension, $\mathfrak{O}$ the principal order of K with respect to $\mathfrak{o}$, and $\mathfrak{O}(x)$ the principal order of $K(x)$ with respect to $\mathfrak{o}(x)$. $\mathfrak{O}(x)$ *consists of the quotients* $\alpha = \beta a^{-1}$ with $\beta \in \mathfrak{O}[x]$ *and primitive* $a \in \mathfrak{o}[x]$.

Proof. Let an $\alpha \in \mathfrak{O}(x)$ satisfy the irreducible equation (1) in $k(x)$. The c_i are elements of $\mathfrak{o}(x)$ and thus have a primitive polynomial a as their common denominator. Then, $\beta = a\alpha$ is a polynomial in the x with coefficients in K. We contend that the coefficients are even in $\mathfrak{O}$. Assume first that $\beta = \beta_0 + \beta_1 x_1 + \beta_2 x_1{}^2 + \cdots$ depends on one variable only. Inserting $x_1 = 0$ we see that β_0 depends integrally on $\mathfrak{o}$, so $\beta_0 \in \mathfrak{o}$. Then $x_1{}^{-1}(\beta - \beta_0) = \beta_1 + \beta_2 x_1 + \cdots$ is in $\mathfrak{O}(x)$, so that $\beta_1 \in \mathfrak{o}$, etc. If β depends on n variables we can consider β as a polynomial in the first variable the coefficients of which are polynomials in the others and thus proceed by induction on n.

Hence, every $\alpha \in \mathfrak{O}(x)$ can be written as

$$\alpha = \beta/a, \qquad \beta \in \mathfrak{O}[x], \quad a \in \mathfrak{o}[x]$$

with primitive a. Conversely, such α are contained in $\mathfrak{O}(x)$. ▼

$\mathfrak{O}(x)$ *is a principal ideal domain.*

Proof. Let $\mathfrak{a} \subseteq \mathfrak{O}(x)$ be a finite $\mathfrak{O}(x)$-ideal and α_i a system of generators of $\mathfrak{a}$ as an $\mathfrak{o}(x)$-module. We take just as many indeterminates x_i not occurring in the α_i, form $\alpha = \sum \alpha_i x_i$, and maintain $\mathfrak{a} = \alpha\mathfrak{O}(x)$.

Let β be an arbitrary element of $\mathfrak{a}$. For some indeterminate ξ consider the quotient

$$f(\xi) = \frac{n_{K/k}(\alpha\xi - \beta)}{n_{K/k}(\alpha)} = n_{K/k}(\xi - \gamma), \qquad \gamma = \beta/\alpha .$$

It is a polynomial in ξ, in fact the characteristic polynomial for γ in the sense of I, §4,1. For, with a field basis ω_i of K/k and elements δ_{ij} of the unit matrix,

$$\omega_i(\xi - \gamma) = \sum_j (\delta_{ij}\xi - c_{ij})\omega_j$$

holds, and thus

$$n_{K/k}(\xi - \gamma) = |E\xi - (c_{ij})|.$$

If we also represent β in the system of generators of $\mathfrak{a}$ as $\beta = \sum \alpha_i b_i$, then

$$f(\xi) = \frac{n_{K/k}(\sum \alpha_i u_i)}{n_{K/k}(\sum \alpha_i x_i)}, \qquad u_i = x_i\xi - b_i .$$

Both denominator and numerator here are polynomials, their only difference being the variables x_i and u_i, and these are independent of each other as of the variables occurring in the α_i. Dividing by the g.c.d. of the coefficients yields a rational function, whose numerator polynomial has coefficients in $\mathfrak{o}(x, \xi)$ and whose denominator polynomial is primitive.

As a consequence all coefficients of $f(\xi)$ are in $\mathfrak{o}(x)$.

The two facts, that $f(\xi)$ is the characteristic polynomial of γ and that its coefficients are in $\mathfrak{o}(x)$, the highest being 1, show that $\gamma \in \mathfrak{O}(x)$. Hence $\beta = \alpha\gamma$ with $\gamma \in \mathfrak{O}(x)$, so that $\mathfrak{a} \subseteq \alpha\,\mathfrak{O}(x)$. The opposite inclusion is obvious.

The possibility of prime decomposition arises by norm formation with respect to k. The norm of elements of $\mathfrak{O}(x)$ lie in $\mathfrak{o}(x)$. Now, if an $\alpha(x) \in \mathfrak{O}(x)$ were decomposable without limit, the possibility of prime decomposition in $\mathfrak{o}(x)$ would assure that, with a finite number of exceptions, all the norms are units. But then the factors themselves would be units. ▼

The theorem just proven permits us, because of the uniqueness of prime factor decomposition in principal ideal domains, to draw the consequence: *The factor group*

$$\mathfrak{K}_o = K(x)^\times/\mathfrak{E}(x)$$

of the multiplicative group $K(x)^\times$ *of* $K(x)$ *with respect to the group of units* $\mathfrak{E}(x)$ *of* $\mathfrak{O}(x)$ *is the direct product of the cyclic subgroups generated by the prime elements of* $\mathfrak{O}(x)$. $\mathfrak{K}_0$ contains the factor group

$$\mathfrak{H}_0 = K^\times/\mathfrak{E}, \qquad \mathfrak{E} = K \cap \mathfrak{E}(x),$$

where $\mathfrak{E}$ is the unit group of $\mathfrak{O}$. We call the elements of $\mathfrak{K}_0$ *Kronecker divisors of* $\mathfrak{O}(x)$ *with respect to* $\mathfrak{o}$, the elements of $\mathfrak{H}_0$ (Kronecker) *principal divisors with respect to* $\mathfrak{o}$.

Later another concept of divisor will be introduced and placed more in the foreground. Kronecker divisors play a small part in the current literature and we too will only apply them twice. But then they prove themselves superior to other conceptual aids. The first application will be the construction of ideal theory, to which we now turn.

We conclude with a remark which is not necessary in what follows.

After a single variable z is adjoined, $\mathfrak{O}(z)$ *is already a principal ideal domain.*

Proof. We maintain that, if

$$\alpha(x) = \sum a_{i_1,\ldots} x_1^{i_1} \cdots$$

is a polynomial in the variables $x_1, \ldots, x_m$, one can find exponents $n_i = n_{i_1,\ldots,i_m}$ such that, setting

$$\beta(z) = \sum a_{i_1,\ldots} z^{n_i},$$

the quotient $\alpha(x)\beta(z)^{-1}$ is a unit in $\mathfrak{O}(x_1, \ldots, x_m, z)$. This statement yields the proposition. For, as we showed above, the g.c.d. of a finite number of $\alpha_i(x) \in \mathfrak{O}(x)$ can be written as a polynomial $\alpha(x)$. But, by our construction, it is also equal to $\beta(z)$. If the $\alpha_i(x)$ depend only upon a single variable x their g.c.d. is also equal to $\beta(x)$, as $\beta(z)\beta(x)^{-1}$ is, of course, a unit.

Choose n_i in such a manner that the polynomials $\alpha(x)$ and $\beta(z)$ have the same coefficients. Also, choose $d(x) = d(x_{m+1}, \ldots)$ so that it represents the g.c.d. of all the $a_{i_1,\ldots}$. Then, $\alpha(x)\, d(x)^{-1}$ is a primitive polynomial in $x_1, \ldots, x_m$ with coefficients in $\mathfrak{O}(x_{m+1}, \ldots)$. Its norm with respect to k is a unit in $\mathfrak{o}(x_1, \ldots, x_{m+1}, \ldots)$. For, the norm is a product of elements, conjugate with respect to k, all of which are primitive in this case. By Gauss's theorem, their product is also primitive. As a consequence, $\alpha(x)\, d(x)^{-1}$ is a unit in $\mathfrak{O}(x_1, \ldots, x_{m+1}, \ldots)$. The same argument shows that $\beta(z)\, d(x)^{-1}$ is a unit in $\mathfrak{O}(z, x_{m+1}, \ldots)$. The quotient $\alpha(x)\beta(z)^{-1}$ is therefore a unit in $\mathfrak{O}(z, x_1, \ldots, x_m)$, which was to be proved. ▼

4. Ideals

$\mathfrak{O}$ will from here on in be the principal order of K with respect to $\mathfrak{o}$.

An $\mathfrak{O}$-ideal (simply called an ideal when no confusion can result) is defined as a finite $\mathfrak{o}$-module $\mathfrak{a}$ of elements of K not consisting only of the 0, and such that for every pair $\omega \in \mathfrak{O}$ and $\alpha \in \mathfrak{a}$ the product $\omega\alpha \in \mathfrak{a}$. This last condition can be written symbolically as

$$\mathfrak{O}\mathfrak{a} = \mathfrak{a}. \tag{2}$$

This concept of ideal is somewhat more general than that of abstract algebra, where $\mathfrak{a} \subseteq \mathfrak{O}$ is required.

If $\mathfrak{a}$, $\mathfrak{b}$ are two ideals, then the finite sums $\sum \alpha_i \beta_i$ with arbitrary $\alpha_i \in \mathfrak{a}$, $\beta_i \in \mathfrak{b}$ also form an ideal called the *product* of $\mathfrak{a}$ and $\mathfrak{b}$ and written $\mathfrak{ab}$. If α_i, β_i are systems of generators of $\mathfrak{a}$ and $\mathfrak{b}$, then $\alpha_i \beta_j$ is a system of generators of $\mathfrak{ab}$. In particular, (2) represents a product, so that it becomes plausible to call $\mathfrak{O}$ the unit ideal, as ideals remain unchanged when multiplied by it. By §1,2, $\mathfrak{O}$ is a finite $\mathfrak{o}$-module, and thus an ideal.

If $\mathfrak{a}$, $\mathfrak{b}$ and $\mathfrak{c}$ are three ideals, the associative law

$$(\mathfrak{ab})\mathfrak{c} = \mathfrak{a}(\mathfrak{bc})$$

holds.

Special ideals, those of the form $\mathfrak{O}\alpha$ with $\alpha \neq 0$, are called principal ideals. The short notation $\mathfrak{O}\alpha = (\alpha)$ is used. Clearly

$$(\alpha)(\beta) = (\alpha\beta). \tag{3}$$

This can be genralized to $\mathfrak{a}(\alpha) = \mathfrak{a}\alpha$. Because of (3) the principal ideals form a group, the unit element of which is $\mathfrak{O} = (1)$. This group is a homomorphic image of the multiplicative group $K^{\times}$ of K. The units of $\mathfrak{O}$, i.e., those elements $\varepsilon \in \mathfrak{O}$ whose inverse is contained in $\mathfrak{O}$, are mapped onto the unit ideal under this homomorphism.

An ideal $\mathfrak{a}$ is said to be *divisible* by an ideal $\mathfrak{b}$ if

$$\mathfrak{a} \subseteq \mathfrak{b} \tag{4}$$

holds. If $\mathfrak{a}$ is divisible by $\mathfrak{O}$ it is called an *integral ideal.* Multiplying (4) by the inverse ideal $\mathfrak{b}^{-1}$ (which we shall soon prove to exist) shows that the divisibility of $\mathfrak{a}$ by $\mathfrak{b}$ is equivalent to the fact that $\mathfrak{ab}^{-1}$ is integral, as is familiar from numbers.

An element α is also said to be divisible by $\mathfrak{a}$ if $\alpha \in \mathfrak{a}$.

If $\mathfrak{a}$ and $\mathfrak{b}$ are arbitrary ideals, the $\mathfrak{o}$-module consisting of their union is an ideal written $(\mathfrak{a}, \mathfrak{b})$. It is a divisor of $\mathfrak{a}$ and $\mathfrak{b}$ in the sense of (4), and is called the g.c.d. of $\mathfrak{a}$ and $\mathfrak{b}$. In apparent contradiction to its name, the g.c.d. is the smallest ideal containing both $\mathfrak{a}$ and $\mathfrak{b}$. The g.c.d. of several ideals is defined accordingly. For principal ideals $\mathfrak{a} = (\alpha)$, $\mathfrak{b} = (\beta)$, etc., one writes $(\alpha, \beta, \ldots)$ for short.

As is familiar, an integral ideal $\mathfrak{p}$ is called a *prime ideal* if its residue class ring $\mathfrak{O}/\mathfrak{p}$ contains no divisors of 0. *Our hypothesis being that $\mathfrak{o}$ is a principal ideal domain, the residue class ring $\mathfrak{O}/\mathfrak{p}$ is even a field.*

Proof. All elements in $\mathfrak{o}$ divisible by $\mathfrak{p}$ form an $\mathfrak{o}$-ideal. By hypothesis it is a principal ideal (p). The fact that $\mathfrak{p}$ is a prime ideal implies that p is a prime element (i.e., not decomposable into proper factors). By I, §1,1, the residue classes of a principal ideal domain with respect to a prime element form a field.

Now let $\alpha \in \mathfrak{O}$, $\alpha \notin \mathfrak{p}$, and $\alpha^r + c_1\alpha^{r-1} + \cdots + c_r \equiv 0 \bmod \mathfrak{p}$ with $c_i \in \mathfrak{o}$ be the congruence of least degree that α satisfies. Certainly $c_r \not\equiv 0 \bmod p$, for otherwise we would have $\alpha(\alpha^{r-1} + \cdots + c_{r-1}) \equiv 0 \bmod \mathfrak{p}$ and, because of our assumption, the second factor would be $\equiv 0 \bmod \mathfrak{p}$. Since the residue class ring $\mathfrak{o}/p$ is a field, there exists a $c \in \mathfrak{o}$ with $cc_r \equiv 1 \bmod p$. But then

$$-\alpha c(\alpha^{r-1} + \cdots + c_{r-1}) \equiv 1 \bmod \mathfrak{p},$$

and the assertion is proved. ▼

A prime ideal $\mathfrak{p}$ *is divisible only by itself and* $\mathfrak{O}$.

Proof. Let $\mathfrak{a}$ lie between $\mathfrak{p}$ and $\mathfrak{O}$. The residue classes of the elements in $\mathfrak{o}$ form an ideal in $\mathfrak{O}/\mathfrak{p}$. But in this field there can be only a zero ideal and unit ideal, corresponding to whether $\mathfrak{a} = \mathfrak{p}$ or $\mathfrak{a} = \mathfrak{O}$. ▼

Principal Theorem, *If* $\mathfrak{O}$ *is the principal order of* K *with respect to a principal ideal domain* $\mathfrak{o}$ *of* k, *then the* $\mathfrak{O}$*-ideals form a group. Furthermore, all ideals can be written uniquely as power products of prime ideals.*

The principal ideals form a subgroup of the group of all ideals. The cosets of this subgroup are called *classes of ideals*, and they also form a group, the factor group. Ideals of the same coset, or simply class, are called *equivalent*:

$$\mathfrak{a} \sim \mathfrak{b} \quad \text{if} \quad \mathfrak{b} = \mathfrak{a}\alpha, \qquad \alpha \in K.$$

5. Proof of the Principal Theorem

As in §1,3 we adjoin infinitely many indeterminates x. Then every element $\alpha(x) \in \mathfrak{O}(x)$ can be written as quotient $a_n(x)a_d(x)^{-1}$ of polynomials, the denominator $a_d(x)$ being a primitive polynomial in $\mathfrak{o}[x]$. The $\mathfrak{O}$-ideal generated by the coefficients of the numerator $a_n(x)$ is called the *content* of $\alpha(x)$. It will turn out that the content does not change when the quotient suffers cancellation or extension with primitive polynomials in $\mathfrak{o}[x]$. Conversely, it is possible to associate an element $\alpha(x) = \sum \alpha_i x^i$ with a given ideal $\mathfrak{a}$ having the system of generators α_i; its content is $\mathfrak{a}$.

To every ideal $\mathfrak{a}$ there exists an inverse ideal $\mathfrak{a}^{-1}$:

$$\mathfrak{a}\mathfrak{a}^{-1} = \mathfrak{O}.$$

Thus the ideals form a group.

Proof. Introducing another set of indeterminates x' consider the $\alpha(x)$ as rational functions with coefficients in the principal ideal domain $\mathfrak{O}(x')$. As such, they (in particular the polynomials in x) have contents which are Kronecker divisors, i.e., cosets of the form $\alpha_1(x')\mathfrak{E}(x')$, where $\mathfrak{E}(x')$ is the group of units of $\mathfrak{O}(x')$. By Gauss's theorem the content of a product is equal to the product of the contents.

Let an ideal $\mathfrak{a}$ be given, and let α_i be a system of generators of $\mathfrak{a}$ as a $\mathfrak{o}$-module. We form $\alpha(x) = \sum \alpha_i x_i$ and the inverse $\beta(x) = \alpha(x)^{-1} = b_n(x)b_d(x)^{-1}$. Then

$$\alpha(x)b_n(x) = b_d(x), \qquad b_d(x) \in \mathfrak{o}[x]$$

with primitive $b_d(x)$. Now we consider the contents of these polynomials as Kronecker divisors with respect to the variables x'. Because $b_d(x)$ is primitive the contents $\alpha_1(x')$ and $b_{n1}(x')$ of $\alpha(x)$ and $b_n(x)$ are inverse to each other, and the products of the coefficients of $\alpha(x)$ and $b_n(x)$ are in $\mathfrak{O}(x')$. As they are also in K, they must be contained in $\mathfrak{O}$.

Furthermore, it becomes clear that the coefficients of $b_d(x)$ can be written as sums of products of coefficients of $\alpha(x)$ and $b_n(x)$. Since $\mathfrak{o}$ is a principal ideal domain and the coefficients of $b_d(x)$ have g.c.d. 1, it is possible to represent 1 as the sum of products of coefficients of $\alpha(x)$ and of $b_n(x)$ and of elements of $\mathfrak{O}$. This finally implies that every element of $\mathfrak{O}$ can be so written.

Denoting the ideal generated by coefficients of $b_n(x)$ with $\mathfrak{a}^{-1}$, the result attained is $\mathfrak{a}\mathfrak{a}^{-1} = \mathfrak{O}$, which is what was to be proven. ▼

The point should not be missed that the assumptions of $\mathfrak{o}$ being a principal ideal domain was necessary. An example in §1,6 will demonstrate this clearly.

The group of ideals is isomorphic to the group of Kronecker divisors, i.e., the factor group $K(x)^\times/\mathfrak{E}(x)$ of the multiplicative group of $K(x)$ with respect to the group of units of $\mathfrak{O}(x)$.

Proof. Comparison of coefficients shows that the rational functions $\alpha(x)$, $\beta(x)$, $\gamma(x)$ with $\alpha(x)\beta(x) = \gamma(x)$ have contents $\mathfrak{a}$, $\mathfrak{b}$, $\mathfrak{c}$ satisfying $\mathfrak{a}\mathfrak{b} \subseteq \mathfrak{c}$. Using $\beta(x) = \alpha(x)^{-1}\gamma(x)$ leads to $\mathfrak{a}^{-1}\mathfrak{c} \subseteq \mathfrak{b}$, and multiplying this with $\mathfrak{a}$ yields $\mathfrak{c} \subseteq \mathfrak{a}\mathfrak{b}$. This proves the assertion. ▼

The earlier assertion, that the contents of $\alpha(x)$ remain invariant under cancellation and extension, also follows directly.

The unique prime decomposition of Kronecker divisors yields, finally: *Any ideal can be written uniquely as the product of prime ideals.*

6*. Extension of Divisibility Theory

To conclude this paragraph we show that the concept of Kronecker divisors leads further than the concept of ideals. We return to this possibility in Chapter V.

If $\mathfrak{o}$ is a noetherian domain in which every element can be represented uniquely as the product of prime elements, then $\mathfrak{o}(z)$ is a principal ideal domain.

Proof. As in §1,3, we first adjoin an infinite number of indeterminates, x_i. Given $\alpha_i(x) \in \mathfrak{o}(x)$, their g.c.d. can be represented as $\alpha(x) = \sum \alpha_i(x)x_i$, taking variables x_i not occurring in the $\alpha_i(x)$. For, cancelling by the g.c.d. of the denominators in $\alpha_i(x)\alpha(x)^{-1}$ leaves elements with g.c.d. 1 in $\mathfrak{o}(x)$

remaining. A second step now shows that adjunction of a single variable suffices. This is done, e.g., by associating the polynomials

$$\alpha(x) = \sum a_{i_1,\ldots} x_1^{i_1} \cdots \quad \text{and} \quad \alpha(z) = \sum a_{i_1,\ldots} z^{m_i}$$

where

$$m_i = m_{i_1,\ldots} = i_1 + n_1 i_2 + n_1 n_2 i_3 + \cdots,$$

the n_ϱ being the degrees of the $\alpha(x)$ in the x_ϱ.

The feasibility of prime decomposition of a polynomial is a consequence of the fact that $\mathfrak{o}[z]$ is also a noetherian ring (cf. I,§1,4). The prime decomposition of an element of $\mathfrak{o}(z)$, i.e., of a quotient of polynomials of $\mathfrak{o}[z]$ with primitive denominator, comes from the decomposition of the numerator. ▼

The principal order of $K(z)$ with respect to $\mathfrak{o}(z)$ is a principal ideal domain.

The proof is word for word that of §1,3. Feasibility of prime decomposition is, as in §1,3, demonstrated by the formation of norms with respect to k.

Again as in §1,3, the group $\mathfrak{K}_\mathfrak{o} = K(z)^\times/\mathfrak{E}(z)$ of Kronecker divisors with respect to $\mathfrak{o}$ and its subgroup $\mathfrak{H}_\mathfrak{o} = \mathfrak{K}^\times/\mathfrak{E}$ of principal divisors can be formed.

Ideal theory does not lead this far. Take the example

$$k = k_0(y_1, y_2), \qquad K = k(\sqrt{y_1 y_2}), \qquad \mathfrak{o} = k_0[y_1, y_2].$$

A basis of $\mathfrak{O}$ with respect to $\mathfrak{o}$ is $1, \sqrt{y_1 y_2}$. Now, let $\mathfrak{a}$, $\mathfrak{b}$ be ideals with bases $y_1, \sqrt{y_1 y_2}$ and $y_2, \sqrt{y_1 y_2}$ with respect to $\mathfrak{o}$. Then, $\mathfrak{ab}$ can be generated by $y_1\sqrt{y_1 y_2}, y_2\sqrt{y_1 y_2}, y_1 y_2$ with respect to $\mathfrak{o}$, but has no two-element basis. There can be no inverse to $\mathfrak{a}$. Such an ideal would, at the most, contain elements of $\mathfrak{b}(\sqrt{y_1 y_2})^{-1}$, but $\mathfrak{ab}\,(\sqrt{y_1 y_2})^{-1} \subset \mathfrak{O}$. The calculation with Kronecker divisors is the following:

$$\alpha(x) = y_1 + \sqrt{y_1 y_2}\, x, \qquad \beta(x) = y_2 + \sqrt{y_1 y_2}\, x, \qquad \alpha(x)\beta(x) = \sqrt{y_1 y_2}\; \varepsilon(x)$$

with the unit $\varepsilon(x) = (y_1 + y_2)x + \sqrt{y_1 y_2}(1 + x^2)$. The fact that $\varepsilon(x)$ is a unit follows from the norm

$$n_{K/k}(\varepsilon(x)) = (y_1 + y_2)^2 x^2 - y_1 y_2 (1 + x^2)^2,$$

which is a primitive polynomial in x.

The theory of divisibility based on divisors is due to Kronecker. He, however, uses them only in the proof that the contents of the product of two polynomials is equal to the product of their contents. This yields, in principle, the theorem that the ideals form a group, and from there it is not far to the unique prime decomposition of ideals. As was seen here, divisors attain prime decomposition quicker, and the ideals would be superfluous. The concept of ideals turns out to be indispensible for other reasons, however. Nevertheless, multiplication of ideals could often be omitted from the theory, which would be a simplification.

Another possibility of extending the theory of divisibility is the theory of quasi-equal ideals of van der Waerden and Artin.† The assumptions made are weaker yet, but with them it is easy to show that if $\mathfrak{o}$ is an integral domain in which the theory of quasi-equal ideals holds then $\mathfrak{o}(z)$ is a principal ideal domain. Thus, both extensions of the theory of divisibility turn out to have the same effect. Without doubt, the Kronecker divisors are easier to handle than quasi-equal ideals in concrete problems.

§2. Local Rings

1. Basic Concepts

In arithmetic, the local considerations indicated in the introduction play the same part as in analysis, where, for example, a curve is considered in relation to its tangent. Since the basic operations and concepts of arithmetic do not have the same visual meaning in the immediate sense of analysis and geometry, the meaning of local considerations only becomes apparent step by step as the theory is built up. Our first task is to fix in a conceptual way the integral domain $\mathfrak{i}_\xi$ constructed in the introduction.

An integral domain $\mathfrak{o}$ is called a *special local ring*‡ if it has the following properties: (1) the quotient field k of $\mathfrak{o}$ is not equal to $\mathfrak{o}$; (2) if $\alpha \in k$ then either $\alpha \in \mathfrak{o}$ or $\alpha^{-1} \in \mathfrak{o}$. If k is a function field, we will also require: (3) $\mathfrak{o}$ contains the exact field of constants.

A special local ring contains a single maximal prime ideal $\mathfrak{p}$; *it consists of all* $\alpha \in \mathfrak{o}$ *such that* $\alpha^{-1} \notin \mathfrak{o}$.

Proof. Let $\alpha \in \mathfrak{o}$, $\alpha^{-1} \notin \mathfrak{o}$. Then, if $\gamma \in \mathfrak{o}$ so is $\alpha\gamma$, but $(\alpha\gamma)^{-1} \notin \mathfrak{o}$. Thus, $\alpha \in \mathfrak{p}$, $\gamma \in \mathfrak{o}$ implies $\alpha\gamma \in \mathfrak{p}$. Further, let $\alpha_1, \alpha_2 \in \mathfrak{p}$. Suitable indexing gives $\alpha_1\alpha_2^{-1} \in \mathfrak{o}$. Then, $\alpha_1 - \alpha_2 = (\alpha_1\alpha_2^{-1} - 1)\alpha_2 \in \mathfrak{p}$, proving $\mathfrak{p}$ to be an $\mathfrak{o}$-ideal.

Those $\alpha \in \mathfrak{o}$ not in $\mathfrak{p}$ have, by definition, the property $\alpha^{-1} \in \mathfrak{o}$. Hence, the residue class ring is a field, and $\mathfrak{p}$ is therefore prime. We emphasize: *the residue class ring of a special local ring with respect to its prime ideal is a field.*

Now let $\mathfrak{q}$ be some maximal prime $\mathfrak{o}$-ideal. If $\kappa \in \mathfrak{q}$, then $\kappa^{-1} \notin \mathfrak{o}$. Thus, $\mathfrak{q} \subseteq \mathfrak{p}$, and as $\mathfrak{q}$ is maximal, $\mathfrak{q} = \mathfrak{p}$. ▼

A finite ideal $\mathfrak{a}$ *for a special local ring* $\mathfrak{o}$, *that is, the union of a finite number of principal ideals* $\mathfrak{o}\beta$, *is a principal ideal* $\mathfrak{o}\alpha$.

Proof. For any two nonzero elements α, $\beta \in k$, we define an ordering $(\alpha) \leqq (\beta)$ if $\alpha\beta^{-1} \in \mathfrak{o}$. Then we have: (1) either $(\alpha) \leqq (\beta)$ or $(\beta) \leqq (\alpha)$, with

† B. L. van der Waerden, *Algebra*, 4th ed., Vol. II, §131. Berlin-Göttingen-Heidelberg, 1959.

‡ A local ring as such is defined as an integral domain having a single maximal prime ideal.

both certainly holding if $\alpha = \beta$, we write $(\alpha) = (\beta)$; (2) if $(\alpha) \leqq (\beta)$ and $(\beta) \leqq (\gamma)$ then $(\alpha) \leqq (\gamma)$; (3) if $(\alpha) \leqq (\beta)$ and $\gamma \neq 0$ then $(\alpha\gamma) \leqq (\beta\gamma)$; and, finally, (4) $\mathfrak{o}$ consists of all α with $(\alpha) \leqq (1)$ and $\alpha = 0$.

A finite ideal $\mathfrak{a}$ has a largest element. Assume this to be false, and there to exist an infinite series α_ν with $(\alpha_\nu) < (\alpha_{\nu+1})$; i.e., $(\alpha_\nu) \leqq (\alpha_{\nu+1})$ but not $(\alpha_{\nu+1}) \leqq (\alpha_\nu)$. For no subscript n, then, can we write α_n as

$$\alpha_n = \gamma_1\alpha_1 + \cdots + \gamma_{n-1}\alpha_{n-1}, \quad \gamma_\nu \in \mathfrak{o},$$

for this equation would imply

$$1 = \gamma_1(\alpha_1/\alpha_n) + \cdots + \gamma_{n-1}(\alpha_{n-1}/\alpha_n).$$

By hypothesis $\alpha_\nu/\alpha_n = \alpha_\nu/\alpha_{\nu+1} \cdots \alpha_{n-1}/\alpha_n$ are contained in $\mathfrak{o}$, but not their inverse elements. Thus, the $\alpha_\nu/\alpha_n \in \mathfrak{p}$, yet 1 cannot be in $\mathfrak{p}$. We have thus shown that $\mathfrak{a}$ cannot be a finite ideal, which is a contradiction.

For a maximal $\alpha \in \mathfrak{a}$ we have $(\beta) \leqq (\alpha)$ for all $\beta \in \mathfrak{a}$, which means that $\beta \in \mathfrak{o}\alpha$. Thus, $\mathfrak{a} = \mathfrak{o}\alpha$, as maintained. ▼

The discussion also implies: *The finite $\mathfrak{o}$-ideals form an ordered group under the definition $\mathfrak{o}\alpha \leqq \mathfrak{o}\beta$ if $(\alpha) \leqq (\beta)$.*

Exercise. The triangle inequality $\mathfrak{o}(\alpha + \beta) \leqq \max(\mathfrak{o}\alpha, \mathfrak{o}\beta)$ holds; i.e., the mapping $\alpha \to \mathfrak{o}\alpha$ is a valuation of k.

A special local ring $\mathfrak{o}$ will be called *discrete* if its prime ideal $\mathfrak{p}$ is finite and if to every $\alpha \in k$ there exists a power of $\mathfrak{p}$ such that $\alpha\mathfrak{p}^h \subseteq \mathfrak{o}$. Taking the least power h, one has $\alpha\mathfrak{p}^h = \mathfrak{o}$. In fact, let $\mathfrak{p} = \mathfrak{o}\pi$. Were $\alpha\mathfrak{p}^h \neq \mathfrak{o}$, then $\alpha\mathfrak{p}^h \subseteq \mathfrak{o}\pi$ and $\alpha\pi^{-1}\mathfrak{p}^h = \alpha\mathfrak{p}^{h-1} \subseteq \mathfrak{o}$, and h would not be minimal. *Thus, for these discrete local rings all finite ideals are powers of the prime ideal.*

Further (maximality): *An integral domain $\mathfrak{o}'$ properly containing a discrete local ring $\mathfrak{o}$ and having the same quotient field k coincides with k.*

Proof. Let $\kappa \in \mathfrak{o}'$, $\kappa \notin \mathfrak{o}$. Then, $\pi = \kappa^{-1}$ is in $\mathfrak{p}$. As $\mathfrak{o}$ is discrete, there exists to every $\alpha \in k$ an exponent h, such that $\alpha\pi^h = \beta \in \mathfrak{o}$. Clearly one may choose $h \geqq 0$. Then, every α can be represented $\alpha = \beta\pi^{-h}$. This, however, implies that adjoining $\kappa = \pi^{-1}$ to $\mathfrak{o}$ yields the field k. ▼

Example. Let the field k have the property that every element can be written essentially as a power product of prime elements, $a = ep_1^{\alpha_1} \cdots p_r^{\alpha_r}$, with a unit e. These units and the prime elements belong to an integral domain $\mathfrak{o}$ in k. Distinguishing some prime p, all a in which p occurs with nonnegative powers form a discrete local ring $\mathfrak{o}_p$. The i_ξ mentioned in the introduction are of this variety.

2. Local Rings in Algebraic Extensions

Let a discrete local ring $\mathfrak{o}$ with quotient field k be given, K be a finite extension of k, and $\mathfrak{O}$ the principal order of K with respect to $\mathfrak{o}$. Then, $\mathfrak{O}$ is a principal ideal domain with only finitely many prime ideals.

These prime ideals have the form $\mathfrak{p}_\varrho = \mathfrak{O}\pi_\varrho$, where π_ϱ comes from a prime decomposition

$$p = \varepsilon\pi_1^{e_1} \cdots \pi_r^{e_r} \tag{1}$$

of a prime element p of $\mathfrak{o}$. (p is determined up to a unit factor by $\mathfrak{p} = \mathfrak{o}p$.) ε is a unit of $\mathfrak{O}$. The π_ϱ are called the prime elements for the $\mathfrak{p}_\varrho$.

Proof. As $\mathfrak{o}$ is a principal ideal ring the ideal theory holds for $\mathfrak{O}$. Let $\mathfrak{q}$ be a prime $\mathfrak{O}$-ideal, then $\mathfrak{o} \cap \mathfrak{q}$ is a prime $\mathfrak{o}$-ideal; thus it must be $\mathfrak{p}$, so that $\mathfrak{q}$ is a divisor of $\mathfrak{O}p$. Let

$$\mathfrak{O}p = \mathfrak{p}_1^{e_1} \cdots \mathfrak{p}_r^{e_r}$$

be the prime decomposition in $\mathfrak{O}$. All ideals are power products of the $\mathfrak{p}_\varrho$.

We now maintain: there exists a $\pi_1 \in \mathfrak{p}_1$ such that $\pi_1\mathfrak{p}_1^{-1}$ is relatively prime to $\mathfrak{p}_1 \cdots \mathfrak{p}_s$ $(s = 1, 2, \ldots, r)$. This assertion holds for $s = 1$, as $\mathfrak{p}_1^2 \subset \mathfrak{p}_1$. Further, $\mathfrak{p}_1\mathfrak{p}_2 \subset \mathfrak{p}_1$ implies the existence of a $\pi_1{}' \in \mathfrak{p}_1$ not divisible by $\mathfrak{p}_2$. Therefore, either π_1 or $\pi_1 + \pi_1'^2$ is divisible by $\mathfrak{p}_1$ exactly once and not at all by $\mathfrak{p}_2$, proving the assertion for $s = 2$. We proceed by induction, denoting an element of $\mathfrak{p}_i$ as $\pi_{i;j,\ldots}$ if it is such that $\pi_{i;j,\ldots}\mathfrak{p}_i^{-1}$ is relatively prime to $\mathfrak{p}_i\mathfrak{p}_j \cdots$. The inductive hypothesis gives us $\pi_{1;2,\cdots,s-1}$. We have finished if this element is not divisible by $\mathfrak{p}_s$. But if it is, then

$$\pi_{1;2,\ldots,s} = \pi_{1;2,\ldots,s-1} + \pi_{1;s}^2\pi_{2;s} \cdots \pi_{s-1;s}$$

is an element with the property asserted.

In particular, taking $s = r$ and forming the analogous elements $\pi_2, \pi_3, \ldots \pi_r$, one has $\mathfrak{p}_\varrho = \mathfrak{O}\pi_\varrho$, proving the theorem. ▼

Those $\alpha \in K$ for which a fixed $\mathfrak{p}_\varrho$ does not occur with a negative power in the prime decomposition of $\mathfrak{O}\alpha = (\alpha)$ form a discrete local ring $\mathfrak{O}_\varrho$. It contains $\mathfrak{o}$, and

$$\mathfrak{O} = \mathfrak{O}_1 \cap \cdots \cap \mathfrak{O}_r$$

holds. Conversely, these are all the special local rings $\mathfrak{B}$ containing $\mathfrak{o}$.

Proof. The first statements are obvious (see the example at the end of §1). Keeping the notation from above, let $\mathfrak{B}$ be such a ring. Any $\alpha \in \mathfrak{O}$ satisfies an equation $a^n + c_1\alpha^{n-1} + \cdots + c_n = 0$, $c_i \in \mathfrak{o}$. If α were not contained in $\mathfrak{B}$ we could write $\alpha = -c_1 - c_2\alpha^{-1} - \cdots - \alpha^{-n+1}$, the right side of which would be in $\mathfrak{B}$. Thus, $\alpha \in \mathfrak{B}$ and $\mathfrak{O} \subseteq \mathfrak{B}$.

Let $\mathfrak{q}$ be the prime ideal of $\mathfrak{B}$ and set $\mathfrak{O} \cap \mathfrak{q} = \mathfrak{r}$; it is an integral $\mathfrak{O}$-ideal. In order to show that it cannot be divisible by two different prime $\mathfrak{O}$-ideals, we assume the contrary, that it is divisible by $\mathfrak{p}_1, \ldots, \mathfrak{p}_s$. Then, π_1 is not in $\mathfrak{r}$, nor in $\mathfrak{q}$. Thus, $\pi_1^{-1} \in \mathfrak{q}$; similarly $\pi_2^{-1}, \ldots \in \mathfrak{q}$. Altogether, $\alpha = (\pi_1^{-1} \cdots \pi_s^{-1})^h \in \mathfrak{q}$ for any natural h. Choosing a sufficiently large h, by hypothesis $\alpha^{-1} \in \mathfrak{r}$, and thus in $\mathfrak{q}$. This is impossible.

The possibility remaining is $\mathfrak{O} \cap \mathfrak{q} = \mathfrak{p}_1{}^h$, with suitable indexing. Certainly $h \neq 0$; otherwise $1 \in \mathfrak{q}$. $h > 1$ drops out, as then $\pi_1 \notin \mathfrak{q}$, but $\pi_1^{-1} \in \mathfrak{q}$ is just as impossible.

By elimination, then, we have

$$\mathfrak{O} \cap \mathfrak{q} = \mathfrak{p}_1.$$

Proceeding, $\pi_2, \ldots, \pi_r \in \mathfrak{O} \subseteq \mathfrak{B}$. They are not in $\mathfrak{p}_1$, thus cannot belong to $\mathfrak{q}$, and are units of $\mathfrak{B}$. Therefore, all elements $\alpha\pi_2^{-h_2} \ldots \pi_r^{-h_r} \in \mathfrak{B}$, if $\alpha \in \mathfrak{O}$ for arbitrary h_ϱ. These are exactly the elements that form the ring $\mathfrak{O}_1$. ▼

In conclusion we draw attention to the fact that these theorems can be proved without resort to the ideal theory of §1, which is, in fact, accomplished by the use of *valuation theory* in several texts. This leads to another foundation of arithmetic meeting with ours in §5. It borrows its basic concepts from classical function theory and is thus quite intuitive. On the negative side, certain difficulties are encountered when it comes to inseparable field extensions, as we saw in §1,2, and these can only be overcome by limitation of the generality. It can, in fact, be shown that the important equation §3 ,(14) is incorrect without such limitations. Our method is preferable for these reasons, as well as because it is shorter and makes fewer assumptions.

The results of this section, incidentally, can be achieved without reference either to §1 or to valuation theory, by simply deriving them from the properties of $\mathfrak{o}$, as long as the residue class field $\bar{k} = \mathfrak{o}/\mathfrak{p}$ contains infinitely many elements. We will do this in III, §2 for the field $k = k_0(x)$, k_0 algebraically closed, thus choosing the path otherwise arrived at by valuation theory. For $\bar{k}$ with only finitely many elements the procedure of §1 can be shortened considerably. It is a worthwhile exercise (to be contemplated, say, after reading Chapters II and III, §2) to carry this out.

3. Local Rings in Algebraic Number and Function Fields

Let $k = Q$ or $k = k_0(x)$, $\mathfrak{i}$ being as in the introduction. Let p denote a prime element and $\mathfrak{i}_p$ be the integral domain of all $\alpha \in k$, in whose prime decomposition p does not occur with a negative power. In the function case, let i_∞ be the integral domain of functions with nonpositive degree, which we can now write in the form $\mathfrak{i}_\infty = \mathfrak{i}'_{p'}$, with $x' = x^{-1}$, $\mathfrak{i}' = k_0[x']$, and p' being the prime element x'.

The $\mathfrak{i}_p$ *(including* $\mathfrak{i}_\infty$*) are all the special local rings of* k.

Proof. A local ring $\mathfrak{o}$ always contains 1, and thus $-1, \pm 2, \ldots$. In the number case, therefore, $\mathfrak{o} \supseteq \mathfrak{i}$. In the function case let us assume first that $x \in \mathfrak{o}$. The constant field is always contained in a local ring, hence again, $\mathfrak{i} \subseteq \mathfrak{o}$.

Not all $a \in \mathfrak{i}$ can be units of $\mathfrak{o}$, for then $\mathfrak{o}$ would contain k, the quotient field of $\mathfrak{i}$. If some a is a nonunit of $\mathfrak{o}$, then one of its prime factors p must be a nonunit. Thus, the prime ideal $\mathfrak{p}$ contains at least one prime element p, but it also contains only one. For, if it contained two, say p and q, $\mathfrak{i}_p$ would also contain 1, because of the representation $pu + qv = 1$. As all primes of $\mathfrak{i}$ except for p are now units in $\mathfrak{o}$, this ring contains the ring $\mathfrak{i}_p$. But $\mathfrak{i}_p$ is a discrete local ring, so that due to the maximality property we have $\mathfrak{o} = \mathfrak{i}_p$.

We have yet to discuss the possibility that k is a function field but $x \notin \mathfrak{o}$. But then $x' = x^{-1} \in \mathfrak{o}$, and the same argument holds: $\mathfrak{o} = \mathfrak{i}'_{p'}$ with some prime polynomial $p' = x'^n + \cdots + c_n$. Now, if $p' \neq x'$ we would have $p' = x^{-n}p$, where $p = c_n x^n + \cdots + 1$. It is easy to see that this would lead to $\mathfrak{i}'_{p'}, = \mathfrak{i}_p$, and then $x \in \mathfrak{o}$. Hence we have $p' = x'$. ▼

By now applying the last theorem of §2,2, a complete view of all the special local rings of an algebraic number or function field has been obtained. This result is especially remarkable for functions in view of the following consideration.

An algebraic function field K is always given as a finite extension of a rational function field $k = k_0(x)$ over the exact constant field k_0. The pair K, $k_0(x)$ is called a *model* of K. Given the function field K, a model is determined by an element x. Clearly, any $x \notin k_0$ can be used. In our construction of all the special local rings of K, we used a particular model, but the local rings are actually invariant of the model. We shall return to this fact in §5.

We here extend the concept of a *place* of a field by associating a place with every discrete local ring. The remark of I, §3,1 holds here in that our definition is essentially only a manner of speaking to be used, for example, in the sense: an element is integral at a place if it is contained in the corresponding local ring.

We extend the terminology by saying a place of a finite extension K/k lies over a place of k if the local ring of K contains that of k. This expression takes on geometric meaning in the case of classical function theory, but this will not appear until IV, §1.

4. The Component Decomposition of Ideals

We now consider some methodical preparations for §§3 and 4. Let k and $\mathfrak{o}$ satisfy the assumptions of §1, and L be an intermediate field between k and the finite extension K, the cases where L coincides with K or k not being excluded. The ideals of K are not only $\mathfrak{o}$-modules, but also $\mathfrak{O}_L$-modules, where $\mathfrak{O}_L$ is the principal order of L with respect to $\mathfrak{o}$. In order to study the correspondence between the ideals of K and L, the ideals of K must be taken as $\mathfrak{O}_L$-modules. The fact that $\mathfrak{O}_L$ is not a principal ideal domain is a hindrance here, but it is overcome by the use of local considerations.

To avoid the complication of subscripts we write k and $\mathfrak{o}$ for L and $\mathfrak{O}_L$, and we will use the result of §1, that all $\mathfrak{o}$-ideals and all $\mathfrak{O}$-ideals form a group in which unique prime decomposition always holds.

First we form the local rings $\mathfrak{o}_\mathfrak{p}$; for all prime ideals $\mathfrak{p}$ of k. These consist of those $\alpha \in k$ which are not divisible by a negative power of $\mathfrak{p}$. Clearly

$$\mathfrak{o} = \bigcap_{\mathfrak{p}} \mathfrak{o}_\mathfrak{p}. \tag{2}$$

We list further results as (a), ...

(a) *Every* $\alpha \in k$ *lies in almost all* $\mathfrak{o}_\mathfrak{p}$ (i.e., in all $\mathfrak{o}_\mathfrak{p}$ with at most finitely many exceptions). For $\alpha \neq 0$ the same holds for α^{-1}, so that any nonzero $\alpha \in k$ is a unit of $\mathfrak{o}_\mathfrak{p}$ for almost all $\mathfrak{p}$.

For an $\mathfrak{O}$-ideal $\mathfrak{a}$ we now introduce the $\mathfrak{p}$-*components*

$$\mathfrak{a}_\mathfrak{p} = \mathfrak{a}\mathfrak{o}_\mathfrak{p}.$$

(b) $\mathfrak{O}_\mathfrak{p} = \mathfrak{O}\mathfrak{o}_\mathfrak{p}$ *is the principal order of* K *with respect to* $\mathfrak{o}_\mathfrak{p}$.

Proof. Denote this principal order by $\mathfrak{O}_\mathfrak{p}$, and assume $\alpha \in \mathfrak{O}_\mathfrak{p}$. Let α satisfy the irreducible equation $\alpha^n + c_1\alpha^{n-1} + \cdots + c_n = 0$ in k, the c_i lying in $\mathfrak{o}_\mathfrak{p}$ by assumption. The common denominator $\mathfrak{b}$ of the ideals (c_i) is relatively prime to $\mathfrak{p}$. As in §2,2, choose a $b \in \mathfrak{b}$ not divisible by $\mathfrak{b}\mathfrak{p}$, and in addition a $c \in \mathfrak{o}$ such that $bc \equiv 1 \bmod \mathfrak{p}$. Then, $\alpha_1 = \alpha bc$ satisfies the equation $\alpha_1{}^n + c_1 bc\alpha_1^{n-1} + \cdots + c_n(bc)^n = 0$. Thus, $\alpha_1 \in \mathfrak{O}$. By construction $(bc)^{-1} \in \mathfrak{o}_p$, and thus $\alpha = \alpha_1(bc)^{-1} \in \mathfrak{O}\mathfrak{o}_\mathfrak{p}$. ▼

(c) *Every* $\alpha \in K$ *is contained in almost all* $\mathfrak{O}_\mathfrak{p}$.

Proof. Once more let $\alpha^n + \cdots + c_n = 0$. By (a) almost all $\mathfrak{o}_\mathfrak{p}$ contain all the c_i, and thus, by (b), α lies in almost all $\mathfrak{O}_\mathfrak{p}$. ▼

(d) *If* $\mathfrak{a}$ *is an* $\mathfrak{O}$-*ideal, then* $\mathfrak{a}_\mathfrak{p} = \mathfrak{O}_\mathfrak{p}$ *almost always.*

Proof. Let $\alpha_1, \ldots, \alpha_m$ be a system of generators of $\mathfrak{a}$ as an $\mathfrak{O}$-module. As the α_i and α_i^{-1} lie in almost all $\mathfrak{O}_\mathfrak{p}$, they are almost always units of $\mathfrak{O}_\mathfrak{p}$. For these $\mathfrak{O}_\mathfrak{p}$, $\mathfrak{a}_\mathfrak{p} = \mathfrak{O}_\mathfrak{p}$. ▼

(e) *Two ideals,* $\mathfrak{a}$, $\mathfrak{b}$ *satisfy*

$$(\mathfrak{a}\mathfrak{b})_\mathfrak{p} = \mathfrak{a}_\mathfrak{p}\mathfrak{b}_\mathfrak{p}, \qquad (\mathfrak{a}^{-1})_\mathfrak{p} = (\mathfrak{a}_\mathfrak{p})^{-1} = \mathfrak{a}_\mathfrak{p}^{-1}.$$

The *proof* is an immediate consequence of the definitions.

(f) *Every* $\mathfrak{O}$-*ideal is the intersection of its* $\mathfrak{p}$-*components.*

Proof. First, two ideals $\mathfrak{a}$, $\mathfrak{b}$ satisfy

$$\mathfrak{a} \bigcap_{\mathfrak{p}} \mathfrak{b}_\mathfrak{p} \subseteq \bigcap_{\mathfrak{p}} \mathfrak{a}\mathfrak{b}_\mathfrak{p}. \tag{3}$$

Applying (3) to $\mathfrak{a}^{-1}$, $\mathfrak{a}\mathfrak{b}$ in place of $\mathfrak{a}$, $\mathfrak{b}$ yields, with (e),

$$\bigcap_{\mathfrak{p}} \mathfrak{b}_\mathfrak{p} = \bigcap_{\mathfrak{p}} \mathfrak{a}_\mathfrak{p}^{-1}\mathfrak{a}_\mathfrak{p}\mathfrak{b}_\mathfrak{p} = \bigcap_{\mathfrak{p}} \mathfrak{a}^{-1}\mathfrak{a}_\mathfrak{p}\mathfrak{b}_\mathfrak{p} \supseteq \mathfrak{a}^{-1} \bigcap_{\mathfrak{p}} \mathfrak{a}_\mathfrak{p}\mathfrak{b}_\mathfrak{p} = \mathfrak{a}^{-1} \bigcap_{\mathfrak{p}} \mathfrak{a}\mathfrak{b}_\mathfrak{p}.$$

Multiplication by $\mathfrak{a}$ and comparison with (3) yields

$$\mathfrak{a} \bigcap_{\mathfrak{p}} \mathfrak{b}_{\mathfrak{p}} = \bigcap_{\mathfrak{p}} \mathfrak{a}\mathfrak{b}_{\mathfrak{p}} = \bigcap_{\mathfrak{p}} \mathfrak{a}_{\mathfrak{p}}\mathfrak{b}_{\mathfrak{p}}.$$

Now, set $\mathfrak{b} = \mathfrak{O}$ to get the result

$$\mathfrak{a} = \bigcap_{\mathfrak{p}} \mathfrak{a}_{\mathfrak{p}}. \tag{4}$$

▼

(g) *Let there be given an arbitrary $\mathfrak{O}_{\mathfrak{p}}$-ideal $\mathfrak{a}_{\mathfrak{p}}$ to every $\mathfrak{p}$, with the provision that almost always $\mathfrak{a}_{\mathfrak{p}} = \mathfrak{O}_{\mathfrak{p}}$. Then, the intersection* (4) *is an $\mathfrak{O}$-ideal.*

Proof. The equation $\mathfrak{a}\mathfrak{O} = \mathfrak{O}$ is obvious, and what must be shown is that $\mathfrak{a}$ is finite in the sense required. Let $\mathfrak{a}_1 \subseteq \mathfrak{a}_2 \subseteq \cdots \subseteq \mathfrak{a}$ be a sequence of finite $\mathfrak{O}$-ideals leading to $\mathfrak{a}$. By assumption $\mathfrak{a} \subseteq \mathfrak{a}_{\mathfrak{p}}$, and therefore

$$\mathfrak{a}\mathfrak{o}_{\mathfrak{p}} \subseteq \mathfrak{a}_{\mathfrak{p}}$$

holds for the $\mathfrak{p}$-components of $\mathfrak{a}$. By hypothesis in conjunction with (d) we have, for almost all $\mathfrak{p}$,

$$\mathfrak{O}_{\mathfrak{p}} = \mathfrak{a}_{1\mathfrak{p}} = \mathfrak{a}_1\mathfrak{o}_{\mathfrak{p}} \subseteq \mathfrak{a}\mathfrak{o}_{\mathfrak{p}} \subseteq \mathfrak{a}_{\mathfrak{p}} = \mathfrak{O}_{\mathfrak{p}}$$

and then $\mathfrak{a}_{i\mathfrak{p}} = \mathfrak{a}_{\mathfrak{p}}$ for all i. Because of $\mathfrak{a}_n\mathfrak{o}_{\mathfrak{p}} \subseteq \mathfrak{a}_{\mathfrak{p}}$ the ascending chain theorem can be applied to the modules $\mathfrak{a}_n\mathfrak{o}_p$. For each of the finitely many exceptional $\mathfrak{p}$ there exists an index n, such that $\mathfrak{a}_n\mathfrak{o}_{\mathfrak{p}} = \mathfrak{a}_{n+1}\mathfrak{o}_{\mathfrak{p}} = \cdots$. The maximal of these n applies to all, and for this index is $\mathfrak{a}_n = \mathfrak{a}_{n+1} = \cdots = \mathfrak{a}$. ▼

(h) *Let $\mathfrak{a}$ be generated as in* (g). *Then, the $\mathfrak{p}$-components of $\mathfrak{a}$ are $\mathfrak{a}\mathfrak{o}_{\mathfrak{p}} = \mathfrak{a}_{\mathfrak{p}}$.*

Proof. Distinguish $\mathfrak{p}$ and note that, by §2,2, $\mathfrak{a}_{\mathfrak{p}} = \mathfrak{O}_{\mathfrak{p}}\alpha_{\mathfrak{p}}$ is a principal ideal. By (c), $\alpha_{\mathfrak{p}}$ lies in almost all the other $\mathfrak{a}_{\mathfrak{p}}$. Say $\alpha_{\mathfrak{p}} \notin \mathfrak{a}_{\mathfrak{p}_i}$ for a finite number of $\mathfrak{p}_i \neq \mathfrak{p}$. Because $\mathfrak{p}\mathfrak{p}_1\mathfrak{p}_2 \cdots \subset \mathfrak{p}_1\mathfrak{p}_2 \cdots$ there exists an $a \in k$ which is divisible by all the $\mathfrak{p}_i$ but not by $\mathfrak{p}$. With this a we have $\mathfrak{a}_{\mathfrak{p}} = \mathfrak{O}_{\mathfrak{p}}\alpha_{\mathfrak{p}}a^h$ with any h and, with suitably large h, $\alpha_{\mathfrak{p}}a^h \in \mathfrak{a}_{\mathfrak{p}_i}$. Thus, we can set $\alpha_{\mathfrak{p}} \in \mathfrak{a}_{\mathfrak{p}'}$ for all $\mathfrak{p}' \neq \mathfrak{p}$. But, with this choice we have $\mathfrak{a}\mathfrak{o}_{\mathfrak{p}} = \mathfrak{O}_{\mathfrak{p}}\alpha_{\mathfrak{p}}$, completing the proof. ▼

(i) *Let $\mathfrak{q}$ be a prime $\mathfrak{O}$-ideal. Then, $\mathfrak{p} = \mathfrak{q} \cap \mathfrak{o}$ is a prime $\mathfrak{o}$-ideal, and we have the residue class isomorphism*

$$\mathfrak{O}/\mathfrak{q} \cong \mathfrak{O}_{\mathfrak{p}}/\mathfrak{q}_{\mathfrak{p}}.$$

Proof. The residue classes of $\mathfrak{o}$ mod $\mathfrak{p}$ form a subring in the residue class field $\mathfrak{O}/\mathfrak{q}$. As such a subring cannot contain divisors of zero, $\mathfrak{p}$ is a prime ideal. Each residue class of $\mathfrak{O}/\mathfrak{q}$ naturally determines a residue class of $\mathfrak{O}_{\mathfrak{p}}/\mathfrak{q}_{\mathfrak{p}}$. On the other hand, let $\alpha \in \mathfrak{O}_{\mathfrak{p}}$. As in the proof of (h), find an $a \in \mathfrak{o}$ not divisible by $\mathfrak{p}$ but with $\alpha a \in \mathfrak{O}$. For a $b \in \mathfrak{o}$ with $ab \equiv 1 \bmod \mathfrak{p}$ one has $\alpha' = \alpha ab \in \mathfrak{O}$, and $\alpha' \equiv \alpha \bmod \mathfrak{q}$. Thus each residue class of $\mathfrak{O}_{\mathfrak{p}}/\mathfrak{q}_{\mathfrak{p}}$ is really determined by one of $\mathfrak{O}/\mathfrak{q}$. It is obvious that the correspondence $\mathfrak{O}/\mathfrak{q} \leftrightarrow \mathfrak{O}_{\mathfrak{p}}/\mathfrak{q}_{\mathfrak{p}}$ behaves properly under addition and multiplication. ▼

§3. Ideals in Different Fields; the Norm

1. Extension of an Ideal

Let $\mathfrak{o}$ be a principal ideal domain with quotient field k and K be a finite extension of k. In case the extension is inseparable, we again make the limiting assumption of §1,2 regarding $\mathfrak{o}$. Thus, the theory developed in §1 holds for ideals in K. In addition to K we consider an intermediate field L, denoting the principal orders by $\mathfrak{O}_K$, $\mathfrak{O}_L$, respectively. The theorems of §1 hold, of course, for the ideals of $\mathfrak{O}_L$, too. The subject matter of §3 will be two homomorphic mappings of the ideal groups of $\mathfrak{O}_L$ and $\mathfrak{O}_K$, each onto a subgroup of the other.

The first of these mappings is an *extension* of an $\mathfrak{O}_L$-ideal $\mathfrak{a}_L$ to an $\mathfrak{O}_K$-ideal $\mathfrak{a}_K = \mathfrak{O}_K\mathfrak{a}_L$. For another ideal, $\mathfrak{b}_L$,

$$(\mathfrak{a}\mathfrak{b})_K = \mathfrak{O}_K\mathfrak{a}_L\mathfrak{b}_L = \mathfrak{a}_K\mathfrak{b}_K \tag{1}$$

clearly holds. Thus, we have a homomorphism. But, the mapping is even an isomorphism. In fact, if $\mathfrak{a}_K = \mathfrak{O}_K$ then all elements of $\mathfrak{a}_K$ depend integrally on $\mathfrak{O}_L$, especially those of $\mathfrak{a}_L$, and so $\mathfrak{a}_L \subseteq \mathfrak{O}_L$. Now we compute $\mathfrak{O}_K\mathfrak{a}_L^{-1}\mathfrak{a}_K$ in two ways, using (1) and the hypothesis:

$$\mathfrak{O}_K\mathfrak{a}_L^{-1}\mathfrak{a}_K = \mathfrak{O}_K\mathfrak{a}_L^{-1}\mathfrak{O}_K\mathfrak{a}_L = \mathfrak{O}_K = \mathfrak{O}_K\mathfrak{a}_L^{-1}\mathfrak{O}_K = \mathfrak{O}_K\mathfrak{a}_L^{-1}.$$

By the previous argument $\mathfrak{a}_L^{-1} \subseteq \mathfrak{O}_L$. Therefore $\mathfrak{a}_L = \mathfrak{O}_L$.

Whenever only multiplicative relations between ideals are dealt with, it is permissible to identify the ideal $\mathfrak{a}_L$ of L with its extension $\mathfrak{a}_K$ of K. Then, as a rule, their distinction by subscripts becomes superfluous. In particular, the unit ideal (1) is designated with the same symbol in K and L. Furthermore, an ideal $\mathfrak{a}$ of K is said to already belong to L if it can be generated by extension of an ideal of L. This convention is based on the conception that the notion ideal is a concrete realization of an already existent idea, guaranteeing the possibility of unique prime decomposition.

2. The Norm

Our starting point is the component decomposition

$$\mathfrak{O}_L = \bigcap_{\mathfrak{p}} \mathfrak{O}_{L\mathfrak{p}},$$

where $\mathfrak{p}$ runs through all prime $\mathfrak{O}_L$-ideals. The $\mathfrak{O}_{L\mathfrak{p}}$ are principal ideal domains.

$\mathfrak{O}_K$ is the principal order of K with respect to $\mathfrak{O}_L$. Let $\mathfrak{O}_{K\mathfrak{p}}$ denote the principal orders of K with respect to the $\mathfrak{O}_{L\mathfrak{p}}$. By §2,2 these are also principal ideal domains, along with the $\mathfrak{O}_{L\mathfrak{p}}$.

An $\mathfrak{O}_K$-ideal $\mathfrak{a}$ can be represented, according to §2,4,(f), as the intersection of its $\mathfrak{p}$-components $\mathfrak{a}_\mathfrak{p}$, $\mathfrak{p}$ running through the places of L. As the $\mathfrak{O}_{K\mathfrak{p}}$ are principal ideal domains there exist $\alpha_\mathfrak{p} \in K$ such that

$$\mathfrak{a}_\mathfrak{p} = \mathfrak{O}_{K\mathfrak{p}}\alpha_\mathfrak{p} \tag{2}$$

and the $\alpha_\mathfrak{p}$ are almost always units of $\mathfrak{O}_{K\mathfrak{p}}$ by §2,4,(c). The $\mathfrak{O}_K$-ideal $\mathfrak{a}$ is now assigned the $\mathfrak{O}_L$-ideal

$$n_{K/L}(\mathfrak{a}) = \bigcap_\mathfrak{p} \mathfrak{O}_{L\mathfrak{p}} n_{K/L}(\alpha_\mathfrak{p}) \tag{3}$$

as its *norm* with respect to L. The norms of units of $\mathfrak{O}_{K\mathfrak{p}}$ with respect to L are units of $\mathfrak{O}_{L\mathfrak{p}}$. Thus, the $n_{K/L}(\alpha_\mathfrak{p})$ are units almost everywhere, so that (3) defines an ideal according to §2,4,(g).

By §2,4,(e), the components $\mathfrak{a}_\mathfrak{p}$ are multiplicative functions of $\mathfrak{a}$. Thus, along with the multiplicativity of the element norm, we see that the ideal norm is a multiplicative function, i.e.,

$$n_{K/L}(\mathfrak{a}\mathfrak{b}) = n_{K/L}(\mathfrak{a})n_{K/L}(\mathfrak{b}). \tag{4}$$

For $\alpha \neq 0$ we have, in particular,

$$n_{K/L}(\mathfrak{O}_K\alpha) = \mathfrak{O}_L n_{K/L}(\alpha). \tag{5}$$

If $\mathfrak{O}_L$ is a principal ideal domain, then $\mathfrak{O}_K$ and $\mathfrak{a}$ each have bases with respect to $\mathfrak{O}_L$ consisting of $[K:L]$ elements ω_i and α_i, respectively. Let

$$\alpha_j = \sum \omega_i a_{ij}$$

with $a_{ij} \in L$. Then

$$n_{K/L}(\mathfrak{a}) = \mathfrak{O}_L|a_{ij}|. \tag{6}$$

When the bases ω_i and α_i are changed, the determinant $|a_{ij}|$ is multiplied by a unit and the ideal on the right in (6) remains invariant. This remark will be used in the proof, which follows. By §2,4,

$$\mathfrak{a}_\mathfrak{p} = \mathfrak{O}_{L\mathfrak{p}}\mathfrak{a}.$$

Hence, the α_i are also a basis of $\mathfrak{a}_\mathfrak{p}$ with respect to $\mathfrak{O}_{L\mathfrak{p}}$. Moreover, the ω_i form a basis of $\mathfrak{O}_{K\mathfrak{p}}$ with respect to $\mathfrak{O}_{L\mathfrak{p}}$. Using the elements $\alpha_\mathfrak{p}$ defined by (2), then, the $\alpha_\mathfrak{p}\omega_i$ form another basis of $\mathfrak{a}_\mathfrak{p}$ with respect to $\mathfrak{O}_{L\mathfrak{p}}$. Set

$$\alpha_\mathfrak{p}\omega_j = \sum \omega_i a'_{ij},$$

so that the determinants $|a_{ij}|$ and $|a'_{ij}|$ differ but by a unit factor in $\mathfrak{O}_{L\mathfrak{p}}$. But the latter determinant is exactly $n_{K/L}(\alpha_\mathfrak{p})$. This shows that $|a_{ij}|$ is divisible by the exact same power of each prime ideal $\mathfrak{p}$ of L as is $n_{K/L}(\alpha_\mathfrak{p})$, which by definition (3) above is exactly as often as is $n_{K/L}(\mathfrak{a})$. ▼

An ideal $\mathfrak{a}$ of L satisfies

$$n_{K/L}(\mathfrak{a}) = \mathfrak{a}^{[K:L]}. \tag{7}$$

For if in L, $\mathfrak{a}_\mathfrak{p} = \mathfrak{O}_{L\mathfrak{p}}\alpha_\mathfrak{p}$ holds with certain $\alpha_\mathfrak{p} \in L$, then in K correspondingly $\mathfrak{a}_\mathfrak{p} = \mathfrak{O}_{K\mathfrak{p}}\alpha_\mathfrak{p}$, and $n_{K/L}(\alpha_\mathfrak{p}) = \alpha_\mathfrak{p}{}^{[K:L]}$ yields (7).

3. The Prime Ideals

Assume the prime ideal $\mathfrak{p}$ of L decomposes in K to

$$\mathfrak{O}_K\mathfrak{p} = \mathfrak{p}_1^{e_1} \cdots \mathfrak{p}_r^{e_r} \tag{8}$$

with prime ideals $\mathfrak{p}_\varrho$ of K. The exponents $e_\varrho = e_{K/L}(\mathfrak{p}_\varrho)$ are called the *ramification indices* of the $\mathfrak{p}_\varrho$ with respect to L.

Another natural number, $f_\varrho = f_{K/L}(\mathfrak{p}_\varrho)$, called the *residue class degree* or simply the *degree* of $\mathfrak{p}_\varrho$ with respect to L is defined by the following theorem: *The residue class field $\mathfrak{O}_K/\mathfrak{p}_\varrho$ is an extension of finite degree*

$$f_{K/L}(\mathfrak{p}_\varrho) = [\mathfrak{O}_K/\mathfrak{p}_\varrho : \mathfrak{O}_L/\mathfrak{p}] \tag{9}$$

over the residue class field $\mathfrak{O}_L/\mathfrak{p}$. According to §2,4,(i) the residue class fields in question are isomorphic to the local residue class fields $\mathfrak{O}_{K\mathfrak{p}_\varrho}/\mathfrak{p}_{\varrho\mathfrak{p}}$ and $\mathfrak{O}_{L\mathfrak{p}}/\mathfrak{p}_\mathfrak{p}$, respectively. It is therefore also possible to define the residue class degree locally as

$$f_{K/L}(\mathfrak{p}_\varrho) = [\mathfrak{O}_{K\mathfrak{p}_\varrho}/\mathfrak{p}_{\varrho\mathfrak{p}} : \mathfrak{O}_{L\mathfrak{p}}/\mathfrak{p}_\mathfrak{p}]. \tag{10}$$

Proof. The residue classes of the elements of $\mathfrak{O}_L$ mod $\mathfrak{p}_\varrho$ and mod $\mathfrak{p}$ coincide. Thus the former field of the theorem is an extension of the latter. And, there can be at most as many linearly independent residue classes in $\mathfrak{O}_K/\mathfrak{p}_\varrho$ with respect to $\mathfrak{O}_L/\mathfrak{p}$ as there are linearly independent elements in $\mathfrak{O}_K$ with respect to $\mathfrak{O}_L$, i.e., $[K:L]$. ▼

Given a prime ideal $\mathfrak{p}_K$ of K, there exists exactly one prime ideal $\mathfrak{p}_L$ of L divisible by $\mathfrak{p}_K$, and then

$$n_{K/L}(\mathfrak{p}_K) = \mathfrak{p}_L{}^f, \qquad f = f_{K/L}(\mathfrak{p}_K). \tag{11}$$

Proof. Start with the residue class field $\mathfrak{O}_K$ mod $\mathfrak{p}_K$. The residue classes representable by elements of $\mathfrak{O}_L$ form a subring without divisors of zero, isomorphic to $\mathfrak{O}_L/\mathfrak{p}_L$, where $\mathfrak{p}_L = \mathfrak{O}_L \cap \mathfrak{p}_K$. Hence $\mathfrak{p}_L$ is a prime ideal of L divisible by $\mathfrak{p}_K$.

Moreover, for a prime ideal $\mathfrak{q}_L \neq \mathfrak{p}_L$ we have $(\mathfrak{p}_L, \mathfrak{q}_L) = \mathfrak{O}_L$ and thus $(\mathfrak{q}_L, \mathfrak{p}_K) = \mathfrak{O}_K$, i.e., $\mathfrak{p}_{K\mathfrak{q}_L} = \mathfrak{O}_{K\mathfrak{q}_L}$. This implies that $n_{K/L}(\mathfrak{p}_{K\mathfrak{q}_L}) = \mathfrak{O}_{L\mathfrak{q}_L}$, so that (11) holds with some power f. It remains to show that f coincides with the residue class degree.

For this aim we observe that the $\mathfrak{p}_L$-components satisfy the corresponding equation

$$n_{K/L}(\mathfrak{p}_{K\mathfrak{p}_L}) = \mathfrak{p}_{L\mathfrak{p}_L}^f$$

with the same exponent. Therefore (11) need only be proved for the $\mathfrak{p}_L$ components which is done with the use of Eq. (10) for the residue class degree. Apply the elementary divisor theorem (I, §1,2) to the $\mathfrak{O}_{L\mathfrak{p}_L}$-modules $\mathfrak{O}_{K\mathfrak{p}_L}$ and $\mathfrak{p}_{K\mathfrak{p}_L}$, arriving at a basis ω_i of the former and elements $t_i \in \mathfrak{O}_{L\mathfrak{p}_L}$ such that $t_i\omega_i$ form a basis of the latter. As $\mathfrak{O}_{K\mathfrak{p}_L}\mathfrak{p}_L$ is contained in $\mathfrak{p}_{K\mathfrak{p}_L}$, some of the t_i are units in $\mathfrak{O}_{L\mathfrak{p}_L}$ while the rest are divisible exactly once by $\mathfrak{p}_{L\mathfrak{p}_L}$. There are exactly as many of the latter as the rank of $\mathfrak{O}_{K\mathfrak{p}_L}/\mathfrak{p}_{K\mathfrak{p}_L}$ over $\mathfrak{O}_{L\mathfrak{p}_L}/\mathfrak{p}_{L\mathfrak{p}_L}$. Thus the norm of $\mathfrak{p}_{K\mathfrak{p}_L}$ with respect to L is, by (6), equal to $\mathfrak{p}_{L\mathfrak{p}_L}^f$, where f is the number in (11), which is verified. ▼

The definitions of the residue class degree and the ramification index lead immediately to the *transitivity formulas*

$$f_{K/k}(\mathfrak{p}_K) = f_{K/L}(\mathfrak{p}_K) f_{L/k}(\mathfrak{p}_L), \qquad e_{K/k}(\mathfrak{p}_K) = e_{K/L}(\mathfrak{p}_K) e_{L/k}(\mathfrak{p}_L), \tag{12}$$

where $\mathfrak{p}_K$ is a prime ideal in K and $\mathfrak{p}_L$ denotes the prime ideal in L divisible by $\mathfrak{p}_K$. Furthermore, arbitrary ideals $\mathfrak{a}$ satisfy the transitivity formula for the norm

$$n_{K/k}(\mathfrak{a}) = n_{L/k}(n_{K/L}(\mathfrak{a})). \tag{13}$$

It suffices to prove this for prime ideals, and then apply (4). For them, (13) follows from a double application of (11) taking into account the first of Eqs. (12).

A most important formula is

$$\sum_{\varrho=1}^{r} e_{K/L}(\mathfrak{p}_\varrho) f_{K/L}(\mathfrak{p}_\varrho) = [K : L], \tag{14}$$

where $\mathfrak{p}_\varrho$ runs through all prime ideals appearing in (8).

Proof. Take the norm on both sides of (8). Because of (7) the left side gives $\mathfrak{p}^{[K:L]}$, while the right side gives $\prod \mathfrak{p}^{e_\varrho f_\varrho}$ because of (4) and (11). ▼

Exercise. Build up the theory of the norm of ideals developed in §3 on the basis of Kronecker divisors, in particular proving (14). To do so prove and use

$$[\mathfrak{O}_K/\mathfrak{p}_\varrho : \mathfrak{O}_L/\mathfrak{p}] = [\mathfrak{O}_K(z)/\mathfrak{O}_K(z)\mathfrak{p}_\varrho : \mathfrak{O}_L(z)/\mathfrak{O}_L(z)\mathfrak{p}].$$

This method is even a short cut. Our detour serves, though, to familiarize one with the arguments of local considerations.

§4. The Complement, Different, and Discriminant

1. The Complement

Retain the assumptions and notation of §§1 and 3. The elements of K can be taken as vectors in a $[K:L]$-dimensional vector space over L. If σ is an arbitrary pseudotrace of K/L (cf. I, §4,4), then $\alpha\beta = \sigma(\alpha\beta)$ defines a scalar product of these vectors with values in L. In particular, it has the properties of I, §1,3, if K is simultaneously identified with its dual vector space.

Now let $\mathfrak{a}$ be an $\mathfrak{O}_L$-module in K and $\mathfrak{a}_{K/L}^*$ be the complementary $\mathfrak{O}_L$-module in the sense of I, §1,3. Then $\mathfrak{a}_{K/L}^*$ is called the *pseudocomplement* of $\mathfrak{a}$ with respect to the pseudotrace used. For separable K/L the trace may be chosen for σ. One then speaks of the *complement* of $\mathfrak{a}$. Here, as for all other derived concepts, we use or omit the prefix "pseudo" according to whether a pseudotrace or the trace is being used. The concept of the complement is due to R. Dedekind.

As was determined in I, §1,3, if $\mathfrak{a}$ and $\mathfrak{b}$ are modules,

$$\mathfrak{a} \subseteq \mathfrak{b} \quad \text{implies} \quad \mathfrak{a}_{K/L}^* \supseteq \mathfrak{b}_{K/L}^*. \tag{1}$$

Further, the following equation is evident:

$$(\mathfrak{a}\alpha)_{K/L}^* = \mathfrak{a}_{K/L}^*\alpha^{-1}. \tag{2}$$

The pseudocomplement of an ideal is an ideal.

Proof. If $\mathfrak{a}$ has a basis with respect to $\mathfrak{O}_L$ consisting of $[K:L]$ elements α_i, the complementary basis α_i^* can be calculated by the method of I, §1,3. It is a basis of $\mathfrak{a}_{K/L}^*$, which is then, along with $\mathfrak{a}$, a finite $\mathfrak{O}_L$-module, and even a finite $\mathfrak{o}$-module because $\mathfrak{O}_L$ is finite. But, in general, $\mathfrak{a}$ has no such basis. Yet it has the rank $[K:L]$ with respect to $\mathfrak{O}_L$, for if $\alpha_1, \ldots, \alpha_{[K/L]}$ are elements of K linearly independent with respect to L, then there is an $a \in \mathfrak{O}_L$ so that $a\alpha_i \in \mathfrak{a}$. Without loss of generality, then, let all the $\alpha_1 \in \mathfrak{a}$. They span a submodule $\mathfrak{b}$ of $\mathfrak{a}$, but then the α_i^* span the module $\mathfrak{b}_{K/L}^*$ containing $\mathfrak{a}_{K/L}^*$. The former is finite so that, by the finiteness criterion of I, §1,1, we have the finiteness of $\mathfrak{a}_{K/k}^*$.

Arbitrary elements α, α^*, ω of $\mathfrak{a}$, $\mathfrak{a}_{K/L}^*$, $\mathfrak{O}_K$, respectively, satisfy

$$\sigma(\alpha \cdot \alpha^* \omega) = \sigma(\alpha\omega \cdot \alpha^*) \in \mathfrak{O}_L.$$

Hence, as maintained, we have $\mathfrak{O}_K \mathfrak{a}_{K/L}^* = \mathfrak{a}_{K/L}^*$. ▼

As α traverses an $\mathfrak{O}_K$-ideal $\mathfrak{a}$, *$\sigma(\alpha)$ traverses an $\mathfrak{O}_L$-ideal in L.* It can be denoted $\sigma(\mathfrak{a})$, the *pseudotrace* of $\mathfrak{a}$. With this notation

$$\sigma(\mathfrak{a}\mathfrak{a}_{K/L}^*) = \mathfrak{O}_L \tag{3}$$

holds. For certainly $\sigma(\mathfrak{a}\mathfrak{a}_{K/L}^*) = \mathfrak{b}$ is an integral $\mathfrak{O}_L$-ideal. If we had $\mathfrak{b} \subset \mathfrak{O}_L$, then $\mathfrak{a}_{K/L}^*\mathfrak{b}^{-1} \supset \mathfrak{a}_{K/L}^*$. But, also

$$\sigma(\mathfrak{a}\mathfrak{a}_{K/L}^*\mathfrak{b}^{-1}) = \sigma(\mathfrak{a}\mathfrak{a}_{K/L}^*)\mathfrak{b}^{-1} = \mathfrak{O}_L,$$

and thus $\mathfrak{a}_{K/L}^* \supseteq \mathfrak{a}_{K/L}^*\mathfrak{b}^{-1} \supset \mathfrak{a}_{K/L}^*$. ▼

Let pseudotraces be given satisfying the transitivity formulas I, §4,4,(10). Then the pseudocomplements with respect to the extensions K/k, K/L, and

L/K satisfy the transitivity formula:

$$\mathfrak{O}^*_{K/k} = \mathfrak{O}^*_{K/L}\mathfrak{O}^*_{L/k}. \tag{4}$$

Proof. Apply (3) to k instead of L, so that

$$\sigma_{L/k}(\sigma_{K/L}(\mathfrak{O}^*_{K/k})) = \sigma_{K/k}(\mathfrak{O}^*_{K/k}) = \mathfrak{o}.$$

Thus $\sigma_{K/L}(\mathfrak{O}^*_{K/k})$ is an $\mathfrak{O}_L$-ideal of pseudotrace $\mathfrak{o}$, i.e.,

$$\sigma_{K/L}(\mathfrak{O}^*_{K/k}) \subseteq \mathfrak{O}^*_{L/k}.$$

This implies that

$$\sigma_{K/L}(\mathfrak{O}^*_{K/k}\mathfrak{O}^{*-1}_{L/k}) = \sigma_{K/L}(\mathfrak{O}^*_{K/k})\mathfrak{O}^{*-1}_{L/k} \subseteq \mathfrak{O}_L$$

and thus, on the one hand, $\mathfrak{O}^*_{K/k}\mathfrak{O}^{*-1}_{K/k} \subseteq \mathfrak{O}^*_{K/L}$ or

$$\mathfrak{O}^*_{K/k} \subseteq \mathfrak{O}^*_{K/L}\mathfrak{O}^*_{L/k}. \tag{5}$$

By (3), on the other,

$$\sigma_{K/L}(\mathfrak{O}^*_{K/L}\mathfrak{O}^*_{L/k}) = \sigma_{K/L}(\mathfrak{O}^*_{K/L})\mathfrak{O}^*_{L/k} = \mathfrak{O}^*_{L/k}$$

so that, by taking $\sigma_{L/k}$ on both sides,

$$\sigma_{K/k}(\mathfrak{O}^*_{K/L}\mathfrak{O}^*_{L/k}) = \sigma_{L/k}(\mathfrak{O}^*_{L/k}) = \mathfrak{o}.$$

This gives

$$\mathfrak{O}^*_{K/L}\mathfrak{O}^*_{L/k} \subseteq \mathfrak{O}^*_{K/k}.$$

This, along with (5), demonstrates the assertion (4). ▼

We finally mention the two formulas

$$\mathfrak{a}^*_{K/L} = \mathfrak{O}^*_{K/L}\mathfrak{a}^{-1}, \tag{6}$$

$$(\mathfrak{a}^*_{K/L})^*_{K/L} = \mathfrak{a}. \tag{7}$$

Because of (3) we have $\sigma_{K/L}(\mathfrak{O}^*_{K/L}\mathfrak{a}^{-1}\mathfrak{a}) = \mathfrak{O}_L$; thus $\mathfrak{O}^*_{K/L}\mathfrak{a}^{-1} \subseteq \mathfrak{a}^*_{K/L}$. On the other hand, $\sigma_{K/L}(\mathfrak{O}_K\mathfrak{a}^*_{K/L}\mathfrak{a}) = \mathfrak{O}_L$ so that $\mathfrak{a}^*_{K/L}\mathfrak{a} \subseteq \mathfrak{O}^*_{K/L}$. Together this yields (6), while (7) is proved in the next section.

2. Different and Discriminant

The ideal inverse to the pseudocomplement is called the *pseudodifferent*;

$$\mathfrak{d}_{K/L} = \mathfrak{O}^{*-1}_{K/L}. \tag{8}$$

From (4) we carry over the *transitivity formula*

$$\mathfrak{d}_{K/k} = \mathfrak{d}_{K/L}\mathfrak{d}_{L/k}. \tag{9}$$

The norm of the pseudodifferent is called the *pseudodiscriminant*:

$$\mathfrak{D}_{K/L} = n_{K/L}(\mathfrak{d}_{K/L}). \tag{10}$$

It is an ideal in L. The transitivity formula

$$\mathfrak{D}_{K/k} = n_{L/k}(\mathfrak{D}_{K/L})\mathfrak{D}_{L/k}^{[K:L]} \tag{11}$$

follows from (9) and §3,(4) and (7). It is quite analogous to Eqs. I,§4,(7) and (11), and stands in connection with these. In fact, we have the theorem:

If $\mathfrak{O}_K$ has a basis with $[K:L]$ elements ω_i with respect to $\mathfrak{O}_L$, then

$$\mathfrak{D}_{K/L} = \mathfrak{O}_L\Delta(\omega_i), \tag{12}$$

that is, it is the $\mathfrak{O}_L$-ideal generated by the pseudodiscriminant of the ω_i.

Proof. As was seen in §4,1, the complementary basis generates the pseudocomplement. This basis can be computed as

$$\omega_i^* = \sum c_{ij}\omega_j \qquad \text{with} \quad (c_{ij})(\sigma(\omega_i\omega_j)) = (\delta_{ij}) = \text{unit matrix},$$

by I,§1,3. From the definition of the norm in §3, then

$$n_{K/L}(\mathfrak{d}_{K/L})^{-1} = n_{K/L}(\mathfrak{O}_{K/L}^*) = \mathfrak{O}_L|c_{ij}| = \mathfrak{O}_L\Delta(\omega_i)^{-1},$$

as was to be proved. ▼

For separable K/L, choose the trace for the pseudotrace. Then $s_{K/L}(\mathfrak{O}_K) \subseteq \mathfrak{O}_L$, and thus $\mathfrak{O}_{K/L}^* \supseteq \mathfrak{O}_K$.

As a consequence the different and discriminant are integral ideals.

As in §2,4, to all prime ideals $\mathfrak{p}$ of L we form the $\mathfrak{p}$-components

$$\mathfrak{a}_\mathfrak{p} = \mathfrak{O}_{L\mathfrak{p}}\mathfrak{a}$$

of ideals of K. The following theorem then holds:

The pseudocomplement of the $\mathfrak{p}$-component of an ideal $\mathfrak{a}$ is the $\mathfrak{p}$-component of its pseudocomplement:

$$(\mathfrak{a}_\mathfrak{p})_{K/L}^* = (\mathfrak{a}_{K/L}^*)_\mathfrak{p}. \tag{13}$$

Proof. The definition of pseudocomplement gives

$$\mathfrak{a}_{K/L}^* \supseteq \bigcap_\mathfrak{p} (\mathfrak{a}_\mathfrak{p})_{K/L}^*.$$

Using (g) and (h) of §2,4, the right side is an $\mathfrak{O}_K$-ideal with $\mathfrak{p}$-components $(\mathfrak{a}_\mathfrak{p})_{K/L}^*$. Taking $\mathfrak{p}$-components on the left as well, one has

$$(\alpha_{K/L}^*)_\mathfrak{p} \supseteq (\mathfrak{a}_\mathfrak{p})_{K/L}^*. \tag{14}$$

On the other hand, the definition of $\mathfrak{p}$-components shows that $(\mathfrak{a}_{K/L}^*)_\mathfrak{p}\mathfrak{a}_\mathfrak{p} = \mathfrak{a}_{K/L}^*\mathfrak{a}\mathfrak{O}_{L\mathfrak{p}}^*$; taking pseudotraces this is

$$\sigma((\mathfrak{a}_{K/L}^*)_\mathfrak{p}\mathfrak{a}_\mathfrak{p}) = \mathfrak{O}_{L\mathfrak{p}}.$$

The definition of $(\mathfrak{a}_\mathfrak{p})_{K/L}^*$ therefore assures $(\mathfrak{a}_{K/L}^*)_\mathfrak{p} \subseteq (\mathfrak{a}_\mathfrak{p})_{K/L}^*$, which, together with (14), gives the desired equation (13). ▼

We can now prove (7) as follows. The equation certainly holds for all $\mathfrak{a}_\mathfrak{p}$, as these have bases with respect to $\mathfrak{O}_{L\mathfrak{p}}$. By (13), then,

$$((\mathfrak{a}^*_{K/L})^*_{K/L})_\mathfrak{p} = ((\mathfrak{a}^*_{K/L})_\mathfrak{p})^*_{K/L} = ((\mathfrak{a}_\mathfrak{p})^*_{K/L})^*_{K/L} = \mathfrak{a}_\mathfrak{p},$$

which in turn gives (7).

3. The Dedekind Discriminant Theorem

Let K now be a separable extension of L.

Let the decomposition of the prime ideal $\mathfrak{p}$ of L in K into prime ideals of residue class degrees f_ϱ be given by §3,(8). The discriminant $\mathfrak{D}_{K/L}$ is divisible by $\mathfrak{p}$ at least to the power $f_1(e_1 - 1) + \cdots + f_r(e_r - 1)$.

Furthermore, $\mathfrak{D}_{K/L}$ is always divisible by $\mathfrak{p}$ if one of the residue class fields $\mathfrak{O}_K/\mathfrak{p}$ is an inseparable extension of the residue class field $\mathfrak{O}_L/\mathfrak{p}$.

Proof. From (13) it is known that the $\mathfrak{p}$-components of the different $\mathfrak{d}_{K/L}$ are the differents of $\mathfrak{O}_{K\mathfrak{p}}$ with respect to $\mathfrak{O}_{L\mathfrak{p}}$. This, in turn, implies the corresponding statement concerning discriminants. The assertion can therefore be restated identically for all the discriminants of $\mathfrak{O}_{K\mathfrak{p}}$ with respect to $\mathfrak{O}_{L\mathfrak{p}}$. By (12), these can be computed as discriminants of bases of $\mathfrak{O}_{K\mathfrak{p}}$ with respect to $\mathfrak{O}_{L\mathfrak{p}}$.

To compute these discriminants a special basis of $\mathfrak{O}_{K\mathfrak{p}}$ with respect to $\mathfrak{O}_{L\mathfrak{p}}$ is constructed for each $\mathfrak{p}$. It will be used again in III, §2,4. Using the decomposition of $\mathfrak{p}$ mentioned in the theorem, §2,2 gives

$$\mathfrak{O}_{K\mathfrak{p}} = \bigcap_\varrho \mathfrak{O}_{K\mathfrak{p}_\varrho}. \tag{15}$$

Using the elements π_ϱ given in §2,2, we first form

$$\kappa_{\varrho j} = \pi_\varrho{}^j(\pi_1^{e_1} \cdots \pi_r^{e_r}\pi_\varrho^{-e_\varrho})^h, \qquad j = 0, \ldots, e_\varrho - 1 = e_{K/L}(\mathfrak{p}_\varrho) - 1, \tag{16}$$

where the exponent $h > 0$ will be fixed later. Then, let $\lambda_{\varrho i}$ $(i = 1, \ldots, f_\varrho = f_{K/L}(\mathfrak{p}_\varrho))$ be a system of elements of $\mathfrak{O}_{K\mathfrak{p}_\varrho}$, such that their residue classes mod $\mathfrak{p}_{\varrho\mathfrak{p}}$ form a basis of $\mathfrak{O}_{K\mathfrak{p}_\varrho}/\mathfrak{p}_{\varrho\mathfrak{p}}$ with respect to its subfield $\mathfrak{O}_{L\mathfrak{p}}/\mathfrak{p}_\mathfrak{p}$. We maintain that the products

$$\lambda_{\varrho i}\kappa_{\varrho j}, \qquad \varrho = 1, \ldots, r;\ i = 1, \ldots, f_\varrho;\ j = 0, \ldots, e_\varrho - 1 \tag{17}$$

form a basis of $\mathfrak{O}_{K\mathfrak{p}}$ with respect to $\mathfrak{O}_{L\mathfrak{p}}$. By first choosing h sufficiently large, it can be assured that the elements (17) lie in all the $\mathfrak{O}_{K\mathfrak{p}_\varrho}$, and thus, by (15), in $\mathfrak{O}_{K\mathfrak{p}}$. Because of §3,(14) there are exactly $[K:L]$ elements in (17), as is required.

To demonstrate the basis property, start with an $\alpha \in \mathfrak{O}_{K\mathfrak{p}}$. To each index ϱ there are elements $a_{\varrho i0} \in \mathfrak{O}_{L\mathfrak{p}}$ such that

$$\alpha \equiv \sum_i a_{\varrho i0}\lambda_{\varrho i}\kappa_{\varrho 0} \bmod \mathfrak{p}_{\varrho\mathfrak{p}}.$$

Remembering what the $\kappa_{\varrho 1}$ are, we have $(\alpha - \sum a_{\varrho i0}\lambda_{\varrho i}\kappa_{\varrho 0})\kappa_{\varrho 1}^{-1} \in \mathfrak{O}_{K\mathfrak{p}_\varrho}$, so that new elements $a_{\varrho i1} \in \mathfrak{O}_{L\mathfrak{p}}$ can be found with

$$(\alpha - \sum a_{\varrho i0}\lambda_{\varrho i}\kappa_{\varrho 0})\kappa_{\varrho 1}^{-1} \equiv \sum a_{\varrho i1}\lambda_{\varrho i} \bmod \mathfrak{p}_{\varrho\mathfrak{p}}.$$

The same argument applied e_ϱ times yields the congruence

$$\alpha \equiv \sum_{i,j} a_{\varrho ij}\lambda_{\varrho i}\kappa_{\varrho j} \bmod \mathfrak{p}_{\varrho\mathfrak{p}}^{e_\varrho}.$$

Simultaneously, remembering definition (16), we have for $h > e_\sigma$

$$0 \equiv \sum_{i,j} a_{\varrho ij}\lambda_{\varrho i}\kappa_{\varrho j} \bmod \mathfrak{p}_{\sigma\mathfrak{p}_\sigma}^{e_\sigma}, \qquad \sigma \neq \varrho.$$

Finally, with §3,(8), both congruences together yield

$$\alpha \equiv \sum_{\varrho,i,j} a_{\varrho ij}\lambda_{\varrho i}\kappa_{\varrho j} \bmod \mathfrak{O}_{K\mathfrak{p}}\mathfrak{p}.$$

It is to be noted, in this argument, that the $a_{\varrho ij}$ are uniquely determined mod $\mathfrak{p}_\mathfrak{p}$. This assures the linear independence of the elements (17) mod $\mathfrak{p}_\mathfrak{p}$, and then certainly the linear independence as such. If ω_ν is a basis of $\mathfrak{O}_{K\mathfrak{p}}$ with respect to $\mathfrak{O}_{L\mathfrak{p}}$, then a system of linear equations

$$\omega_\nu = c^{-1}\sum_{\varrho,i,j} c_{\nu,\varrho ij}\lambda_{\varrho i}\kappa_{\varrho j}$$

holds, where c, $c_{\nu,\varrho ij} \in \mathfrak{O}_{L\mathfrak{p}}$, c being the common denominator. If c were divisible by $\mathfrak{p}_\mathfrak{p}$, multiplication would give $\sum c_{\nu,\varrho ij}\lambda_{\varrho i}\kappa_{\varrho j} \equiv 0 \bmod \mathfrak{O}_{K\mathfrak{p}}\mathfrak{p}_\mathfrak{p}$, contradicting the linear independence mod $\mathfrak{p}_\mathfrak{p}$. Thus, indeed, the $\lambda_{\varrho i}\kappa_{\varrho j}$ are a basis of $\mathfrak{O}_{K\mathfrak{p}}$.

We now arrange these basis elements, according to the index ϱ, into r groups, each containing $f_\varrho e_\varrho$ elements. The product of any two, with differing ϱ, is, because of (16), divisible by $\mathfrak{p}_\mathfrak{p}^h$. By increasing h we can increase the powers of $\mathfrak{p}_\mathfrak{p}$ by which the elements of the discriminant matrix are divisible so long as they are not in the r boxes with $f_\varrho e_\varrho$ rows along the diagonal. The consequence is that the power of $\mathfrak{p}_\mathfrak{p}$ by which the discriminant $\mathfrak{D}_{K/L,\mathfrak{p}}$ is divisible is the same as that which goes into the product of the $f_\varrho e_\varrho$ row subdeterminants

$$|s_{K/L}(\lambda_{\varrho i}\kappa_{\varrho j}\lambda_{\varrho i'}\kappa_{\varrho j'})|, \qquad \varrho = 1, \ldots, r.$$

We now show that $s_{K/L}(\alpha\kappa_{\varrho j}) \equiv 0 \bmod \mathfrak{p}_\mathfrak{p}$ for $j > 0$ and all $\alpha \in \mathfrak{O}_{K\mathfrak{p}}$. This shown, all rows and columns of the above matrix would be divisible by $\mathfrak{p}_\mathfrak{p}$, except for the f_ϱ first ones. This yields the first assertion.

For this last argument we use the method of I, §4,1 to compute the traces:

$$\alpha\kappa_{\varrho j}\cdot\lambda_{\sigma i}\kappa_{\sigma l} = \sum_{\sigma',i',l'} \lambda_{\sigma' i'}\kappa_{\sigma' l'}a_{\sigma' i' l',\sigma i l}.$$

The $f_\varrho e_\varrho$ by $f_\varrho e_\varrho$ submatrices of the matrix $(a_{\sigma' i' l',\sigma i l})$ which stand along the diagonal are almost all $\equiv 0 \bmod p_\mathfrak{p}$, except those of index $\sigma = \varrho$;

the other elements are not interesting in connection with the trace. The exceptional submatrices are of the form

$$\begin{pmatrix} 0 & \cdots & 0 & \cdots & 0 \\ \cdot & \cdots & \cdot & \cdots & \cdot \\ * & \cdots & 0 & \cdots & 0 \\ \cdot & \cdots & \cdot & \cdots & \cdot \\ 0 & \cdots & * & \cdots & 0 \end{pmatrix},$$

where 0 stands for f_ϱ row matrices divisible by $\mathfrak{p}_\mathfrak{p}$ and $*$ for certain matrices with elements in $\mathfrak{O}_{K\mathfrak{p}}$, $*$ taking the jth place in the first column. This completes the last argument of the proof.

In proving the second assertion of the discriminant theorem only the case where all $e_\varrho = 1$ need be considered, for otherwise the discriminant is divisible by $\mathfrak{p}_\mathfrak{p}$ anyway. By the hypothesis, let $\mathfrak{O}_{K\mathfrak{p}\varrho}/\mathfrak{p}_{\varrho\mathfrak{p}}$ be inseparable over $\mathfrak{O}_{L\mathfrak{p}}/\mathfrak{p}_\mathfrak{p}$. This makes the residue class of $|s_{K/L}(\lambda_{\varrho i}\lambda_{\varrho i'}\kappa_{\varrho 0}^2)|$ mod $\mathfrak{p}_\mathfrak{p}$ the discriminant of $\mathfrak{O}_{K\mathfrak{p}_\varrho}/\mathfrak{p}_{\varrho\mathfrak{p}}$ with respect to $\mathfrak{O}_{L\mathfrak{p}}/\mathfrak{p}_\mathfrak{p}$, that is 0. ▼

Another statement concerning the power of $\mathfrak{p}$ by which the discriminant $\mathfrak{D}_{K/L}$ is divisible will be arrived at in §6.

It is not difficult to rearrange the above considerations so as to get information concerning the powers of $\mathfrak{p}_\varrho$ by which the different $\mathfrak{d}_{K/L}$ is divisible. For the *Dedekind different theorem* which results see Hasse† where, however, a different method is used.

The theory of the different can be carried over to more general classes of rings, as was done in papers by E. Noether and E. Kaehler, which introduce two independent new notions of the different. In our special case they of course coincide with the Dedekind different. For the latest results see Berger and Kunz.‡

§5. Divisors

1. The Rational Function Field

In §5 we collect what has been said in I, §3 and II, §2 concerning the local point of view, and complete the development of this tool. We first illustrate our procedure using the example of the rational function field $k = k_0(x_i) = k_0(x_1, \ldots, x_m)$.

With each of the places p of the introduction, i.e., the prime polynomials of $\mathfrak{i}$ and $p = \infty$ a *prime divisor* $\mathfrak{p} = \mathfrak{z}_p$ is formally associated. These prime

† H. Hasse, *Zahlentheorie*, p. 310 ff., Berlin, 1949.

‡ R. Berger, *Über verschiedene Differentenbegriffe*, S.-B. Heidelberger Akad. Wiss., Math. Nat. Kl. pp. 3–44 (1960). E. Kunz, *Die Primidealteiler der Differente in allgemeinen Ringen*, J. Reine Angew. Math. **204**, 165–182 (1959).

divisors are taken as the generators of a free abelian group $\mathfrak{k}$, whose elements are called *divisors*, and, by definition, are power products

$$\mathfrak{a} = \prod_{\mathfrak{p}} \mathfrak{p}^{v_{\mathfrak{p}}(\mathfrak{a})} \tag{1}$$

of prime divisors. The *exponents* or *order functions* $v_{\mathfrak{p}}(\mathfrak{a})$ are all supposed to be 0, with a finite number of exceptions.

The *multiplication* of divisors implies addition of the exponents. The *unit divisor* is written $\mathfrak{a} = (1)$. A divisor is said to be *integral* if all the $v_{\mathfrak{p}}(\mathfrak{a}) \geqq 0$, and $\mathfrak{a}$ is called *divisible* by $\mathfrak{b}$ if $\mathfrak{a}\mathfrak{b}^{-1}$ is integral.

These pure formalities being set, let $f = f(x_i) \neq 0$ be an element of k and

$$f = p_1^{v_1} \cdots p_h^{v_h}$$

its decomposition into prime polynomials p_i of $\mathfrak{i} = k_0[x_1, \ldots, x_m]$. From now on $G(f)$ will denote the degree of a rational function in the sense of the variables x_i. With f associate the *principal divisor*

$$(f) = \mathfrak{z}_{p_1}^{v_1} \cdots \mathfrak{z}_{p_h}^{v_h} \mathfrak{z}_{\infty}^{v_\infty} \qquad \text{with} \quad v_\infty = -G(f). \tag{2}$$

In particular, for a prime polynomial p we have

$$(p) = \mathfrak{z}_p \mathfrak{z}_\infty^{-G(p)}.$$

The exponent v_∞ with which $\mathfrak{p}_\infty$ occurs in (2) can be interpreted in yet another way. The principal ideal domain $\mathfrak{i}_\infty$ has only one prime ideal $\mathfrak{i}_\infty(1/y)$, where y is any first degree function. All $\mathfrak{i}_\infty$-ideals are powers of the ideal, and in particular

$$\mathfrak{i}_\infty f = \mathfrak{i}_\infty(1/y)^{-G(f)},$$

as $fy^{-G(f)}$ is of degree 0 and must therefore be a unit of $\mathfrak{i}_\infty$. This last equation brings $p = \infty$ into line with the other places, for they also satisfy

$$\mathfrak{i}_p f = \mathfrak{i}_p p^{v_p(f)}.$$

The divisor associated with the product of two functions f, g is clearly the product of their divisors,

$$(fg) = (f)(g). \tag{3}$$

Hence the principal divisors form a subgroup $\mathfrak{h}$ of $\mathfrak{k}$. It is isomorphic to the factor group $k^\times/k_0^\times$ of the multiplicative groups of k and the constant field.

Each divisor is assigned a *degree* $g(\mathfrak{p})$ by setting $g(\mathfrak{z}_\infty) = 1$ and $g(\mathfrak{z}_p) = G(p)$, the degree of the prime polynomial with which $\mathfrak{z}_p$ is associated. For a general divisor one can then define

$$g(\mathfrak{a}) = \sum_{\mathfrak{p}} v_{\mathfrak{p}}(\mathfrak{a}) g(\mathfrak{p}). \tag{4}$$

These degrees satisfy

$$g(\mathfrak{a}\mathfrak{b}) = g(\mathfrak{a}) + g(\mathfrak{b}). \tag{5}$$

Principal divisors are always of degree 0, as follows immediately from (2). Conversely, every divisor of degree 0 is a principal divisors. To prove this, let the divisor on the right side of (2) be of degree 0. Then, $-v_\infty = \sum v_i g(\mathfrak{z}_{p_i})$, so that the divisor is a product $\prod (p_i^{v_i})$ of principal divisors:

$$\prod (p_i^{v_i}) = \prod (\mathfrak{z}_{p_i} \mathfrak{z}_\infty^{-g(\mathfrak{z}_{p_i})})^{v_i}. \quad \blacktriangledown$$

In §3 the concept of divisors will be carried over to algebraic number and function fields. There, however, the simple distinction between principal divisors and other divisors, expressed by the degree, will not generally hold.

2*. Projective Invariance

The question as to the generality of this notion of divisor becomes natural. The generating indeterminates x_i of the field k are not distinguished by any inner properties. They might, for example, undergo a projective transformation

$$x_i' = \frac{\alpha_{i0} + \alpha_{i1}x_1 + \cdots + \alpha_{im}x_m}{\alpha_{00} + \alpha_{01}x_1 + \cdots + \alpha_{0m}x_m}. \tag{6}$$

The inverse transformation is

$$x_i = \frac{\beta_{i0} + \beta_{i1}x_1' + \cdots + \beta_{im}x_m'}{\beta_{00} + \beta_{01}x_1' + \cdots + \beta_{0m}x_m'}, \tag{7}$$

where β_{ij} is calculated from the matrix identity

$$(\alpha_{ij})(\beta_{ij}) = (\delta_{ij}).$$

For later use we point out the special relation

$$(\alpha_{00} + \alpha_{01}x_1 + \cdots + \alpha_{0m}x_m)(\beta_{00} + \beta_{01}x_1' + \cdots + \beta_{0m}x_m') = 1. \tag{8}$$

We will show that the integral domains $\mathfrak{i}_p$ permute under the projective transformation (6). *The divisors are thus projective invariants of the generators x_i of the field.*

For the case of functions of only one variable x, it is easy to observe that $x' = (\alpha_{10} + \alpha_{11}x)/(\alpha_{00} + \alpha_{01}x)$ are the only functions which also generate k. If $m > 1$ indeterminates are given, there are other "birational transformations" such as $x_1' = x_1$, $x_2' = x_1x_2$. The divisors are thus birationally invariant for $m = 1$ only. For the one variable case though, with which this book is primarily concerned, an essentially further reaching invariance of the divisor concept will appear in §5,5.

Under (7), a polynomial $f(x_i)$ of degree G becomes

$$f(x_i) = f'(x_i')(\beta_{00} + \beta_{01}x_1' + \cdots + \beta_{0m}x_m')^{-G} \tag{9}$$

with a polynomial $f'(x_i')$ of degree $G' \leqq G$. Replacing the x_i' by their values from (6) gives

$$f'(x_i') = f''(x_i)(\alpha_{00} + \alpha_{01}x_1 + \cdots + \alpha_{0m}x_m)^{-G'} \tag{10}$$

with a polynomial $f''(x_i)$ of degree $G'' \leqq G'$. If $f(x_i)$ was a prime polynomial, then (8), (9), and (10) show that only two possibilities exist:

(1) $f'' = f$ and $G'' = G' = G$;
(2) $f'' =$ constant, $f = \alpha_{00} + \alpha_{01}x_1 + \cdots + \alpha_{0m}x_m$, $G = 1$, $G' = 0$.

In the first case $f'(x_i')$ is also a prime polynomial in the x_i'. For, if

$$f'(x_i') = f_1'(x_i')f_2'(x_2')$$

with two factors of degrees G_1', $G_2' > 0$, and the equations corresponding to (10) were

$$f_1'(x_i') = f_1''(x_i)(\alpha_{00} + \alpha_{01}x_1 + \cdots)^{-G_1'},$$

$$f_2'(x_i') = f_2''(x_i)(\alpha_{00} + \alpha_{01}x_1 + \cdots)^{-G_2'},$$

the first of the two possibilities applying in both cases. This would necessitate $f_1''f_2'' = f'' = f$, the degrees of the factors being $G_1'' = G_1'$, $G_2'' = G_2'$, contrary to the assumed irreducibility of f.

These considerations show that every prime polynomial, with the single exception of $f(x_i) = \alpha_{00} + \alpha_{01}x_1 + \cdots + \alpha_{0m}x_m$, corresponds under (9) to another prime polynomial $f'(x_i')$ of the same degree. The rest is obvious, for if

$$f = f(x_i) = (\alpha_{00} + \alpha_{01}x_1 + \cdots + \alpha_{0m}x_m)^{\nu_0}p_1(x_i)^{\nu_1}\cdots$$

is the decomposition of a function into irreducible factors, and $p_1'(x_i'), \ldots$ are the corresponding prime polynomials in the x_i', then

$$f = p_1'(x_i')^{\nu_1}\cdots(\beta_{00} + \beta_{01}x_1' + \cdots + \beta_{0m}x_m')^{-\nu_0-\nu_1 G(p_1)-\cdots}$$

is the decomposition of the same function into irreducible factors in the x_i'.

Thus the integral domains $\mathfrak{i}_p$ and $\mathfrak{i}_{p'}$, correspond for corresponding p, p', etc., and further $\mathfrak{i}_p$ and $\mathfrak{i}_\infty$, correspond when $p = \alpha_{00} + \alpha_{01}x_1 + \cdots$. Finally, $\mathfrak{i}_\infty$ and $\mathfrak{i}_{p'}$ also correspond for $p' = \beta_{00} + \beta_{01}x_1' + \cdots$. Hence even the exponents remain invariant, which is clear for all p except ∞ and is true also for this exception because the degree of the divisor of f is 0. This is exactly the assertion that the divisors are projectively invariant. ▼

3. Divisors in Algebraic Number and Function Fields

Let K be a finite extension of the rational fields $k = Q$ or $k_0(x_i)$. The $\mathfrak{i}_p$ are the same as in the introduction, and the $\mathfrak{I}_p$ are the principal orders of K

with respect to the $\mathfrak{i}_p$. Let the prime ideal $(p) = \mathfrak{i}_p p$ of k decompose in K as

$$(p) = \mathfrak{I}_p p = \mathfrak{p}_1^{e_1} \cdots \mathfrak{p}_r^{e_r}$$

into prime ideals $\mathfrak{p}_\varrho$. As before, choose an arbitrary function of degree -1 for the element p of the place $p = \infty$.

With every prime ideal $\mathfrak{p}_\varrho$, for all places p of k, one associates formally a *place* $\mathfrak{p}_\varrho$ and a *prime divisor* $\mathfrak{p}_\varrho$ of K. Introduction of new special designations is unnecessary. The places $\mathfrak{p}_\varrho$ of K are said to lie over the place p of k. The geometric import of this convention in the classical case will become apparent in III, §2 and IV, §1.

As before, form the free abelian group $\mathfrak{K}$ of all divisors from the prime divisors. If $\alpha \neq 0 \in K$, and if for each p

$$\mathfrak{I}_p \alpha = \mathfrak{p}_1^{\nu_{\mathfrak{p}_1}(\alpha)} \cdots \mathfrak{p}_r^{\nu_{\mathfrak{p}_r}(\alpha)},$$

we associate with the element α the *principal divisor*

$$(\alpha) = \prod_{\mathfrak{p}} \mathfrak{p}^{\nu_\mathfrak{p}(\alpha)}, \tag{11}$$

where the product is taken over all prime divisors, or, in other words, all prime ideals $\mathfrak{p}_\varrho$ for all p. Assurance is still needed that only a finite number of the exponents $\nu_\mathfrak{p}(\alpha)$ in (11) are nonzero. By §2,4,(c) almost all the $\mathfrak{I}_p$ contain α, so that $\nu_\mathfrak{p}(\alpha) \geqq 0$, almost always. The same argument applies to α^{-1}.

The principal divisors form a subgroup $\mathfrak{H} \subset \mathfrak{K}$. By its definition it is a homomorphic image of the multiplicative group $K^\times$ of K.

The importance of divisors is, that they permit the global statement of the divisibility properties for all places p, as they are expressed in ideal theory for $\mathfrak{I}_p$. Moreover, it will turn out that for the case of fields of functions of a single variable the divisors represent a concept which does not depend upon the particular model of the field K. For the case of algebraic number fields K/Q, though, divisor theory cannot yield more than ideal theory for the principal order $\mathfrak{I}$ with respect to $\mathfrak{i} = Z$, since ideal theory of $\mathfrak{I}$ already contains the ideal theories of all the $\mathfrak{I}_p$, as $\mathfrak{I} = \bigcap \mathfrak{I}_p$.

At times it is practical to split up the product representation of a divisor in the manner

$$\mathfrak{a} = \prod_p \mathfrak{a}_p, \qquad \mathfrak{a}_p = \prod_{\varrho=1}^{r} \mathfrak{p}_\varrho^{\nu_{\mathfrak{p}_\varrho}(\alpha)};$$

i.e., the places $\mathfrak{p}_\varrho$ of K lying over a place p of k are grouped together. This procedure will furnish a bridge to the concept of a linear divisor of I, §3.

Let us now investigate the nature of the homomorphic mapping $K^\times \to \mathfrak{H}$ somewhat more exactly. We have

$$\mathfrak{H} \cong K^\times/\mathfrak{E},$$

where $\mathfrak{E}$ is the group of those $\varepsilon \in K$ with principal divisor $(\varepsilon) = (1)$. These ε are called the *units* of K. For the case of a number field, these units will be considered more closely in the appendix to this chapter; they form a finitely generated group. The result for a function field is quite simple: $\mathfrak{E} = k_0{}^\times$, as long as we assume that k_0 is the exact constant field.

Proof. That $k_0{}^\times \subseteq \mathfrak{E}$ is obvious. Now let $\varepsilon \in \mathfrak{E}$ and thus be a unit in all $\mathfrak{J}_p$. If $k = K$, then clearly $\varepsilon \in k_0$. Otherwise, let $\varepsilon^n + c_1\varepsilon^{n-1} + \cdots + c_n = 0$ be the equation in k which ε satisfies. By §1,1, $\varepsilon \in \mathfrak{J}_p$ if and only if all the $c_i \in \mathfrak{i}_p$. This, as we have just seen, is equivalent to $c_i \in k_0$, if it is to hold for all p. Thus ε belongs to the exact field of constants which, by hypothesis, is k_0. ▼

A consequence of the above is: *an element is determined by its divisor, up to a factor from the exact constant field.*

In the proof we also showed that for a function field

$$\bigcap_p \mathfrak{i}_p = k_0, \qquad \bigcap_p \mathfrak{J}_p = k_0. \tag{12}$$

The distinction from I§3,(1) should be noted.

The elements of the factor group

$$\mathfrak{C} = \mathfrak{K}/\mathfrak{H}$$

are called *divisor classes*. Divisors of the same class are called *equivalent*, written $\mathfrak{a}\alpha = \mathfrak{b} \sim \mathfrak{a}$.

Automorphisms S of K/k act on the divisor by their action on the ideals. As $(\alpha^S) = (\alpha)^S$ is obvious, the S are also automorphisms of the group of divisor classes.

4. The Behavior of Divisors under Field Extensions

Let an intermediate field L between k and K be given. The group $\mathfrak{L}$ of divisors of L is embedded as follows into the divisor group $\mathfrak{K}$ of K. Say a prime ideal $\mathfrak{p}$ for $\mathfrak{J}_p = \mathfrak{J}_{L,p}$ of L decomposes in K into

$$\mathfrak{J}_{K,p}\mathfrak{p} = \mathfrak{p}_1^{e_1} \cdots \mathfrak{p}_r^{e_r}. \tag{13}$$

Proceed by identifying the divisor of L with the divisor in K of the same notation,

$$\mathfrak{p} = \mathfrak{p}_1^{e_1} \cdots \mathfrak{p}_r^{e_r}, \tag{14}$$

and expand this embedding into an isomorphism of $\mathfrak{L}$ onto a subgroup of $\mathfrak{K}$. It is unnecessary to distinguish between a divisor in $\mathfrak{L}$ and its image in $\mathfrak{K}$.

This embedding, which is formally possible, carries with it an embedding of the group $\mathfrak{H}_L$ of principal divisors of L into the corresponding group $\mathfrak{H}_K$ of K. It is to be expected that this embedding corresponds with the natural embedding of $\mathfrak{H}_L \cong L^*/k_0^\times$ into $\mathfrak{H}_K \cong K^\times/k_0^\times$. Indeed, this is so. For, if

$$(\alpha) = \prod \mathfrak{p}^{\nu_\mathfrak{p}(\alpha)}$$

is a principal divisor in L, and the decompositions (14) are substituted for the $\mathfrak{p}$, its image in K is

$$(\alpha) = \prod_\mathfrak{p} \prod_\varrho \mathfrak{p}_\varrho^{e_\varrho \nu_\mathfrak{p}(\alpha)}.$$

Considering (α) as a principal ideal in L and K, however, we also have

$$\mathfrak{J}_{L,p}\alpha = \prod \mathfrak{p}^{\nu_\mathfrak{p}(\alpha)}, \qquad \mathfrak{J}_{K,p}\alpha = \prod_\mathfrak{p} \prod_\varrho \mathfrak{p}_\varrho^{e_\varrho \nu_\mathfrak{p}(\alpha)},$$

where here the $\mathfrak{p}_\varrho$ simultaneously denotes the prime factors of $(\mathfrak{p})$ in $\mathfrak{J}_{L,p}$ and these $\mathfrak{p}$ are decomposed in $\mathfrak{J}_{K,p}$ according to (13). Thus, here also, divisor theory can be seen as the formal global summarization of ideal theory of the individual places p.

In embedding $\mathfrak{H}_L$ in $\mathfrak{H}_K$ one must carefully avoid the error of thinking that $\mathfrak{H}_L = \mathfrak{L} \cap \mathfrak{H}_K$. It is well possible for a divisor of L that is not principal to become a principal divisor in K.

The residue class degree and ramification index of a prime divisor $\mathfrak{p} \in \mathfrak{R}$ with respect to a subfield L are fixed as the same numbers as for the corresponding ideals. Further, the *norm* $n_{K/L}(\mathfrak{a})$ of an $\mathfrak{a} \in \mathfrak{R}$ of the form (1) is defined as the product $\prod n_{K/L}(\mathfrak{p})^{\nu_\mathfrak{p}(\mathfrak{a})}$, and the norms of the $\mathfrak{p}$, in turn, by setting them equal to the ideal norms of §3; they are taken as divisors in L. The product formula §3,(4) holds, and in particular,

$$n_{K/L}((\alpha)) = (n_{K/L}(\alpha)) \tag{15}$$

the norm of a principal divisor (α) *is equal to the principal divisor of the element norm.*

In the case of a function field a *degree* $g(\mathfrak{p})$ of a prime divisor is defined as follows. Say $\mathfrak{p}$ corresponds to a prime factor in K of the ideal $\mathfrak{J}_p p$ of residue class degree $f_{K/k}(\mathfrak{p})$, p being an irreducible polynomial in the generating variables. Then, set $g(\mathfrak{p}) = f_{K/k}(\mathfrak{p})G(p)$, where $G(p)$ is the degree as polynomial, for $p \neq \infty$. The case $p = \infty$ is handled similarly; $\mathfrak{p}$ corresponds to a prime factor of $\mathfrak{J}_\infty(1/y)$, y being some rational functions of the first degree. Thus use $G(\infty) = 1$.

Define the degrees of composite divisors (1) by again using (4). Equation (5) still holds, *and the divisors of degree* 0 *form a subgroup.*

The degree of a divisor was here defined with respect to the field K. If one is dealing with a divisor of a subfield L, one must distinguish between the degrees $g_L(\mathfrak{a})$ and $g_K(\mathfrak{a})$ of $\mathfrak{a}$ as a divisor of L and K.

Every divisor $\mathfrak{a}$ of K satisfies

$$g_K(\mathfrak{a}) = g_L(n_{K/L}(\mathfrak{a})). \tag{16}$$

Clearly we need only demonstrate this for prime divisors. Let, therefore, $\mathfrak{a} = \mathfrak{p}_K$, and $\mathfrak{p}_L$ be the prime divisor in L divisible by $\mathfrak{p}_K$. On the one hand, §3,(11) gives $n_{K/L}(\mathfrak{p}) = \mathfrak{p}_L^{f_{K/L}}$, and on the other, by definition and §3,(12), the degrees of p_K, p_L are

$$g_K(\mathfrak{p}_K) = f_{K/L} f_{L/k} G(p), \qquad g_L(\mathfrak{p}_L) = f_{L/k} G(p),$$

which proves (16).

A consequence of (15) and (16) applied to $L = k_0(x_i) = k$ and the fact that the principal divisors of k have degree 0 is the important theorem:

The degree of a principal divisor is 0; *divisors of the same class have the same degree.*

Formula §3,(7) carries over word for word to divisors. From it, in connection with (16), we have the important relation: *the degree of a divisor of L in K is* $[K:L]$ *times its degree in L.*

5. The Prime Divisors

The introduction of divisors concludes the common treatment of algebraic number and function fields, so that it now seems fitting to compare these two sorts of fields and consider them in relation to each other.

As mentioned in §5,3, the group of divisors of an algebraic number field is isomorphic to its group of $\mathfrak{I}$-ideals. The reason for this can be said to be, that the intersection of all the integral domains $\mathfrak{I}_p$ for all primes p of $\mathfrak{i}$ is equal to $\mathfrak{I}$. This is no longer so for function fields. There, indeed, we still have

$$\mathfrak{I} = \bigcap_{p \neq \infty} \mathfrak{I}_p,$$

but by (12) the intersection of all the $\mathfrak{I}_p$ is the exact constant field. As a consequence, there is no integral domain whose ideal theory can reflect the properties of all the $\mathfrak{I}_p$. Thus the divisors are needed to sum up globally the divisibility properties of all the places.

There is another difference between number and function fields. In contrast to number fields, function fields can be generated in various manners. Any maximal set of algebraically independent functions x_i in K defines a model (cf. §2,3). K is a finite algebraic extension of the rational function field $k_0(x_i)$. For fields where the number of variables $m > 1$, it was shown in §5,2 with an example that our notion of divisor depends upon the generation of the field. In other words, the divisors are not invariant under birational transformations.

For fields in one variable we prove the following:

Invariance Theorem. *The prime divisors* $\mathfrak{p}$ *of an algebraic function field K of one variable with constant field* k_0 *correspond, one-to-one, with the special local* *rings* $\mathfrak{o}_\mathfrak{p}$ *of* K *(which contain* k_0*) in such a manner that* $\mathfrak{o}_\mathfrak{p}$ *consists of all those functions,* α*, whose principal divisors do not contain negative powers of* $\mathfrak{p}$.

The residue class rings $\mathfrak{o}_\mathfrak{p}/\mathfrak{p}_\mathfrak{p}$ *are extensions of degree* $g(\mathfrak{p})$ *of the exact constant field.*

The first assertion follows directly from the characterization of local rings in §2,2 and §2,3. Now let x be some nonconstant element. Then K is a finite extension of $k = k_0(x)$. Say the prime divisor $\mathfrak{p}$ divides the prime divisor $\mathfrak{p}_0$ of k, and assume, for the time, that $\mathfrak{p}_0$ is the numerator divisor of a prime polynomial $p(x)$ of degree g_0. Further, say $\mathfrak{p}$ divides $\mathfrak{p}_0$ with residue class degree f. Then the residue class ring $\mathfrak{i}/p(x)$ is an extension of k_0 of degree g_0. By §3,3, $\mathfrak{o}_\mathfrak{p}/\mathfrak{p}_\mathfrak{p}$ is an extension of $\mathfrak{i}/p(x)$ of degree f, and thus an extension of k_0 of degree $g(\mathfrak{p}) = g_0 f$. For the exception, where $\mathfrak{p}_0$ is the denominator divisor of x, replace $\mathfrak{i}$ by $\mathfrak{i}_\infty$ in the above argument. ▼

In this connection we incidentally mention another possibility of defining divisors, based on §1,3. We limit ourselves to the simple case in which K is generated by simple adjunction to a rational subfield. (This generation is always possible if k_0 is perfect, as follows from III, §4,1. The generalization possibilities of the following considerations lie close at hand.) Let, therefore, $K = k_0(x, y)$ with an equation $f(x, y) = 0$. Using three "homogeneous" variables ξ, η, ζ, set $x = \xi\zeta^{-1}$, $y = \eta\zeta^{-1}$. A homogeneous equation $F(\xi, \eta, \zeta) = 0$ then holds, and it defines a finite extension K_h over $k_h = k_0(\xi, \zeta)$. Also, set $\mathfrak{o}_h = k_0[\xi, \zeta]$. Now adjoin another indeterminate z and define, as *divisors of* K, the cosets $\alpha(z)\mathfrak{E}_h(z)$, where $\alpha(z)$ are functions of $K_h(z)$, homogeneous in ξ, η, and ζ, while $\mathfrak{E}_h(z)$ is the group of units of the principal order $\mathfrak{O}_h(z)$ of $K_h(z)$ with respect to $\mathfrak{o}_h(z)$, homogeneous in ξ, η, and ζ.

The divisors in this and the previous sense must be shown to correspond one-to-one. Let $\pi(z) \in K_h(z)$ be a homogeneous prime element. A local ring $\mathfrak{O}_{\pi(z)} \subset K_h(z)$ corresponds to $\pi(z)$. Certainly $\mathfrak{O}_{\pi(z)}(\pi z)$ contains a homogeneous polynomial $f(\xi, \zeta) \in k_0[\xi, \zeta]$. This means that $\pi(z)$ maps the rational subfield $k_0(\xi\zeta^{-1})$ onto a finite extension of k_0, and thus does the same to K. Hence $\pi(z)$ belongs to a prime divisor $\mathfrak{p}$, with local ring $\mathfrak{O}_\mathfrak{p} = \mathfrak{O}_{\pi(z)} \cap K$. Conversely, let a prime divisor in the previous sense, $\mathfrak{p}$, be given. We may assume that $\mathfrak{p}$ does not divide the denominator of x, for otherwise we could use x^{-1}. Now, a prime element $\pi_0(z)$, integrally dependent upon $\mathfrak{i}(z)$ ($\mathfrak{i} = k_0[x]$), corresponds to $\mathfrak{p}$. Choose a power of ζ such that $\pi(z) = \pi_0(z)\xi^l$ is integrally dependent upon $\mathfrak{o}_h(z)$. We will even show that if l is minimal, then this element is also a prime element. Once this is known the first correspondence again gives $\mathfrak{p}$, so that we are finished.

$\pi(z)$ is homogeneous. Represent this element as the quotient of polynomials, the denominator being a primitive polynomial of $\mathfrak{o}_h[z]$. After any possible cancellations, homogeneous functions remain in the denominator and numerator. If the denominator is primitive, the prime decomposition of an element amounts to the prime decomposition of the numerator. Now, it is very easy to see that the decomposition of a polynomial in z homogeneous in ξ, η, and ζ gives homogeneous factors. Then, if contrary to our assertion, $\pi_0(z)\zeta^l = \alpha(z)\beta(z)$ with functions $\alpha(z)$, $\beta(z)$ of degrees a, b, homogeneous in ξ, η, ζ, and integrally dependent upon $\mathfrak{o}_h(z)$, we should have a decomposition $\pi_0(z) = \alpha(z)\zeta^{-a}\cdot\beta(z)\zeta^{-b}$ into factors integral at all places of K not divisors of the denominator of x. But, because of the relation of $\pi_0(x)$ to the prime divisor $\mathfrak{p}$, any decomposition of this sort must be trivial. Thus, one of the factors is the product of a unit with a power of ζ, which is what was to be shown.

6. Divisors and Linear Divisors

The time has come to clarify the relation to the concept of linear divisor introduced in I, §3. Our starting point is §2 of this chapter, but specialized with k as rational number or function field and $\mathfrak{o}_\mathfrak{p} = \mathfrak{i}_\mathfrak{p}$, $\mathfrak{O}_\mathfrak{p} = \mathfrak{I}_p$. An arbitrary basis ω_i of K/k is given. By associating with every $\alpha = \sum a_i\omega_i \in K$ the vector (a_i), we have a one-to-one mapping $K \leftrightarrow k^n$. If a pseudotrace $\sigma_{K/k}$ is given, then K is simultaneously mapped onto the dual space k^{n*}, as in §4.

We now decompose an arbitrary divisor $\mathfrak{a}$ to a product

$$\mathfrak{a} = \prod_p \mathfrak{a}_p, \tag{17}$$

whose factors $\mathfrak{a}_p$ consist only of power products of the finitely many $\mathfrak{p}_\varrho$ lying over p, in the sense of §5,3. As became clear there, the groups $\mathfrak{K}_p$ of these $\mathfrak{a}_p$ are isomorphic with the groups of $\mathfrak{I}_p$-ideals carrying the same notation $\mathfrak{a}_p$, the latter being exactly all $\alpha \in K$ with exponents $\nu_{\mathfrak{p}_\varrho}(\alpha) \geqq \nu_{\mathfrak{p}_\varrho}(\mathfrak{a}_p) = \nu_{\mathfrak{p}_\varrho}(\mathfrak{a})$.

These $\mathfrak{I}_p$-ideals $\mathfrak{a}_p$ are simultaneously finite $\mathfrak{i}_p$-modules of rank $n = [K:k]$. *They are the p-components of a linear divisor in the sense of* I, §3. All we need show for this is that the above field basis ω_i forms a basis of $\mathfrak{a}_p$ with respect to $\mathfrak{i}_p$ for almost all p.

We show this first under the assumption that K is a number field or function field of one variable. By §1,1 the intersection $\mathfrak{I}$ of all $\mathfrak{I}_p$ (except $p = \infty$) has a basis ι_i consisting of n elements. Set $\omega_j = \sum \iota_i c_{ij}$; the matrix (c_{ij}) is unimodular for almost all p. Furthermore, as components of an $\mathfrak{I}$-ideal the $\mathfrak{a}_p$ are almost all $\mathfrak{I}_p$ (§2,4,(d)). Thus, the ω_i form a basis of $\mathfrak{a}_p$ for almost all p. ▼

Using the basis theorem of I, §1,5, one can carry this argument over to function fields in two variables. We shall not need this theorem for functions of several variables, and therefore omit the proof of the general case. It can be carried out by use of the methods of §1,1 and §1,2, under the same limiting assumptions made there in the case of inseparable field extensions.

The linear divisors corresponding to equivalent divisors, $\mathfrak{a}$ *and* $\mathfrak{a}\alpha$, *are equivalent in the sense of* I, §3,1. For, if $\mathfrak{a}_p$ has the basis

$$\alpha_{pi} = \sum_j a_{p,ij}\omega_j$$

with respect to $\mathfrak{i}_p$, then $\alpha_{pi}\alpha$ is a basis of $\mathfrak{a}\alpha$. Setting

$$\alpha\omega_i = \sum_j a_{ij}\omega_j$$

we have

$$\alpha_{pi}\alpha = \sum_{j,l} a_{p,ij}a_{jl}\omega_l .$$

The following theorem is important in later applications.

There exists a divisor $\mathfrak{d}_{K/k}$, *such that the linear divisors corresponding to an arbitrary divisor* $\mathfrak{a}$ *and to*

$$\mathfrak{a}^* = \mathfrak{d}_{K/k}^{-1}\mathfrak{a}^{-1} \qquad (\mathfrak{a}^{**} = \mathfrak{a}) \tag{18}$$

are complementary in the sense of I, §3,1. We call $\mathfrak{d}_{K/k}$ the *pseudodifferent divisor* with respect to k; its p-components $\mathfrak{d}_{K/k,p}$, taken as $\mathfrak{J}_p$-ideals, are the pseudodifferent ideals of $\mathfrak{J}_p$ in the sense of §4. Our assertion is quite the same as formula §4,(6), applied in the case $L = k$ to the components of the divisors and using the abbreviation of §4,(8).

For separable K/k, we can use the trace as pseudotrace, $\mathfrak{d}_{K/k}$ is then the *different divisor*, and the correspondence $\mathfrak{a} \leftrightarrow \mathfrak{a}^*$ is unique. For inseparable extensions K/k, (18) does not define a unique *pseudocomplementary divisor. At least, though, the class of* $\mathfrak{a}^*$ *is uniquely determined by* $\mathfrak{a}$.

The distinction between two pseudotraces σ, σ' is, in fact, that $\sigma'(\alpha) = \sigma(\delta\alpha)$ for all $\alpha \in K$ with some fixed $\delta \neq 0 \in K$, according to I, §4,4. For pseudocomplements this gives

$$\mathfrak{a}_p^{*\prime} = \mathfrak{a}_p{}^*\delta^{-1} .$$

And, for the divisors defined by (17), we then have

$$\mathfrak{a}^{*\prime} = \mathfrak{a}^*\delta^{-1}.$$

7. The Linear Degree

The Riemann–Roch theorem for linear divisors of I, §3,3 can now easily be formulated as a theorem for divisors. As it will turn out, it is the central theorem for the theory of algebraic functions of one variable, and therefore its formulation will belong to Chapter III. Nevertheless, it is practical to calculate the linear degree $G(\mathfrak{a})$ of a divisor at this point.

Let K be a function field in one variable.† Fix a basis ω_i of the principal

† All we really need know is that $\mathfrak{J}$ has a basis with respect to $\mathfrak{i}$. By a remark in the last section this also holds for function fields in two variables.

order $\mathfrak{J}$ with respect to $\mathfrak{i}$, and use this basis for the computation of the linear norm according to the procedure prescribed in $\mathfrak{J}$ §3,2. For all places $p \neq \infty$ of k the p-components of $N(\mathfrak{a})$ and of $n_{K/k}(\mathfrak{a})$ are the same, as follows immediately from §3,(6). Further, if $\omega_{\infty i} = \sum c_{ij}\omega_j$ is a basis of $\mathfrak{J}_\infty$ with respect to $\mathfrak{i}_\infty$, the same reasoning shows that, for the ∞-component,

$$N(\mathfrak{a})_\infty = (n_{K/k}(\mathfrak{a})|c_{ij}|)_\infty$$

holds. It follows from I,§3,2, then, that the linear degree of $\mathfrak{a}$ is

$$G(\mathfrak{a}) = g_k(n_{K/k}(\mathfrak{a})) + c$$

with some rational integer c, which can best be computed by the use of (18) and I,§3,(12). With (16) it is also possible to write

$$G(\mathfrak{a}) = g(\mathfrak{a}) + c. \tag{19}$$

Because of (18), the divisor associated with $\mathfrak{a}$ by the definition I,§3,(11), is

$$\mathfrak{a}' = \mathfrak{a}^{-1}\mathfrak{d}_{K/k}^{-1}\mathfrak{z}_\infty{}^2,$$

where $\mathfrak{d}_{K/k}$ denotes the pseudodifferent divisor and $\mathfrak{z}_\infty$ is the denominator divisor of (x). The degree of $\mathfrak{z}_\infty$ as a divisor of K is n because of the theorem at the end of §5,4. On the one hand, this yields

$$g(\mathfrak{a}) + g(\mathfrak{a}') = 2n - g(\mathfrak{d}_{K/k}).$$

On the other hand, we saw in I,§3,(12) that

$$G(\mathfrak{a}) + G(\mathfrak{a}') = 2n.$$

Comparison of these two equations finally gives $c = \frac{1}{2}g(\mathfrak{d}_{K/k})$ in (19). Altogether we thus have

$$G(\mathfrak{a}) = g(\mathfrak{a}) + \tfrac{1}{2}g(\mathfrak{d}_{K/k}). \tag{20}$$

§6*. Decomposition of Prime Ideals in Galois Extensions

In §6 K will always be a separable Galois extension of k, of degree $[K:k] = n$. The behavior of a prime $\mathfrak{o}$-ideal $\mathfrak{p}_0$ of k is obviously completely determined by the behavior of the corresponding prime $\mathfrak{o}_{\mathfrak{p}_0}$-ideal, $\mathfrak{o}_{\mathfrak{p}_0}$ being the local ring associated with $\mathfrak{p}_0$ by §2,2. As $\mathfrak{o}_{\mathfrak{p}_0}$ is a principal ideal domain, there is no loss of generality in assuming that $\mathfrak{o}$ is a principal ideal domain.

In §6 we assume throughout that the residue fields $\mathfrak{O}/\mathfrak{p}$ of the principal order $\mathfrak{O}$ of K with respect to all prime ideals $\mathfrak{p}$ dividing $\mathfrak{p}_0$ are separable extensions of $\mathfrak{o}/\mathfrak{p}_0$. This assumption is actually dispensable, but simplifies matters.†

The theory which we will now develop is due to Hilbert. Among the most important results to be attained is a strengthening of the Dedekind discriminant theorem.

1. The Decomposition Group and Inertia Group

The elements of the Galois group $\mathfrak{G}$ of K/k are denoted $S, T, \cdots$. Application of S to an element or set of elements is written $\alpha \to \alpha^S$. Application of S to an ideal $\mathfrak{a}$ yields another ideal $\mathfrak{a}^S$.

One prime ideal $\mathfrak{p}$ of K will be fixed throughout. The totality of all $S \in \mathfrak{G}$ which map $\mathfrak{p}$ into itself form a subgroup $\mathfrak{Z} \subseteq \mathfrak{G}$, the *decomposition group* of $\mathfrak{p}$. If

$$\mathfrak{G} = \mathfrak{Z}S_1 \cup \cdots \cup \mathfrak{Z}S_r \qquad (S_1 \in \mathfrak{Z}) \tag{1}$$

is a decomposition of $\mathfrak{G}$ into left cosets, then $\mathfrak{p}_1 = \mathfrak{p}^{S_1} = \mathfrak{p}, \ldots, \mathfrak{p}_r = \mathfrak{p}^{S_r}$ are the prime ideals *conjugate* to $\mathfrak{p}$. The decomposition groups of the $\mathfrak{p}_\varrho$ are $S_\varrho^{-1}\mathfrak{Z}S_\varrho$.

Let $\mathfrak{p}_0$ be the prime ideal of k divisible by $\mathfrak{p}_1$. We maintain that

$$\mathfrak{p}_0 = (\mathfrak{p}_1 \cdots \mathfrak{p}_r)^e, \qquad e \geqq 1. \tag{2}$$

Proof. Let $\mathfrak{p}_0 = \mathfrak{p}_1^{e_1} \cdots \mathfrak{p}_r^{e_r}\mathfrak{p}_1'^{e_1'} \cdots \mathfrak{p}_s'^{e_s'}$, where the $\mathfrak{p}_\sigma'$ and their conjugates differ from the $\mathfrak{p}_\varrho$. This decomposition continues to hold under automorphisms in $\mathfrak{G}$. Thus we already have $e_1 = \cdots = e_r = e$. By §2,2, there exists an $\alpha \in K$ divisible by $\mathfrak{p}_1$ but by none of the $\mathfrak{p}_\sigma'$ or their conjugates. Then the conjugates of α are divisible neither by the $\mathfrak{p}_\sigma'$ nor by their conjugates, and consequently neither is the product of the conjugates of α which is $n_{K/k}(\alpha)$. On the other hand, $n_{K/k}(\alpha)$ is divisible by the $\mathfrak{p}_\varrho$, and thus by $\mathfrak{p}$. Hence we have $e_1' = \cdots = 0$. ▼

The *inertia group* $\mathfrak{T}$ of $\mathfrak{p}$ is the totality of all $\mathfrak{T} \in \mathfrak{G}$ for which

$$\omega^T \equiv \omega \bmod \mathfrak{p} \qquad \text{for all} \quad \omega \in \mathfrak{O} \tag{3}$$

holds. Certainly (3) holds for $\omega \in \mathfrak{p}$, so that $\mathfrak{T} \subseteq \mathfrak{Z}$.

The subfields K_z and K_t belonging to $\mathfrak{Z}$ and $\mathfrak{T}$ in the sense of Galois theory are called the *decomposition field* and *inertia field* of $\mathfrak{p}$, respectively. We will denote the principal orders and prime ideals divisible by $\mathfrak{p}$, of these fields, $\mathfrak{O}_z, \mathfrak{O}_t, \mathfrak{p}_z, \mathfrak{p}_t$.

---

† The theory is carried out under more general conditions in O. Zariski and P. Samuel, *Commutative Algebra I*, p. 290 ff., London, Toronto, New York, 1958. (Tr. note: It has become customary to use the symbols $\mathfrak{Z}$ and $\mathfrak{T}$, standing for the German words Zerlegung and Trägheit, in English as well.)

The inertia group is a normal subgroup of the decomposition group. If the residue class field $\mathfrak{O}/\mathfrak{p}$ *is a separable extension of* $\mathfrak{o}/\mathfrak{p}_0$, *then every element of* $\mathfrak{O}$ *is congruent* mod $\mathfrak{p}$ *to an element of* $\mathfrak{O}_t$, *so that* $\mathfrak{O}/\mathfrak{p} \cong \mathfrak{O}_t/\mathfrak{p}_t$. *Under that assumption, the field* $\mathfrak{O}/\mathfrak{p}$ *is a Galois extension of* $\mathfrak{o}/\mathfrak{p}_0$ *with the group* $\mathfrak{Z}/\mathfrak{T}$.

Proof. Application of (3) on ω^S in place of ω with an arbitrary $S \in \mathfrak{Z}$ yields $\omega^{STS^{-1}} \equiv \omega \bmod \mathfrak{p}$. This is equivalent to the first assertion.

Let T run through all the elements of $\mathfrak{T}$. Again because of (3), any element $\alpha \in \mathfrak{O}$ satisfies the congruence,

$$f(x) = \prod_T (x - \alpha^T) \equiv (x - \alpha)^{[\mathfrak{T}]} \equiv 0 \bmod \mathfrak{p},$$

$[\mathfrak{T}]$ signifying the order of $\mathfrak{T}$ as a group. The coefficients of the polynomial on the left belong to the inertia group; let us consider more closely the polynomial $f(x)$ in K_t. The above congruence assures us that $f(x)$ is a $[\mathfrak{T}]$th power in $\mathfrak{O}/\mathfrak{p}$. But, as we assume $\mathfrak{O}/\mathfrak{p}$ to be separable over $\mathfrak{o}/\mathfrak{p}_0$ and therefore over $\mathfrak{O}_t/\mathfrak{p}_t$, $f(x)$ must already be a $[\mathfrak{T}]$th power in $\mathfrak{O}_t/\mathfrak{p}_t$. Thus, $f(x) \equiv (x - \alpha_t)^{[\mathfrak{T}]}$ mod $\mathfrak{p}_t$ with some $\alpha_t \in \mathfrak{O}_t$, so that $\alpha \equiv \alpha_t \bmod \mathfrak{p}$. For the residue class fields this means that

$$\mathfrak{O}/\mathfrak{p} \cong \mathfrak{O}_t/\mathfrak{p}_t. \tag{4}$$

To prove the last assertion we consider $\mathfrak{O}/\mathfrak{p}$ as an extension of $\mathfrak{O}_z/\mathfrak{p}_z$. For arbitrary S and T in $\mathfrak{S}$ and $\mathfrak{T}$, and all $\alpha \in \mathfrak{O}$ we have, with (3), $\alpha^{TS} \equiv \alpha^S \bmod \mathfrak{p}$. Hence the cosets of $\mathfrak{T}$ in $\mathfrak{Z}$ induce automorphisms of $\mathfrak{O}/\mathfrak{p}$ leaving $\mathfrak{O}_z/\mathfrak{p}_z$ invariant. For an $S \in \mathfrak{Z}$, $S \notin \mathfrak{T}$ there exists, by the definition of $\mathfrak{T}$, an $\alpha \in \mathfrak{O}$ such that $\alpha^S \not\equiv \alpha \bmod \mathfrak{p}$. We have $[\mathfrak{Z}:\mathfrak{T}]$ of these automorphisms of $\mathfrak{O}/\mathfrak{p}$. In other words $[\mathfrak{O}/\mathfrak{p}:\mathfrak{O}_z/\mathfrak{p}_z] \geqq [\mathfrak{Z}:\mathfrak{T}]$.

On the other hand, the symmetric functions of all α^S, $S \in \mathfrak{Z}$ for one $\alpha \in \mathfrak{O}$ are in $\mathfrak{O}_z$. Thus, every $\alpha \in \mathfrak{O}$ satisfies an equation in $\mathfrak{O}_z$ of degree at most $[\mathfrak{Z}:\mathfrak{T}]$. This assures us that

$$f_{K_t/K_z}(\mathfrak{p}_t) = [\mathfrak{O}_t/\mathfrak{p}_t:\mathfrak{O}_z/\mathfrak{p}_z] = [\mathfrak{O}/\mathfrak{p}:\mathfrak{O}_z/\mathfrak{p}_z] = [\mathfrak{Z}:\mathfrak{T}]. \tag{5}$$

Furthermore, $\mathfrak{O}/\mathfrak{p} \cong \mathfrak{O}_t/\mathfrak{p}_t$ is a Galois extension of $\mathfrak{O}_z/\mathfrak{p}_z$ with the group $\mathfrak{Z}/\mathfrak{T}$. The last assertion therefore depends only upon the proof of

$$\mathfrak{O}_z/\mathfrak{p}_z \cong \mathfrak{o}/\mathfrak{p}_0. \tag{6}$$

From the definition of the residue class degree in §3,3, (4) implies $f_{K/K_t}(\mathfrak{p}) = 1$. The only prime ideal of K which divides $\mathfrak{p}_t$ is $\mathfrak{p}$. For, all the prime divisors of $\mathfrak{p}_t$ in K are generated from $\mathfrak{p}$ by the application of elements of the Galois Group of K/K_t [cf. proof of (2)]. But this group is $\mathfrak{T}$, which leaves $\mathfrak{p}$ fixed. Now, as seen in §3,(14), $f_{K/K_t}(\mathfrak{p}) = 1$ implies

$$e_{K/K_t}(\mathfrak{p}) = [K:K_t] \qquad \text{or} \qquad \mathfrak{p}_t = \mathfrak{p}^{[K:K_t]}. \tag{7}$$

Taking norms of (2) according to §3,(11) yields $(p)^n = (p)^{efr}$, so that

$$efr = n, \tag{8}$$

where $r = [\mathfrak{G}:\mathfrak{Z}] = [K_z:k]$. But, as in §3,(12), e is divisible by $e_{K/K_t}(\mathfrak{p})$ and f by $f_{K_t/K_z}(\mathfrak{p}_t)$. By (7) and (5) these numbers are $[K:K_t]$ and $[K_t:K_z]$. This shows that the left side of (8) is divisible by $[K:K_t][K_t:K_z][K_z:k] = [K:k] = n$, which is the right side, so that

$$e = e_{K/k}(\mathfrak{p}) = [K:K_t], \qquad f = f_{K/k}(\mathfrak{p}) = f_{K_t/kz}(\mathfrak{p}_t) = [K_t:K_z]. \tag{9}$$

Equation (9), with §3,(12), yields $f_{K_z/k}(\mathfrak{p}_z) = 1$, which is equivalent to (6). Finally (7) and the first equation of (9), along with §3,(12), yield $e_{K_t/K_z}(\mathfrak{p}_t) = 1$, which is equivalent to

$$\mathfrak{p}_t = \mathfrak{p}_z. \tag{10}$$

This completes the proof.

The result can also be described with other words. The prime ideal $\mathfrak{p}_0$ of k is divisible by a prime ideal $\mathfrak{p}_z$ of degree $f_{K_z/k}(\mathfrak{p}_z) = 1$ in K_z. The other prime divisors of $\mathfrak{p}_0$ in K_z can have larger degree. Stepping up to K_t, (10) assures us that $\mathfrak{p}_z$ remains prime: it is "inert." But, its degree increases to $f = [K_t:K_z]$. In the extension K/K_t, $\mathfrak{p}_t$ becomes the eth power of a prime ideal $\mathfrak{p}$ in K. K/K_t is called *ramified* at the place $\mathfrak{p}$, with index e, if $e > 1$. But $\mathfrak{p}$ retains the degree $f_{K/k}(\mathfrak{p}) = f_{K_t/k}(\mathfrak{p}_t) = f$. Incidentally, in any intermediate field K' between K_t and K the prime ideal $\mathfrak{p}'$ there satisfies $\mathfrak{p}_t = \mathfrak{p}'^{[K':K_t]}$. This is because the degree $f_{K'/K_t}(\mathfrak{p}') = 1$ and the index $e_{K'/K_t}(\mathfrak{p}') = [K':K_t]$, then, using §3,(14). If K'' is another intermediate field, between K and K', we again have $\mathfrak{p}' = \mathfrak{p}''^{[K'':K']}$. Thus, ramification does not cease once it has started.

The norm introduced in §3 for ideals can be written

$$n_{K/k}(\mathfrak{a}) = \prod_{S\in\mathfrak{G}} \mathfrak{a}^S$$

so as to correspond with the norm of elements. The proof need only be brought for a prime ideal. But for these it follows immediately from (1) and (2), together with the above considerations concerning the order of $\mathfrak{Z}$ and $\mathfrak{T}$ and their connection with the degree and ramification index.

Finally, assume that the residue class field $\mathfrak{o}/\mathfrak{p}_0$ has only finitely many elements. The same then holds for $\mathfrak{O}/\mathfrak{p}$. The numbers q and Q of them are then powers of a prime number, with, in particular,

$$Q = q^f.$$

The Galois group of a finite field is cyclic, as is well known, and it is generated by exponentiation with q.

If $\mathfrak{o}/\mathfrak{p}_0$ has $q < \infty$ elements, then $\mathfrak{Z}/\mathfrak{T}$ is a cyclic group and there exists a generating element S such that

$$\omega^S \equiv \omega^q \bmod \mathfrak{p} \qquad (\text{for all } \omega \in \mathfrak{O}). \tag{11}$$

It is uniquely determined up to a factor in $\mathfrak{T}$, and is called the Frobenius automorphism for $\mathfrak{p}$.

2. The Ramification Groups

The first ramification group $\mathfrak{V}_1$ consists of all $V \in \mathfrak{Z}$ such that

$$\omega^V \equiv \omega \bmod \mathfrak{p}^2 \qquad (\text{for all } \omega \in \mathfrak{O})$$

holds. Beside the group $\mathfrak{V}_1$ we are interested in the greatest exponent v_1 ($\geqq 2$) for which

$$\omega^V \equiv \omega \bmod \mathfrak{p}^{v_1} \qquad (\text{for all } V \in \mathfrak{V}_1, \omega \in \mathfrak{O}). \tag{12}$$

$\mathfrak{V}_1$ is an invariant subgroup of $\mathfrak{T}$ and $\mathfrak{Z}$. The factor group $\mathfrak{T}/\mathfrak{V}_1$ is isomorphic to a subgroup of the multiplicative group of the residue class field $\mathfrak{O}/\mathfrak{p}$.

For the proof we introduce the principal ideal domain $\mathfrak{O}_\mathfrak{p}$ of all $\alpha \in K$ without $\mathfrak{p}$ in their denominators (cf. §2,2). All $\mathfrak{O}_\mathfrak{p}$-ideals are of the form $\mathfrak{O}_\mathfrak{p}\pi^h$. An element π of this sort is called a *prime element*. Two prime elements differ but by the factor of a unit of $\mathfrak{O}_\mathfrak{p}$. By §2,4(i) the residue class rings $\mathfrak{O}/\mathfrak{p}$ and $\mathfrak{O}_\mathfrak{p}/\mathfrak{O}_\mathfrak{p}\pi$ are isomorphic.

Comparison of (3) and (12) yields $\mathfrak{V}_1 \subseteq \mathfrak{T}$. If $S \in \mathfrak{Z}$, the application of (12) to ω^S gives $\omega^{SVS^{-1}} \equiv \omega \bmod \mathfrak{p}^2$. Thus $\mathfrak{V}_1$ is an invariant subgroup of $\mathfrak{Z}$.

If $T \in \mathfrak{T}$ and π is a prime element, then so is π^T, and

$$\pi^T = \varepsilon_T \pi \tag{13}$$

with a unit ε_T. If further $T \in \mathfrak{V}_1$, then $\pi^T \equiv \pi \bmod \mathfrak{p}^2$ so that $\varepsilon_T \equiv 1 \bmod \mathfrak{p}$.

We want to show the converse of this: in (13) let $\varepsilon_T \equiv 1 \bmod \mathfrak{p}$. By the last section every $\omega \in \mathfrak{O}_\mathfrak{p}$ is congruent to an $\omega_t \in \mathfrak{O}_{t,\mathfrak{p}} = K_t \cap \mathfrak{O}_\mathfrak{p}$, and thus

$$\omega \equiv \omega_t + \omega_t'\pi \bmod \mathfrak{p}^2$$

with some $\omega_t' \in \mathfrak{O}_{t,\mathfrak{p}}$. Now, (3) implies

$$\omega^T \equiv \omega_t + \omega_t'\pi^T \bmod \mathfrak{p}^2$$

so that, altogether, with (13) and $\varepsilon_T \equiv 1 \bmod \mathfrak{p}$, we have

$$\omega^T \equiv \omega \bmod \mathfrak{p}^2.$$

By definition, then, $T \in \mathfrak{V}_1$.

If $T_1, T_2 \in \mathfrak{T}$, then by (13),

$$\pi^{T_1T_2} = \varepsilon_{T_1T_2}\pi = \varepsilon_{T_1}^{T_2}\varepsilon_{T_2}\pi$$

so that $\varepsilon_{T_1T_2} = \varepsilon_{T_1}^{T_2}\varepsilon^{T_2}$, and with (3),

$$\varepsilon_{T_1T_2} \equiv \varepsilon_{T_1}\varepsilon_{T_2} \bmod \mathfrak{p}. \tag{14}$$

A consequence of (14) is that the mapping of T onto the residue class of ε_T

is a homomorphism of $\mathfrak{T}$ into the multiplicative group of $\mathfrak{O}/\mathfrak{p}$. The kernel is the ramification group $\mathfrak{V}_1$, as was shown.

It appears plausible to define further ramification groups by strengthening the congruence (12). Say the ith *ramification group* $\mathfrak{V}_i$ is defined by the congruence

$$\omega^V \equiv \omega \bmod \mathfrak{p}^{v_i} \qquad (\text{for all } \omega \in \mathfrak{O}) \tag{15}$$

where v_i is maximal. Then, define the $(i+1)$th *ramification group* as the totality of all $V \in \mathfrak{V}_i$ for which

$$\omega^V \equiv \omega \bmod \mathfrak{p}^{v_i+1}$$

always holds. Then choose a *maximal* v_{i+1} for which

$$\omega^V \equiv \omega \bmod \mathfrak{p}^{v_{i+1}} \qquad (\text{for all} \quad V \in \mathfrak{V}_{i+1}, \omega \in \mathfrak{O}).\dagger \tag{16}$$

An argument already used twice shows that:

The ramification groups $\mathfrak{V}_i$ are normal subgroups of the decomposition group $\mathfrak{Z}$.

The invariant fields K_i belonging to the $\mathfrak{V}_i$ are called the *ramification fields*. For the prime ideals $\mathfrak{p}_i$ of these fields which are divisible by $\mathfrak{p}$ we have, by §6,1,

$$\mathfrak{p}_i = \mathfrak{p}_{i+1}^{e_{i+1}} \qquad \text{with} \quad e_{i+1} = [K_{i+1}:K_i] = [\mathfrak{V}_i:\mathfrak{V}_{i+1}] \tag{17}$$

(here $\mathfrak{V}_0 = \mathfrak{T}$, $K_0 = K_t$). The total ramification index of $\mathfrak{p}$ is then

$$e = e_{K/k}(\mathfrak{p}) = e_1 e_2 \cdots . \tag{18}$$

Let us yet consider the factor groups $\mathfrak{V}_i/\mathfrak{V}_{i+1}$ for $i \geqq 1$. We know from (15) and (16) that, for a prime element π and a $V \in \mathfrak{V}_i$,

$$\pi^V \equiv \pi + \nu_V \pi^{v_i} \bmod \mathfrak{p}^{v_i+1}$$

with a $\nu_V \in \mathfrak{O}_\mathfrak{p}$, whose residue class mod $\mathfrak{p}$ corresponds to the element V (where π is held fixed). If $V_1, V_2 \in \mathfrak{V}_i$, then

$$\pi^{V_1 V_2} \equiv \pi + (\nu_{V_1}^{V_2} + \nu_{V_2})\pi^{v_i} \bmod \mathfrak{p}^{v_i+1}.$$

That is using (3) and the fact that $V_2 \in \mathfrak{T}$,

$$\nu_{V_1 V_2} \equiv \nu_{V_1} + \nu_{V_2} \bmod \mathfrak{p}. \tag{19}$$

This represents a homomorphism of $\mathfrak{V}_i$ into the additive group of $\mathfrak{O}/\mathfrak{p}$. By definition, the kernel is $\mathfrak{V}_{i+1}$.

The factor groups $\mathfrak{V}_i/\mathfrak{V}_{i+1}$ are isomorphic to subgroups of the additive groups of $\mathfrak{O}/\mathfrak{p}$.

---

† The v_i lie below a bound that depends only upon n, cf. the book quoted on page 91 (Zariski–Samuel).

One speaks of *regular* or *irregular ramification*, depending upon whether $\mathfrak{B}_1$ consists only of the unity or not. *K/k is regularly ramified at the place $\mathfrak{p}$ if and only if the ramification index e is divisible neither by $\mathfrak{p}$ nor by the characteristic of the field.*

Proof. Let K/k be irregularly ramified. Then, $e_2 = [\mathfrak{B}_1 : \mathfrak{B}_2] > 1$. But $\mathfrak{B}_1/\mathfrak{B}_2$ is isomorphic to a subgroup of the additive group of $\mathfrak{O}/\mathfrak{p}$, so that there exists a residue class $\omega \not\equiv 0 \bmod \mathfrak{p}$ and a divisor e_2' of e_2 such that $\omega + \cdots + \omega = e_2'\omega \equiv 0 \bmod \mathfrak{p}$. Then $e_2' \equiv e_2 \equiv 0 \bmod \mathfrak{p}$, so that with (18) either $\mathfrak{p}$ or the characteristic divides e.

Let K/k be regularly ramified, i.e., $\mathfrak{B}_1$ consist of the unity element only, and, contrary to assertion, let e be divisible by some prime $q \equiv 0 \bmod \mathfrak{p}$. † Now e is the order of $\mathfrak{T}$, and $\mathfrak{T}$ is isomorphic to a subgroup of the multiplicative group of $\mathfrak{O}/\mathfrak{p}$. The residue classes mod $\mathfrak{p}$ form a field, and the residue classes of $0, 1, \ldots, q-1$ form a finite subfield Ω because of $q \equiv 0 \bmod \mathfrak{p}$. As the multiplicative group of $\mathfrak{O}/\mathfrak{p}$ has a subgroup of order e and e is divisible by q, there must exist a residue class $\varrho \not\equiv 1 \bmod \mathfrak{p}$ with $\varrho^q \equiv 1 \bmod \mathfrak{p}$. Adjoining ϱ would yield an extension of Ω of the qth degree. An extension of this sort over a field of q elements, defined by $\varrho^q = 1$, is well known not to exist. ▼

3. The Discriminant

We first compute the discriminants $\mathfrak{D}_{K_1}/\mathfrak{D}_{Kt}$ and $\mathfrak{D}_K/\mathfrak{D}_{K_1}$, where K_1 denotes the first ramification field. As we saw in §4, we need only the $\mathfrak{p}$-components.

Every $\omega \in \mathfrak{O}_\mathfrak{p}$ is congruent mod $\mathfrak{p}$ to some $\omega_t \in \mathfrak{O}_{t,\mathfrak{p}} = K_t \cap \mathfrak{O}_\mathfrak{p}$. Thus

$$\omega = \omega_t + \omega'\pi$$

with some $\omega' \in \mathfrak{O}_\mathfrak{p}$ and a prime element π in $\mathfrak{O}_\mathfrak{p}$. Following the same procedure with ω', etc., we find a congruence

$$\omega \equiv \omega_t + \omega_t'\pi + \cdots + \omega_t^{(e-1)}\pi^{e-1} \bmod \mathfrak{p}^e \tag{20}$$

with $\omega_t, \omega_t', \cdots \in \mathfrak{O}_{t,\mathfrak{p}}$. We now maintain that $1, \pi, \ldots, \pi^{e-1}$ are a basis of $\mathfrak{O}_\mathfrak{p}$ with respect to $\mathfrak{O}_{t,\mathfrak{p}}$. In fact, if ω_ν is a basis and

$$\pi^\mu = \sum_\nu c_{\mu\nu}\omega_\nu, \qquad c_{\mu\nu} \in \mathfrak{O}_{t,\mathfrak{p}},$$

one can substitute the representations (20) for the ω_ν and arrive at a system of congruences

$$\pi^\mu \equiv \sum_{\nu,\varrho} c_{\mu\nu}c'_{\nu\varrho}\pi^\varrho \bmod \mathfrak{p}^e, \qquad c'_{\nu\varrho} \in \mathfrak{O}_{t,\mathfrak{p}}.$$

† This includes the case where q is the characteristic.

The elements $1, \pi, \ldots, \pi^{e-1}$ are linearly independent mod $\mathfrak{p}^e$, so that we have the matrix congruence $(c_{\mu\nu})(c'_{\mu\nu}) \equiv E \bmod \mathfrak{p}_t$. But this is only possible if the matrix $(c_{\mu\nu})$ is unimodular in $\mathfrak{O}_{t,\mathfrak{p}}$.

Similarly we find that if π_1 is a prime element of $\mathfrak{O}_{1,\mathfrak{p}} = K_1 \cap \mathfrak{O}_\mathfrak{p}$, then $1, \pi_1, \ldots, \pi_1^{e_1-1}$ is a basis of $\mathfrak{O}_{1,\mathfrak{p}}$ with respect to $\mathfrak{O}_{t,\mathfrak{p}}$. Further, $1, \pi, \ldots, \pi^{e_1'-1}$ $(e_1' = e/e_1)$ is a basis of $\mathfrak{O}_\mathfrak{p}$ with respect to $\mathfrak{O}_{1,\mathfrak{p}}$.

By §4,2, the $\mathfrak{p}_t$-component of $\mathfrak{D}_{K_1/K_t}$ is the principal ideal generated by

$$|s_{K_1/K_t}(\pi_1{}^\mu \pi_1{}^\nu)| = \left| \sum_{\varrho=1}^{e_1} (\pi_1{}^\mu \pi_1{}^\nu)^{T_\varrho} \right|$$

$$= \begin{vmatrix} 1 & \cdots & 1 \\ \pi_1^{T_1} & \cdots & \pi_1^{T_{e_1}} \\ \cdot & \cdots & \cdot \\ (\pi_1^{e_1-1})^{T_1} & \cdots & (\pi_1^{e_1-1})^{T_{e_1}} \end{vmatrix} \begin{vmatrix} 1 & \pi_1^{T_1} & \cdots & (\pi_1^{e_1-1})^{T_1} \\ \cdot & \cdot & \cdots & \cdot \\ \cdot & \cdot & \cdots & \cdot \\ 1 & \pi_1^{T_{e_1}} & \cdots & (\pi_1^{e_1-1})^{T_{e_1}} \end{vmatrix}$$

$$= \pm \prod_{\varrho \neq \sigma} (\pi_1^{T_\varrho} - \pi_1^{T_\sigma})$$

where T_ϱ runs through a system of representatives for the cosets of $\mathfrak{V}_1$ in $\mathfrak{T}$. By (13) we have here

$$\pi_1^{T_\varrho} - \pi_1^{T_\sigma} = \pi_1^{T_\sigma}(\varepsilon_{T_\varrho T_\sigma^{-1}}^{T_\sigma} - 1) = \pi_1 \varepsilon_{T_\sigma}(\varepsilon_{T_\varrho T_\sigma^{-1}}^{T_\sigma} - 1),$$

and because we assume $T_\varrho T_\sigma^{-1} \notin \mathfrak{V}_1$ we know $\varepsilon_{T_\varrho T_\sigma^{-1}}^{T_\sigma} \not\equiv 1 \bmod \mathfrak{p}$. The $\mathfrak{p}_t$-component of $\mathfrak{D}_{K_1/Kt}$ is thus equal to $\mathfrak{p}_1^{e_1(e_1-1)} = \mathfrak{p}_t^{e_1-1}$.

The $\mathfrak{p}_1$-component of $\mathfrak{D}_{K/K_1}$ is similarly found to be the principal ideal generated by

$$\prod_{\varrho \neq \sigma} (\pi^{V_\varrho} - \pi^{V_\sigma}), \tag{21}$$

where V_ϱ, V_σ traverse the entire group $\mathfrak{V}_1$. The V_ϱ, V_σ are represented as products

$$V_\varrho = \cdots V_{2\varrho_2} V_{1\varrho_1}, \qquad V_\sigma = \cdots V_{2\sigma_2} V_{1\sigma_1},$$

the $V_{i\varrho_i}$, $V_{i\sigma_i}$ traversing a system of representatives of the cosets of $\mathfrak{V}_{i+1}$ in $\mathfrak{V}_i$. We have

$$V_\varrho V_\sigma^{-1} = \cdots V_{2\varrho_2} V_{1\varrho_1} V_{1\sigma_1}^{-1} V_{2\sigma_2}^{-1} \cdots.$$

We here assume that $V_{1\varrho_1} = V_{1\sigma_1}, \ldots, V_{i-1,\varrho_{i-1}} = V_{i-1,\sigma_{i-1}}, V_{i\varrho_i} \neq V_{i\sigma_i}$. The $V_\varrho V_\sigma^{-1}$ then lie in $\mathfrak{V}_i$ but not in $\mathfrak{V}_{i+1}$, all the subgroups occurring being normal. Then §6,2 give us;

$$\pi^{V_\varrho} - \pi^{V_\sigma} = (\pi^{V_\varrho V_\sigma^{-1}} - \pi)^{V_\sigma} \equiv (v_{V_\varrho V_\sigma^{-1}} \pi^{v_i})^{V_\sigma} \bmod \mathfrak{p}^{v_i+1};$$

i.e., every term is divisible by $\mathfrak{p}$ exactly v_i times. Thus in the product (21), holding the V_σ fixed, there are exactly $[\mathfrak{V}_i] - [\mathfrak{V}_{i+1}] = (e_{i+1} - 1)(e/e_1 \cdots e_{i+1})$ factors equal to $(\pi^{v_i})^{V_\sigma}$ up to units, this holding for $i = 1, 2, \ldots$. $[V_i]$ denotes the order of $\mathfrak{V}_i$. Remembering still that $(\prod \pi^{V_\sigma}) = \prod \mathfrak{p}^{V_\sigma} = \mathfrak{p}_1$, it has been shown that the $\mathfrak{p}_1$-components of $\mathfrak{D}_{K/K_1}$ is $\mathfrak{p}_1$ to the power

$$v_1([\mathfrak{B}_1] - [\mathfrak{B}_2]) + v_2([\mathfrak{B}_2] - [\mathfrak{B}_3]) + \cdots$$
$$= v_1(e_2 - 1)(e/e_1 e_2) + v_2(e_3 - 1)(e/e_1 e_2 e_3) + \cdots .$$

Finally we apply the transitivity formula §4,(11), noting that, by the Dedekind discriminant theorem of §4,3, $\mathfrak{D}_{K_t/k}$ is relatively prime to $\mathfrak{p}$ whenever the residue class field $\mathfrak{O}/\mathfrak{p}$ is separable over $\mathfrak{o}/p$. This gives the final result:

Let the residue class field $\mathfrak{O}/\mathfrak{p}$ be separable over $\mathfrak{o}/\mathfrak{p}_0$. $\mathfrak{D}_{K/k}$ is then divisible by $\mathfrak{p}_0{}^{\delta}$, where

$$\delta = f((e_1 - 1)(e/e_1) + v_1(e_2 - 1)(e/e_1 e_2) + v_2(e_3 - 1)(e/e_1 e_2 e_3) + \cdots); \qquad (22)$$

(18) *should be taken into account. For regular ramification at $\mathfrak{p}$ this exponent is $\delta = f(e - 1)$; otherwise it is larger.*

Appendix to Chapter II*

Topics from the Theory of Algebraic Number Fields

The theorem of the unique decomposition into prime ideals gives us a basis for the theory of algebraic numbers. The main task remains to find as close as possible a connection between elements and ideals, and this is carried out primarily in two theorems (§1): first, that the set of ideals decomposes into a finite number of classes (II, §1,4), and second, an explanation of the structure of the group of units (II, §5,3), so that the connection between elements and principal ideals becomes clear. The results of these two theorems can be interpreted as saying that the concept of an ideal leads only a finite step beyond the concept of an element. In §2 the information thus obtained is briefly illustrated on two types of fields, as important as they are simple. The space here does not permit a deeper penetration of algebraic number fields. The interested reader will find it in the following texts:

H. HASSE, *Vorlesungen über Zahlentheorie*, 2nd ed. Springer Verlag, Berlin-Göttingen-Heidelberg, 1964.

H. HASSE, *Zahlentheorie*. Akademie Verlag, Berlin, 1949; rev. ed., East Berlin, 1963.

E. HECKE, *Vorlesungen über die Theorie der algebraischen Zahlen*. Leipzig, 1923, reprinted, Chelsea Press, New York, 1948.

§1. The Finiteness Theorems

1. THE FINITENESS OF THE IDEAL CLASS NUMBER

As was discussed in II, §5,3, the divisor concept can offer nothing more for the study of algebraic number fields than the ideal concept of II, §1,4, where $\mathfrak{o}$ is taken to be the principal order $\mathfrak{i} = Z$ of the rational base field $k = Q$. We will stay with these assumptions throughout.

Let the degree $[K\colon Q]$ be n. By the fundamental theorem of algebra, K can be considered as a subfield of the field C of complex numbers. Let ϑ be a primitive element of K and $\vartheta_1 = \vartheta, \vartheta_2, \ldots, \vartheta_n$ be the algebraic conjugate elements to ϑ. The nonreal elements among these occur in pairs, for if ϑ_ν is conjugate to ϑ, so is $\bar{\vartheta}_\nu$. The number of real conjugates will be denoted r_1 throughout, the number of pairs of complex conjugates with r_2. Then,

$$n = r_1 + 2r_2. \tag{1}$$

These elements ϑ_ν generate n number fields $K_\nu = k(\vartheta_\nu)$ conjugate to K which, however, need not be distinct.

The number of ideal classes is finite. In many cases, in fact, this number is one. This means that then every ideal is a principal ideal, and every element can be written uniquely, except for units, as a product of prime elements.

The calculation of the number of ideal classes or, what is more, the determination of the structure of the group of ideal classes is a difficult problem, insofar as one desires to make general statements.†

The proof of the theorem consists of finding a certain finite set of ideals containing a representative of every ideal class. Let ι_ν be a basis of the principal order $\mathfrak{J}$ of K, and A some ideal class; let it contain the ideal $\mathfrak{a}$. Furthermore, let ϱ_ν be a basis of the inverse ideal $\mathfrak{a}^{-1}$. Then there is a linear substitution

$$\varrho_i = \sum_j m_{ij}\iota_j$$

with rational m_{ij}. By II, §3,2, its determinant has absolute value

$$\|m_{ij}\| = n(\mathfrak{a})^{-1}.$$

By I, §2,1 there exist rational integers x_i, not all vanishing, such that the inequalities

$$\Bigl|\sum_i x_i m_{ij}\Bigr| < c_j = 2\sqrt[n]{n(\mathfrak{a})^{-1}} \tag{2}$$

hold for all j. We now set

$$\beta = \sum \varrho_i x_i = \sum \iota_j y_j, \qquad y_j = \sum_i x_i m_{ij},$$

and majorize the norm of β. As defined in I, §4,1 it is $|n(\beta)| = |f(y_j)|$ with a homogeneous polynomial in the y_j of degree n. If $2^{-n}\tau$ denotes the maximum of $|f(z_j)|$ for $|z_j| \leqq 1$, then by (2), $|n(\beta)| < cn(\mathfrak{a})^{-1}$.

Now, the ideal $\mathfrak{a}\beta$ belongs to the class A, is integral, and has norm

$$n(\mathfrak{a}\beta) = n(\mathfrak{a})|n(\beta)| < c.$$

† H. Hasse, *Über die Klassenzahl abelscher Zahlkörper*, Akademie Verlag, East Berlin, 1952. C. Meyer, *Die Berechnung der Klassenzahl abelscher Körper über quadratischen Zahlkörpern*, Akademie Verlag, East Berlin, 1957.

There can exist but finitely many integral ideals whose norm is thus bounded, namely, those occurring in the prime decomposition of natural numbers $< c$.

2. The Discriminant

The discriminant $\mathfrak{D}_{K/Q}$ of the field K is, according to II, §4,(12), equal to the principal ideal $\mathfrak{i}D(\iota_i)$, where ι_i is, again, a basis of $\mathfrak{J}$. It can be minorized

$$|D(\iota_i)| > (\pi/2)^{2r_2} \geqq 1. \tag{3}$$

Hence, using the Dedekind discriminant theorem of II, §4,3, we know that there are always rational prime numbers which ramify in K.

The *proof* depends upon refining the argument of the last section. Let $\iota_{i\nu}$ ($\nu = 1, \ldots, n$) be the conjugates of the ι_i. Then

$$D(\iota_i) = |s(\iota_i \iota_j)| = \left| \sum_{\nu=1}^{n} \iota_{i\nu} \iota_{j\nu} \right| = |\iota_{i\nu}|^2.$$

Assume, contrary to the assertion, that (writing $\| \cdots \|$ for the absolute value of the determinant)

$$\|\iota_{i\nu}\| \leqq (\pi/2)^{r_2}. \tag{4}$$

If the sign $<$ holds we can choose positive numbers c_ν with the product

$$\|\iota_{i\nu}\|(\pi/2)^{-r_2} < \prod_{\nu=1}^{n} c_\nu < 1.$$

Then, by I, §2,1 there exist rational integers x_i, not all vanishing such that

$$\left| \sum_{i=1}^{n} x_i \iota_{\nu i} \right| < c_\nu \tag{5}$$

holds. The norm of $\xi = \sum x_i \iota_i$ is then, taken absolutely,

$$|n(\xi)| = \prod_{\nu=1}^{n} \left| \sum_{i=1}^{n} x_i \iota_{i\nu} \right| < \prod_{\nu=1}^{n} c_\nu \leqq 1. \tag{6}$$

If (4) is an equality let $c_\nu = c_{\nu 0} + t$ with positive t and positive $c_{\nu 0}$ whose product is 1. (5) defines a convex body changing continuously with t, and for all $t > 0$ it contains a lattice point in its interior aside from the origin. This lattice point cannot change as $t \to 0$, and therefore

$$\left| \sum_{i=1}^{n} x_i \iota_{i\nu} \right| \leqq c_{\nu 0}.$$

If we assign $c_{\nu 0}$ such irrational values that the corresponding equations cannot hold for rational x_i, we have again the inequality sign and so (6).

But $\xi \in \mathfrak{J}$, so that $n(\xi)$ is an integer, and by our inequality it would have to be 0. On the other hand, our construction shows that ξ is not 0, so that $n(\xi) = 0$ is impossible. This proves (3). ▼

The approximation (3) can easily be refined to†

$$|D(\iota_i)| > (\pi/4)^n(n^n/n!)^2.$$

This is done by finding a convex symmetric domain as large as possible within the surface $|n(\sum x_i\iota_i)| = 1$ of affine space with coordinates x_i and then, in connection with the general Minkowski lattice point theorem, finding a lattice point aside from the origin. The inequality (5) does, in fact, define such a convex region, but it leads to the approximation (3). There exist more favorable regions that, at the cost of their simplicity, permit the improved inequality.

3. The Dirichlet Unit Theorem

As already noted, the concept of divisors and of ideals ($\mathfrak{J}$-ideals) is essentially identical here, so that the units of K in the sense of II, §5,3 coincide with the units of $\mathfrak{J}$. Dirichlet could already analyze the structure of the group of units $\mathfrak{E}$:

Let there be r_1 real and $2r_2$ pairwise complex conjugate number fields algebraically conjugate to K. Then, there exist

$$r - 1 = r_1 + r_2 - 1 \tag{7}$$

units ε_i in K, such that every unit of K can be written uniquely as a power product

$$\varepsilon = \zeta\varepsilon_1^{h_1} \cdots \varepsilon_{r-1}^{h_{r-1}} \tag{8}$$

up to a root of unity ζ.

Such units are said to form a *system of base units* of K. Thus, drawing on the fact normally proved in algebra that all the roots of unity of a finite algebraic number field form a cyclic group, the Dirichlet unit theorem asserts that the unit group $\mathfrak{E}$ has a basis of r elements.

The proof must be preceded by several auxiliary considerations. In them a surface $\mathfrak{N}$, defined by

$$\left|n\left(\sum_{i=1}^{n} x_i\iota_i\right)\right| = 1 \tag{9}$$

and to be considered as in the n-dimensional affine space R^n with real coordinates x_i called the "norm one surface," plays an important part. The points are simply denoted $\xi = \sum x_i\iota_i$.

Those points having rational coordinates are simultaneously elements of K, and form a group with respect to multiplication. If α is an element of this sort, then

$$\xi\alpha = \sum x_i\iota_i\alpha = \sum y_j\iota_j = \eta, \qquad y_j = \sum x_i a_{ij}, \tag{10}$$

† Cf., for example H. Hasse, *Zahlentheorie*, p. 454 Berlin, 1949; p. 590, 1963.

defines an affine coordinate transformation, with determinant $|a_{ij}| = n(\alpha) = \pm 1$. This gives a faithful representation of the multiplicative group of these α by affine transformations.

As multiplication and inverse formation are continuous operations in $K \cap \mathfrak{R}$, the points of $\mathfrak{R}$ taken naturally as limit points of elements with rational coordinates become elements of an abelian group, which is also faithfully represented by affine transformations.

Lemma 1. *In the bounded subset $\mathfrak{R}_c$ of $\mathfrak{R}$, defined by $|x_i| < c$, there exist at most a finite number of $\varepsilon \in \mathfrak{E}$ such that the corresponding affine transformations map a point of $\mathfrak{R}_c$ into the same or another pont of $\mathfrak{R}_c$.*

Proof. Let ξ, η be a pair of such points, with $\eta = \varepsilon\xi$ or $\varepsilon = \eta\xi^{-1}$. This product can be calculated in the representation group of affine transformations. It is a consequence of the continuity that the coefficients of the matrices of $\eta, \xi, \xi^{-1}, \varepsilon$ are bounded, the bounds depending only upon c. Thus, ε also lies in a bounded part of $\mathfrak{R}$, while on the other hand its coordinates are rational integers. This leaves only a finite number of values for ε. ▼

Lemma 2. *There exists a constant c, such that to every point $\alpha \in \mathfrak{R}$ there exists a unit ε with $\varepsilon\alpha \in \mathfrak{R}_c$.*

Proof. Use the notation of (10). As $|a_{ij}| = \pm 1$, I,§2,1 assures us that we can find rational integers x_i, not all vanishing, such that

$$\left|\sum_i x_i a_{ij}\right| = |y_j| < 2$$

holds for all j. Thus, $\eta \in \mathfrak{R}_2$. Let γ denote the maximum of all $|n(\eta)|$ with $\eta \in \mathfrak{R}_2$.

We now have $\xi = \sum x_i \iota_i \in \mathfrak{J}$ and $|n(\xi)| = |n(\eta)| \leqq \gamma$. Thus $\mathfrak{J}\xi$ is an integral ideal, with norm in a finite set. Using the argument of the last section, $\mathfrak{J}\xi$ also belongs to a finite set of principal ideals $\mathfrak{J}\beta_\nu$. Then $\xi = \beta_\nu\varepsilon$ with a unit ε and β_ν belongs to a finite set.

Using (10), finally, gives $\varepsilon\alpha = \beta_\nu^{-1}\eta$. As η belongs to $\mathfrak{R}_1$ and β_ν to a finite set, there clearly exists a constant c, dependent only upon $\mathfrak{J}$, with $\varepsilon\alpha \in \mathfrak{R}_c$. ▼

Based on the fact that the determinant $|\iota_{i\nu}| \neq 0$ (cf. §1,2), we have the (otherwise obvious) following lemma:

Lemma 3. *Using the algebraic conjugate bases $\iota_{i\nu}$, form*

$$\xi_\nu = \sum_{i=1}^{n} x_i \iota_{i\nu},$$

numbered so that $\bar{\xi}_{r_1+\nu} = \xi_{r_1+r_2+\nu}$ $(\nu = 1, \ldots, r_2)$. Then, with every vector ξ associate the vector $\mathfrak{l}(\xi)$ with components

$$\log|\xi_\nu|, \qquad \nu = 1, \ldots, r-1.$$

This maps every bounded subset of $\mathfrak{N}$ onto a bounded subset of the $(r-1)$-*dimensional space* R^{r-1}. *As*

$$\log|n(\xi)| = \sum_{\nu=1}^{r_1} \log|\xi_\nu| + 2\sum_{\nu=1}^{r_2} \log|\xi_{r_1+\nu}| = 0,$$

the points of $\mathfrak{N}$ mapped onto a bounded subset of R^{r-1} belong, conversely, to a bounded set.

Proof of the Dirichlet unit theorem. Multiplication of the ξ becomes addition of the $\mathfrak{l}(\xi) \in R^{r-1}$. Lemma 3 carries Lemmas 1 and 2 over to R^{r-1}, in the manner to be expected. By Lemma 1 the $\mathfrak{l}(\varepsilon)$ of units ε have no limit point at the origin, thus they have no limit point at all. Construct the base units as follows: First choose an ε_1 giving the vector $\mathfrak{l}(\varepsilon_1)$ minimal length. If $\varepsilon_2, \ldots, \varepsilon_{\varrho-1}$ have already been chosen ($\varrho < r-1$) find an ε_ϱ such that $\mathfrak{l}(\varepsilon_1)$, $\ldots$, $\mathfrak{l}(\varepsilon_\varrho)$ span a parallelepiped of minimal positive volume. If such an ε_ϱ exists, then there can exist no $\varepsilon \in \mathfrak{E}$ for which

$$\mathfrak{l}(\varepsilon) = s_1\mathfrak{l}(\varepsilon_1) + \cdots + s_\varrho\mathfrak{l}(\varepsilon_\varrho) \tag{11}$$

holds with nonintegral s_i. For otherwise, such an ε would exist with $0 \leqq s_i < 1$. Say $s_\sigma \neq 0$, $s_{\sigma+1} = \cdots = 0$. Then $\mathfrak{l}(\varepsilon_1), \ldots, \mathfrak{l}(\varepsilon_{\sigma-1}), \mathfrak{l}(\varepsilon)$ would span a parallelepiped of volume smaller than $\mathfrak{l}(\varepsilon_1), \ldots, \mathfrak{l}(\varepsilon_\sigma)$, contrary to assumption.

The nonexistence of such an ε_ϱ could have two possibilities as its cause. First, there could exist an infinite sequence of such $\varepsilon_{\varrho j}$, for which the volume decreases monotonely. To see that this is impossible, decompose R^{r-1} into the subspace R_1 spanned by $\mathfrak{l}(\varepsilon_1), \ldots, \mathfrak{l}(\varepsilon_{\varrho-1})$ and its perpendicular space R_2. In the first space, the $\mathfrak{l}(\varepsilon_\nu)$ ($\nu = 1, \ldots, \varrho-1$) also span a point lattice, and every vector of R_1 can be brought into the base parallelepiped defined by the same vectors by translation with a vector of the form (11) with integral s_i and $\varrho - 1$ in place of ϱ. This is done, in particular, for the components in R_1 of the vectors $\mathfrak{l}(\varepsilon_{\varrho j})$, and these vectors are then replaced by $\mathfrak{l}(\varepsilon_{\varrho j}) - \sum s_\nu \mathfrak{l}(\varepsilon_\nu)$. This does not change the volume of $\mathfrak{l}(\varepsilon_1), \ldots, \mathfrak{l}(\varepsilon_{\varrho j})$. Thus there is no loss of generality in assuming the components of $\mathfrak{l}(\varepsilon_{\varrho j})$ to be bounded in R_1. The volume sought is equal to the volume of $\mathfrak{l}(\varepsilon_1), \ldots, \mathfrak{l}(\varepsilon_{\varrho-1})$, multiplied by the length of the component of $\mathfrak{l}(\varepsilon_{\varrho j})$ in R_2, which must also be bounded. Thus, $\mathfrak{l}(\varepsilon_{\varrho j})$ must itself belong to a bounded subset of R^{r-1}, so that by Lemma 1 there can only be finitely many such $\varepsilon_{\varrho j}$.

The second possible reason for the nonexistence of such an ε_ϱ is that the volume of $\mathfrak{l}(\varepsilon_1), \ldots, \mathfrak{l}(\varepsilon_\varrho)$ is always zero. From what has already been shown, this would mean that $\mathfrak{l}(\varepsilon_1), \ldots, \mathfrak{l}(\varepsilon_{\varrho-1})$ already generate all the translations associated with units. But these translations leave the vectors of R_2 invariant. Thus there would exist vectors that could not be translated into a bounded set in R^{r-1} by translations $\mathfrak{l}(\varepsilon)$. This would, however, contradict our lemma 2.

Thus it has been shown that there exist $r-1$ units ε_i, such that for every unit ε Eq. (11) holds with rational integral s_i and $\varrho = r-1$. Then,

$$\varepsilon = \zeta \varepsilon_1^{s_1} \cdots \varepsilon_{r-1}^{s_{r-1}}$$

is satisfied with a unit ζ which, along with all its conjugates ζ_ν, has absolute value 1. This property is also held by all powers of ζ. By Lemma 3 they all belong to a bounded subset of $\mathfrak{N}$, and then by Lemma 1 there can only be a finite number of them. Thus ζ is a root of unity, and the proof is completed.

4. The Regulator

Choosing a system of base units ε_i we define the absolute value of the determinant

$$R = 2^\varrho \left\| \begin{matrix} \log|\varepsilon_{1,1}| & \cdots & \log|\varepsilon_{1,r-1}| \\ \cdot & \cdots & \cdot \\ \log|\varepsilon_{r-1,1}| & \cdots & \log|\varepsilon_{r-1,r-1}| \end{matrix} \right\| \tag{12}$$

where

$$\varrho = \begin{cases} 0 & \text{for} \quad r_2 = 0 \\ r_2 - 1 & \text{for} \quad r_2 > 0 \end{cases}$$

and ε_{ij} denotes the jth conjugate of ε_i, numbered by the convention of lemma 3 above, as the *regulator* of K. The fact that it is independent of the particular choice of base units or the numbering of the conjugates is almost self-evident.

§2. Quadratic Number Fields and Cyclotomic Fields

1. Quadratic Number Fields

Quadratic number fields are discussed in detail in numerous elementary and algebraic number theory texts, so that we may limit ourselves to only the simplest facts, considering them primarily as exercises to the theory developed thus far.

A quadratic number field arises when the square root of a nonsquare $v \in Q$ is adjoined to Q. There is no loss of generality in assuming v to be integral and free of square divisors. *A basis of* $\mathfrak{J}$ *is given by* $\{1, \frac{1}{2}(1+\sqrt{v})\}$ *if* $v \equiv 1 \bmod 4$ and $\{1, \sqrt{v}\}$ *otherwise.*

Proof. Let $\alpha = a + b\sqrt{v} \in \mathfrak{J}$, with $a, b \in Q$. By II,§1,1 we know that $\alpha \in \mathfrak{J}$ if and only if the trace $s(\alpha) = 2a$ and the norm $n(\alpha) = a^2 - vb^2$ are contained in $\mathfrak{i} = Z$. It follows that either a and b are contained in Z, or that $a \equiv b \equiv \frac{1}{2}$ mod 1 and $(2a)^2 - (2b)^2 v \equiv 0 \bmod 4$; this latter case is thus only possible when $v \equiv 1 \bmod 4$. The totality of the α is represented by the bases given. ▼

The discriminant of II, §4,(12) *is* $\mathfrak{D}_{K/Q} = \mathfrak{i}D$, *where*

$$D = D(\iota_i) = \begin{cases} v & \text{for} \quad v \equiv 1 \bmod 4, \\ 4v & \text{for} \quad v \not\equiv 1 \bmod 4. \end{cases} \tag{1}$$

Decomposition of the rational prime ideals (prime numbers) $\mathfrak{I}p$:

(a) If $(D/p) = -1$ (Legendre symbol), then $\mathfrak{I}p$ remains a prime ideal.

(b) If $(D/p) = 1$, then $\mathfrak{I}p = \mathfrak{p}_1\mathfrak{p}_2$, where $\mathfrak{p}_1, \mathfrak{p}_2$ have the bases $\{p, \frac{1}{2}(x + \sqrt{v})\}$, $\{p, \frac{1}{2}(x - \sqrt{v})\}$ respectively, or $\{p, x + \sqrt{v}\}$, $\{p, x - \sqrt{v}\}$, respectively, depending upon whether $v \equiv 1 \bmod 4$ or $v \not\equiv 1 \bmod 4$. Here x is a solution of $x^2 \equiv v \bmod p$ if $p \neq 2$ and $x^2 \equiv v \bmod 8$ if $p = 2$.

(c) If $(D/p) = 0$, then $\mathfrak{I}p = \mathfrak{p}^2$, where $\mathfrak{p}$ has the basis $\{p, \frac{1}{2}(p + \sqrt{v})\}$ if $v \equiv 1 \bmod 4$, $\{p, \sqrt{v}\}$ if $\pm p \equiv v \not\equiv 1 \bmod 4$, and $\{2, 1 + \sqrt{v}\}$ in the final case that $v \equiv 3 \bmod 4$ and $p = 2$.

Proof. Let the decomposition be II, §3,(8), with $e_1 f_1 + \cdots + e_r f_r = 2$ because of II, §3,(14). Three cases remain:

(a) $r = 1,\ e = 1,\ f = 2$; (b) $r = 2,\ e_i = f_i = 1$; (c) $r = 1,\ e = 2,\ f = 1$.

In case (a), $\mathfrak{I}p$ is a prime ideal, as $n(\mathfrak{I}p) = p^2$. In case (c), $\mathfrak{I}p = \mathfrak{p}^2$, and the Dedekind discriminant theorem gives $(D/p) = 0$. Conversely, $(D/p) = 0$ implies the case (c). For the case (b), certainly $\mathfrak{I}p = \mathfrak{p}_1\mathfrak{p}_2$. Now, if $\alpha = \frac{1}{2}(a + b\sqrt{v})$ (with $a, b \in Z$) is not divisible by p but exactly once by $\mathfrak{p}_1$, then its norm $n(\alpha) = \frac{1}{4}(a^2 - b^2 v)$ must be divisible by p exactly once. Thus a and b are relatively prime to p if p is odd, and thus v, and with it D, are squares $\bmod\, p$; for $v \not\equiv 1 \bmod 4$, replace $\frac{1}{2}a, \frac{1}{2}b$ in this argument by a, b. For $v \equiv 1 \bmod 4$ and $p = 2$, v and thus D are even squares mod 8. Thus then, $(D/p) = 1$. Conversely, the existence of an $\alpha = \frac{1}{2}(x + \sqrt{v})$ or $= x + \sqrt{v} \in \mathfrak{I}$, with norm divisible by p, follows from $(D/p) = 1$. Then α is divisible by a proper divisor of $\mathfrak{I}p$, and case (a) cannot apply, nor can (c), for then we would have $(D/p) = 0$. All has been proven. ▼

Exercise. Confirm the equations $\mathfrak{I}p = \mathfrak{p}_1\mathfrak{p}_2$ and $\mathfrak{I}p = \mathfrak{p}^2$ by carrying out the product on the right.

The units:

(a) $v < 0$. The number r from the Dirichlet unit theorem is 1. Thus all units are roots of unity. Save ± 1, the only such roots which satisfy a quadratic equation are those of order 3 and 4 in the fields with discriminants -3 and -4.

(b) $v > 0$. The number r is 2. No roots of unity can occur, aside from ± 1. By the Dirichlet unit theorem, all units are powers of a single base unit ε. It is even uniquely determined if one requires $\varepsilon > 1$. For small values of v the most rapid way of finding it is by trial and error, seeking the smallest unit $\varepsilon > 1$.

2. Special Cyclotomic Fields

We proceed to the consideration of the pth cyclotomic field $K = Q(\zeta)$, where ζ is a primitive pth root of unity, and p an odd prime. It is well known that $[K: Q] = p - 1$. K is a Galois extension of Q, the group being the multiplicative group of residue classes mod p prime to p. It is cyclic.

A basis of $\mathfrak{J}$ *is* $1, \zeta, \ldots, \zeta^{p-2}$. *The discriminant of* II, §4,(12) *is* $\mathfrak{D}_{K/Q} = \mathfrak{i}D$, *where*

$$D = D(1, \zeta, \ldots, \zeta^{p-2}) = (-1)^{\frac{1}{2}p(p-1)} p^{p-2}. \tag{2}$$

Proof. The traces can be found directly from the defining equations:

$$s(1) = p - 1, \qquad s(\zeta^i) = -1 \qquad \text{for} \quad i \not\equiv 0 \bmod p.$$

Thus,

$$D(1, \zeta, \ldots, \zeta^{p-2})$$
$$= \begin{vmatrix} p-1 & -1 & -1 & \cdots & -1 \\ -1 & -1 & -1 & \cdots & -1 \\ -1 & -1 & -1 & \cdots & p-1 \\ \cdot & \cdot & \cdot & \cdots & \cdot \\ -1 & -1 & p-1 & \cdots & -1 \end{vmatrix} = \begin{vmatrix} p & 0 & 0 & \cdots & 0 \\ -1 & -1 & -1 & \cdots & -1 \\ 0 & 0 & 0 & \cdots & p \\ \cdot & \cdot & \cdot & \cdots & \cdot \\ 0 & 0 & p & \cdots & 0 \end{vmatrix},$$

so that

$$D(1, \zeta, \ldots, \zeta^{p-2}) = (-1)^{\frac{1}{2}p(p-1)} p^{p-2}.$$

The ζ^i are contained in $\mathfrak{J}$, so that $\mathfrak{D}_{K/Q}$ is a divisor of $D(1, \zeta, \ldots)$. The element $\lambda = \zeta - 1$ satisfies the equation

$$f(x) = \frac{(x+1)^p - 1}{x} = x^{p-1} + px + \cdots + p = 0.$$

Hence, $n(\lambda) = p$, and $\mathfrak{J}\lambda$ is a prime ideal. Furthermore, the quotients

$$(\zeta^i - 1)/(\zeta - 1) = \zeta^{i-1} + \cdots + 1$$

lie in $\mathfrak{J}$, so that all the prime ideals $\mathfrak{J}(\zeta^i - 1)$ conjugate to $\mathfrak{J}\lambda$ are identical with $\mathfrak{J}\lambda$, and

$$n(\mathfrak{J}\lambda) = (\mathfrak{J}\lambda)^{p-1} = \mathfrak{J}p.$$

Thus, $\mathfrak{p} = \mathfrak{J}\lambda$ ramifies with index $e = p - 1$. The Dedekind discriminant theorem now states that $\mathfrak{D}_{K/Q}$ is divisible by p exactly $p - 2$ times. This proves (2). ▼

Decomposition of the rational prime ideals (prime numbers) $\mathfrak{J}q$: The decomposition of $q = p$ has been given. Let $q \neq p$. Using the Dedekind discriminant theorem, no ramification can occur, so that

$$\mathfrak{J}q = \mathfrak{q}_1 \cdots \mathfrak{q}_r$$

with r different prime ideals $\mathfrak{q}_i$ of degree f, satisfying

$$rf = p - 1.$$

We maintain that f is the smallest natural number for which

$$q^f - 1 \equiv 0 \bmod p \tag{3}$$

holds.

Proof. Let $\mathfrak{q}$ be one of the $\mathfrak{q}_i$. The residue class field $\mathfrak{J}/\mathfrak{q}$ is an extension of $\mathfrak{i}/q$ of degree f, and thus has q^f elements. Its multiplicative group has $q^f - 1$ elements, so that every $\iota \in \mathfrak{J}$ prime to $\mathfrak{q}$ satisfies $\iota^{q^f-1} \equiv 1 \bmod \mathfrak{q}$. In particular,

$$\zeta^{q^f-1} \equiv 1 \bmod \mathfrak{q}.$$

But then, were $q^f - 1 \equiv 0 \bmod p$ not satisfied, all roots of unity would be congruent 1 mod $\mathfrak{q}$. Then $\mathfrak{q}$ would divide $\lambda = \zeta - 1$, which is impossible as $n(\lambda) = p$.

Conversely, let $q^{f'} - 1 \equiv 0 \bmod p$ with a minimal $f' > 0$. Let

$$(q^f - 1) - (q^{tf'} - 1) = q^{tf'}(q^{f-tf'} - 1) \equiv q^{f-tf'} - 1 \equiv 0 \bmod p.$$

From this it is not difficult to see that the minimal f' is a divisor of f.

It remains to be shown that, if $q^{f'} - 1 \equiv 0 \bmod p$ with minimal f', then $f = f'$. Now, let f' be a proper divisor of f. It is well known from the theory of finite fields that the residue class field $\mathfrak{J}/\mathfrak{q}$ is generated by adjunction of a primitive $(q^f - 1)$th root of unity to the prime field $\mathfrak{i}/q$. But, by our assumption, ζ is a primitive $(q^{f'} - 1)$th root of unity, and thus not a primitive $(q^f - 1)$th root. This contradicts the fact that $\mathfrak{J}/\mathfrak{q}$ is generated by adjunction of ζ to $\mathfrak{i}/q$. ▼

Comparison of the decomposition laws of this and the last sections yields a proof of the quadratic reciprocity law. It is well known that (cf. §2,1 of the Appendix to Chapter I)

$$\left(\sum_{r=0}^{p-1} \zeta^{r^2}\right)^2 = (-1)^{\frac{1}{2}(p-1)} p = \varepsilon p.$$

Now, $K_1 = Q(\sqrt{\varepsilon p})$ is a quadratic subfield of K. If $(\varepsilon p/q) = 1$, then q decomposes to two prime ideals in K_1. These are, in turn, decomposed into prime ideals of some degree f, which divides $[K:K_1] = \frac{1}{2}(p-1)$. Thus by (3) we have

$$q^{\frac{1}{2}(p-1)} \equiv 1 \bmod p.$$

The consequence is that the residue class of $q \bmod p$ is a square in the multiplicative residue class group. This means that $(q/p) = 1$. The reciprocity law can be completely derived from this with the use of the first complementary theorem.

CHAPTER III

Algebraic Functions and Differentials

The theory of algebraic functions can be approached in three ways. The first, which we followed in Chapter II, erects the foundation of the theory of algebraic numbers simultaneously. The second proceeds along the framework of classical function theory, and can therefore only deal with fields of characteristic 0. We will develop this method in the next two paragraphs. This is done for various reasons.

First, it is natural to connect our theory with the framework of classical function theory for the general educational value of doing so.

Second, the simplest fields of characteristic $p > 0$ are generated from the rational integers by residue class formation with a prime p. It is possible and often advantageous to generate more general fields of characteristic p, and in particular algebraic function fields by similar residue formation. This process, which shall be carried out in §6, will be explicitly based on this second approach.

Third, this approach is characterized by an intuitive clarity which serves to illustrate the abstract concepts.

The third approach comes forth from the second; in principle it consists of the same formal arguments, but abstracted from the special assumptions of the classical case. It is the quickest, beyond doubt. We intentionally postpone it until the end of §2, because we believe the student can only appreciate the novelty fully when he has the means to associate concrete formations with the abstract concepts.

A study of this book can be started with Chapter III, as was already mentioned in the preface. At the beginning of §2 Sections I, §1,1 and 3 as well as II, §1,1 and 2 must be inserted, though. Further, I, §3 is needed for the proof of the Riemann–Roch theorem in §3.

The central problem of the theory is developed and solved in §3. It is the question as to the existence and multiplicity of functions with given zeros and poles. What remains then is the development of a differential and integral calculus. Even if only the classical case is to be considered, it is profitable here to develop first a formal calculus. In other words, differential and integral calculus can be pursued without the continuity concept. Not until Chapter IV will that concept occur.

§1. Power Series Expansions of Algebraic Functions

1. The Field of Power Series

First let k_0 be an arbitrary field; later we shall have to assume it to be of characteristic 0. By a *power series* with coefficients $a_\mu \in k_0$ a formal expression

$$f(x) = 0 \qquad \text{or} \qquad f(x) = \sum_{\mu=\nu}^{\infty} a_\mu x^\mu, \qquad a_\nu \neq 0, \tag{1}$$

is meant. One way of defining calculation with power series is to state the rules for determining the coefficients of the sum and product of two power series, as is done for polynomials. Alternatively, the power series can be taken as limits of finite sums in the sense of the valuation of the rational function field $k_0(x)$ by which a polynomial $f(x) \neq 0$ of (1) has value $|f(x)| = \sigma^\nu$ with some constant σ between 0 and 1, fixed once and for all.

$f(x)$ is said to have a *zero of order* ν at $x = 0$ if $\nu > 0$; if $\nu < 0$ then $f(x)$ has a *pole of order* $|\nu|$ at $x = 0$. In that case

$$\sum_{\mu=\nu}^{-1} a_\mu x^\mu \tag{2}$$

is called the *principal part* of the pole.

If k_0 is contained in the field C of complex numbers the *radius of convergence* is defined for the power series. A concept formally related to this is the following: Let the ideal theory of II,§1 hold in k_0. The largest ideal $\mathfrak{a}$ such that $\mathfrak{a}^\mu a_\mu$ is almost always an integral ideal is called the *arithmetic convergence radius*. For example, the series $\sum x^\mu, \sum 1/\mu! x^\mu$ have the arithmetic convergence radii (1) and (0).

In the following we shall often make statements of the sort: *A power series has a (respective) convergence radius* $\neq 0$. This means that if $k_0 \subseteq C$ then the convergence radius is positive, and if the ideal theory holds for k_0 then the arithmetic convergence radius is not the zero ideal.

All power series with coefficients in k_0 form a field. The same is true of those power series with nonzero convergence radius.

The simple steps necessary for the proof need not be carried out here. We will only show that the reciprocal power series $g(x)$ of a power series $f(x)$ with radius of convergence $\neq 0$ also has radius of convergence $\neq 0$. Evidently it suffices to do this for $f(x) = 1 + c_1 x + c_2 x^2 + \cdots$, and verification consists of the application with $g(x) = 1 + y$ and $cx + \varphi(x,y) = (1 - f(x))(1 + y)$ of the lemma which follows. This lemma, due to Cauchy, not only will serve us on several later occasions, but its underlying principle has even proved useful in other branches of mathematics.

Lemma. *Let*

$$\varphi(x, y) = \sum_{i+j>1} c_{ij} x^i y^j$$

be a power series, all of whose terms have degree $\geqq 2$. *The equation*

$$y = cx + \varphi(x, y) \tag{3}$$

then has, as its solution, a power series

$$y = \sum_{\mu=1}^{\infty} c_\mu x^\mu \qquad \text{with} \quad c_1 = c. \tag{4}$$

If the ideal theory holds in k_0 *and there exists a nonzero* $a \in k_0$ *such that* $a^{i+j-1}c_{ij}$ *are integral, then* (4) *has an arithmetic radius of convergence* $\neq 0$. *If* $\varphi(x, y)$ *has a positive radius of convergence, so does* (4).

Proof. Equation (3) is solved by iteration. Set

$$y_1 = cx, \ldots, y_{\mu+1} = cx + \varphi(x, y_\mu).$$

Here y_μ and $y_{\mu+1}$ do not differ in their first μ terms. Indeed, by hypothesis, this is so for $\mu = 1$, and assume it to be so for $\mu - 1$. The inductive hypothesis thus states that $y_\mu = y_{\mu-1} + x^\mu z_\mu$ with a power series z_μ without negative exponents. Iteration yields

$$\begin{aligned} y_{\mu+1} &= cx + \varphi(x, y_{\mu-1} + x^\mu z_\mu) \\ &= cx + \varphi(x, y_{\mu-1}) + x^{\mu+1} z_{\mu+1} = y_\mu + x^{\mu+1} z_{\mu+1}, \end{aligned}$$

where the hypothesis concerning $\varphi(x, y)$ assures that $z_{\mu+1}$ also has no negative exponents.

Hence the formal limit $y = \lim y_\mu$ exists; it solves (3) and has the property (4).

To demonstrate a nonzero arithmetic radius of convergence under the given assumptions, replace x and y by ax and ay. Then (3) becomes the corresponding equation with $a^{i+j-1}c_{ij}$ in place of c_{ij}. Thus there is no loss of generality in assuming the c_{ij} to be integral. In the process of the iteration it turns out that the arithmetic radii of convergence of all the series y_μ, and therefore also that of y, are divisors of the denominator of c.

Only the question of the positive radius of convergence remains. Let $\varphi(x, y)$ converge absolutely for $|x|, |y| < \gamma$. Then, with some further constant γ_0, we have

$$|c| < \gamma_0 \gamma^{-1} = C, \qquad |c_{ij}| < \gamma_0 \gamma^{-i-j} = C_{ij}. \tag{5}$$

From their calculation above the c_μ of (4) turn out as polynomials $P_\mu(c, c_{ij})$ with positive integral coefficients. Replacing c and the c_{ij} by their majorants (5) yields the majorants $C_\mu = P_\mu(C, C_{ij})$ of the c_μ. The C_μ are coefficients of the power series

$$z = \sum_{\mu=1}^{\infty} C_\mu x^\mu, \tag{6}$$

which solves the equation

$$z = Cx + \sum_{i,j} C_{ij} x^i z^j.$$

Using (5) this equation becomes

$$z = \gamma_0 \left(\frac{\gamma^2}{(\gamma - x)(\gamma - z)} - 1 - \frac{z}{\gamma} \right).$$

This is a quadratic equation with solution

$$z = \frac{\gamma^2}{2(\gamma + \gamma_0)} \left[1 - \left(1 - \frac{4\gamma_0(\gamma + \gamma_0)}{\gamma^3} \frac{x}{\gamma - x} \right)^{1/2} \right].$$

It can be expanded in the power series (6) and it is from this nature of (6) that we know it has a positive radius of convergence. The power series (4) then has a radius of convergence at least as large. ▼

Exercise. Show similarly that if $f(x)$ and $q(x)$ have radii of convergence $\neq 0$ and $q(0) = 0$, then $f(q(x))$ also has a nonzero radius of convergence.

2. Divisibility, Rearranging of Power Series

Power series (1) with first exponent $\nu \geqq 0$ are called *integral* or holomorphic. They form an integral domain $\mathfrak{J}$. The power series with first exponents $\nu \geqq \nu_0$ form an $\mathfrak{J}$-ideal. The ideal for $\nu_0 = 1$ is a prime ideal $\mathfrak{Q}$, and the elements of k_0 are a complete system of representatives of the residue classes of $\mathfrak{J}$ mod $\mathfrak{Q}$. Thus the residue class ring $\mathfrak{J}/\mathfrak{Q}$ is isomorphic to k_0, the proofs being obvious.

A power series $q(x) = q$ with first exponent $\nu = 1$ is called a *prime element.* For simplicity sake we assume that the coefficient of x in $q(x)$ is 1. Letting $f(x)$ be as in (1), we have

$$f_1(x) = f(x) - a_\nu q(x)^\nu = a_{1,\nu+1} x^{\nu+1} + \cdots$$

and further

$$f_2(x) = f_1(x) - a_{1,\nu+1} q(x)^{\nu+1} = a_{2,\nu+2} x^{\nu+2} + \cdots.$$

This process finally leads to the identity

$$f(x) = a_\nu q(x)^\nu + a_{1,\nu+1} q(x)^{\nu+1} + a_{2,\nu+2} q(x)^{\nu+2} + \cdots, \tag{7}$$

which is still to be considered purely formal, that is, in the sense that the coefficients coincide. In the next section we will show that in the special case $f(x) = x$ this series has a radius of convergence $\neq 0$ if this is so for $q(x)$. Then, by the exercise at the end of the last section, this holds for the series (7) in general.

We will illuminate the theory of power series briefly in the light of the results of Chapter II.† Let $\mathfrak{J}$ be an integral domain with quotient field k. Let

† The reader starting with Chapter III may skip these remarks.

there exist a single prime $\mathfrak{J}$-ideal $\mathfrak{Q}$, which is a principal ideal, $\mathfrak{Q} = \mathfrak{J}q$. Further, let the residue classes of $\mathfrak{J}$ mod $\mathfrak{Q}$ all be represented by elements of a field k_0 contained in $\mathfrak{J}$.

Any $f \in \mathfrak{J}$ can then be expanded in a power series (7) with coefficients in k_0: let a_0 be that uniquely determined element of k_0 which satisfies the congruence $f \equiv a_0 \bmod \mathfrak{Q}$. Then $f_1 = (f - a_0)q^{-1} \in \mathfrak{J}$; and say $f_1 \equiv a_1 \bmod \mathfrak{Q}$. Again let $f_2 = (f_1 - a_1)q^{-1} \equiv a_2 \bmod \mathfrak{Q}$, etc. Call $a_0 + a_1 q + \ldots$ the power series expansion of f. If f is not contained in $\mathfrak{J}$, then with some exponent $-\nu$ certainly $fq^{-\nu} \in \mathfrak{Q}$, and this element can be treated as described. Then f is associated with the power series beginning with $a_\nu q^\nu$.

It is now easy to show that the correspondence between the elements $f \in k$ and their power series is an isomorphism.

The situation assumed always occurs if k is an algebraic function field, the constant field k_0 being assumed algebraically closed. For $\mathfrak{J}$ choose the integral domain of functions of k whose denominator divisor is not divisible by some prime divisor $\mathfrak{q}$. (Compare this with the example at the end of II, §2,1.) The prime ideal $\mathfrak{Q}$ is the totality of those functions whose numerator divisor is divisible by $\mathfrak{q}$. As k_0 is algebraically closed all prime divisors, and in particular $\mathfrak{q}$, have degree 1, so that all residue classes of $\mathfrak{J}$ mod $\mathfrak{Q}$ are represented by elements of k_0. Therefore every algebraic function can be expanded with respect to a prime element q for an arbitrary place $\mathfrak{q}$. From this point of view divisibility comes first, and power series are a consequence. The foundation of the theory which we are now developing, though, starts with power series and bases divisibility upon them.

3. Inversion of a Power Series

We start with the following lemma:

Lemma. *Let*

$$f(x) = 1 + a_1 x + a_2 x^2 + \cdots = 1 + \varphi(x) \tag{8}$$

and $e > 1$ be a natural number not divisible by the characteristic of k_0. There then exists exactly one power series

$$g(x) = 1 + c_1 x + c_2 x^2 + \cdots \tag{9}$$

satisfying the equation $g(x)^e = f(x)$. If $f(x)$ has a convergence radius $\neq 0$, then so does $g(x)$.

Proof. In (8) let $a_1 = \cdots = a_{r-1} = 0$, $a_r \neq 0$, and put $g(x) = 1 + x^{r-1}y$. Then our equation

$$g(x)^e - f(x) = ex^{r-1}(y - (a_r/e)x - \cdots) = 0$$

is a special case of (3), and our lemma is implied in that of III, §1,1. ▼

With this preparation we can show:

Let e be a natural number not divisible by the field characteristic and

$$y = f(x) = a_e x^e + a_{e+1} x^{e+1} + \cdots, \qquad a_e \neq 0. \tag{10}$$

There then exists a power series

$$x = \varphi(\eta) = c_1\eta + c_2\eta^2 + \cdots, \qquad c_1 = (\sqrt[e]{a_e})^{-1}, \tag{11}$$

with coefficients in the field $k_0(\sqrt[e]{a_e})$ *which, substituted for x in* (10), *yields the identity* $y = \eta^e$. *After determination of an* eth *root of* a_e *all the coefficients of* (11) *are uniquely determined. If* (10) *has a nonzero radius of convergence, then so does* (11).

For the case $e = 1$ where y is a prime element, (11) is the expansion of x with respect to it, which was shown to exist in the last section. For this case the coefficients c_i of (11) can even be explicitly stated.† They are the coefficients of x^i in the power series expansions of

$$(x/f(x))^{i+1} \cdot df(x)/dx$$

Proof. Choose the value given in (11) for c_1 and set indeterminate coefficients in a power series for $\varphi(\eta)$. Substitution of (11) into (10) yields a power series in η for y, beginning with η^e and with

$$e a_e c_1^{e-1} c_\mu + p_\mu(c_1, \ldots, c_{\mu-1}; a_e, \ldots, a_{e+\mu-1})$$

as the coefficient of $\eta^{e+\mu-1}$, the p_μ being polynomials in the variables indicated. Set these coefficients from $\mu = 2$ onwards to zero; this gives recursion formulas for the c_μ. They are solvable because e is not divisible by the characteristic, and the solution is unique.

No generality is lost in assuming $a_e = c_1 = 1$ for the convergence proof. Using the above lemma the right side of (10) may be written

$$y = g(x)^e = (x + a_2'x^2 + \cdots)^e,$$

where $g(x)$ has a radius of convergence $\neq 0$. The series (11) must now be found so as to fulfill the identity $\eta = g(x)$. This means that the convergence proof is only necessary for the case $e = 1$. Now write (10) as

$$y = x + \varphi(x)$$

which is a special case of (3), and which was shown in III,§1,1 to be solvable with a power series of radius of convergence $\neq 0$. ▼

---

† M. Deuring, *Eine Bemerkung über die Bürmann–Lagrangesche Reihe*, Nachr. Akad. Wiss. Göttingen Math.-Phys. Kl., 33–35 (1946).

4. Algebraic Functions; Regular Places

Let K be a finite separable extension of the rational function field $k = k_0(x)$. (For k_0 of characteristic 0 the assumption of separability is trivial.) The theorem of the primitive element states that $K = k(y)$ with some $y \in K$ satisfying the irreducible equation

$$f(y, x) = y^n + c_1(x)y^{n-1} + \cdots + c_n(x) = 0 \tag{12}$$

with coefficients $c_i(x) \in k$. Its degree is $n = [K : k]$.

If $h(x)$ is the common denominator of the rational functions $c_i(x)$, then $y_1 = h(x)y$ satisfies the equation

$$y_1{}^n + h(x)c_1(x)y_1^{n-1} + h(x)^2c_2(x)y_1^{n-2} + \cdots + h(x)^n c_n(x) = 0,$$

with coefficients that are polynomials in x. Replace y by y_1 so that now y is a primitive element of K satisfying Eq. (12) with polynomials $c_i(x)$.

We want to expand all the functions of K in power series in $x - \xi$ or x^{-1}, where ξ is an arbitrary element of k_0. We state beforehand that this is not always possible, but it is in a great majority of the cases. Clearly it suffices to so expand the single function y. For x trivially has such an expansion, so that then §1,1 assures the same for all rational functions in x and y, that is for all functions of K.

In a field extension K_1 of k the polynomial (12) decomposes to factors of the first degree

$$f(y, x) = \prod_{i=1}^{n} (y - w_i)$$

whose zeros, by our assumption, do not coincide. The discriminant

$$D(f) = \prod_{i<j} (w_i - w_j)^2 = d(x)$$

can, as is familiar, be calculated as a polynomial in the $c_i(x)$, so that it is itself a polynomial in x not identically 0. This polynomial has, at most, a finite number of zeros $x = \xi \in k_0$, called the *critical places*† of Eq. (12), other $\xi \in k_0$ being called *regular places*. At a regular place the equation $f(y, \xi) = 0$ has exactly n distinct zeros $y = \eta_i$ in an extension of k_0, and we maintain:

For a regular place ξ there exists, to every zero $y = \eta_i$ of $f(y, \xi)$, exactly one power series

---

† Tr. note: The term "place" here, for the German "Stelle," is not the place defined in Chapter II, but rather simply means point. As the development of the subject proceeds it will turn out, though, that the two senses of "place" coincide in effect and require the same word, thus justifying the terminology of Chapter II anew.

$$y = \sum_{\mu=0}^{\infty} \eta_{i\mu}(x - \xi)^{\mu} \qquad \text{with} \quad \eta_{i0} = \eta_i, \tag{13}$$

satisfying (12) (*identically in x*). *The coefficients* $\eta_{i\mu}$ *are elements of the field* $k_0(\eta_i)$.

The respective radii of convergence of these series are not zero.

Proof. To (12) apply the variable substitution

$$x \to x' = x - \xi, \qquad y \to y' = y - \eta_i \tag{14}$$

leading to the equation $f'(y', x') = 0$ which has the simple zero $y' = 0$ for $x' = 0$. This means that the last two coefficients of $f'(y', x')$ have the properties $c_n'(0) = 0$, $c'_{n-1}(0) \neq 0$. To be proved is the existence of a power series $y'(x')$ with first exponent $\nu > 0$ which solves $f'(y', x') = 0$. To simplify the notation we omit the primes and consider Eq. (12) under the assumption

$$c_{n-1}(0) \neq 0, \qquad c_n(0) = 0.$$

It can now be written as

$$y = \frac{1}{c_{n-1}(0)}(c_{10}x + \varphi(x, y)), \qquad \varphi(x, y) = \sum_{i+j>1} c_{ij}x^i y^j$$

which is of the form (3). Thus the solvability now follows from the lemma in III, §1,1, as does the existence of a nonzero radius of convergence, provided k_0 satisfies the necessary hypothesis.

Remember that we started with the substitution (14), we see that the coefficients of the equation solved belong to the field $k_0(\eta_i)$. Following the solution of the equation in the proof of the lemma it is seen that the coefficients of the solution (13) also belong to that field. ▼

5. Continuation; Critical Places

We must now assume that the field of constants has characteristic 0. Let ξ be a critical place. Again substitute as in (14), so that $c_n(0) = 0$. In general, though, we now no longer have $c_{n-1}(0) = 0$, but can only prove the following: *There exists a natural number e and a power series with coefficients in a finite field extension of* k_0,

$$y = \sum_{\mu=1}^{\infty} \eta_{\mu}(\sqrt[e]{x})^{\mu}, \tag{15}$$

which solves (12). *It has a convergence radius* $\neq 0$

Proof. Replace (12) by the more general equation

$$f(y, x) = \sum_{r=h}^{H} f_r(y, x) = 0 \tag{16}$$

which can be decomposed into its homogeneous parts

$$f_r(y, x) = \sum_{i=0}^{r} c_{ri} y^i x^{r-i}$$

and about which we assume nothing further than that it contain y and have no multiple solutions .The sum (16) is even permitted to be infinite. Now let h be the smallest degree for which some $f_r(y, x)$ is nonzero. This h is called the *subdegree* of $f(y, x)$, and the existence of the solution (15) is proved by induction on it.

One first remark will be useful. Assume a power series

$$x = a_e(\sqrt[e']{y})^e + a_{e+1}(\sqrt[e']{y})^{e+1} + \cdots, \qquad a_e \neq 0,$$

satisfying (16) has been found. The theorem of §1,3 then gives the inverted series

$$\sqrt[e']{y} = b_1\sqrt[e]{x} + b_2(\sqrt[e]{x})^2 + \cdots,$$

and

$$y = c_{e'}(\sqrt[e]{x})^{e'} + c_{e'+1}(\sqrt[e]{x})^{e'+1} + \cdots$$

is a solution of (16). With these manipulations a respective radius of convergence persists, so that we may interchange x and y in the course of the proof.

This immediately settles the beginning of induction. By the last section $h = 1$ implies that a power series expansion exists either for y in x or for x in y, with first exponent $v > 0$. We may therefore assume the contention to hold for all smaller subdegrees than h, and let $h > 1$.

By interchanging x and y if necessary, we may assume that

$$f_h(y, x) \neq ax^h.$$

In (16) set $y = xy_1'$ and divide by x^h, arriving at

$$\begin{aligned} f_1'(y_1', x) &= x^{-h}f(xy_1', x) \\ &= f_h'(y_1') + xf_{h+1}'(y_1') + \cdots + x^{n-h}f_n'(y_1') + \cdots = 0, \end{aligned}$$

with

$$f_r'(y_1') = x^{-r}f_r(xy_1', x),$$

polynomials in y_1' alone. Our hypothesis assures that $f_h'(y_1')$ is not a constant, so that this polynomial must have a zero ζ_1 in some extension of k_0. Set

$$y_1' = y_1 + \zeta_1$$

so that in the resulting equation,

$$f_1'(y_1 + \zeta_1, x) = f_1(y_1, x) = f_h(y_1) + xf_{h+1}(y_1) + \cdots + x^{n-h}f_n(y_1) + \cdots = 0,$$

$f_h(0) = 0$ holds. Now, the degree of $f_h(y_1)$ is at most h, so that the subdegree h_1 of $f_1(y_1, x)$ must lie between 1 and h. If, indeed, we have attained $h_1 < h$, then we know by inductive hypothesis that y is a power series in $(x)^{1/e}$ with some natural number e, starting with $\zeta_1 x$. This would complete the proof.

If, on the other hand, we still have $h_1 = h$, we repeat the argument. It must be shown that the subdegree h, if $\geqq 2$, cannot remain unchanged infinitely often. If $h_1 = h$, then $f_h(y_1)$ has a zero of multiplicity h at the place $y_1 = 0$. This implies (writing partial derivation with respect to the first variable as an overbar)

$$f_1(0, x) \equiv 0 \bmod x, \qquad \bar{f}_1(0, x) \equiv 0 \bmod x.$$

Because

$$f_1(y_1, x) = x^{-h} f(x(y_1 + \zeta_1), x), \qquad \bar{f}_1(y_1, x) = x^{1-h} \bar{f}(x(y_1 + \zeta_1), x), \tag{17}$$

it follows that

$$f(\zeta_1 x, x) \equiv 0 \bmod x^{h+1}, \qquad \bar{f}(\zeta_1 x, x) \equiv 0 \bmod x^h.$$

As we have assumed that the subdegree does not decrease at the next step either, etc., these same congruences also hold for $f_1(y_1, x)$ and the polynomials obtained from them in like manner. We maintain that the congruences

$$\begin{aligned} f(\zeta_1 x + \cdots + \zeta_q x^q, x) &\equiv 0 \bmod x^{qh+1}, \\ \bar{f}(\zeta_1 x + \cdots + \zeta_q x^q, x) &\equiv 0 \bmod x^{qh}, \end{aligned} \tag{18}$$

hold for $q = 1, 2, \ldots$ and also for the polynomials $f_1(y_1, x)$, etc., formed as described above. We prove this by induction on q, the first step being already taken. Our inductive assumption, which we formulate for f_1 in place of f, is $f_1(\zeta_2 x + \cdots + \zeta_q x^{q-1}, x) \equiv 0 \bmod x^{(q-1)h+1}$, $\bar{f}_1(\zeta_2 x + \cdots + \zeta_q x^{q-1}, x) \equiv 0 \bmod x^{(q-1)h}$. Using (17) this gives (18).

But, the congruences (18) would show that the infinite power series $\zeta_1 x + \zeta_2 x^2 + \cdots$ is a double root of (16), which would contradict the hypothesis that (16) is irreducible. ▼

6. Puiseux's Theorem

Let K be an algebraic extension of the rational function field $k = k_0(x)$ of degree $n = [K:k]$, where k_0 is the exact constant field of K and of characteristic 0. Then, with every element ξ of k_0 there are associated $r = r(\xi) \leqq n$ natural numbers $e_\varrho = e_\varrho(\xi)$ whose sum is

$$e_1 + \cdots + e_r = n; \tag{19}$$

similar numbers $e_\varrho(\infty)$ are associated with the symbol $\xi = \infty$. Save for a finite number of exceptions, $r(\xi) = n$ and $e_\varrho(\xi) = 1$ always hold.

These numbers have the following meaning: Setting

$$z_\xi = x - \xi, \qquad z_\infty = 1/x, \tag{20}$$

the irreducible equation (12) *satisfied by an arbitrary function y of K over k has for solutions the $r = r(\xi)$ power series*

$$y_\varrho = \sum_{\mu=\nu_\varrho}^{\infty} a_{\varrho\mu}(\sqrt[e_\varrho(\xi)]{z_\xi})^\mu, \qquad a_{\varrho\nu_\varrho} \neq 0, \qquad \varrho = 1, 2, \ldots, r(\xi). \tag{21}$$

*With a primitive e_ϱ*th *root of unity ζ form*

$$y_{\varrho i} = \sum_\mu a_{\varrho\mu}\zeta^{i\mu}(\sqrt[e_\varrho(\xi)]{z_\xi})^\mu, \qquad i = 0, \ldots, e_\varrho(\xi) - 1; \tag{22}$$

then the left side of (12) *is identical with*

$$f(y, x) = \prod_{i,\varrho} (y - y_{\varrho i}). \tag{23}$$

The coefficients $a_{\varrho\mu}$ are elements of a finite field extension k_1 of k_0, and their images under isomorphisms of k_1 give permutations of the $y_{\varrho i}$ in (23). *The power series* (21) *have respective radii of convergence* $\neq 0$.

There is not much left to prove. By §1,4 every $y \in K$ is of the form $y = h(x)^{-1}y_1$, where $h(x)$ is a polynomial and y_1 satisfies an equation (12) with polynomials $c_i(x)$. But, for such y_1 we have shown the existence of solutions (21) in the last two sections, and this then also holds for y by what was shown in §1,1.

As for the symbol $\xi = \infty$, the variable x is replaced there by $z_\infty = x^{-1}$. The place $z_\infty = 0$ can turn out to be regular or critical, leading to power series (21) with $e_\varrho(\infty) = 1$ or $e_\varrho(\infty) > 1$.

It has already been shown that the $a_{\varrho\mu}$ belong to a finite extension k_1 of k_0. Now, the fact that (21) solves (12) is expressed in algebraic equations in the $a_{\varrho\mu}$ with coefficients in k_0, which continue to hold under an automorphism. This means that automorphisms of k_1/k_0 transform solutions of (12) into solutions of (12).

Only Eqs. (19) and (23) remain to be proved, the first of these being an immediate consequence of the second. At regular places we found n differing solutions in §1,4, and, since there cannot be more, (23) is obvious for that case.

Now let ξ be a critical place. The partial products

$$\prod_{i=0}^{e_\varrho - 1} (y - y_{\varrho i})$$

of (23) are polynomials in y of degree e_ϱ. The coefficients are power series in $\sqrt[e_\varrho]{z_\xi}$, but because of the symmetry of their construction they remain invariant under multiplication of the variable by an e_ϱth root of unity. Hence, the coefficients are power series in z_ξ. The same is true of the right side of

(23), as well as for the quotient of both sides of (23). If, further, the quotient still depends on y, the process of the last two sections can be used to find further solutions which will turn out to be of the sort (21). (Note that in the last section we explicitly permitted (16) to be an infinite power series.) But, under the assumption that (21) and (22) are all the possible solutions, (23) must hold. ▼

Corollary. *If the coefficients of* (12) *can be expanded in integral power series at* $x = \xi$, *then all the power series* (21) *turn out integral.* The proof is contained in §1,5.

§2. Algebraic Function Fields

1. Divisors in Rational Function Fields

In §2,1–6 the constant field k_0 will be assumed to be algebraically closed and of characteristic 0.

As is familiar, every rational function $w \neq 0 \in k = k_0(x)$ can be decomposed to a product of powers of first degree factors

$$w = \gamma \prod_{\xi} (x - \xi)^{v_\xi(w)}, \qquad \gamma \neq 0 \in k_0, \tag{1}$$

where the ξ are a finite number of elements of k_0. With every element $\xi \in k_0$ we formally associate a *place*, without considering these to be in a space, the geometry of which we want to consider. It is true that in the classical case $k_0 = C$ the ξ are visualized to be in the complex number plane, which is even closed by a point ∞. This visualization is without consequence in the algebraic theory, though.

The exponent $v_\xi(w)$ of (1) is called the *order of* w *at the place* ξ. Now, w has an order at every place ξ, but as only a finite number of factors occur in (1) all but a finite number of these orders are 0. Finally, formally introduce the place $\xi = \infty$ where the order is defined to be

$$v_\infty(w) = -\sum_{\xi \neq \infty} v_\xi(w). \tag{2}$$

Note, incidentally, that $-v_\infty(w)$ is the degree of the rational function w.

With the places we further associate *prime divisors* $\mathfrak{p} = \mathfrak{z}_\xi$. Let these generate the free abelian group $\mathfrak{k}$, the elements of which are called *divisors*. By definition these are thus products

$$\mathfrak{a} = \prod_{\mathfrak{p}} \mathfrak{p}^{v_\mathfrak{p}(\mathfrak{a})} = \prod_{\xi} \mathfrak{z}_\xi^{v_\xi(\mathfrak{a})}, \tag{3}$$

with but a finite number of orders $v_\mathfrak{p}(\mathfrak{a}) \neq 0$. Add orders to multiply divisors;

the unit element of the group is then the *unit divisor*, written (1), whose order is $v_\mathfrak{p}(1) = 0$ for all $\mathfrak{p}$.

A divisor $\mathfrak{a}$ is called *integral* if all its orders $v_\mathfrak{p}(\mathfrak{a}) \geqq 0$. $\mathfrak{a}$ is said to be *divisible* by $\mathfrak{b}$ if $\mathfrak{a}\mathfrak{b}^{-1}$ is integral, or, what is the same, $v_\mathfrak{p}(\mathfrak{a}) \geqq v_\mathfrak{p}(\mathfrak{b})$ for all $\mathfrak{p}$.

With an element $w \neq 0 \in k$ we can associate a special divisor by setting

$$(w) = \prod_{\mathfrak{p}} \mathfrak{p}^{v_\mathfrak{p}(w)} = \prod_{\xi} \mathfrak{z}_\xi^{v_\xi(w)}, \tag{4}$$

with $v_\xi(w)$ as in (1) and $v_\infty(w)$ from (2). Note that now, as opposed to (1), the place $\xi = \infty$ occurs in the product. These divisors are called *principal divisors*. Comparison of (1) and (4) shows

$$(w_1 w_2) = (w_1)(w_2); \tag{5}$$

that is, the product of two principal divisors is the principal divisor associated with the product of the two elements. Hence the principal divisors form a subgroup $\mathfrak{h} \subset \mathfrak{k}$.

The elements of the factor group $\mathfrak{k}/\mathfrak{h}$ are called *divisor classes*. Two divisors $\mathfrak{a}$, $\mathfrak{b}$ are called *equivalent*, written $\mathfrak{a} \sim \mathfrak{b}$, if their quotient is a principal divisor. There is no danger of ambiguity in writing $\mathfrak{b} = w\mathfrak{a}$ for $b = (w)\mathfrak{a}$.

A principal divisor (w) is the unit divisor (1) if and only if $w \in k_0$, as is seen with a glance at (4). This implies that the group $\mathfrak{h}$ of principal divisors is isomorphic to the factor group $k^\times/k_0{}^\times$ of the multiplicative groups of k and k_0.

The divisor $\mathfrak{a}$ is said to have the *degree*

$$g(\mathfrak{a}) = \sum_{\mathfrak{p}} v_\mathfrak{p}(\mathfrak{a}), \tag{6}$$

and then

$$g(\mathfrak{a}\mathfrak{b}) = g(\mathfrak{a}) + g(\mathfrak{b}). \tag{7}$$

Our definition (2) assures us of the important fact

$$g(w) = 0; \tag{8}$$

the degree of a principal divisor is 0.

For rational functions, no more is to be learned from divisors as can be seen from formal consideration of (1). The importance lies in the fact that divisors can be carried over to the theory of algebraic functions, where the decomposition (1) is no longer possible.

Exercise. All $a \in k$ with $v_\xi(a) \geqq v_\xi(\mathfrak{a})$ form the ξ-component of a linear divisor in the space k^1, in the sense of I, §3. Its linear degree is equal to $g(\mathfrak{a})$.

2. Divisors in Algebraic Function Fields

Now let K be a finite extension of $k = k_0(x)$ generated, say, by adjoining a root y of Eq. §1,(12). The contents of the last section can be carried over literally as soon as we have defined places, their prime divisors, and orders in K.

First, with each of the $r(\xi)$ power series of §1,(21) associate a *place* ξ_ϱ *of* $K(\rho = 1, \ldots, r(\xi))$. The *places* ξ_ϱ are said to *lie over the place* ξ *of* k. This is to be taken as a convention of speech, the application of which is yet to be explained. The e_ϱ-fold ambiguous root $\sqrt[e_\varrho]{z_\xi}$ of §1,(21) is to be fixed arbitrarily.

Any element $w \in K$ can be written as a polynomial

$$w = P(y, x)$$

in y whose coefficients are rational functions in x. Inserting the power series §1,(21) for y and expanding the coefficients of $P(y, x)$ in power series in z_ξ we obtain power series

$$w = \sum_{\mu=\nu_\varrho}^{\infty} b_{\varrho\mu}(\sqrt[e_\varrho]{z_\xi})^\mu, \qquad b_{\varrho\nu_\varrho} \neq 0. \tag{9}$$

Evidently (9) depends only on w, not on the particular polynomial $P(y, x)$ representing w; the only ambiguity is the e_ϱth root of unity with which $\sqrt[e_\varrho]{z_\xi}$ may be multiplied. Equation (9) is called the *power series expansion of* w *at the place* ξ_ϱ. Equation (9) can be understood as an isomorphic mapping of K into the field of power series in a variable $\sqrt[e_\varrho]{z}$ which varies with the point ξ_ϱ. We do not, however, consider w as a continuous function in x or ξ, even if the power series occurring have positive radii of convergence. It is only after the question as to the correspondence between the ξ_ϱ and the series §1,(21) has been settled that continuous functions are found. This matter will come up in IV, §1. By §1,(23) the $y_{\varrho i}$ are conjugate among themselves and with y, with respect to k. The conjugates of $w = P(y, x)$ are thus $P(y_{\varrho i}, x)$, and w satisfies the equation

$$g(w, x) = \prod_{\varrho,i} (w - P(y_{\varrho i}, x)) = 0$$

in k. The second and last coefficients of this polynomial in w are called the *trace* and *norm* of w:

$$s_{K/k}(w) = \sum_{\varrho,i} P(y_{\varrho i}, x), \qquad n_{K/k}(w) = \prod_{\varrho,i} P(y_{\varrho i}, x). \tag{10}$$

Remembering that for a primitive eth root of unity

$$\sum_{\mu=1}^{e} \zeta^{r\mu} = \begin{cases} e & \text{for} \quad r \equiv 0 \bmod e \\ 0 & \text{for} \quad r \not\equiv 0 \bmod e \end{cases}$$

holds, (9) and (10) give

$$s_{K/k}(w) = \sum_{\varrho=1}^{r(\xi)} \sum_{\mu} e_\varrho b_{\varrho, e_\varrho \mu} z_\xi{}^\mu \tag{11}$$

The arbitrariness left in our choice of an e_ϱth root of z_ξ has no effect on the first exponent $v_\varrho(w) = v_{\xi_\varrho}(w)$ of series (9). It is called the *order of* w *at the place* ξ_ϱ. We now assert that *the order of a function* $w \neq 0$ *is* 0 *except at a finite number of places.*

Proof. Let

$$w = P(y, x) = a_0(x) + a_1(x)y + \cdots + a_{n-1}(x)y^{n-1}$$

with $a_i(x) \in k$. Except for a finite number of places ξ of k the $a_i(x)$ can be expanded as integral power series in $z_\xi = x - \xi$. In the proof of Puiseux's theorem in §1 it was further seen that the series §1,(21) contains no negative exponents if y is taken as a solution of an equation of the type §1,(12) whose coefficients are polynomials. (It was shown there that this last assumption is no loss of generality.) Thus integral power series (9) are found at almost all places for w. The same is also true of w^{-1}, which gives the assertion. ▼

The reader can verify that our definition of orders based on power series corresponds, in the case $K = k$, to the orders defined in the last section.

The definition (10) of the norm yields

$$v_\xi(n_{K/k}(w)) = \sum_{\varrho=1}^{r(\xi)} v_{\xi_\varrho}(w). \tag{12}$$

Together with (2) we have

$$\sum_{\xi_\varrho} v_{\xi_\varrho}(w) = 0. \tag{13}$$

The orders also behave as expected under multiplication:

$$v_{\xi_\varrho}(w_1 w_2) = v_{\xi_\varrho}(w_1) + v_{\xi_\varrho}(w_2). \tag{14}$$

With every place ξ_ϱ we now associate a *prime divisor* $\mathfrak{p} = \mathfrak{z}_{\xi_\varrho}$, proceeding as above to define the group $\mathfrak{K}$ of all *divisors* and its subgroup $\mathfrak{H} \subset \mathfrak{K}$ of *principal divisors*

$$(w) = \prod_{\mathfrak{p}} \mathfrak{p}^{v_\mathfrak{p}(w)} = \prod_{\xi_\varrho} \mathfrak{z}_{\xi_\varrho}^{v_{\xi_\varrho}(w)},$$

whose orders we have just defined. The group property of $\mathfrak{H}$ and its embedding in $\mathfrak{K}$ stem from Eq. (14).

The adaptation of the concepts of integral divisors, divisibility and equivalence of divisors, and divisor classes is obvious.

A proof is needed, though, for the following theorem: *A principal divisor* (w) *is the unit divisor* (1) *if and only if* $w \in k_0$. *Thus,*

$$\mathfrak{H} \cong K^\times / k_0^\times.$$

Proof. If $w \in k_0$, clearly $(w) = (1)$. Now, conversely, let $(w) = (1)$, that is, let all power series (9) for w start with the constant term. If $\gamma \in k_0$, then all the orders of $\gamma - w$ must be $\geqq 0$, and if further γ is equal to the constant term of one of the series (9), then at least one is even > 0. But this would contradict (13) were not $\gamma - w = 0$

The divisor $\mathfrak{a}$ again written in the form (3) is said to be of degree

$$g(\mathfrak{a}) = \sum_{\mathfrak{p}} v_{\mathfrak{p}}(\mathfrak{a}). \tag{15}$$

As before Eq. (7) holds, and (13) can be restated as (8): *every principal divisor has degree* 0. This further implies that *equivalent divisors have the same degree*, so that *the degree is a function of the divisor class.*

3. Decomposition of Rational Divisors

The concepts of divisors in different fields are logically independent. We are, nevertheless, free to embed the divisor group $\mathfrak{k}$ of k into the divisor group $\mathfrak{K}$ of K. Care must be taken, though, to be sure that such an embedding coincides with the natural embedding of $k^{\times}/k_0^{\times}$ into $K^{\times}/k_0^{\times}$. For every place ξ of k we set

$$\mathfrak{z}_{\xi} = \mathfrak{z}_{\xi_1}^{e_1} \cdots \mathfrak{z}_{\xi_r}^{e_r} \qquad (r = r(\xi)), \tag{16}$$

where the e_{ϱ} and ξ_{ϱ} are as in the last section. To prove the consistency, let $w \neq 0$ be an element of k and

$$(w) = \prod_{\xi} \mathfrak{z}_{\xi}^{v_{\xi}(w)}, \qquad (w) = \prod_{\xi_{\varrho}} \mathfrak{z}_{\xi_{\varrho}}^{v_{\xi_{\varrho}}(w)},$$

be its decomposition into prime divisors of k and K; we must show the latter to result from the former by the substitution (16). Now, the rational function w is divisible exactly $v_{\xi}(w)$ times by the function $z_{\xi} = x - \xi$. But in K the order of z_{ξ} at ξ_{ϱ} is exactly e_{ϱ}. Thus we arrive at the exponent $v_{\xi_{\varrho}}(w) = e_{\varrho} v_{\xi}(w)$ of $\mathfrak{z}_{\xi_{\varrho}}$ for w, as was to be shown.

We call $\mathfrak{z}_{\xi}$ the *norm* of the prime divisors $\mathfrak{z}_{\xi_{\varrho}}$, written $\mathfrak{z}_{\xi} = n_{K/k}(\mathfrak{Z}_{\xi\varrho})$. In general, this norm is defined as

$$n_{K/k}\left(\prod_{\xi_{\varrho}} \mathfrak{z}_{\xi_{\varrho}}^{v_{\xi_{\varrho}}(\mathfrak{a})}\right) = \prod_{\xi_{\varrho}} \mathfrak{z}_{\xi}^{v_{\xi_{\varrho}}(\mathfrak{a})}.$$

Then,

$$n_{K/k}(\mathfrak{a}\mathfrak{b}) = n_{K/k}(\mathfrak{a}) n_{K/k}(\mathfrak{b}) \tag{17}$$

holds, giving a homomorphic mapping of $\mathfrak{K}$ into $\mathfrak{k}$. Furthermore,

$$n_{K/k}((w)) = (n_{K/k}(w)); \tag{18}$$

that is, the *norm of a principal divisor is the principal divisor of the element norm*. This follows immediately from (12).

The degrees of a divisor are defined differently in k and K, so that they must be distinguished by subscripts. We have

$$g_K(\mathfrak{a}) = g_k(n_{K/k}(\mathfrak{a})). \tag{19}$$

This is obvious for prime divisors, and carries over to all divisors by (17). A special case of (19) is

$$g_K(\mathfrak{a}) = [K:k]g_k(\mathfrak{a}) \qquad \text{if} \quad \mathfrak{a} \in \mathfrak{k}, \tag{20}$$

which will be needed below.

4. The Principal Orders

From now on we must try to translate statements concerning power series into the language of linear algebra. The knowledge to be gained is already contained in II, §2 and 5, but here we follow a different path. As it turns out it is advantageous to make as little use as possible of the special properties of the power series.

Those functions of k which are integral (i.e., can be expanded in an integral power series) at a place ξ (including $\xi = \infty$) form an integral domain $\mathfrak{i}_\xi$. Those functions y of K, which are integrally dependent on $\mathfrak{i}_\xi$ in the sense of II, §1,1, or in other words which satisfy an equation §1,(12) with $c_i(x)$ in $\mathfrak{i}_\xi$, form an integral domain $\mathfrak{J}_\xi$ in K, the *principal order* of K with respect to $\mathfrak{i}_\xi$. $\mathfrak{J}_\xi$ is of rank $n = [K:k]$ over k. For, if $y \in K$ and h is the smallest exponent such that, for all i, $z_\xi^{ih}c_i(x) \in \mathfrak{i}_\xi$, then $y' = z_\xi{}^h y$ satisfies the equation $y'^n + \cdots + z_\xi^{ih}c_i(x)y'^{n-i} + \cdots = 0$, so that $z_\xi{}^h y \in \mathfrak{J}_\xi$. Now, if w_i are any n elements of K linearly independent with respect to k we can thus find an h such that all the $z_\xi{}^h w_i \in \mathfrak{J}_\xi$. These are of course still linearly independent.

$\mathfrak{J}_\xi$ is a principal ideal domain.

Proof. Let $\mathfrak{a} \subseteq \mathfrak{J}_\xi$ be an $\mathfrak{J}_\xi$-ideal. Find an element a_1 of $\mathfrak{a}$ whose norm with respect to k is divisible by a minimal power of z_ξ. Such an a_1 exists, as the norms of all elements of $\mathfrak{J}_\xi$ lie in $\mathfrak{i}_\xi$. Now let $a \in \mathfrak{a}$ and $\alpha \in k_0$ be arbitrarily given. Then

$$f(\alpha) = n_{K/k}\left(\alpha - \frac{a}{a_1}\right) = \frac{n_{K/k}(\alpha a_1 - a)}{n_{K/k}(a_1)} = \alpha^n + b_1(x)\alpha^{n-1} + \cdots + b_n(x)$$

is a polynomial in α. Using the assumed property of a_1 and the fact that $\alpha a_1 - a \in \mathfrak{a}$, we know that $f(\alpha)$ lies in $\mathfrak{i}_\xi$ for all α. We assert that even all the $b_i(x)$ are in $\mathfrak{i}_\xi$. Say this were not so, and the largest negative power z_ξ^{-h} occurred in $b_{i_1}(x), \ldots, b_{i_m}(x)$. These negative powers would have to cancel out when any $\alpha \in k_0$ is inserted into $f(\alpha)$. In other words, if we set $z_\xi = 0$ in

$$z_\xi{}^h b_{i_1}(x)\alpha^{n-i_1} + \cdots + z_\xi{}^h b_{i_m}(x)\alpha^{n-i_m},$$

then this expression vanishes for all $\alpha \in k_0$. Hence the coefficients of this polynomial in α must all vanish when set $z_\xi = 0$, which in turn contradicts the assumption that all the $b_{i_\nu}(x)$ have denominator $z_\xi{}^h$.

Now, since $f(a/a_1) = 0$, and the coefficients of this polynomial are integral at ξ, we have $a/a_1 \in \mathfrak{J}_\xi$. Thus $\mathfrak{a} = \mathfrak{J}_\xi a_1$, and the proof is completed. ▼

As is well known, the theorem of unique decomposition into prime elements holds in a principal ideal domain, or, more precisely, two given decompositions of an element into prime elements can differ at most by factors of units.

From II,§1,2 we see: $\mathfrak{J}_\xi$ *is a finite* $\mathfrak{i}_\xi$*-module of rank n.*

$\mathfrak{J}_\xi$ *has only a finite number of prime elements* $q_1, \ldots, q_r$ *which all occur in the decomposition*

$$z_\xi = u q_1^{\varepsilon_1} \ldots q_r^{\varepsilon_r}, \tag{21}$$

u being a unit.

Proof. Let q be a prime element. As we saw above, we can find an h such that $z_\xi^h q^{-1} \in \mathfrak{J}_\xi$. Thus q is a divisor of z_ξ. ▼

In (21) $r = r(\xi)$ *and a suitable indexing gives*

$$v_{\xi_\sigma}(q_\varrho) = \delta_{\sigma\varrho} = \begin{cases} 1 & \text{if} \quad \sigma = \varrho, \\ 0 & \text{if} \quad \sigma \neq \varrho, \end{cases} \tag{22}$$

the ε_ϱ *of* (21) *being equal to the* e_ϱ.

Proof. By the corollary of §1,6, the elements of $\mathfrak{J}_\xi$ have nonnegative orders, this being also true, in particular, of the unit u and u^{-1}. Thus, $v_{\xi_\sigma}(u) = 0$ for all σ. Similarly, $v_{\xi_\varrho}(q_\varrho) \geqq 0$. Taking orders of (21) leads to the equation

$$\sum_\varrho \varepsilon_\varrho v_{\xi_\sigma}(q_\varrho) = e_\sigma \qquad \text{for} \quad \sigma = 1, \ldots, r(\xi). \tag{23}$$

As the q_ϱ are mutually prime there must exist, to any pair of indices $\varrho_1 \neq \varrho_2$, elements $a, b \in \mathfrak{J}_\xi$ with

$$a q_{\varrho_1} + b q_{\varrho_2} = 1.$$

But this implies that $v_{\xi_\sigma}(q_{\varrho_1}) > 0$ and $v_{\xi_\sigma}(q_{\varrho_2}) > 0$ cannot hold simultaneously. Hence the sums (23) are reduced to one term each, with $\varrho = \varrho_\sigma$, and we have

$$e_{\varrho_\sigma} v_{\xi_\sigma}(q_{\varrho_\sigma}) = e_\sigma, \qquad v_{\xi_\sigma}(q_\varrho) = 0 \qquad \text{for} \quad \varrho \neq \varrho_\sigma.$$

Furthermore, holding one ϱ fixed, $v_{\xi_\sigma}(q_\varrho) = 0$ cannot hold for all σ. For then q_ϱ and q_ϱ^{-1} would each satisfy an equation with coefficients in $\mathfrak{i}_\xi$, so that q_ϱ would be a unit of $\mathfrak{J}_\xi$. Thus $r \leqq r(\xi)$, and we can renumber the q_ϱ or the ξ_ϱ so that

$$\varepsilon_\varrho v_{\xi_\varrho}(q_\varrho) = e_\varrho, \qquad \varrho = 1, \ldots, r, \tag{24}$$

holds. The rest of the proof depends upon the additional assertion:

A basis of $\mathfrak{J}_\xi$ *with respect to* $\mathfrak{i}_\xi$ *is given by* $q_{\varrho i} = q_1^{\varepsilon_1} \cdots q_r^{\varepsilon_r} q_\varrho^{i - \varepsilon_\varrho}$ $(\varrho = 1, \ldots, r;$ $i = 0, \ldots, \varepsilon_\varrho - 1)$, *and we have*

$$\sum \varepsilon_i = n = [K:k]. \tag{25}$$

For, assuming (25) to hold, then (24) and §1,(19) yield

$$n = \sum_{\varrho=1}^{r(\xi)} e_\varrho \geqq \sum_{\varrho=1}^{r} e_\varrho = \sum_{\varrho=1}^{r} \varepsilon_\varrho v_{\xi_\varrho}(q_\varrho) \geqq \sum_{\varrho=1}^{r} \varepsilon_\varrho = n.$$

If $r(\xi) > r$ we would have $>$ on the left, and $>$ would also have to hold on the right if only one $v_{\xi_\varrho}(q_\varrho) > 1$. Thus the proof of the former assertion depends upon (25).

To prove the last assertion and (25), consider some q_ϱ and some residue class $\bar{a}$ of $\mathfrak{J}_\xi$ mod q_ϱ. It is not difficult to see that it satisfies an equation $f(\bar{a}) = 0$ whose coefficients are residue classes of $\mathfrak{i}_\xi$ mod z_ξ. As k_0 is algebraically closed, these residue classes can all be represented by elements of k_0. Now, the residue class ring of $\mathfrak{J}_\xi$ mod q_ξ can have no divisors of zero, so that $\bar{a}$ must even satisfy an irreducible equation of this sort, and, k_0 being algebraically closed, this equation will be of degree 1. Hence each $a \in \mathfrak{J}_\xi$ is congruent mod q_ϱ to some $a_0 \in \mathfrak{i}_\xi$.

The elements $q_{\varrho 0}$ are $\not\equiv 0$ mod q_ϱ.

Now let some $a \in \mathfrak{J}_\xi$ be given, and $aq_{\varrho 0}^{-1} \equiv a_{\varrho 0} \in \mathfrak{i}_\xi$ mod q_ϱ. Further, let

$$(a - a_{\varrho 0} q_{\varrho 0}) q_\varrho^{-1} q_{\varrho 0}^{-1} = (a - a_{\varrho 0} q_{\varrho 0}) q_{\varrho 1}^{-1} \equiv a_{\varrho 1} \bmod q_\varrho,$$

etc. This yields elements $a_{\varrho i} \in \mathfrak{i}_\xi$ satisfying the congruences

$$a \equiv a_{\varrho 0} q_{\varrho 0} + a_{\varrho 1} q_{\varrho 1} + \cdots + a_{\varrho, \varepsilon_\varrho - 1} q_{\varrho, \varepsilon_\varrho - 1} \bmod q_\varrho^{\varepsilon}.$$

The construction of the $q_{\varrho i}$ entails the right side of this congruence to be $\equiv 0$ mod $q_\sigma^{\varepsilon_\sigma}$ for $\varrho \neq \sigma$. Thus

$$a \equiv \sum_{\varrho=1}^{r} \sum a_{\varrho i} q_{\varrho i} \bmod \mathfrak{J}_\xi \prod_\varrho q_\varrho^{\varepsilon_\varrho} = \mathfrak{J}_\xi z_\xi. \tag{26}$$

Now let ω_ν be a basis of $\mathfrak{J}_\xi$ with respect to $\mathfrak{i}_\xi$, which we know to exist from I, §1, 1, and let

$$q_{\varrho i} = \sum_\nu c_{\varrho i, \nu} \omega_\nu, \qquad c_{\varrho i, \nu} \in \mathfrak{i}_\xi.$$

On the other hand, apply (26) to the ω_ν, so that

$$\omega_\nu \equiv \sum_{\varrho, i} c'_{\nu, \varrho i} q_{\varrho i} \bmod \mathfrak{J}_\xi z_\xi.$$

and then

$$\omega_\nu \equiv \sum_{\varrho, i, \mu} c'_{\nu, \varrho i} c_{\varrho i, \mu} \omega_\mu \bmod \mathfrak{J}_\xi z_\xi.$$

The ω_ν must also be linearly independent mod z_ξ, so that

$$\sum_{\varrho, i} c'_{\nu, \varrho i} c_{\varrho i, \mu} \equiv \delta_{\nu\mu} \bmod \mathfrak{J}_\xi z_\xi$$

must hold. Hence $(c_{\varrho i, \nu})$ is a unimodular matrix over $\mathfrak{i}_\xi$, and the $q_{\varrho i}$ also form a basis of $\mathfrak{J}_\xi$.

As $\mathfrak{J}_\xi$ has the rank n with respect to k, Eq. (25) is simply a matter of counting. ▼

This has brought us to our goal. The basis $q_{\varrho i}$ will soon play an important role. Noting that these calculations made no use of the power series (9), one might conjecture that they could be dispensed with completely. This will be confirmed in III, §2,7. We also point out the fact that a basis similar to $q_{\varrho i}$, but even more complicated, was used in the proof of the Dedekind discriminant theorem in II, §4,3.

One last theorem serves to bridge the gap between the older concepts, based on power series, and the newer, more abstract, algebraic ones: $\mathfrak{I}_\xi$ *coincides with that integral domain of functions* $f \in K$ *whose power series are integral at all places* ξ_ϱ *lying over* ξ. To show this one need only choose the representation $f = \sum q_{\varrho i} f_{\varrho i}$ with $f_{\varrho i} \in k_0(x)$, and prove that all the power series for f at the ξ_ϱ are integral if and only if the power series of all the $f_{\varrho i}$ are integral at ξ. This is not difficult, when the construction of the $q_{\varrho i}$ and (22) are used (exercise).

As was already predicted in the introduction to this chapter, the power series now fade into the background of the development of the algebraic theory.

5. Divisors and Linear Divisors

Our path meets here with that of Chapter II at II, §5,6. As we do not want to presuppose that chapter now, we must consider some material anew. A divisor $\mathfrak{a}$ is written as

$$\mathfrak{a} = \prod_\xi \mathfrak{a}_\xi, \qquad \mathfrak{a}_\xi = \prod_{\varrho=1}^{r(\xi)} \mathfrak{z}_{\xi_\varrho}^{v_{\xi\varrho}(\mathfrak{a})}, \tag{27}$$

with ξ traversing all the places of k. The ξ-component $\mathfrak{a}_\xi$ of $\mathfrak{a}$ thus defined can be associated with the set of all $a \in K$ with $v_{\xi\varrho}(a) \geqq v_{\xi\varrho}(\mathfrak{a})$, $\varrho = 1, \ldots, r(\xi)$. This is clearly an $\mathfrak{I}_\xi$-ideal, and no misunderstanding can occur if we denote it also by $\mathfrak{a}_\xi$.

Choose an $h \geqq \max(0, -v_{\xi_\varrho}(\mathfrak{a}))$, then $z_\xi{}^h \mathfrak{a}_\varrho \subseteq \mathfrak{I}_\xi$. The finiteness criterion of I, §1,1 together with the fact that $\mathfrak{I}_\xi$ is a finite $\mathfrak{i}_\xi$-module implies that $\mathfrak{a}_\xi$ is finite. Given an arbitrary $b \in K$, we can clearly also require $z^h b \in \mathfrak{a}_\xi$; certainly then $\mathfrak{a}_\xi$ has rank n with respect to k. Finally, as almost all $v_{\xi_\varrho}(\mathfrak{a}) = 0$ we have $\mathfrak{a}_\xi = \mathfrak{I}_\xi$ almost always.

A basis representation $a = \sum a_\nu \omega_\nu$ with $a_\nu \in k$ of the elements $a \in K$ is now interpreted as an identification $a \leftrightarrow (a_\nu)$ of these elements with the vectors of an n-dimensional space over k. We can then assert that the $\mathfrak{a}_\xi$ are the components, in the sense of I, §3, of a linear divisor. To show this we form the intersection

$$\mathfrak{I} = \bigcap_{\xi \neq \infty} \mathfrak{I}_\xi. \tag{28}$$

It is an $\mathfrak{i}$-module, where

$$\mathfrak{i} = \bigcap_{\xi \neq \infty} \mathfrak{i}_\xi = k_0[x].$$

By definition $\mathfrak{I}$ consists of all elements of K integrally dependent upon $\mathfrak{i}$ and by II, §1,1 (where now $\mathfrak{o} = \mathfrak{i}$ and $\mathfrak{O} = \mathfrak{I}$) we see that $\mathfrak{I}$ is a finite $\mathfrak{i}$-module of rank n, and therefore has a basis ω_ν with respect to $\mathfrak{i}$ (I, §1,1). The ω_ν are also a basis of $\mathfrak{J}_\xi$ with respect to $\mathfrak{i}_\xi$ for all $\xi \neq \infty$. For, were $a = \sum \omega_\nu a_\nu \in \mathfrak{J}_\xi$ with $a_\nu \in k$ but not all $a_\nu \in \mathfrak{i}_\xi$, we could find a common denominator $h = h_1 z_\xi{}^t$ of the a_ν, with h_1 prime to z_ξ. Then $ah_1 = \sum \omega_\nu a_\nu h_1 \in \mathfrak{J}_\xi$, while not all the $a_\xi h_1 \in \mathfrak{i}_\xi$. On the other hand, $a_\nu h_1 \in \mathfrak{i}_{\xi'}$ for all $\xi' \neq \xi$, so that by our hypothesis, $ah_1 \in \mathfrak{J}$. But some of the $a_\nu h_1$ are nonintegral, which contradicts the basis property of the ω_ν.

As the ω_ν now form a basis of $\mathfrak{J}_\xi$ with respect to $\mathfrak{i}_\xi$ for every ξ, and as $\mathfrak{a}_\xi = \mathfrak{J}_\xi$ almost always, the ω_ν are almost always a basis of $\mathfrak{a}_\xi$. Thus all the properties of a linear divisor have been confirmed.

The linear divisors corresponding to equivalent divisors $\mathfrak{a}$ *and* $\mathfrak{a}a$ *are equivalent in the sense of* I, §3. The proof can be taken exactly from II, §5,6, if only the symbol p there is replaced by ξ.

The field K has been associated in the natural manner with the n-dimensional vector space k^n. Simultaneously we may associate K with the dual space k^{n*} if we can find a bilinear form aa^* of K with values in k which possesses the property (3) in the list of I, §1,3. We choose the trace: $aa^* = s_{K/k}(aa^*)$. The property (3) required is then the fact that the discriminant $D(\omega_\nu) = |s_{K/k}(\omega_\mu \omega_\nu)|$ of a basis ω_ν of K/k is not 0. Let complementary modules and complementary linear divisors be defined now in this sense.

There exists a divisor $\mathfrak{d}_{K/k}$ *such that an arbitrary divisor* $\mathfrak{a}$ *and its "complement"*

$$\mathfrak{a}^* = \mathfrak{d}_{K/k}^{-1}\mathfrak{a}^{-1} \tag{29}$$

correspond to linear divisors which are complementary in the sense of I, §3.

Proof. As opposed to II, §5,6, we can now explicitly describe $\mathfrak{d}_{K/k}$. In the last section it was seen that $\mathfrak{J}_\xi$ is a principal ideal domain, so that for every ξ there exists an $a_\xi \in K$ such that

$$\mathfrak{a}_\xi = \mathfrak{J}_\xi a_\xi. \tag{30}$$

Define the module $\mathfrak{J}_\xi{}^*$ by

$$s_{K/k}(\omega\omega^*) \in \mathfrak{i}_\xi \qquad \text{for all} \quad \omega \in \mathfrak{J}_\xi, \quad \omega^* \in \mathfrak{J}_\xi{}^*;$$

then clearly

$$\mathfrak{a}_\xi{}^* = \mathfrak{J}_\xi{}^* a_\xi^{-1}. \tag{31}$$

Using the fact that $\mathfrak{J}_\xi$ is an integral domain and applying obvious notation we have

$$s_{K/k}(\mathfrak{J}_\xi \cdot \mathfrak{J}_\xi{}^*) = s_{K/k}(\mathfrak{J}_\xi \mathfrak{J}_\xi \cdot \mathfrak{J}_\xi{}^*) - s_{K/k}(\mathfrak{J}_\xi \cdot \mathfrak{J}_\xi \mathfrak{J}_\xi{}^*).$$

This means that $\mathfrak{J}_\xi{}^*$ is an $\mathfrak{J}_\xi$-ideal, and (31) becomes

$$\mathfrak{a}_\xi{}^* = \mathfrak{J}_\xi{}^* \cdot \mathfrak{J}_\xi a_\xi^{-1} = \mathfrak{J}_\xi{}^* \mathfrak{a}_\xi^{-1}.$$

This is Eq. (29) if the local components of $\mathfrak{d}_{K/k}$ are given by

$$(\mathfrak{d}_{K/k})_\xi = \mathfrak{J}_\xi^{*-1}. \tag{32}$$

We can determine these ideals explicitly:

$$\mathfrak{J}_\xi^* = \mathfrak{J}_\xi q_1^{1-e_1} \cdots q_r^{1-e_r}, \tag{33}$$

with the q_ϱ and e_ϱ of the last section. From (22) it is not difficult to derive the power series expansions

$$q = q_1^{1-e_1} \cdots q_r^{1-e_r} = \sum_{\mu=1-e_\varrho}^{\infty} b_{\varrho,\mu}(\sqrt[e_\varrho]{z_\xi})^\mu, \qquad b_{\varrho,1-e_\varrho} \neq 0, \tag{34}$$

for all ϱ. The elements of $\mathfrak{J}_\xi$ have integral power series at all places ξ_ϱ. Then (11) and (34) show that the elements of $\mathfrak{J}_\xi q$ all have traces in $\mathfrak{i}_\xi$. This means that $\mathfrak{J}_\xi q \subseteq \mathfrak{J}_\xi^*$. But, if $\mathfrak{J}_\xi q \subset \mathfrak{J}_\xi^*$, say $\mathfrak{J}_\xi^* = \mathfrak{J}_\xi q_1^{-f_1} \cdots$, where some $f_\varrho \geqq e_\varrho$, then choose some $a \in \mathfrak{J}_\xi$ whose series expansion at ξ_ϱ starts with $(\sqrt[e_\varrho]{z_\xi})^{f_\varrho - e_\varrho}$, while that at $\xi_\sigma (\sigma \neq \varrho)$ starts with $(\sqrt[e_\varrho]{z_\xi})^{f_\sigma - e_\sigma + \varepsilon}$, $\varepsilon > 0$. By (11), $s_{K/k}(aq_1^{-f_1} \cdots)$ would not be in $\mathfrak{i}_\xi$. This contradiction proves (33).

According to Puiseux's theorem almost all the $e_\varrho = 1$, so that the product

$$\mathfrak{d}_{K/k} = \prod_\xi \mathfrak{z}_{\xi_1}^{e_1 - 1} \cdots \mathfrak{z}_{\xi_r}^{e_r - 1} \tag{35}$$

is finite and defines a divisor of K, whose ξ-components, understood as $\mathfrak{J}_\xi$-ideals, are

$$(\mathfrak{d}_{K/k})_\xi = \mathfrak{J}_\xi q_1^{e_1 - 1} \cdots q_r^{e_r - 1}.$$

These are just the inverses of the ideals (33). Thus (32) gives the local components of a divisor. ▼

Simultaneously, the formula

$$g(\mathfrak{d}_{K/k}) = \sum_{\xi,\varrho} (e - 1) \tag{36}$$

has been found. We call (36) the *ramification index* of K with respect to k. (The arithmetic justification of this terminology is to be found in II, §3, while its geometric justification will arise in IV, §1.) $\mathfrak{d}_{K/k}$ is called the *different divisor* of K with respect to k.

We must finally calculate the linear degree of a divisor, in the sense of I, §3. This is done exactly as in II, §5, 6 and 7. The assertion made there, that with a basis ω_ν of $\mathfrak{J}$ the ξ-components of $N(\mathfrak{a})$ and $n_{K/k}(\mathfrak{a})$ coincide for all $\xi \neq \infty$, follows easily from (30). For

$$\omega_\mu a_\xi = \sum_\nu a_{\mu\nu}\omega_\nu, \qquad a_{\mu\nu} \in k,$$

is a basis of $\mathfrak{a}_\xi$, and the determinant $|a_{\mu\nu}|$ is the ξ-component of $N(\mathfrak{a})$. But this is also true to $n_{K/k}(\mathfrak{a})$ by the definition in I, §4. Altogether, by II, §5, (20)

$$G(\mathfrak{a}) = g(\mathfrak{a}) + \tfrac{1}{2}g(\mathfrak{d}_{K/k}) = g(\mathfrak{a}) + \tfrac{1}{2}\sum_{\xi,\varrho}(e_\varrho - 1). \tag{37}$$

6. The Invariance of the Concept of Divisors

We introduced the concept of divisors with the help of a specific model of the field K as a finite extension $k(y)$ of $k = k_0(x)$. The element x was thus distinguished. K could just as well be considered a finite extension of $k' = k_0(y)$; this gives $K = k'(x)$ where x satisfies the same equation §1,(12). Clearly there are even infinitely many possible models.

However, each prime divisor $\mathfrak{p}$ can be associated with the integral domain $\mathfrak{I}_\mathfrak{p}$ of those functions w with $v_\mathfrak{p}(w) \geqq 0$. Now, such an $\mathfrak{I}_\mathfrak{p}$ is a discrete local ring in the sense of II,§2,2, and in II,§2,3 it was shown that there exist no other discrete local rings in K than the $\mathfrak{I}_\mathfrak{p}$, where the $\mathfrak{p}$ are the $\mathfrak{z}_{\xi_\varrho}$ defined in §2,2. Hence the concept of the discrete local ring is independent of the specific model of K, and then the same holds true for prime divisors and, finally, for the concept of divisors.

7*. Extension to More General Constant Fields

From now on let k_0 be an arbitrary algebraically closed field. Further, let there exist what is called a separable model of K, that is, let K be a separable extension of $k = k_0(x)$ by suitable choice of x. In §4,1 it will be shown that this is certainly possible whenever k_0 is algebraically closed. The definition of the $\mathfrak{i}_\xi$ remains unchanged, and the prime divisors $\mathfrak{z}_\xi$ of k, and then the general divisors of k, can be introduced as in §2,1. To introduce the divisors of K, though, we can now no longer use the power series (9); they need not exist. Instead, we use the method of §2,4 to form the principal orders $\mathfrak{I}_\xi$. There is no change in the proof that these are principal ideal domains.

Equation (21) also holds. The exponents ε_ϱ and the number r occurring there must be left in place of the e_ϱ and $r(\xi)$, for there is now no other way of defining these. Prime divisors $\mathfrak{z}_{\xi\varrho}$ are associated with the q_ϱ. Everything else in §2,2, §2,3, §2,5, and §2,6 may be taken over verbatim, with the sole exception of the computation of the complement (33). For here we no longer can use formula (11) and, in fact, (33) is no longer correct if any one of the e_ϱ is divisible by the characteristic. Nevertheless, a complementary ideal $\mathfrak{I}_\xi{}^*$ to $\mathfrak{I}_\xi$ exists, for the separability of K/k assures us that the discriminant of K with respect to k does not vanish. Set

$$\mathfrak{I}_\xi{}^* = \mathfrak{I}_\xi q_1^{-\delta_1} \cdots q_r^{-\delta_r};$$

then $\delta_\varrho = 0$ for almost all ξ, as here again $\mathfrak{I}_\xi$ is the ξ-component of a linear divisor. In place of (35) define

$$\mathfrak{d}_{K/k} = \prod_\xi \mathfrak{z}_{\xi_1}^{\delta_1} \cdots \mathfrak{z}_{\xi_r}^{\delta_r}, \tag{38}$$

so that Eq. (32) is restored.

All we now have for the relation between linear degree and divisor degree is

$$G(\mathfrak{a}) = g(\mathfrak{a}) + \tfrac{1}{2}g(\mathfrak{d}_{K/k}), \tag{39}$$

and the ramification index $g(\mathfrak{d}_{K/k})$ is, in general, larger than the sum (36). It can be computed with the help of (20), for

$$g(\mathfrak{d}_{K/k}) = g_k(\mathfrak{D}_{K/k}), \qquad \mathfrak{D}_{K/k} = n_{K/k}(\mathfrak{d}_{K/k}).$$

The latter is the discriminant divisor. By II, §4,(12) its ξ-components are the discriminants of bases of $\mathfrak{J}_\xi$ with respect to $\mathfrak{i}_\xi$.

A similar procedure is difficult when k_0 is not algebraically closed. It is possible, though, by a trick used in algebraic geometry. K is extended to $\bar{K}$ by adjoining the algebraic closure $\bar{k}_0$ of the exact constant field k_0, and then the above arguments hold in $\bar{K}$. An integral divisor $\mathfrak{a}$ of $\bar{K}$ is said to be a divisor of K (or $\mathfrak{a}$ is said to be defined for k_0) whenever $\mathfrak{a}$ is the g.c.d. of numerators of functions a_i of K. In a fully cancelled representation one then has $(a_i) = \mathfrak{a}\mathfrak{b}_i\mathfrak{c}_i^{-1}$, with integral $\mathfrak{b}_i$, $\mathfrak{c}_i$, the $\mathfrak{b}_i$ being mutually prime. To these divisors of K we annex their quotients, so that the divisors of K form a subgroup $\mathfrak{K} \subset \bar{\mathfrak{K}}$.

Whenever two integral divisors of $\mathfrak{K}$ have a common divisor in $\bar{\mathfrak{K}}$ it is, by definition, also in $\mathfrak{K}$. Thus every divisor of $\mathfrak{K}$ can be represented uniquely as a power product of indecomposable divisors of $\mathfrak{K}$. These are then called the prime divisors of K.

The elements $w \in K$ generate the group $\mathfrak{H}$ of principal divisors of K, and the factor group $\mathfrak{K}/\mathfrak{H}$ consists of the divisor classes of K.

It is finally easy to see that the complement (29) of a divisor $\mathfrak{a}$ also belongs to $\mathfrak{K}$.

§3. The Riemann–Roch Theorem

1. Dimension of a Divisor Class

No assumptions are made concerning the exact field of constants k_0. Whenever K is a separable extension of the rational subfield $k_0(x)$, and thus in particular whenever the characteristic is 0, the trace is taken for the pseudotrace, so that the prefix "pseudo" also drops out for the different and discriminant.

Let $\mathfrak{a}$ be an arbitrary divisor. The dimension of the linear divisor corresponding to $\mathfrak{a}^{-1}$, all in the sense of I, §3, is called the *dimension* of $\mathfrak{a}$, written

$$\dim(\mathfrak{a}) = l(\mathfrak{a}^{-1}). \tag{1}$$

The linear divisor to $\mathfrak{a}^{-1}$ is a system of finite $\mathfrak{i}_\mathfrak{p}$-modules $\mathfrak{a}_p^{-1}$ or $\mathfrak{i}_\xi$-modules $\mathfrak{a}_\xi^{-1}$ in K. Thus there are exactly $\dim(\mathfrak{a})$ elements $b_i \in K$ linearly independent with respect to k_0 and lying in all the $\mathfrak{a}_p^{-1}$ or $\mathfrak{a}_\xi^{-1}$, respectively. For all p (or ξ) this means that $\mathfrak{a}_p b_i \subseteq \mathfrak{J}_p$ (or $\mathfrak{a}_\xi b_i \subseteq \mathfrak{J}_\xi$). This can also be expressed by saying that the divisors $\mathfrak{a}b_i = \mathfrak{b}_i$ are integral. Then

$$(b_i) = \mathfrak{b}_i \mathfrak{a}^{-1}$$

with integral divisors $\mathfrak{b}_i$, and it is plausible to call the b_i *multiples* of $\mathfrak{a}^{-1}$.

Thus dim($\mathfrak{a}$) can also be defined as the number of multiples of $\mathfrak{a}^{-1}$ linearly independent with respect to k_0. As we are assuming that k_0 is the exact constant field of K, the formula $\mathfrak{H} \cong K^\times / k_0^\times$ of II, §5,3 (or, respectively, III, §2,2) states that the b_i are uniquely determined up to a nonvanishing factor in k_0 by the (b_i) and therefore by the $\mathfrak{b}_i$. Changing these factors does not affect the linear independence, though, and it becomes clear that we can speak of *linearly independent* (or linearly dependent) *divisors* $\mathfrak{b}_i$ of a divisor class.

In this connection it is also customary to multiply linearly independent divisors of a class with constants $\varrho_i \in k_0$ and to add them. An expression of the sort $\mathfrak{b} = \sum \varrho_i \mathfrak{b}_i$ is then the numerator divisor of $b = \sum \varrho_i b_i$. This notation has a unique meaning only after the b_i for which $(b_i) = \mathfrak{b}_i \mathfrak{a}^{-1}$ have been fixed, and then $(b) = \mathfrak{b}\mathfrak{a}^{-1}$.

By I, §3,3 *the dimension as well as the degree is a function of the divisor class.* Classes will always be denoted by upper-case roman letters from now on.

The unit class of $\mathfrak{a} = (1)$ is usually called the *principal class*, as it consists of the principal divisors.

2. The Riemann–Roch Theorem

There is a distinguished divisor class W and a rational number $g \geqq 0$ such that the formula

$$\dim(A) = g(A) + \dim(W/A) + 1 - g \tag{2}$$

holds for every divisor class A. If $g(A) < 0$ then $\dim(A) = 0$. *The principal class $A = (1)$ has the dimension dim*$(1) = 1$, *while all other classes A of degree* 0 *have dim*(A) = 0.

One calls g the *genus* and W the *canonical class* of K. The term differential class used in the classical case is, in general, misunderstandable and we will not use it. In this regard cf. §4,8.

Proof. Let $\mathfrak{a}$ be a divisor of the class A. Denote the linear divisor associated with $\mathfrak{a}^{-1}$ by $\mathfrak{b}$. The divisor $\mathfrak{b}'$ associated with $\mathfrak{b}$ by I, §3(11) is, according to II, §5,6 and III, §2,5 (or §2,7),

$$\mathfrak{b}' = \mathfrak{b}^{-1}\mathfrak{d}_{K/k}^{-1}\mathfrak{z}_\infty{}^2 = \mathfrak{a}\mathfrak{d}_{K/k}^{-1}\mathfrak{z}_\infty{}^2, \tag{3}$$

where $\mathfrak{d}_{K/k}$ is the pseudodifferent and $\mathfrak{z}_\infty$ the denominator divisor of (x). We maintain the canonical class is that of the divisor of $\mathfrak{d}_{K/k}\mathfrak{z}_\infty^{-2}$:

$$W = (\mathfrak{d}_{K/k}\mathfrak{z}_\infty^{-2}). \tag{4}$$

By I, §3,(12) we have

$$l(\mathfrak{b}) - l(\mathfrak{b}') = n - G(\mathfrak{b}),$$

where $G(\mathfrak{b})$ denotes the linear degree. Compiling (3), (4), and II, §5,(20) or III, §2,(37) or (39), respectively, we have

$$\dim A - \dim(W/A) = g(A) + n - \tfrac{1}{2}g(\mathfrak{d}_{K/k}), \qquad n = [K:k],$$

which, using the abbreviation

$$g = \tfrac{1}{2}g(\mathfrak{d}_{K/k}) - [K:k] + 1, \tag{5}$$

is identical to (2).

The degree of an integral divisor is nonnegative, so that a class of negative degree can contain no integral divisors. This means $\dim(A) = 0$ for $g(A) < 0$. Further, clearly (1) is the only integral divisor in the principal class, so that $\dim(1) = 1$. If $g(A) = 0$ and $\dim(A) < 0$ there is an integral divisor $\mathfrak{a}$ in A, and it must be $\mathfrak{a} = (1)$.

When A is the principal class (2) goes into

$$\dim W = g. \tag{6}$$

This assures that $g \geqq 0$, completing the proof. ▼

Setting $A = W$ in (2) and using (6), we have

$$g(W) = 2g - 2. \tag{7}$$

An expression in terms of the degree of the pseudodiscriminant in k for the genus is found by combining (5) with II, §5,(16) for $L = k$ or III, §2,(20):

$$g = \tfrac{1}{2}g_k(\mathfrak{D}_{K/k}) - [K:k] + 1. \tag{8}$$

If L is an intermediate field between k and K the transitivity formula II, §4,(11) gives

$$g_k(\mathfrak{D}_{K/k}) = g_k(n_{L/k}(\mathfrak{D}_{K/L})) + [K:L]g_k(\mathfrak{D}_{L/k}) = g_L(\mathfrak{D}_{K/L}) + [K:L]g_L(\mathfrak{d}_{L/k}),$$

where g_L denotes the divisor degree in L. Substitution into (8) yields

$$g = \tfrac{1}{2}g_L(\mathfrak{D}_{K/L}) - [K:L] + 1 + [K:L](\tfrac{1}{2}g_L(\mathfrak{d}_{L/k}) - [L:k] + 1).$$

We introduce the *relative genus*

$$g(K/L) = \tfrac{1}{2}g_L(\mathfrak{D}_{K/L}) - [K:L] + 1 = \tfrac{1}{2}g(\mathfrak{d}_{K/L}) - [K:L] + 1 \tag{9}$$

and thus can finally write

$$g = g(K/k) = g(K/L) + [K:L]g(L/k). \tag{10}$$

The pair of equations (9) and (10) is called the *Hurwitz genus formula.* Note that for the relative genus $g(K/L) \geqq 0$ need not necessarily hold.

3. Questions of Invariance

As already mentioned in II, §5,5 and III, §2,6, the element x, for which the field K is considered a finite extension of $k_0(x)$, is actually arbitrary. The concept of divisor and, in particular, of prime divisor is invariant under a change of this element. This immediately gives the invariance of the divisor degree under the special assumptions of §2 from its definition in connection with prime divisor decomposition §2,(3). Under the more general assumptions of Chapter II the invariance theorem of II, §5,5 states that the degree of a prime divisor is independent of x, and then so is the degree of every divisor. Finally, the concepts of divisor class, integral divisor, and dimension are invariant of the choice of the model of K.

With these considerations (2) implies the *invariance of the genus* g, as for a class A of sufficient degree we can write (2) as

$$g = g(A) - \dim(A) + 1 \qquad \text{for} \quad g(A) > g(W) \tag{11}$$

(Riemann's part of the Riemann–Roch theorem). If the constant field $k_0 \subseteq C$, then we shall see in IV, §1 that g is even a topological invariant of a certain two-dimensional manifold associated with the field K independently of the model used. This would be another proof of the invariance of the genus. As for the canonical class, we finally have:

W is the only class with the properties (6) *and* (7). For, if W' is some class, satisfying these equations, then (2) reads $\dim(W'/W) = 1$. But as $g(W'/W) = 0$ the Riemann–Roch theorem shows that $W'/W = (1)$.

4. Extension of the Constant Field

Let k_0' be an extension of k_0 obtained by the successive adjunction of finitely many elements. We want to adjoin them as well to K and thus obtain a function field K' over k_0', and to study the behavior of the divisors under the extension. But this meets with difficulties if the characteristic is a prime. We shall avoid these difficulties by imposing the following restriction on the field K which does not exclude the most important applications of the theory.†

K has the exact constant field k_0 and is separably generated over k_0. This means, there exists at least one "separating" element x such that K is a finite separable extension of $k = k_0(x)$.

† For a more thorough treatment see the book by C. Chevalley cited on page 142 and J. Tate, *Genus change in inseparable extensions of function fields*, Proc. Amer. Math. Soc. **3**, 400–406 (1952).

Given a finitely generated extension k_0'/k_0 there exists exactly one minimal field

$$K' = Kk_0'$$

over k_0 containing both K and k_0' as subfields with $K \cap k_0' = k_0$. K' is a finite algebraic function field with the exact constant field k_0'. If k_0'/k_0 is finite then

$$[K' : K] = [k_0' : k_0]. \tag{12}$$

Proof. The adjunction of a single element to k_0 and K restores the situation and it suffices to treat such an extension. The contention is obvious for the adjunction of a transcendental element ϑ.

Now let ϑ satisfy the irreducible equation $f(\vartheta) = 0$ in k_0. If it remains irreducible in K we can adjoin ϑ to k_0 and K without ambiguity, and (12) holds. If $f(\vartheta) = f_1(\vartheta)f_2(\vartheta)$ were a proper decomposition in K, then K would contain the coefficients of $f_1(\vartheta)$. These are algebraic over k_0 and thus, as k_0 is the exact constant field, are contained in k_0. So, contrary to the hypothesis, the decomposition would already hold in k_0.

In order to show that the exact constant field of K' is k_0' we assume at first that k_0' is separable over k_0. Let

$$\alpha = \beta_0 + \beta_1\vartheta + \cdots + \beta_{m-1}\vartheta^{m-1} \qquad \text{with} \quad \beta_i \in K \tag{13}$$

be a constant in $K(\vartheta)$, *i.e.*, algebraic over k_0' and therefore over k_0, where m is the degree of $f(\vartheta)$. Replacing here ϑ by its conjugates $\vartheta_1 = \vartheta, \ldots, \vartheta_m$ with respect to k_0, we get m constants $\alpha_i = \beta_0 + \beta_1\vartheta_1 + \cdots$. Using the Vandermonde determinant we can compute the β_i as

$$\beta_i = \sum_j c_{ij}\alpha_j \qquad \text{with} \quad c_{ij} \in k_0.$$

Thus the β_i are algebraic over k_0 and therefore lie in k_0, it being the exact constant field of K. Thus, also $\alpha \in k_0(\vartheta) = k_0'$, which is what was to be proved.

Finally, let ϑ be an arbitrary algebraic element over k_0 and α a constant in $K(\vartheta)$ satisfying an algebraic equation $g(\alpha) = 0$ in k_0'. Furthermore let k_1' be the largest subfield of k_0' which is separable over k_0. With some power p^e of the characteristic the p^eth powers of the coefficients of $g(\alpha)$ lie in k_1'. Now α^{p^e} satisfies the equation $g^{p^e}(\alpha^{p^e}) = (g(\alpha))^{p^e} = 0$ in k_1'. Using what already has been proved α^{p^e} must lie in k_1' and a fortiori in k_0'. So α lies on one hand in k_0' or a purely inseparable extension of k_0' and so in a purely inseparable extension of $k_0'(x)$, also. On the other hand, α lies in $k_0'(x, y)$, which is, by assumption, a separable extension of $k_0'(x)$. So α lies in $k_0'(x)$ and, being a constant, in k_0'. This completes the proof. ▼

The divisors of K occur among those of K' in a natural manner. We now study the behavior of the degrees and dimensions of the divisors of K under such an extension.

Under the assumptions made above the degree of a divisor of K does not change with the extension of the constant field, and the dimension does at least not decrease. If k_0' is separable over k_0, the dimension is an invariant.

Proof. Let x be a transcendental element of K. Then, by II, §5,(16) or III, §2,(20), the degrees of a divisor $\mathfrak{a}$ in K' and K are those of both sides of

$$n_{K'/k_0'(x)}(\mathfrak{a}) = n_{K/k_0(x)}(\mathfrak{a}),$$

whence the invariance of the degree.

Next we study the dimension of $\mathfrak{a}$ under a finite extension of k_0. Let a_ν be multiples of $\mathfrak{a}^{-1}$ in K, linearly independent, with respect to k_0. If they are also linearly independent with respect to k_0', the dimension of $\mathfrak{a}$ in K' is at least as large as that dimension on K. This is indeed the case, for a linear relation in k_0' would mean $\sum a_\nu \alpha_\nu = 0$ with $\alpha_\nu = \sum w_i \alpha_{i\nu}$, $\alpha_{i\nu} \in k_0$ and w_i being a basis of k_0'/k_0. This would lead to $\sum w_i \sum \alpha_{i\nu} a_\nu = 0$. But because of (12) the w_i also form a basis of K'/K. So we find $\sum \alpha_{i\nu} a_\nu = 0$ and, because the a_ν are linearly independent with respect to k_0, $\alpha_{i\nu} = 0$ or $\alpha_\nu = 0$, as contended.

If k_0'/k_0 is even separable, all multiples of $\mathfrak{a}^{-1}$ in K' are linear combinations of those multiples in K. In fact, let $k_0' = k_0(\vartheta)$ and

$$a = b_0 + b_1\vartheta + \cdots + b_{m-1}\vartheta^{m-1}, \qquad b_j \in K,$$

be a multiple of $\mathfrak{a}^{-1}$ in K'. The divisor $a\mathfrak{a}$ remains integral if ϑ is replaced by its conjugates ϑ_i with respect to k_0; in this way we get m integral divisors $a_i\mathfrak{a}$ in some further extension field with $a_i = b_0 + b_1\vartheta_i + \cdots$. Using the Vandernomde determinant we see that then the $b_i\mathfrak{a}$ are integral, which is what we proposed to show.

Lastly we adjoin an indeterminate ϑ to k_0 and K. The substitutions $\vartheta \to \vartheta + \gamma$ with constants $\gamma \in k_0$ are automorphisms of K' which leave K fixed and take integral divisors into integral divisors. Thus, if $a(\vartheta)$ is a multiple of $\mathfrak{a}^{-1}$ in K', so is $a(\vartheta + \gamma)$. $a(\vartheta)$ is a quotient of polynomials in $[K\vartheta]$. Now, if k_0 is an infinite field, and $a(\vartheta)$ were not the product of an element of K (which could be taken into the numerator) and a polynomial in $k_0[\vartheta]$ we should get infinitely many multiples $a(\vartheta + \gamma)$ of $\mathfrak{a}^{-1}$ linearly independent with respect to $k_0(\vartheta)$, which contradicts the Riemann–Roch theorem. If k_0 is a finite field we only need extend it finitely which leaves the dimension of $\mathfrak{a}$ unchanged, in order to arrive at the same contradiction. So $a(\vartheta)$ has as its denominator a polynomial in $k_0[\vartheta]$. Multiplying $a(\vartheta)$ by it we may assume $a(\vartheta)$ to be in $K[\vartheta]$. Because $a(\vartheta + \gamma)$ is always a multiple of a^{-1}, all the coefficients of $a(\vartheta)$ must be multiples of $\mathfrak{a}^{-1}$. With this the proof is complete. ▼

Under the assumptions made above the genus of K' is at most that of K. If k_0'/k_0 is separable, K and K' have the same genus.

Proof. Take a divisor $\mathfrak{a}$ of sufficiently large degree such that (11) holds both in K and K'. Because $g(\mathfrak{a})$ is invariant and $\dim(\mathfrak{a})$ does not decrease, the genus g does not increase. It even remains invariant if the dimension of divisors does also, and this is the case with a separable extension of constants. ▼

If the above assumptions hold, and if K and K' have the same genus, then the dimension of a divisor of K is invariant under the extension.

Proof. Let $\mathfrak{a}$ be some divisor and $\mathfrak{b}$ an integral divisor of so large a degree that, for $\mathfrak{ab}$, (11) holds both in K and K'. Since the degree is an invariant and the genus g is supposed the same, the dimension of $\mathfrak{ab}$ does not change with the extension. The multiples of $\mathfrak{a}^{-1}$ are among those of $(\mathfrak{ab})^{-1}$. Now, if the extension would lead to an increase of the number of multiples of $\mathfrak{a}^{-1}$ it would do the same for those of $(\mathfrak{ab})^{-1}$. This is not the case. ▼

The genus can, indeed, decrease under extension of constants as is shown by the following example. Let k_0 be a nonperfect field of characteristic $p > 2$ and α an element of k_0 not a pth power. Let $K = k_0(x, \sqrt{x^p - \alpha})$, a separable extension of $k = k_0(x)$ as $p \neq 2$. The numerator divisor of $x^p - \alpha$ and the denominator divisor of x are squares of divisors in K. By the Dedekind discriminant theorem the discriminant $\mathfrak{D}_{K/k}$ is then divisible by the product of these divisors (and is, in fact, exactly the product). By (5) then, $g \geqq (p-1)/2$. If the pth root β of α is adjoined to k_0 then $K' = k_0'(x, \sqrt{x - \beta})$ $= k_0'(\sqrt{x - \beta})$ is the rational function field in the variable $\sqrt{x - \beta}$. Its genus is $g' = 0$.

We conclude this section with an example where K' has an exact constant field larger than k_0'. Let k_{00} be a field of characteristic $p > 0$ and λ_1, λ_2 algebraically independent indeterminates over k_{00}. Set $k_0 = k_{00}(\lambda_1, \lambda_2)$ and $K = k_0(x, \sqrt[p]{\lambda_1 x^p + \lambda_2})$. Adjoin $\vartheta = \sqrt[p]{\lambda_1}$. The element

$$\sqrt[p]{\lambda_1 x^p + \lambda_2} - \vartheta x = \sqrt[p]{\lambda_2}$$

of K' is a constant, but not contained in $k_0(\vartheta)$.

An alternative treatment of the questions of this section would use the following assumptions: $K = k_0(x; y_1, \ldots, y_m)$ where y_i is defined by an irreducible equation $f_1(y_i; x, y_1, \ldots, y_{i-1}) = 0$ with coefficients in k_0 which remains irreducible under any algebraic extension of k_0. Such equations are called absolutely irreducible. But it can be proved that this assumption amounts to the same as made above.

5. The Fields of Genus 0†

If and only if K is of genus $g = 0$ and has a divisor class of degree 1 is $K = k_0(x)$ with a suitable $x \in K$.

Proof. The fact that $k_0(x)$ really has genus $g = 0$ follows from (8) with $K = k$. Both denominator and numerator of x are divisors of degree 1. Conversely, let $g = 0$ and A have degree 1. As $\dim(A) = 2$ by (2), there must exist two linearly independent integral divisors $\mathfrak{z}$ and $\mathfrak{n}$ in A. This independence assures that the principal divisor $(x) = \mathfrak{z}/\mathfrak{n}$ is not the unit divisor, and then $x \notin k_0$. Thus the field K is a finite extension of $k = k_0(x)$ of degree, say, n. By II, §5,4 or III, §2,3 the numerator and denominator divisors of (x) in K have degree n. But these divisors $\mathfrak{z}$ and $\mathfrak{n}$ have degree 1, so that $n = 1$ and $K = k_0(x)$, as was to be shown.

Either a function field K of genus $g = 0$ with the exact constant field k_0 contains a class of degree 1 and is then $K = k_0(x)$ with suitable x, or only divisors of even degree occur. In the latter case $K = k_0(x, y)$ where an equation with coefficients in k_0 of the sort

$$\alpha_{11}x^2 + \alpha_{12}xy + \alpha_{22}y^2 + \alpha_{01}x + \alpha_{02}y + \alpha_{00} = 0 \tag{14}$$

holds between x and y, it having no solutions in k_0.

Proof. Now assume there to be no class of degree 1 in K. Clearly all degrees are multiples of some smallest degree g_0. But $g(W) = 2g - 2 = -2$ assures $g_0 = 2$. By (2) a divisor class A of degree 2 has dimension 3; let $\mathfrak{x}$, $\mathfrak{y}$, $\mathfrak{n}$ be three linearly independent divisors of A, and $(x) = \mathfrak{x}\mathfrak{n}^{-1}$, $(y) = \mathfrak{y}\mathfrak{n}^{-1}$. We maintain that, for any two elements $\alpha, \beta \in k_0$ except $\alpha = \beta = 0$, K is a second degree extension of $k_0(\alpha x + \beta y)$. Indeed, the denominator of the principal divisor $(\alpha x + \beta y)$ is $\mathfrak{n}$, and is of degree 2. By the same argument as above, then, $[K : k_0(\alpha x + \beta y)] = 2$.

Thus, every linear combination $\alpha' x + \beta' y \neq 0$ satisfies an equation of degree 1 or 2 over $k_0(\alpha x + \beta y)$,

$$(\alpha' x + \beta' y)^2 + (\alpha' x + \beta' y) f_1(\alpha x + \beta y) + f_2(\alpha x + \beta y) = 0.$$

All of these equations must originate in a single equation, which is only possible when it is of the form (14).

In particular, $\alpha_{22} \neq 0$. For $\alpha_{22} = 0$ would imply that either $K = k_0(x)$ or x and y were independent. The latter is impossible, and the former contradicts the hypothesis that K has no divisor of the first degree.

We must finally show that (14) has no solutions in k_0. If ξ, η were a

† To gain practice in the basic concepts it is advisable to read IV, §2,1 after the present section. There fields of genus 1 are similarly considered.

solution and $\mathfrak{p}$ a divisor by which the numerator of $x - \xi$ were divisible, then $\mathfrak{p}$ would also divide the numerator of

$$\alpha_{22}y^2 + (\alpha_{12}\xi + \alpha_{02})y + \alpha_{11}\xi^2 + \alpha_{01}\xi + \alpha_{00} = 0.$$

But by assumption this polynomial in y decomposes into two factors of the first degree $\alpha_{22}(y - \eta)(y - \eta')$, with $\alpha_{22} \neq 0$. The numerator of $(y - \eta)$ or $(y - \eta')$ would be divisible by $\mathfrak{p}$, say $(y - \eta)$. Now, the numerators of $(x - \xi)$ and $(y - \eta)$ are both of degree 2 and, as there are no divisors of degree 1, the divisors of both these numerators would be $\mathfrak{p}$ and $g(\mathfrak{p}) = 2$. On the other hand, $(x - \xi)$ and $(y - \eta)$ have denominator $\mathfrak{n}$, which would imply $(x - \xi) = (y - \eta)$, contradicting the linear independence of $\mathfrak{x}$, $\mathfrak{y}$, $\mathfrak{n}$.

Conversely, let (14) be unsolvable in k_0. Then either $\alpha_{22} \neq 0$ or $\alpha_{22} = \alpha_{12} = \alpha_{02} = 0$. The latter would make y independent of x. Now form the field $K = k_0(x, y)$ with this equation. Let us first determine the genus of K. The principal order $\mathfrak{J}$ of K with respect to

$$\mathfrak{i} = \bigcap_{p \neq \infty} \mathfrak{i}_p = k_0[x]$$

has the basis $1, y$, as $\alpha_{22} \neq 0$. The same argument, after (14) has been divided by x^2, shows that $1, y/x$ is a basis of $\mathfrak{J}_\infty$, the principal order of K with respect to $\mathfrak{i}_\infty$. Define a pseudotrace

$$\sigma(a_1 + a_2 y) = a_2 \qquad (a_1, a_2 \in k_0(x)).$$

The pseudodiscriminant of $\mathfrak{J}$ is then

$$\mathfrak{D}_\mathfrak{i} = \mathfrak{i}\,\Delta(1, y) = \mathfrak{i}$$

and the pseudodiscriminant of $\mathfrak{J}_\infty$ is a divisor of

$$\mathfrak{i}_\infty\,\Delta(1, y/x) = \mathfrak{i}_\infty \mathfrak{n}^2,$$

where $\mathfrak{n}$ is the denominator divisor of (x) in $k = k_0(x)$. Then

$$g_k(\mathfrak{D}_{K/k}) \leqq 2.$$

Equation (5) for the genus then assures $g = 0$, for $g < 0$ is impossible.

There can be no divisor class in K of degree 1, for then we would have $K = k_0(z)$ with some z. This would make $x = f(z)$ and $y = g(z)$ with rational functions over k_0. Setting this into (14) with an arbitrary z in k_0 would lead to a solution of (14) in k_0, contrary to hypothesis. The assertion is thus proved in full. ▼

6. Lüroth's Theorem

Let K have the exact constant field k_0 and genus 0, and L be an intermediate field not equal to k_0. Then L is a function field with exact constant field k_0 of genus 0. If $K = k_0(z)$ then $L = k_0(t)$ with a suitable $t \in L$.

Proof. Let x be some element of L not in k_0, and $k = k_0(x)$. L and K are then finite extensions of k. Applying the Hurwitz genus formulas (9) and (10) with $g = g(K/k) = 0$ gives

$$[K:L]g(L/k) = [K:L] - \tfrac{1}{2}g_L(\mathfrak{D}_{K/L}) - 1 < [K:L].$$

The genus $g(L/k)$ of L must therefore be smaller than 1, i.e., 0.

If $K = k_0(z)$, and were L not generated in this manner, then $L = k_0(x, y)$ would hold with x, y satisfying Eq. (14). Now, x and y are rational functions of z with coefficients in k_0. Setting these into (14) with an arbitrary z in k_0 leads to a solution of (14) in k_0, contrary to hypothesis. ▼

The extensions k_0'/k_0 in which (14) becomes solvable coincide with the splitting fields of a certain division algebra of degree 2 with center k_0. There is a similar connection between division algebras of degree $m > 2$ and center k_0 and certain function fields K/k_0 in $m - 1$ variables. See P. ROQUETTE, *On the Galois cohomology of the projective linear group and its application to the construction of generic splitting fields of algebras*, Math. Ann. **150** 411–439 (1963).

7. FURTHER PROOFS AND GENERALIZATIONS OF THE RIEMANN–ROCH THEOREM

It is impossible to give a complete survey of all proofs and generalizations here. Of course, we may omit all extensions to functions of several variables, especially as the notion of a generalization of the Riemann–Roch theorem is interpreted with some generosity there. We also omit methods leading from the theory of algebraic curves (found, for example, in [11]†).

Basically, three proofs then remain. The very first proof dates back to Riemann. It stays in the framework of classical function theory, and is based on the so-called Dirichlet principle, that is, on existence theorems for solutions of the partial differential equation $\Delta U = 0$. Modern treatments are found in [1] and [12]. From the point of view of the algebraist today, this access is rather a field of application and exercise for certain methods of analysis than a constructive proof, as is wanted for any development of the theory. These intentions also appear to motivate Teichmueller's proof [2], based on the uniformization of algebraic functions and the theory of Poincaré series. Otherwise one would have to call this foundation of the theory the longest detour thus far to the Riemann–Roch theorem.

Two substantial extensions of the Riemann–Roch theorem, based on generalizations of the concept of divisor, fall into connection here. They are related to each other, but each surpasses the other in one direction. One of them is due to Weil [10] and the other to Petersson; the latter will come up in IV, §4,9.

† The numbers in brackets refer to the bibliography at the end of this section.

While the proof of Dedekind and Weber consists of substituting the function-theoretical concepts into an already standing general statement, Weil's proof [9] starts with a new idea (cf. also [3], [4], [7], [8]). The residue theorem $\sum \operatorname{res}_{\mathfrak{p}}(\alpha_1\, du) = 0$ (cf. §4) gives, for every differential du, a linear relation between the local power series expansions of any function $\alpha \in K$. Inverting this fact, define a system of such relations (for all α) to be a differential. It is easy to associate a divisor with a differential in this sense; the divisors of all differentials lie in the same class, the canonical class. The Riemann–Roch theorem is a statement concerning the number of solutions of two associated problems of linear algebra. These problems are, despite formally different presentation, the same here and in the proof of Dedekind and Weber. Thus, seen from a higher vantage point, the proofs do not differ essentially.

It should still be mentioned that the Dedekind–Weber proof is to be found in Schmidt [6] in somewhat modernized form.

The Riemann–Roch theorem has been generalized in three senses. Rosenlicht [5] based his divisor concept on ideals of more general rings of K than our $\mathfrak{J}_p$ (cf. also [4]). Witt [13] considered the theory for hypercomplex systems in place of fields. The third generalization has already been mentioned. All these generalizations can easily be derived from the theorem in I, §3,3.

References

[1] M. Schiffer and D. C. Spencer, *Functionals on finite Riemann surfaces.* Princeton, 1954.

[2] H. Behnke and F. Sommer, *Theorie der analytischen Funktionen*, p. 530. Berlin-Göttingen-Heidelberg, 1955.

[3] C. Chevalley, *Introduction to the theory of algebraic functions of one variable.* New York, 1951.

[4] P. Roquette, *Über den Riemann–Rochschen Satz in Funktionenkörpern vom Transzendenzgrad* 1, Math. Nachr. **19**, 375–404 (1958).

[5] M. Rosenlicht, *Equivalence relations on algebraic curves*, Ann. of Maths. **56**, 169–191 (1952).

[6] F. K. Schmidt, *Zur arithmetischen Theorie der algebraischen Funktionen I*, Math. Z. **41**, 415–438 (1936).

[7] T. Tamagawa, *On the theorem of Riemann–Roch*, J. Fac. Sci. Univ. Tokyo Sect. I, **6**, 133–144 (1951).

[8] B. L. van der Waerden, *Algebra II*, 4th ed. Berlin-Göttingen-Heidelberg, 1959.

[9] A. Weil, *Zur algebraischen Theorie der algebraischen Funktionen*, J. Reine Angew. Math. **179**, 129–133 (1938).

[10] A. Weil, *Généralisation des fonctions Abéliennes*, J. Math. Pures Appl. [IX], **17**, 47–87 (1938).

[11] A. Weil, *Sur les courbes algébriques et les variétés qui s'en déduisent*, Actualités Sci. Ind. 1041 (1948).

[12] H. Weyl, *Die Idee der Riemannschen Fläche*, 3rd ed. Berlin, 1955.

[13] E. Witt, *Riemann–Rochscher Satz und Zetafunktion im Hyperkomplexen*, Math. Ann. **110**, 12–28 (1934).

§4. Differentials

1. Differential Quotients

From now until the end of Chapter IV we will always assume K to be separably generated. If a field has the characteristic 0 then of course all its elements are separating. But, even in the case of prime characteristic, we have:

If the field of constants is perfect, then K is separably generated.

Proof. Let the characteristic be $p > 0$, and let K be generated by adjoining the finite number of elements $x, y_1, \ldots, y_h$ to k_0, among which the equations $f_i(y_i, x)$ hold. First, let $h = 1$. If $f(y, x)$ is not a polynomial in y^p and x^p, then $f(y, x)$ is separable either with respect to x or with respect to y. Thus either $K/k_0(y)$ or $K/k_0(x)$ is separable. If, however, $f(y, x) = F(y^p, x^p)$, let $F^{p^{-1}}$ be that polynomial whose coefficients are the pth roots of those of F. By hypothesis these coefficients lie in k_0. Then $f(y, x) = (F^{p^{-1}}(y, x))^p = 0$, so that $F^{p^{-1}}(y, x) = 0$. If this polynomial again contains only powers of x and y divisible by p, the same argument can be repeated. After a finite number of steps an equation $f'(y, x)$ results, separable in y or in x. Thus the equation of minimal degree between y and x is separable.

Now, say we have already proved that $K_1 = k_0(x, y_1, \ldots, y_{h-1})$ is separably generated, and x_1 is a separating variable. Another irreducible equation $f(y_h, x_1) = 0$ holds, and is separable in either y_h or x_1 by the above argument. Depending upon which of these two is the case, $k_0(x_1, y_h)/k_0(x_1)$ or $k_0(x_1, y_h)/k_0(y_h)$ is separable. But, as $K_1/k_0(x_1)$ is separable, $K = K_1(y_h)$ is separable over either $k_0(x_1)$ or $k_0(y_h)$. This completes the proof.

Using the theorem of the primitive element, a separable extension $K/k_0(x)$ can be generated as $k_0(x, y)$. We will always assume this to be the case, with the irreducible equation

$$f(y, x) = 0$$

relating x and y. Let $x = \xi \neq \infty$ be a regular place of k_0 in the sense of §1,4. Then y and with it every function $u \in K$ can be expanded in a power series

$$u = \sum_{\mu=\nu}^{\infty} a_{\varrho\mu} z_\xi^{\mu} \qquad (z_\xi = x - \xi)$$

at a place ξ_ϱ over ξ. This expansion is, by §1, an isomorphic mapping of K onto a subfield of the field of power series in z_ξ.

A *differential quotient* is defined by

$$du/dx = \sum_{\mu=\nu}^{\infty} \mu a_{\varrho\mu} z_\xi^{\mu-1}.$$

The familiar sum, product, and quotient rules hold. In particular,

$$\frac{\partial f}{\partial y}\frac{dy}{dx}+\frac{\partial f}{\partial x}=f_y\frac{dy}{dx}+f_x=0.$$

Because x is a separating element and f is irreducible, we have $f_y \neq 0$, and thus

$$dy/dx = -f_x/f_y. \tag{1}$$

This equation shows that the differential quotient again belongs to the field K, and that it can always be determined by (1). Formula (1) is correct, by its derivation, for all $y \in K$. The detour to power series was necessary only to demonstrate that the formal rules of differentiation hold. Also from (1) it is seen that the differential quotient does not depend upon the place ξ.

We end this section by proving the chain rule

$$\frac{dv}{du}\frac{du}{dx}=\frac{dv}{dx}.$$

To do this we use the equations relating v, u, and x:

$$f_1(v, u)=0, \qquad f_2(u, x)=0.$$

Let $u=u_1, \ldots$ be the critical places of the first equation. Further, let $x=\xi$ be a regular place of the second equation and $f_2(u_1, \xi)\neq 0, \ldots$. Solve $f_2(u, \xi)=0$ for u, the solutions $u=\eta$ being regular places for the first equation due to our assumptions. Thus the two power series

$$v=\sum a_\mu(u-\eta)^\mu, \qquad u-\eta=\sum b_\mu(x-\xi)^\mu$$

solve these equations, and the chain rule holds for these power series. Surely it also holds when the differential quotient is computed by the rule (1).

2. The Differential Calculus with Characteristic p

Several remarks are in order for the case of a field of characteristic $p>0$.

It is familiar that raising to the power p is an isomorphic mapping onto a subfield usually denoted by K^p. We introduce the field

$$K_0 = K^p k_0. \tag{2}$$

For a separating element x of K

$$K = K_0(x) \tag{3}$$

holds.

Proof. As $K/k_0(x)$ is separable there exists a primitive element y, such that $K=k_0(x, y)$ and y satisfies an irreducible equation $f(y, x)=0$ of degree

$n = [K : k_0(x)]$ over $k_0(x)$. Then, over $k_0(x^p)$, y^p satisfies the equation $f(y, x)^p = F(y^p, x^p) = 0$. It is irreducible in $k_0(x^p)$. For otherwise $f(y, x)$ would become reducible, also, after pth roots of certain elements were adjoined to k_0, which would contradict the separability of $K/k_0(x)$. Now, $F(y^p, x^p)$ has the same degree n in y^p, so that

$$[K_0 : k_0(x^p)] = [k_0(y^p, x^p) : k_0(x^p)] = [K : k_0(x)] = n. \tag{4}$$

As x satisfies an irreducible equation of degree p over $k_0(x^p)$, we have

$$[k_0(x) : k_0(x)^p] = p. \tag{5}$$

Equation (4) leads to

$$[K : k_0(x^p)] = [K : K_0][K_0 : k_0(x^p)] = [K : K_0][K : k_0(x)]$$

and (5) leads to

$$[K : k_0(x^p)] = [K : k_0(x)][k_0(x) : k_0(x^p)] = p[K : k_0(x)].$$

Comparison of the two finally gives

$$[K : K_0] = p. \tag{6}$$

Now, x also satisfies an irreducible equation of degree p over K_0, so that

$$[K_0(x) : K_0] = p.$$

But $K_0(x) \subseteq K$. Comparison of the last equation with (6) then yields the assertion (3). ▼

Because of (3) we can write every element $u \in K$ in the form

$$u = u_0 + u_1 x + \cdots + u_{p-1} x^{p-1}, \qquad u_i \in K_0, \tag{7}$$

uniquely. The concept of differential quotient can now be introduced in yet another way, by setting

$$du/dx = u_1 + 2u_2 x + \cdots + (p-1)u_{p-1} x^{p-2}. \tag{8}$$

The *differential quotient* thus defined also satisfies the familiar sum, product, and quotient rules. The chain rule,

$$\frac{du}{dv}\frac{dv}{dx} = \frac{dv}{dx},$$

follows immediately, even if v is not in K_0. The verification actually remains in the realm of polynomials (7), the key to the argument being that $dx^p/dx = 0$ not only from the definition (8), but also as a result of the product rule and the fact that the characteristic is p.

The derivatives of elements $a \in K_0$ are always 0, and these are the only elements with vanishing derivative. We therefore call them the *p-constants*, while we call a $u \notin K_0$ a *p-variable*. For the case of characteristic p, therefore, the field K_0 defined by (2) plays the part played by the exact constant field when the characteristic is 0. For convenience we also speak of p-constants in that case, setting $K_0 = k_0$.

All p-variable elements are separating elements. If x and y are p-variable then the first partial derivatives f_x and f_y of the irreducible equation $f(y, x)$ between them do not vanish, and dy/dx can be computed by (1).

This also shows that the differential quotient defined by (8) corresponds to that of the last section.

Proof. Let x be a separating variable and u a p-variable of the form (7). Then $K \supseteq K_0(u) \supset K_0$. By (6) we then have

$$K = K_0(u).$$

Now, let $K/k_0(u)$ be inseparable, contrary to our assertion, and K_1 the largest subfield of K separable with respect to $k_0(u)$. Then $K_1 \neq K$, and K is obtained from K_1 by the adjunction of some p^eth roots. Therefore even $K_1 K_0 \neq K$. But $K_1 K_0 \supseteq K_0(u) = K$. This is a contradiction.

Let x and y be p-variables and $f(y, x)$ the irreducible equation between them. As $K/k_0(x)$ is separable so is $k_0(y, x)/k_0(x)$, so that $f_y \neq 0$. By symmetry $f_x \neq 0$, so that (1) must hold. ▼

Later we shall need the following lemma:

If x and y are p-variables, and the expansion

$$\frac{x}{y}\frac{dy}{dx} = g_0 + g_1 x + \cdots + g_{p-1} x^{p-1}, \qquad g_i \in K_0, \tag{9}$$

holds, then

$$g_0 = \left(\frac{x}{y}\frac{dy}{dx}\right)^p = \frac{x^p}{y^p}\frac{dy^p}{dx^p}, \tag{10}$$

where the differential quotient dy^p/dx^p is to be taken in K^p as the derivative of y^p with respect to x^p.

Proof. With p-constant coefficients write

$$y = \alpha_0 x^n + \cdots + \alpha_n, \qquad \alpha_0 \neq 0, \qquad n < p.$$

In some separable (because $n < p$) extension K_0' of K_0 this polynomial decomposes into the product

$$y = \alpha_0 \prod_{\nu=1}^{n} (x - \beta_\nu). \tag{11}$$

We also define a derivation of the elements of $K_0'(x)$ with respect to x by treating the elements of K_0' as constants. As x is not in K_0' (it is inseparable over K_0 while K_0'/K_0 is separable) this is possible. A simple calculation then yields

$$\frac{x}{y}\frac{dx}{dy} = \sum_{\nu=1}^{n} \frac{x}{x-\beta_\nu} = \sum_{\nu=1}^{n}\left(1 - \frac{1}{1 - x/\beta_\nu}\right)$$

$$= \sum_{\nu=1}^{n}\left(1 - \frac{1}{1-(x/\beta_\nu)^p}(1 + x/\beta_\nu + \cdots + (x/\beta_\nu)^{p-1})\right).$$

The asserted equation

$$g_0 = \sum_{\nu=1}^{n}\left(1 - \frac{1}{1-(x/\beta_\nu)^p}\right) = \left(\sum_{\nu=1}^{n}\left(1 - \frac{1}{1-x/\beta_\nu}\right)\right)^p$$

is contained therein. ▼

Exercise. Prove that if K is not separably generated there can exist no element x for which (3) holds. There then always exist several differentiations for which the elements of K_0 have derivatives 0.

3. The Concept of the Differential

By a *differential*

$$dw = u\,dv \tag{12}$$

we mean a pair of elements $u, v \in K$ with $v \notin K_0$, where two such differentials dw and $dw' = u'dv'$ are said to be *equal* whenever

$$u = u'\,dv'/dv.$$

Obviously the differentials form a one-dimensional K-module. A divisor can be made to correspond with each differential, but the procedure depends upon the nature of the constant field k_0. *In this section we assume k_0 to be algebraically closed.* Let $\mathfrak{p}$ be a prime divisor, and q an element of K with the order $v_\mathfrak{p}(q) = 1$, i.e., a prime element for $\mathfrak{p}$. It cannot be a p-constant. This is self-evident for characteristic 0. For characteristic $p > 0$ and algebraically closed constant field, K_0 consists of the pth powers of elements of K. A prime element cannot be a pth power.† Now, set

$$v_\mathfrak{p}(dw) = v_\mathfrak{p}(u\,dv/dq). \tag{13}$$

This definition is independent of the arbitrary choice of q.

Proof. If k_0 is algebraically closed, then by II,§5,5 every prime divisor is of the first degree. By §1,2, then, every element of K can be expanded in a power

† This argument does not hold if k_0 is not closed. For example, $q = x^p - a$ with $a \in k_0$ can be a prime element.

series in q with constant coefficients. In particular, any other prime element can be written as

$$q' = c_1 q + c_2 q^2 + \cdots, \qquad c_1 \neq 0, \quad c_\mu \in k_0 .$$

Using the reasoning of §4,1 we have

$$dq'/dq = c_1 + 2c_2 q + \cdots,$$

so that $v_\mathfrak{p}(dq'/dq) = 0$ and

$$v_\mathfrak{p}\left(u\,\frac{dv}{dq'}\right) = v_\mathfrak{p}\left(u\,\frac{dv}{dq'}\,\frac{dq'}{dq}\right) = v_\mathfrak{p}\left(u\,\frac{dv}{dq}\right).$$

▼

The orders of (13) *are, save for a finite number, all* 0. *Thus*

$$(dw) = \prod_{\mathfrak{p}} \mathfrak{p}^{v_\mathfrak{p}(dw)} \tag{14}$$

defines a divisor.

Given two differentials,

$$\frac{(dw_1)}{(dw_2)} = \frac{(u_1\,dv_1)}{(u_2\,dv_2)} = \left(\frac{u_1}{u_2}\,\frac{dv_1}{dv_2}\right) \tag{15}$$

holds so that, in particular, the quotient of the divisors of the differentials for two p-variables is equal to the principal divisor of their differential quotient.

Proof. The chain rule yields (15). To show that $v_\mathfrak{p}(u\,dv) = 0$ for almost all $\mathfrak{p}$, remember from the last section that K is a separable extension of $k_0(v)$. Let y be a primitive element of K with respect to that subfield, and $f(y, v) = 0$ the irreducible equation between y and v. By §1,4 almost all the places $v = \xi$ are regular, so that y, as well as any function of K, can be expanded in a power series in $v - \xi$ at all places ξ_ϱ over ξ. Thus $v - \xi$ is a prime element for almost all ξ_ϱ and hence for almost all $\mathfrak{p}$. Setting $q = v - \xi$ in (13) gives $dv/dq = 1$. This proves the assertion, as $v_\mathfrak{p}(u) = 0$ almost always. ▼

Prime divisors $\mathfrak{p}$ with $v_\mathfrak{p}(dw) < 0$, or the places corresponding thereto, are called the *poles* of dw. In the corresponding sense we may speak of the *zeros* of dw.

For algebraically closed k_0 any p-variable x satisfies

$$(dx) = \mathfrak{d}_{K/k_0(x)}\mathfrak{z}_\infty^{-2}, \tag{16}$$

where $\mathfrak{z}_\infty$ is the denominator divisor of x. Further, the divisors of all differentials belong to the canonical class.

The proof here will assume the characteristic to be 0; in §4,8 the general case will be considered. By §2,4 there exists, to every place ξ_ϱ lying over a place ξ, a function q_ϱ with Puiseux expansion

$$q_\varrho = \sqrt[e_\varrho]{z_\xi} + \cdots,$$

the dots indicating higher powers of $z_\xi = x - \xi$ or x^{-1}. These q_ϱ are prime elements for the places ξ_ϱ. For the orders this means

$$v_{\xi_\varrho}(dx/dq) = \begin{cases} e_\varrho(\xi) - 1 & \text{for} \quad \xi \neq \infty \\ -e_\varrho(\xi) - 1 & \text{for} \quad \xi = \infty. \end{cases}$$

Altogether,

$$(dx) = \prod_{\xi_\varrho} \mathfrak{z}_{\xi_\varrho}^{e_\varrho(\xi)-1} \prod_{\varrho} \mathfrak{z}_{\infty_\varrho}^{-2e_\varrho(\infty)} = \prod_{\xi_\varrho} \mathfrak{z}_{\xi_\varrho}^{e_\varrho(\xi)-1} \cdot \mathfrak{z}_\infty^{-2}. \tag{17}$$

Using §2,(35) this is equivalent to (16).

In §3,(4) we saw that $\mathfrak{d}_{K/k_0(x)}\mathfrak{z}_\infty^{-2}$ lies in the canonical class W, and thus with (15) all divisors of differentials dw lie in W. ▼

4*. Continuation; Separable and Inseparable Prime Divisors

The residue class field of $\mathfrak{J}_\mathfrak{p}$ modulo a prime divisor $\mathfrak{p}$ of degree $g(\mathfrak{p}) = r$ was seen in II, §5,5 to be an extension of k_0 of degree r. The prime divisor $\mathfrak{p}$ is called *separable* or *inseparable*, depending upon whether this extension is separable or inseparable.

We now want to associate a divisor with any differential in a separably generated function field over an arbitrary field of constants k_0. Let k_0'' be the algebraic closure of k_0 and $K'' = Kk_0''$ (cf. §3,4). We define (dw) as the divisor in K'' of the last section, but show:

There exists a purely inseparable finite extension of constants k_0'/k_0 in which (dw) is defined.

Proof. First let $\mathfrak{p}$ be a separable prime divisor and $\bar{\vartheta}$ a primitive element of the residue class field $\mathfrak{J}_\mathfrak{p}/\mathfrak{p}$ over k_0. Let it satisfy the irreducible equation $f(\bar{\vartheta}) = \bar{\vartheta}^r + \cdots = 0$, $f(\bar{\vartheta})$ being a separable polynomial of degree $r = g(\mathfrak{p})$. Hence there must exist a separable extension k_1/k_0 in which $f(\bar{\vartheta})$ decomposes to r mutually prime factors $\bar{\vartheta} - \alpha_i$. For an element ϑ in the residue class $\bar{\vartheta}$, $f(\vartheta)$ is an element of K whose numerator is divisible by $\mathfrak{p}$. In $K_1 = Kk_1$ we have the decomposition

$$f(\vartheta) = (\vartheta - \alpha_1) \cdots (\vartheta - \alpha_r)$$

into r factors. The difference between any two of them is prime to $\mathfrak{p}$. This leads to the decomposition of $\mathfrak{p}$ to

$$\mathfrak{p} = \mathfrak{p}_1 \cdots \mathfrak{p}_r \tag{18}$$

in K_1. Comparison of degrees shows that the $\mathfrak{p}_\varrho$ are all of degree 1, and must therefore be prime divisors.

A prime element q for $\mathfrak{p}$ in K is a prime element for every $\mathfrak{p}_\varrho$ in K_1, as is seen by (18). For a differential (12) in K and the orders defined by (13) this means

$$\nu_{\mathfrak{p}_1}(dw) = \cdots = \nu_{\mathfrak{p}_r}(dw) = \nu_{\mathfrak{p}}(dw),$$

so that

$$\prod_{\varrho=1}^{r} \mathfrak{p}_\varrho^{\nu_{\mathfrak{p}_\varrho}(dw)} = \mathfrak{p}^{\nu_{\mathfrak{p}}(dw)}.$$

This shows that the contributions of the individual $\mathfrak{p}_\varrho$ to the divisor (dw) altogether yield a power of $\mathfrak{p}$, which is, of course, a divisor of K. *Thus the* $\mathfrak{p}$-*component of* (dw) *for separable prime divisors* $\mathfrak{p}$ *can be computed as in the last section.*

It is well known that, for an inseparable prime divisor $\mathfrak{p}$, a finite purely inseparable extension k_0' of k_0 can be found such that the field composed of k_0' and $\mathfrak{J}_\mathfrak{p}/\mathfrak{p}$ is separable over k_0'. Then $\mathfrak{p}$ is decomposed into separable prime divisors over the extension $K' = Kk_0'$, and the above argument can be applied. This need only be done for finitely many $\mathfrak{p}$ because, as it was seen in the last section, only finitely many $\mathfrak{p}$ contribute to (dw). The assertion is proved. ▼

Simultaneously we have proved the

Lemma. *A finite extension of constants $K' = Kk_0'$ always exists in which a given prime divisor decomposes into prime divisors of the first degree. For separable* $\mathfrak{p}$ *this extension can even be chosen separable.*

By definition we have: *The divisor (dw) of a differential is invariant under extension of constants.*

Formula (15) *is true in the general case, and therefore divisors of all differentials belong to the same class.*

Formula (16), on the other hand, is no longer generally true, and the divisors of differentials need not lie in the canonical class. Consider, for example, the field $K = k_0(x, \sqrt{x^p - a})$ mentioned in §3,4. Now, as the divisors of differentials are defined by closure of the constant field, they are invariant under extension of constants. But in the previous mention of this field we saw that its genus changes under extension of constants, and therefore also its canonical class. §4,8 will again deal with this question.

5. Cartier's Operator

We undertake some considerations in preparation of what follows, assuming the characteristic to be $p > 0$. *A differential of K,*

$$u\,dx/x = (u_0 + u_1 x + \cdots + u_{p-1}x^{p-1})\,dx/x, \qquad u_i \in K_0, \tag{19}$$

is in unique correspondence with the differential $u_0\,dx^p/x^p$ of K_0; that is, the correspondence does not depend upon the element x used. By dx^p we mean the differential of the element x^p in K_0. A linear map S of the differentials of K onto those of K_0 is clearly defined by this theorem; it satisfies

$$S(\alpha_1\,dw_1 + \alpha_2\,dw_2) = \alpha_1 S(dw_1) + \alpha_2 S(dw_2), \qquad \alpha_1, \alpha_2 \in K_0. \tag{20}$$

The proof uses the following not inessential concept. A differential dw is called *exact* if it is the differential of an element of K. For example, $x^{i-1}\,dx = dx^i/i$ for $i \not\equiv 0 \bmod p$. Similarly to all differentials, the exact differentials form a K_0-module, denoted by E, so that we can write

$$dw_1 \equiv dw_2 \bmod E$$

whenever $dw_1 - dw_2$ is exact. Clearly $u_0\,dx^p/x^p$ is a function of the class of $u\,dx/x \bmod E$.

Now let y be another p-variable, and

$$\frac{y}{x}\frac{dx}{dy} = g_0 + g_1 y + \cdots + g_{p-1}y^{p-1}, \qquad g_i \in K_0.$$

Then

$$u\frac{dx}{x} \equiv u_0\frac{dx}{x} = u_0\frac{y}{x}\frac{dx}{dy}\frac{dy}{y} \equiv u_0 g_0\frac{dy}{y} \bmod E.$$

But, by the lemma in §4,2, this can be expressed as

$$u\frac{dx}{x} \equiv u_0\frac{y^p}{x^p}\frac{dx^p}{dy^p}\frac{dy}{y} \bmod E.$$

Therefore, if y is used as the variable instead of x, the differential $S(u\,dx/x)$ corresponding to $u\,dx/x$ in K_0 becomes

$$u_0\frac{y^p}{x^p}\frac{dx^p}{dy^p}\frac{dy^p}{y^p}.$$

If instead, $u_0\,dx^p/x^p$ is immediately calculated out for the variable y^p the result is the same, completing the proof. ▼

The following statements are obvious. *A place $\mathfrak{p}$ of K lies over exactly one place $\mathfrak{p}_0$ of K_0. If (dw) is integral at $\mathfrak{p}$ then $S(dw)$ is integral at $\mathfrak{p}_0$.*

The operator S was discovered by P. Cartier†; it can be defined in the same sense for functions of several variables.

6. Residues of Differentials‡

First consider a place $\mathfrak{p}$ of the first degree. In §1,2 we saw that all functions can be expanded in power series with coefficients in k_0 in terms of powers of a

---

† *Une nouvelle opération sur les formes différentielles*, Compt. Rend. Paris **244**, 426–428 (1957) and *Questions de rationalité des diviseurs en géométrie algébrique*, Bull. Soc. Math. France **86** 177–251 (1958).

‡ Another approach to this question is found in H. Hasse, *Theorie der Differentiale in algebraischen Funktionenkörpern mit vollkommenem Konstantenkörper*, J. Reine Angew. Math. **172**, 55–64 (1934), as well as in the book [3] in §3,7. In the latter a concept of the differential is used which need only coincide with ours for perfect fields of constants.

prime element q. Let a differential

$$dw = u\,dv = u\,\frac{dv}{dq}\,dq = \left(\sum_{\mu} c_\mu q^\mu\right)\frac{dq}{q}, \qquad c_\mu \in k_0, \tag{21}$$

be given; define the *residue of dw at the place* $\mathfrak{p}$ as the coefficient

$$\operatorname{res}_{\mathfrak{p}}(dw) = c_0 \tag{22}$$

of the power series expansion (21). For two differentials dw_1 and dw_2, and two constants $\alpha_1, \alpha_2 \in k_0$ this implies

$$\operatorname{res}_{\mathfrak{p}}(\alpha_1\,dw_1 + \alpha_2\,dw_2) = \alpha_1 \operatorname{res}_{\mathfrak{p}}(dw_1) + \alpha_2 \operatorname{res}_{\mathfrak{p}}(dw_2). \tag{23}$$

It must be shown that the definition (22) does not depend upon the choice of prime element q. But, using (23), we need only prove invariance for differentials $q^\mu\,dq/q$. Let q' be another prime element, and

$$q = a_1 q' + a_2 q'^2 + \cdots, \qquad a_1 \neq 0, \quad a_i \in k_0. \tag{24}$$

Then

$$\frac{dq}{q} = \frac{dq'}{q'}(1 + \cdots).$$

This shows the invariance for $\mu = 0$.

For μ not divisible by the characteristic we have

$$q^\mu \frac{dq}{q} = \frac{1}{\mu}\,dq^\mu.$$

Substituting (24) on the right we get the differential of a power series, which cannot contain a term in dq'/q'.

All that remains is a case where the characteristic $p > 0$. Here we can apply Cartier's operator S. Clearly

$$\operatorname{res}_{\mathfrak{p}}(dw) = \operatorname{res}_{\mathfrak{p}_0}(S(dw)), \tag{25}$$

where $\mathfrak{p}_0$ is the place of K_0 over which $\mathfrak{p}$ lies. Repeated application of S eventually leads to a differential, integral at the respective place, or with a pole of order 1 there. In both cases the residue is proved independent of the prime element. ▼

Now we can let $\mathfrak{p}$ be a separable prime divisor of a higher degree. By the lemma of §4,4 there exists a separable extension k_0'/k_0 such that $\mathfrak{p}$ decomposes into prime ideals $\mathfrak{p}_1', \ldots, \mathfrak{p}_r'$ of the first degree in $K' = Kk_0'$. No generality is lost in assuming the extension to be normal; let the Galois group be $\mathfrak{G}$. By §3,4 k_0' is then the exact constant field of K', and K' has the same Galois group $\mathfrak{G}$ over K. The residue at the place $\mathfrak{p}$ is defined by

$$\operatorname{res}_{\mathfrak{p}}(dw) = \sum_{\varrho=1}^{r} \operatorname{res}_{\mathfrak{p}_\varrho'}(dw) \tag{26}$$

and it will be proved: *the residue of a separable prime divisor lies in the constant field.*

For, let q_1' be a prime element for $\mathfrak{p}_1'$ in the decomposition field of $\mathfrak{p}_1'$ (cf. II, §6,1). Its behavior under automorphisms of $\mathfrak{G}$ will be that of $\mathfrak{p}_1'$. The same holds true for the power series

$$dw = \sum_{\mu} (c_{1\mu} q_1'^{\mu})\, dq_1'/q_1'$$

as well as for their coefficients $c_{1\mu}$. Applying the substitutions S_ϱ of II, §6,(1) yields r such power series. Their coefficients and all the elements Z of the decomposition group of $\mathfrak{p}_1'$ satisfy

$$c_{1\mu}^{ZS_\varrho} = c_{1\mu}^{S_\varrho} = c_{\varrho\mu}.$$

Therefore,

$$\operatorname{res}_p(dw) = \sum_{\varrho=1}^{r} c_{\varrho,0} = \frac{r}{[k_0' : k_0]}\, s_{k_0'/k_0}(c_{1,0}). \quad ▼$$

Now let $\mathfrak{q}$ be a "purely inseparable" prime divisor, that is, let it decompose to $\mathfrak{q} = \mathfrak{q}'^{p^h}$ with $\mathfrak{q}'$ of degree 1 in a purely inseparable extension k_0' of k_0. Then define the residue of dw at $\mathfrak{q}$ to be equal to the residue of dw at the place $\mathfrak{q}'$ of K'. *This residue also lies in the constant field* k_0, *without extension.*

Proof. Apply the operator S h times, remembering (25), and considering $\mathfrak{q} = \mathfrak{q}'^{p^h}$ as a separable divisor of the field $k_0'K^{p^h}$. On the other hand, $S^h(dw)$ is a differential of $k_0K^{p^h}$. Thus the residue lies in k_0. ▼

Exercise. Combining the last two steps define the residue of a differential at an arbitrary place and show that it lies in the field k_0.

Finally we remark that as the orders of a differential are almost always 0 *the residues of a differential are* 0 *at almost every place.*

7. The Residue Theorem

The sum of the residues of a differential in a separably generated field is 0.

The proof, which is an easy corollary of the Cauchy integral theorem in the classical case, takes considerably more effort in general. We may assume we have an algebraically closed constant field, for (26) shows that the sum of residues is unaffected by a separable extension of k_0 and it does not change under an inseparable extension as a matter of definition.

Let us first take care of the rational function field $K = k_0(x)$. Decompose u of the differential $dw = u\,dx$ to the familiar partial sums

$$u = \sum_{\xi \neq \infty} \left(\frac{c_{\xi,-1}}{x - \xi} + \frac{c_{\xi,-2}}{(x-\xi)^2} + \cdots \right) + c_{\infty,2} + c_{\infty,3}x + \cdots.$$

We can now prove the theorem for the individual summands

$$dx/(x-\xi)^i, \qquad x^i\,dx.$$

All residues of the first of these vanish except for $i = 1$; those of the second form all vanish. A quick computation shows

$$\operatorname{res}_\xi \frac{dx}{x-\xi} = 1, \qquad \operatorname{res}_\infty \frac{dx}{x-\xi} = -1$$

completing the proof for the rational case.

Now let K be a separable extension of $k = k_0(x)$. The residue theorem is an immediate consequence of our special case once we have the formula

$$\sum_\varrho \operatorname{res}_{\xi_\varrho}(u\,dx) = \operatorname{res}_\xi(s_{K/k}(u)\,dx) \tag{27}$$

which we proceed to prove. The sum is, of course, taken over all ξ_ϱ lying over ξ. There is no loss of generality in choosing $\xi = 0$. The numerator divisor $\mathfrak{z}_0$ of x decomposes in K, say to

$$\mathfrak{z}_0 = \mathfrak{p}_1^{e_1} \cdots \mathfrak{p}_r^{e_r},$$

so that we can rewrite (27) correspondingly as

$$\sum_{\varrho=1}^{r} \operatorname{res}_{\mathfrak{p}_\varrho}(u\,dx) = \operatorname{res}_0(s_{K/k}(u)\,dx). \tag{28}$$

First, simplify the matter. Choose elements $q_1, \ldots, q_r$ according to §2,4, so that

$$v_{\mathfrak{p}_\sigma}(q_\varrho) = \begin{cases} 1 & \text{for} \quad \sigma = \varrho \\ 0 & \text{for} \quad \sigma \neq \varrho. \end{cases}$$

These q_ϱ can be viewed as primitive elements of K with respect to $k = k_0(x)$. For if they were not they could be replaced by $q_\varrho(1 + \vartheta x^h)$ with a primitive ϑ and sufficiently large h.

At the place $\mathfrak{p}_\varrho$ we can expand $u\,dx$ as

$$u\,dx = \sum_\mu c_{\varrho\mu} q_\varrho^{\mu-1}\,dq_\varrho, \qquad \varrho = 1, \ldots, r, \quad c_{\varrho\mu} \in k_0.$$

It suffices to prove (28) for the special differentials $q_\mathfrak{p}^{\mu-1}\,dq_\varrho$, because of the additivity (23) of residues. Our choice of the q_ϱ assures

$$\operatorname{res}_{\mathfrak{p}_\sigma}(q_\varrho^{\mu-1}\,dq_\varrho) = 0 \qquad \text{for} \quad \sigma \neq \varrho,$$

so that the left side of (28) reduces to a single term for this differential. Only

$$\operatorname{res}_\mathfrak{p}(q_\varrho^{\mu-1}\,dq_\varrho) = \operatorname{res}_0(s_{K/k}(q_\varrho^{\mu-1}\,dq_\varrho/dx)\,dx) \tag{29}$$

remains to be shown. It clearly suffices to do this for $\varrho = 1$.

Let q_1 satisfy the irreducible equation

$$f(q_1, x) = q_1{}^n - c_1(x)q_1^{n-1} + - \cdots + (-1)^n c_n(x) = \prod_{\nu=1}^{n} (q_1 - \kappa_\nu)$$

whose zeros κ_ν we consider to lie in an extension of k having only the elements of k in common with K. Now, $c_n(x)$ is the norm of q_1 with respect to k, and it is divisible by $\mathfrak{z}_0$ exactly f times, f being the residue class degree of $\mathfrak{p}_1$ in K/k. But the degree of $\mathfrak{p}_1$ is 1 by our assumption that k_0 be algebraically closed, so that certainly $f = 1$. Thus

$$\prod_{\nu=1}^{n} \kappa_\nu = c_n(x) \qquad \text{and} \qquad v_0(c_n(x)) = 1.$$

Differentiating this equation logarithmically yields

$$s_{K/k}\left(q_1^{-1}\frac{dq_1}{dx}\right) = \sum_{\nu=1}^{n} \kappa_\nu^{-1}\frac{d\kappa_\nu}{dx} = c_n(x)^{-1}\frac{dc_n(x)}{dx} = x^{-1} + \cdots.$$

This implies

$$\operatorname{res}_0\left(s_{K/k}\left(q_1^{-1}\frac{dq_1}{dx}\right)dx\right) = 1 = \operatorname{res}_{\mathfrak{p}_1}(dq_1/q_1),$$

which is the asserted equation (29) for the case $\mu = 0$.

If μ is not divisible by the characteristic p, then

$$s_{K/k}\left(q_1^{\mu-1}\frac{dq_1}{dx}\right) = \frac{1}{\mu}\, s_{K/k}\left(\frac{dq_1{}^\mu}{dx}\right) = \frac{1}{\mu}\sum_{\nu=1}^{n}\frac{d\kappa_\nu{}^\mu}{dx} = \frac{1}{\mu}\frac{d}{dx}\sum_{\nu=1}^{n}\kappa_\nu{}^\mu.$$

As the power series on the right can contain no expression in $x^{-1}\,dx$, (29) is also proved for this case.

The differential $q_1{}^\mu\, dq_1/q_1$ with exponent $\mu \neq 0$ divisible by the characteristic p remains. We now apply Cartier's operator. Let

$$\frac{x}{q_1}\frac{dq_1}{dx} = \sum_{i=0}^{p-1} g_i x^i, \qquad g_i \in K_0.$$

Then

$$s_{K/k}\left(q_1{}^\mu \frac{x}{q_1}\frac{dq_1}{dx}\right) = \sum_{i=0}^{p-1} s_{K/k}(q_1{}^\mu g_i)x^i,$$

and then $q_1{}^\mu g_i \in K_0$ implies $s_{K/k}(q_1{}^\mu g_i) \in k_0(x^p)$. Thus we have

$$S\left(s_{K/k}\left(q_1{}^\mu\frac{x}{q_1}\frac{dq_1}{dx}\right)\frac{dx}{x}\right) = s_{K/k}(q_1{}^\mu g_0)\frac{dx^p}{x^p} = s_{K/k}\left(\frac{x^p}{dx^p}\, S\left(q_1{}^\mu\frac{x}{q_1}\frac{dq_1}{dx}\frac{dx}{x}\right)\right)\frac{dx^p}{x^p}. \tag{30}$$

This means that the operations S and taking of traces commute. Repeated application of S finally leads either to a differential integral at the place in question, making (29) trivial, or to a differential with a pole of first order which has already been treated. This completes the proof. ▼

At a later point we shall need the following generalization of (27). *Let L be a field between K and $k = k_0(x)$, and $\mathfrak{p}$ a divisor of L whose decomposition* *in K is $\mathfrak{p} = \mathfrak{p}_1^{e_1}\mathfrak{p}_2^{e_2}\cdots$. For any $u \in K$, then*

$$\operatorname{res}_{\mathfrak{p}}(s_{K/L}(u)\,dx) = \sum_i \operatorname{res}_{\mathfrak{p}_i}(u\,dx). \tag{31}$$

Proof. It again suffices to treat first degree prime divisors only. Let the divisor of k divisible by $\mathfrak{p}$ be $\mathfrak{p}_0$, and let $\mathfrak{p}', \mathfrak{p}'', \ldots$ be the other divisors of L which divide $\mathfrak{p}_0$. It is no loss of generality to assume $\mathfrak{p}_0$ not to be the denominator of x. Let $\mathfrak{I}_L$ be the principal order of L with respect to $k_0[x]$. The divisors $\mathfrak{p}, \mathfrak{p}', \ldots$ correspond to $\mathfrak{I}_L$-ideals $\mathfrak{p}_L, \mathfrak{p}_L', \ldots$. Using an argument of II, §2,2 there exists a $w \in \mathfrak{I}_L$ such that $w \equiv 0 \bmod \mathfrak{p}_L$, $w \equiv 1 \bmod \mathfrak{p}_L'\mathfrak{p}_L''\cdots$. The element $v = (1-w)^h$ satisfies the congruences $v \equiv 1 \bmod \mathfrak{p}_L$, $v \equiv 0 \bmod (\mathfrak{p}_L'\mathfrak{p}_L''\cdots)^h$. Sufficiently large choice of h yields residues $\operatorname{res}_{\mathfrak{p}_i}(vu\,dx) = \operatorname{res}_{\mathfrak{p}_i}(u\,dx)$. But also, $\operatorname{res}_{\mathfrak{p}'}(s_{K/L}(vu)\,dx) = \operatorname{res}_{\mathfrak{p}''}(s_{K/L}(u)v\,dx) = 0$, etc.

We now apply (28) twice:

$$\begin{aligned}\sum_i \operatorname{res}_{\mathfrak{p}_i}(vu\,dx) &= \operatorname{res}_{\mathfrak{p}_0}(s_{K/k}(vu)\,dx) = \operatorname{res}_{\mathfrak{p}_0}(s_{L/k}(s_{K/L}(vu))\,dx)\\ &= \operatorname{res}_{\mathfrak{p}}(s_{K/L}(vu)\,dx) + \operatorname{res}_{\mathfrak{p}'}(s_{K/L}(vu)\,dx) + \cdots = \operatorname{res}_{\mathfrak{p}}(s_{K/L}(vu)\,dx)\end{aligned}$$

which, by the above, gives the assertion. ▼

8*. The Differential Class

In §4,4 we saw that all differentials lie in the same class, the *differential class*, which we shall now consider more closely. We first observe:

$\mathfrak{d}_{K/k}\mathfrak{z}_\infty^{-2}$ is always divisible by the divisor (dx). Thus, (16) *is equivalent to the statement that (dx) is in the canonical class.*

Proof. Let q be a prime element in $k_0[x]$ and $\mathfrak{p}_\varrho$ ($\varrho = 1, \ldots, r$) the prime divisors of K by which its numerator $\mathfrak{z}_q$ is divisible. By the procedure of II, §5,6 associate with the divisor $(dx)^{-1}$ the $\mathfrak{I}_q$-ideal $(dx)^{-1}$ of all $u \in K$ satisfying $v_{\mathfrak{p}_\varrho}(u) \geqq v_{\mathfrak{p}_\varrho}((dx)_q^{-1})$. For all $u \in (dx)_q^{-1}$ and all $v_0 \in \mathfrak{I}_q$ we then have $uv_0\,dx/dq \in \mathfrak{I}_q$ and thus $\operatorname{res}_{\mathfrak{p}_\varrho}(uv_0\,dx) = 0$. By (28) then, $\operatorname{res}_{\mathfrak{z}_q}(s_{K/k}(uv_0)\,dx) = 0$. On the other hand, we can say that $\operatorname{res}_{\mathfrak{z}_q}(s_{K/k}(uv_0)\,dx) = 0$ if and only if u belongs to the complement $\mathfrak{I}_q^* = (\mathfrak{d}_{K/k})_q^{-1}$ of $\mathfrak{I}_q$, for this is simply another form of the definition of this complement. Thus, $(dx)_q^{-1} \subseteq (\mathfrak{d}_{K/k})_q^{-1}$. At the place ∞ we must replace x by x^{-1}, arriving at $(dx^{-1})_\infty^{-1} = x^2(dx)_\infty^{-1} \subseteq (\mathfrak{d}_{K/k})_\infty^{-1}$. This verifies the assertion: $\mathfrak{d}_{K/k}\mathfrak{z}_\infty^{-2}$ is divisible by (dx). ▼

Equation (16) *holds generally for separable extensions K of $k = k_0(x)$ if the different $\mathfrak{d}_{K/k}$ is divisible by separable prime divisors only.*

The proof is first produced under the additional assumption that all prime divisors $\mathfrak{p}_\varrho$ of $\mathfrak{d}_{K/k}$ are of degree 1; the exact powers of $\mathfrak{p}_\varrho$ which go into (dx)

and $\mathfrak{d}_{K/k}$ are determined. Our new assumption assures (by §1,2) that every element $u \in K$ can be expanded in a power series in a prime element q_ϱ with coefficients in k_0. As above we assume the q_ϱ not to be divisible by the prime divisors $\mathfrak{p}_\sigma \neq \mathfrak{p}_\varrho$; and also that the q_ϱ are primitive elements. From the previous proof we can now extract the fact that the exact power δ_ϱ of $\mathfrak{p}_\varrho$ which divides (dx) is also the largest integer such that $\operatorname{res}_{\mathfrak{p}_\varrho}(uq_\varrho^{-\delta_\varrho}\,dx) = 0$ for all $u \in K$ integral at $\mathfrak{p}_\varrho$. We now determine this power by setting an arbitrary integral power series in q_ϱ for u; then δ_ϱ is the largest integer such that

$$\operatorname{res}_{\mathfrak{p}_\varrho}(q_\varrho^{-\delta_\varrho}\,dx) = \operatorname{res}_{\mathfrak{p}_\varrho}(q_\varrho^{-\delta_\varrho+1}\,dx) = \cdots = 0.$$

In terms of (28) then, δ_ϱ is the largest integer for which we have $s_{K/k}(q_\varrho^{-\delta_\varrho}), s_{K/k}(q_\varrho^{-\delta_\varrho+1}), \ldots \in \mathfrak{i}_q$. But then, from the definition of the different in II, §4,2, the $\mathfrak{p}_\mathfrak{p}$-component of (dx) and $\mathfrak{d}_{K/k}$ is the same. As the same argument can be applied with x^{-1} in place of x for the place ∞, this case is proved.

If $\mathfrak{d}_{K/k}$ is divisible by separable prime divisors of higher degrees, the lemma of §4,4 states that these decompose to first degree factors in a separable extension of constants. According to §4,3 and §3,4 the divisor (dx) and the genus remain invariant under such extensions. The different divisor can, at most, pass to one of its proper divisors. But Eq. §3,(8) implies, using the invariance of the genus, that its degree remains invariant. Then so does $\mathfrak{d}_{K/k}$. Now, (16) holds after this extension of constants; therefore it must have held before. ▼

A field K is called *conservative* if its genus remains invariant under all finite extensions of constants.

Equation (16) *holds for, and only for, conservative fields.*

Proof. The above argument shows (16) to hold for conservative fields. Conversely, let (16) hold and k_0' be a finite extension of k_0. For the extension $K' = Kk_0'$ we can deduce as follows: first, (dx) remains unchanged, second, as above, it is certainly a divisor of $\mathfrak{d}_{K'/k_0'(x)}\mathfrak{z}_\infty^{-2}$, and third, $\mathfrak{d}_{K'/k_0'(x)}\mathfrak{z}_\infty^{-2}$ divides $\mathfrak{d}_{K/k}\mathfrak{z}_\infty^{-2} = (dx)$. This implies that $\mathfrak{d}_{K'/k_0'(x)}\mathfrak{z}_\infty^{-2} = \mathfrak{d}_{K/k_0(x)}\mathfrak{z}_\infty^{-2}$, so that the invariance of the genus follows from §3,(8). ▼

Combining the last two theorems we have: *a separably generated field $K/k_0(x)$ is conservative if $\mathfrak{d}_{K/k_0}(x)$ is divisible only by separable prime divisors.*

In closing we remark that recently a new notion of differentials, whose divisors always lie in the canonical class, was suggested.† It has the same

† H. J. Nastold, *Zum Dualitätssatz in inseparablen Funktionenkörpern der Dimension* 1, Math. Z. **76**, 75–84 (1961). This paper, in turn, is based on E. Kunz, *Differentialformen inseparabler algebraischer Funktionenkörper*, ibid., pp. 56–74. A similar treatment is given by O. Zariski and P. Falb, *On differentials in function fields*, Amer. J. Math. **83**, 542–556 (1961).

range of applicability as the divisor concept of the book [3] mentioned in §3,7. It fails to coincide with our divisors only in the case that the constant field k_0 is itself a function field in one or more variables, and then it uses differentiation in k_0, too. Thus this notion properly belongs to n-dimensional algebraic geometry.

§5. Differentials and Principal Part Systems

1. Differentials of Higher Degrees

This paragraph is preparatory to Chapter V. Its subject matter can be considered to be an analogy, in an abstract algebraic sense, to integral calculus. We assume throughout that K is separably generated and that the divisors of differentials lie in the canonical class. We saw in §4,8 that this is always true for conservative fields.

Along with the previous differentials, the *differentials of degree* h, where h may be any rational integer, play a part. By such a differential we mean a formal expression

$$dw^h = u\,dv^h,$$

with u and v in K and v a separating element. Set

$$u\,dv^h = u'\,dv'^h \qquad \text{whenever} \quad u' = u(dv/dv')^h.$$

For $h = 1$ this gives the ordinary differentials introduced in §4.

With such a differential of degree h we associate an *order*

$$v_\mathfrak{p}(dw^h) = v_\mathfrak{p}(u\,dv^h) = v_\mathfrak{p}(u(dv/dq)^h)$$

for every place $\mathfrak{p}$, where q is a prime element there. This can again, of course, only be done for separable $\mathfrak{p}$, but a suitable extension of the constant field, as in §4,4, serves to generalize; a finite purely inseparable extension suffices in each case.

Set

$$(dw^h) = \prod_\mathfrak{p} \mathfrak{p}^{v_\mathfrak{p}(dw^h)}$$

to define the *divisor of a differential*. By definition it is invariant under extension of constants. Then the divisors of two differentials as in §4,(15) satisfy

$$\frac{(u_1\,dv_1{}^h)}{(u_2\,dv_2{}^h)} = \left(\frac{u_1}{u_2}\left(\frac{dv_1}{dv_2}\right)^h\right).$$

Thus it is seen that the divisors of all dw^h lie in the hth power of the canonical class, provided the field K satisfies the above assumptions.

Among the differentials of some arbitrary degree h we now distinguish a certain set, whose importance will not become apparent until later. To this end we first fix a divisor $\mathfrak{a}$ once and for all which we stipulate to satisfy one of the following assumptions:

(1) *either* $g(\mathfrak{a}) - 2(g-1)(1-h) \geqq 0$, *and* $\mathfrak{a}$ *does not lie in the* $(1-h)$th *power of the canonical class,*

(2) *or* $\mathfrak{a}$ *lies in the* $(1-h)$th *power of the canonical class.*

We now say that the differential dw^h is of the *first kind* with respect to $\mathfrak{a}$ if $\mathfrak{a}(dw^h)$ is an integral divisor. For $h = 1$ and $\mathfrak{a} = (1)$ we also define differentials of the second and third kinds. In particular, δw is said to be of the *second kind* if to every place $\mathfrak{p}$ there exists a function $w_\mathfrak{p} \in K$ such that $v_\mathfrak{p}(dw - dw_\mathfrak{p}) \geqq 0$ at that place. All other differentials are said to be of the *third kind.* For example, $x^{-1}\,dx$, where x is a separating element, falls in the last class.

The differentials of the first kind with respect to $\mathfrak{a}$ *form a* k_0*-module of rank*

$$g_{h,\mathfrak{a}} = \begin{cases} g(\mathfrak{a}) + (2h-1)(g-1) & \text{in case (1),} \\ g(\mathfrak{a}) + (2h-1)(g-1) + 1 & \text{in case (2).} \end{cases}$$

Proof. All (dw^h) lie in the hth power of the canonical class W, so that the integral divisor $\mathfrak{a}(dw^h)$ lies in the class $\mathfrak{a}W^h$. The number of linearly independent $\mathfrak{a}(\delta w^h)$ can be determined from the Riemann–Roch theorem together with Eq. §3,(7). ▼

2. Principal Part Systems

Hold the divisor $\mathfrak{a}$ fixed throughout §5. By a *principal part system* $d^h\mathfrak{w} = (dw_\mathfrak{p}{}^h)$ *of degree h for the divisor* $\mathfrak{a}$ we mean a set of differentials $dw_\mathfrak{p}{}^h$ of degree h over K, associated in a one-to-one manner with the places $\mathfrak{p}$ of K, and such that the orders $v_\mathfrak{p}(\mathfrak{a}^{-1}w_\mathfrak{p}{}^h) \geqq 0$ for almost all $\mathfrak{p}$. We call two principal part systems $(dw_\mathfrak{p}{}^h)$ and $(dv_\mathfrak{p}{}^h)$ the same if $v_\mathfrak{p}(\mathfrak{a}^{-1}(dv_\mathfrak{p}{}^h - dw_\mathfrak{p}{}^h)) \geqq 0$ for all $\mathfrak{p}$ without exception. For instance, in the case $\mathfrak{a} = (1)$ we are concerned only with the principal parts of the poles of the $dw_\mathfrak{p}{}^h$ at $\mathfrak{p}$. Thus such a system is given by a finite number of data. Furthermore, the terminology is thereby explained.

Under the extension of constants k_0'/k_0, the principal part system $(dw_\mathfrak{p}{}^h)$ of K passes to a principal part system $(dw_{\mathfrak{p}'}^h)$ of $K' = Kk_0'$ defined as follows. Say the places $\mathfrak{p}_\varrho'$ of K' lie over the place $\mathfrak{p}$ of K. Then set

$$dw_{\mathfrak{p}_\varrho'}^h = dw_\mathfrak{p}{}^h \qquad (\varrho = 1, \ldots, r).$$

This convention permits us to identify the principal part systems $(dw_\mathfrak{p}{}^h)$ and $(dw_{\mathfrak{p}'}^h)$ of K and K'.

Principal part systems can be multiplied by elements $\alpha, \beta \in k_0$ and added;

$$\alpha(dw_\mathfrak{p}{}^h) + \beta(dv_\mathfrak{p}{}^h) = (\alpha\, dw_\mathfrak{p}{}^h + \beta\, dv_\mathfrak{p}{}^h).$$

They form a k_0-module.

If to two principal part systems $(dv_\mathfrak{p}{}^h)$ and $(dw_\mathfrak{p}{}^h)$ there exists a differential dw^h in K such that

$$v_\mathfrak{p}(\mathfrak{a}^{-1}(dw_\mathfrak{p}{}^h - dv_\mathfrak{p}{}^h - dw^h)) \geqq 0$$

holds for all $\mathfrak{p}$, then $(dv_\mathfrak{p}{}^h)$ and $(dw_\mathfrak{p}{}^h)$ are called *equivalent*, written $(dw_\mathfrak{p}{}^h) \sim (dv_\mathfrak{p}{}^h)$. That this is really an equivalence relationship is clear. *Principal part system classes* form a k_0-module.

We call a principal part system $d^h\mathfrak{w}$ *exact* if $d^h\mathfrak{w} \sim (0)$, that is, if a differential (dw^h) exists in K such that $v_\mathfrak{p}(\mathfrak{a}^{-1}(dw_\mathfrak{p}{}^h - dw^h)) \geqq 0$ for all $\mathfrak{p}$. Clearly this differential dw^h is uniquely defined by its principal part system $d^h\mathfrak{w}$ up to the addition of another differential dv^h for which $\mathfrak{a}^{-1}(dv^h)$ is an integral divisor, that is, up to the addition of a differential of the first kind with respect to $\mathfrak{a}^{-1}$. For instance, if $h = 0$ and $\mathfrak{a} = (1)$, $dw^0 = w$ is determined up to a constant. For, if simultaneously $v_\mathfrak{p}(w_\mathfrak{p} - w) \geqq 0$ and $v_\mathfrak{p}(w_\mathfrak{p} - v) \geqq 0$ everywhere, then $v_\mathfrak{p}(w - v) \geqq 0$ everywhere, so that $w - v \in k_0$.

3. The Scalar Product

Let dw^h be a differential of the hth degree and first kind with respect to $\mathfrak{a}$, and $d^{1-h}\mathfrak{w}$ a principal part system of degree $1 - h$ for the divisor $\mathfrak{a}$. Then

$$(d^{1-h}\mathfrak{w}, dw^h) = \sum_\mathfrak{p} \operatorname{res}_\mathfrak{p}(dw_\mathfrak{p}^{1-h}\, dw^h) \tag{1}$$

defines a *scalar product*. On the right we have differentials $dv_\mathfrak{p} = dw_\mathfrak{p}^{1-h}\, dw^h$ of degree 1, of which at most a finite number have a pole at $\mathfrak{p}$. If $d^{1-h}\mathfrak{w} = (dw_\mathfrak{p}^{1-h}) = (dv_\mathfrak{p}^{1-h})$, our definition of principal part systems in the last section yields

$$\begin{aligned} v_\mathfrak{p}((dw_\mathfrak{p}^{1-h} - dv_\mathfrak{p}^{1-h})\, dw^h) &= v_\mathfrak{p}(\mathfrak{a}^{-1}(dw_\mathfrak{p}^{1-h} - dv_\mathfrak{p}^{1-h})\mathfrak{a}\, dw^h) \\ &= v_\mathfrak{p}(\mathfrak{a}^{-1}(dw_\mathfrak{p}^{1-h} - dv_\mathfrak{p}^{1-h})) + v_\mathfrak{p}(\mathfrak{a}\, dw^h) \end{aligned}$$

with both summands on the right $\geqq 0$. Thus (1) really depends only upon the principal part system $d^{1-h}\mathfrak{w}$, and not on the representatives $dw_\mathfrak{p}^{1-h}$ used.

Clearly the scalar product is a bilinear function, that is,

$$(\alpha_1\, d^{1-h}\mathfrak{w}_1 + \alpha_2\, d^{1-h}\mathfrak{w}_2, dw^h) = \alpha_1(d^{1-h}\mathfrak{w}_1, dw^h) + \alpha_2(d^{1-h}\mathfrak{w}_2, dw^h),$$

$$(d^{1-h}\mathfrak{w}, \alpha_1\, dw_1{}^h + \alpha_2\, dw_2{}^h) = \alpha_1(d^{1-h}\mathfrak{w}, dw_1{}^h) + \alpha_2(d^{1-h}\mathfrak{w}, dw_2{}^h)$$

hold for $\alpha_1, \alpha_2 \in k_0$.

Using §4,6 we see that the scalar product is always an element of the field of constants k_0.

Let a principal part system $d^{1-h}\mathfrak{w}$ for the divisor $\mathfrak{a}$ be given. If the scalar products with all differentials dw^h of the first kind with respect to $\mathfrak{a}$ vanish, and only then, is $d^{1-h}\mathfrak{w}$ exact.

Some preparations are needed for the proof. As differentials and their divisors, principal part systems and their classes, and finally the scalar product do not change by an extension of the field of constants, we may assume that field to be large enough to assure that all poles $\mathfrak{p}$ occurring are of degree 1, and that there exists at least one further divisor of the first degree $\mathfrak{q}$ by which $\mathfrak{a}$ is not divisible.

Let such a divisor $\mathfrak{q}$ be chosen and held fixed in what follows, and let q be a prime element for $\mathfrak{q}$. A special basis of the differentials dw^h of the first kind must now be produced. By subjecting the power series expansions

$$dw^h = (c_0 + c_1 q + \cdots)\, dq^h, \qquad c_i \in k_0,$$

to a suitable linear transformation, a basis

$$dw_i^h = (q^{\mu_i} + c_{i,\mu_i+1} q^{\mu_i+1} + \cdots)\, dq^h, \qquad c_{ij} \in k_0, \tag{2}$$

such that

$$0 \leqq \mu_1 < \mu_2 < \cdots < \mu_G, \qquad G = g_{h,\mathfrak{a}}, \tag{3}$$

is found, and clearly uniquely determines the μ_i. The number $g_{h,\mathfrak{a}}$ computed in §5,1 is here abbreviated G.

The divisors $\mathfrak{a}(dw_i^h)$ lie in the class $\mathfrak{a}W^h$. By the existence of the basis (2) we can deduce

$$\begin{aligned} \dim(\mathfrak{a}W^h) = \dim(\mathfrak{q}^{-1}\mathfrak{a}W^h) = \cdots &= \dim(\mathfrak{q}^{-\mu_1}\mathfrak{a}W^h) = G, \\ \dim(\mathfrak{q}^{-\mu_1-1}\mathfrak{a}W^h) = \cdots &= \dim(\mathfrak{q}^{-\mu_2}\mathfrak{a}W^h) = G-1, \\ &\cdots \\ \dim(\mathfrak{q}^{-\mu_{G-1}-1}\mathfrak{a}W^h) = \cdots &= \dim(\mathfrak{q}^{-\mu_G}\mathfrak{a}W^h) = 1, \\ \dim(\mathfrak{q}^{-\mu_G-1}\mathfrak{a}W^h) &= \cdots = 0. \end{aligned} \tag{4}$$

Proof. The divisors of all dw^h of the first kind with respect to $\mathfrak{a}$ can be written as $(dw^h) = \mathfrak{a}^{-1}\mathfrak{g}$ with integral $\mathfrak{g}$ which is even divisible by $\mathfrak{q}^{\mu_1}$ because of (2). This divisor $\mathfrak{g}$ is in the class $\mathfrak{a}W^h$, in which there are G linearly independent divisors. The divisors $\mathfrak{g}\mathfrak{q}^{-\nu}$, where $\nu = 1, \ldots, \mu_1$, lie in the classes $\mathfrak{q}^{-\nu}\mathfrak{a}W^h$. As they are all integral, the number linearly independent among them is $\dim(\mathfrak{q}^{-\nu}\mathfrak{a}W^h)$. But there must be just as many linearly independent divisors as there are $\mathfrak{g}$. This verifies the first equation (4). Omitting dw_1^h and thus reducing the dimension by 1 so that the remaining dw^h are of the form $\mathfrak{a}^{-1}\mathfrak{g}$ with integral $\mathfrak{g}$ divisible by $\mathfrak{q}^{\mu_2}$ gives similar arguments for the other equations.

Two lemmas now follow.

Lemma 1. *If for some i $(i = 1, \ldots, G)$ $\mu_i + 2 \leqq \mu_{i+1}$, then to every ν in the sequence $\mu_i + 2, \ldots, \mu_{i+1}$ there exists a differential dv^{1-h} of degree $1-h$ such that the divisor $\mathfrak{q}^\nu\mathfrak{a}^{-1}(dv^{1-h}) = \mathfrak{b}$ is integral and prime to $\mathfrak{q}$.*

Proof. Such a divisor $\mathfrak{b}$ belongs to the class $\mathfrak{q}^{\nu}\mathfrak{a}^{-1}W^{1-h}$. If $\mathfrak{b} = \mathfrak{q}\mathfrak{b}'$, then $\mathfrak{b}'$ lies in the class $\mathfrak{q}^{\nu-1}\mathfrak{a}^{-1}W^{1-h}$. The existence of a differential dv^{1-h} as described is dependent upon the truth of

$$\dim(\mathfrak{q}^{\nu-1}\mathfrak{a}^{-1}W^{1-h}) < \dim(\mathfrak{q}^{\nu}\mathfrak{a}^{-1}W^{1-h}).$$

By the Riemann–Roch theorem this inequality is equivalent to

$$\begin{aligned} \nu - 1 - g(\mathfrak{a}) + (1-2h)(g-1) + \dim(\mathfrak{q}^{1-\nu}\mathfrak{a}W^h) \\ < \nu - g(\mathfrak{a}) + (1-2h)(g-1) + \dim(\mathfrak{q}^{-\nu}\mathfrak{a}W^h). \end{aligned}$$

But this inequality holds whenever

$$\dim(\mathfrak{q}^{1-\nu}\mathfrak{a}W^h) = \dim(\mathfrak{q}^{-\nu}\mathfrak{a}W^h),$$

and this is so, under our hypothesis, according to (4). ▼

Lemma 2. *A principal part system class of degree* $1-h$ *always contains a principal part system* $d^{1-h}\mathfrak{w} = (dv_{\mathfrak{p}}^{1-h})$ *with orders* $\nu_{\mathfrak{p}}(\mathfrak{a}^{-1}(dv_{\mathfrak{p}}^{1-h})) \geqq 0$, *except possibly at* $\mathfrak{p} = \mathfrak{q}$, *where then* $\nu_{\mathfrak{q}}(\mathfrak{a}^{-1}(dv_{\mathfrak{q}}^{1-h})) \geqq -\mu_G - 1$.

Proof. Let $(dw_{\mathfrak{p}}^{1-h})$ be a principal part system of a given class and $\nu_{\mathfrak{p}}(\mathfrak{a}^{-1}(dw_{\mathfrak{p}}^{1-h})) = -l$, $l \geqq 1$, for some $\mathfrak{p} \neq \mathfrak{q}$. We seek a differential dw^{1-h} with

$$\nu_{\mathfrak{p}}(\mathfrak{a}^{-1}(dw^{1-h})) = -l, \qquad \nu_{\mathfrak{q}}(\mathfrak{a}^{-1}(dw^{1-h})) \geqq -\mu_G - 1, \qquad \nu_{\mathfrak{p}'}(\mathfrak{a}^{-1}(dw^{1-h})) \geqq 0$$

for all $\mathfrak{p}' \neq \mathfrak{p}, \mathfrak{q}$. If this dw^{1-h} is multiplied by a suitable constant, then $\nu_{\mathfrak{p}}(\mathfrak{a}^{-1}(dw_{\mathfrak{p}}^{1-h} - dw^{1-h})) \geqq -(l-1)$. The principal part system $(dw_{\mathfrak{p}}^{1-h}) - (dw^{1-h})$ is equivalent to $(dw_{\mathfrak{p}}^{1-h})$, and we operate similarly on it, etc. A finite number of steps leads to an equivalent principal part system $d^{1-h}\mathfrak{w}$ of the sort required.

The differential dw^{1-h} sought above must have as its divisor

$$(dw^{1-h}) = \mathfrak{p}^{-l}\mathfrak{q}^{-\mu_G-1}\mathfrak{a}\mathfrak{c} \tag{5}$$

with some integral divisor $\mathfrak{c}$ prime to $\mathfrak{p}$. It belongs to the class W^{1-h}, so that the class of $\mathfrak{c}$ is $\mathfrak{p}^{l}\mathfrak{q}^{\mu_G+1}\mathfrak{a}^{-1}W^{1-h}$. The Riemann–Roch theorem gives the dimension of this class:

$$\delta_l = l + \mu_G + 1 - g(\mathfrak{a}) + (1-2h)(g-1) + \dim(\mathfrak{p}^{-l}\mathfrak{q}^{-\mu_G-1}\mathfrak{a}W^h).$$

The divisor of a differential (5) for which $\mathfrak{c}$ is divisible by $\mathfrak{p}$ can be written $\mathfrak{p}^{1-l}\mathfrak{q}^{-\mu_G-1}\mathfrak{a}\mathfrak{c}'$ with integral $\mathfrak{c}'$, this $\mathfrak{c}'$ belonging to the class $\mathfrak{p}^{l-1}\mathfrak{q}^{\mu_G+1}\mathfrak{a}^{-1}W^{1-h}$ of dimension

$$\delta_{l-1} = l + \mu_G - g(\mathfrak{a}) + (1-2h)(g-1) + \dim(\mathfrak{p}^{1-l}\mathfrak{q}^{-\mu_G-1}\mathfrak{a}W^h).$$

Thus the existence of a differential (5) with integral $\mathfrak{c}$ not divisible by $\mathfrak{p}$ is assured whenever $\delta_l > \delta_{l-1}$, and this is the case if

$$\dim(\mathfrak{p}^{-l}\mathfrak{q}^{-\mu_G-1}\mathfrak{a}W^h) = \dim(\mathfrak{p}^{1-l}\mathfrak{q}^{-\mu_G-1}\mathfrak{a}W^h) = 0.$$

As $l \geqq 1$ this is an immediate consequence of (4), where we saw that $\dim(\mathfrak{q}^{-\mu_G-1}\mathfrak{a}W^h) = 0$. This proves our lemma. ▼

Little remains to prove the theorem. If $d^{1-h}\mathfrak{w}$ is exact, i.e., if $d^{1-h}\mathfrak{w} = (dv^{1-h})$ then for any differential dw^h we have $(d^{1-h}\mathfrak{w}, dw^h)$ equal to the sum of residues of the differential $dv^{1-h}dw^h$, a differential of the first degree, which sum must be 0 according to the residue theorem of §4,7.

Conversely, let

$$(d^{1-h}\mathfrak{w}, dw^h) = 0 \tag{6}$$

for all dw^h with integral $\mathfrak{a}(dw^h)$. We have already seen that (6) also holds for equivalent principal part systems, so that we may choose for $d^{1-h}\mathfrak{w}$ the system $(dv_{\mathfrak{p}}^{1-h})$ of Lemma 2. For the $\mathfrak{q}$-component we have the power series expansion

$$dv_{\mathfrak{q}}^{1-h} = (c_{-\mu_G-1}q^{-\mu_G-1} + c_{-\mu_G}q^{-\mu_G} + \cdots)\,dq,$$

while for $\mathfrak{p} \neq \mathfrak{q}$ we have $v_{\mathfrak{p}}(\mathfrak{a}^{-1}(dv_{\mathfrak{p}}^{1-h})) \geqq 0$. Now using Lemma 1, we can even improve our principal part system with these properties within the class so that at most the exponents $-\mu_G - 1,\ -\mu_{G-1} - 1,\ \ldots,\ -\mu_1 - 1$ occur with nonvanishing coefficients. Then, by (2) we have

$$((dv_{\mathfrak{q}}^{1-h}), dw^h{}_i) = c_{-\mu_i-1},$$

which must vanish by our assumption. Hence, the $dv_{\mathfrak{q}}^{1-h}$ are integral at $\mathfrak{q}$ and, as $\mathfrak{q}$ does not divide $\mathfrak{a}$, $v_{\mathfrak{q}}(\mathfrak{a}^{-1}(dv_{\mathfrak{q}}^{1-h})) \geqq 0$. We have thus verified that $(dv_{\mathfrak{p}}^{1-h}) = (0)$, proving the theorem. ▼

4. The Relationship to Integral Calculus

For this section we assume the constant field to be algebraically closed, so that all divisors are of the first degree. Further, let $\mathfrak{a} = (1)$.

To any principal part system $d^0\mathfrak{w} = (w_{\mathfrak{p}})$ *there exists in* K *a differential* dv *of the second kind such that* $v_{\mathfrak{p}}(dv - dw_{\mathfrak{p}}) \geqq 0$ *for all* $\mathfrak{p}$. *This* dv *is even uniquely determined up to the addition of a differential of the first kind. The mapping of* $d^0\mathfrak{w}$ *onto the residue class of* dv *modulo the differentials of the first kind is a linear function.*

Conversely, there exists such a $d^0\mathfrak{w}$ *for every* dv *of the second kind. For* K *of characteristic* 0 *there exists only one such* $d^0\mathfrak{w}$.

Proof. Say some $\mathfrak{p}$ has the prime element q and

$$w_{\mathfrak{p}} = c_{-m}q^{-m} + \cdots + c_{-1}q^{-1} + \cdots. \tag{7}$$

We first seek a differential of the second kind, $dv = dv(\mathfrak{p})$, possessing the expansion

$$dv = (-mc_{-m}q^{-m-1} - \cdots - c_{-1}q^{-2} + c_1 + \cdots)\,dq \tag{8}$$

at $\mathfrak{p}_0$ and no other poles, the $\mu c_{-\mu} q^{-\mu-1}$ for which the characteristic divides μ dropping out. Now, a differential dv_μ with a pole of order $\mu \geqq 2$ at $\mathfrak{p}$ and no others has a divisor $\mathfrak{p}^{-\mu}\mathfrak{b}$, with integral $\mathfrak{b}$ not divisible by $\mathfrak{p}$ and in the class $\mathfrak{p}^\mu W$. As in the last section, use

$$\dim(\mathfrak{p}^\mu W) = \mu + g - 1 > \dim(\mathfrak{p}^{\mu-1} W) = \mu + g - 2$$

to show that such a differential exists. Then it is not difficult to construct a differential (8) with the dv_μ. In turn, addition of these $dv(\mathfrak{p})$ for the finite number of places at which $w_\mathfrak{p}$ is singular leads to a differential dv satisfying the theorem. Clearly a differential of the first kind may be added to dv, and the difference between two such dv is of the first kind.

Furthermore, the construction clearly shows the linearity of the mapping.

Conversely, let dv be a differential of the second kind and let (8) represent the principal parts of the finitely many poles. If μ is divisible by the characteristic, no terms $q^{-\mu-1}\,dq$ occur. Thus, as far as negative exponents are concerned, (8) represents the differentials of the elements (7). This shows that the $w_\mathfrak{p}$ are components of a principal part system $d^0\mathfrak{w}$ that maps onto dv.

If the characteristic is 0 the principal part of (7) is uniquely determined by that of (8). In cases with prime characteristic, though, terms $c_{-p} q^{-p} + c_{-2p} q^{-2p} + \cdots$ with arbitrary coefficients may be added. ▼

The relationship of the theorem may be represented symbolically by the indefinite integral

$$d^0\mathfrak{w} = \int dv. \tag{9}$$

Moreover, this permits us to define a scalar product

$$(du, dv) = \sum_{\mathfrak{p}} \operatorname{res}_\mathfrak{p}\left(\left(\int du \cdot dv\right)\right) \tag{10}$$

between any two differentials of the second kind. The power series expansions

$$du = (c_{-m} q^{-m} + \cdots + c_{-1} q^{-1} + c_0 + c_1 q + \cdots)\frac{dq}{q},$$

$$dv = (d_{-m} q^{-m} + \cdots + d_{-1} q^{-1} + d_0 + d_1 q + \cdots)\frac{dq}{q},$$

at a place $\mathfrak{p}$ (in which $c_\mu = d_\mu = 0$ for μ divisible by the characteristic because du and dv are of the second kind) show that

$$\operatorname{res}_\mathfrak{p}\left(\left(\int du \cdot dv\right)\right) = -\frac{c_{-m}\, d_m}{m} - \cdots - \frac{c_{-1}\, d_1}{1} + \frac{c_1\, d_{-1}}{1} + \cdots + \frac{c_m\, d_{-m}}{m}$$

is skew symmetric in du and dv, so that

$$(du, dv) = -(dv, du). \tag{11}$$

This scalar product is clearly 0 if both du and dv are of the first kind. It also vanishes if either du or dv is exact, that is, if du or dv is a differential of a function in K.

Let du_i be the basis (2) of first kind differentials with $h = 1$, and use its expansions at a place $\mathfrak{q}$ to define principal part systems $d^0\mathfrak{v}_i$ by stipulating

$$v_{i\mathfrak{p}} = \begin{cases} q^{-\mu_i - 1} & \text{for } \mathfrak{p} = \mathfrak{q} \\ 0 & \text{for } \mathfrak{p} \neq \mathfrak{q} \end{cases}$$

so that $(d^0\mathfrak{v}_i, du_j) = 0$ or 1, depending upon whether $i \neq j$ or $i = j$. The corresponding differentials of the second kind dv_i are determined only up to differentials $dw_i = \sum c_{ij}\, du_j$. Remembering (11) we can adjust the c_{ij} and thus the dw_i in a manner that the expression

$$\begin{aligned}(dv_i + dw_i, dv_j + dw_j) \\ = (dv_i, dv_j) + \sum_l ((dv_i, du_l)c_{jl} + (dv_j, du_l)c_{il}) = (dv_i, dv_j) + c_{ji} - c_{ij}\end{aligned}$$

always vanishes. Again writing dv_i for $dv_i + dw_i$ we now have

$$(du_i, du_j) = 0, \qquad (dv_i, dv_j) = 0,$$

$$(dv_i, du_j) = -(du_i, dv_j) = \begin{cases} 1 & \text{for } i = j \\ 0 & \text{for } i \neq j. \end{cases} \tag{12}$$

The theorem of the last section together with the correspondence between principal part systems and differentials found here finally give the result: *Every differential of the second kind is a sum of a differential of a function of K and a linear combination of du_i and dv_i.*

5*. The Diagonal

The Green function of classical analysis depends upon two variables and has a singularity at a variable place. The rest of §5 will be devoted to finding an algebraic analog to the Green function.

To this end, consider two isomorphic function fields K and K' in one variable over the exact field of constants k_0. Say $K = k_0(x, y)$ with $f(x, y) = 0$, where $f(x, y)$ is separable in y; take another variable x', algebraically independent of x, and let y' be a solution of $f(x', y') = 0$. Then $K' = k_0(x', y')$ is such a field. $KK' = k_0(x, y; x', y')$ is a field of functions of two variables over k_0, which can be considered a field of functions of one variable over K and also over K' as constant field. In the first case it arises from K' by extension of the field of constants k_0 to K, in the second from K by extension of constants to K'.

KK' has an involutive automorphism, given by interchanging x, y with x', y'. In the following it will always be denoted by priming.

Lemma. *If x_1 and x_2 are any two separating elements of K then there exist two polynomials $h_1(x_1, x_2; x_1', x_2')$ and $h_2(x_1, x_2; x_1', x_2')$, with $h_1(x_1, x_2; x_1, x_2)h_2(x_1, x_2; x_1, x_2) \neq 0$ and such that*

$$\frac{x_2 - x_2'}{x_1 - x_1'} = \frac{h_2(x_1, x_2; x_1', x_2')}{h_1(x_1, x_2; x_1', x_2')}.$$

Proof. A separable equation $f(x_1, x_2) = 0$, irreducible in both variables, holds. Hence, the equation $f(x_1, x_2') = 0$ has $x_2' = x_2$ as its only solution in K, and thus

$$f(x_1, x_2') = (x_2 - x_2')g_2(x_1, x_2; x_2'),$$

where g_2 is a polynomial in x_2' with functions rational in x_1, x_2 as coefficients, and for which $g_2(x_1, x_2; x_2) = -\partial f(x_1, x_2)/\partial x_2 \neq 0$. The same argument shows that

$$f(x_1, x_2') = (x_1 - x_1')g_1(x_1; x_1', x_2'),$$

with a polynomial g_1 in x_1 whose coefficients lie in K' and such that $g_1(x_1; x_1, x_2) \neq 0$. Division of these equations yields

$$\frac{x_2 - x_2'}{x_1 - x_1'} = \frac{g_1(x_1; x_1', x_2')}{g_2(x_1, x_2; x_2')}$$

which goes into the representation desired when the fraction is extended by the denominators $m_1(x_1', x_2')$ and $m_2(x_1, x_2)$ of g_1 and g_2. ▼

Corollary. *If the fields K and K' are identified in the sense of the isomorphism given above, then*

$$\left(\frac{x_2 - x_2'}{x_1 - x_1'}\right)_{K'=K} = \frac{dx_2}{dx_1}.$$

Proof. The assertion follows immediately from Taylor's theorem for polynomials, which states that

$$g_2(x_1, x_2; x_2) = -f_{x_2}(x_1, x_2) \text{ and } g_1(x_1; x_1, x_2) = f_{x_1}(x_1, x_2). \quad ▼$$

This taken care of, consider the integral domain $K \times K'$ of all finite sums $x_1y_1' + x_2y_2' + \cdots$ with arbitrary $x_i \in K$ and $y_i' \in K'$. The expressions $(x - x')Y$ with arbitrary $x \in K$ and $Y \in K \times K'$ form an ideal $\mathfrak{D}_{K \times K'}$, which is even a prime ideal as the residue class ring is isomorphic to the field K.

KK' is the quotient field of $K \times K'$, so that in view of the lemma every $A \in KK'$ can be represented as

$$A = (x - x')^l(X/Y)$$

with X and Y in $K \times K'$ but not divisible by $\mathfrak{D}_{K \times K'}$. The A with $l > 0$ form an integral domain $\mathfrak{J}_{\mathfrak{D}}$, which is even a discrete local ring in the sense of II, §2.

The prime divisor $\mathfrak{D}$ of K corresponding to the ideal $\mathfrak{D}_{K\times K'}$ is called the *diagonal divisor*, or simply the *diagonal* of KK'. If KK' is taken as a field of functions of one variable over the constant field K', then $\mathfrak{D}$ becomes a prime divisor of the first degree, for the residue class field of $\mathfrak{J}_{\mathfrak{D}}$ mod $\mathfrak{D}$ is the constant field K'. Interchanging K and K' leads to corresponding statements. By the lemma $x - x'$ is a prime element for $\mathfrak{D}$, where x is any separating element of K.

6*. The Analog of the Green Function

Let dw_i^h be the basis (2) of degree h differentials of the first kind with respect to $\mathfrak{a}$. Then

$$dw_{i\mathfrak{p}}^{1-h} = \begin{cases} q^{-\mu i-1}\, dq^{1-h} & \text{for } \mathfrak{p} = \mathfrak{q} \\ 0 & \text{for } \mathfrak{p} \neq \mathfrak{q} \end{cases}, \quad i = 1, \ldots, G = g_{h,\mathfrak{a}},$$

defines $g_{h,a}$ principal part systems $d^{1-h}\mathfrak{w}_i$ for $\mathfrak{a}$, with

$$(d^{1-h}\mathfrak{w}_i, dw_j^h) = \begin{cases} 1 & \text{for } i = j \\ 0 & \text{for } i \neq j \end{cases} \tag{13}$$

according to (2). In view of §5,3 these $d^{1-h}\mathfrak{w}_i$ form a basis of all the principal part system classes, the *complementary basis* to dw_i^h.

Contragredient linear transformations of both bases leave the relation (13) intact; the bases remain complementary. Such transformations also leave the bilinear form

$$dd'^{1-h}\mathfrak{G}_0^h = \sum_i d'^{1-h}\mathfrak{w}_i'\, dw_i^h \tag{14}$$

invariant. Here we denoted by $d'^{1-h}\mathfrak{w}_i'$ the principal part systems carried from K to K' by isomorphism; priming the d will turn out to be practical. Equation (14) is to be read as between principal part systems of degree $1-h$ in KK' taken as a function field over K, multiplied by a differential of degree h of K.

According to the second lemma of §5,3 there now exists an exact principal part system $d'^{1-h}\, d\mathfrak{G}_1^h$ of degree $1-h$ in KK'/K, multiplied by a differential of degree h in K, such that the equivalent principal part system

$$d'^{1-h}\, d\mathfrak{G}^h = d'^{1-h}\, d\mathfrak{G}_0^h - d'^{1-h}\, d\mathfrak{G}_1^h = (d'G_{\mathfrak{p}'}^{1-h}\, dx^h) \tag{15}$$

(with some separating x of K) has the property $v_{\mathfrak{p}}'(\mathfrak{a}'^{-1}\, d'G_{\mathfrak{p}}^{1-h}) \geqq 0$ at all places $\mathfrak{p}'$ of KK'/K except for $\mathfrak{p}' = \mathfrak{D}$, where that order is $\geqq -\mu_G - 1$. The notation μ_G is that of §5,3 with $\mathfrak{q} - \mathfrak{D}$. At the place $\mathfrak{p}' = \mathfrak{D}$,

$$d'G_{\mathfrak{D}}^{1-h}\, dx^h = (c_{-\mu_G-1}(x'-x)^{-\mu_G-1} + \cdots + c_{-1}(x'-x)^{-1})\, d'x'^{1-h}\, dx^h \tag{16}$$

with coefficients $c_{-1}, \ldots \in K$. As in §5,3, the first lemma even permits a choice with coefficients $c_j = 0$ with the possible exception of the $c_{-\mu_i-1}$. Moreover, we now assert that all $c_j = 0$ with the exception $c_{-1} = 1$.

To prove this we note that $d'^{1-h}d\mathfrak{G}_1{}^h$ is exact, so it does not contribute to the scalar product; by (15), (14), and (13), then

$$(d'^{1-h}\, d\mathfrak{G}^h, d'w_i'^h) = (d'^{1-h}\, d\mathfrak{G}_0{}^h, d'w_i'^h) = dw_i{}^h. \tag{17}$$

Clearly (17) holds for all differentials of the first kind, for the $dw_i{}^h$ form a basis of these.

In order to evaluate (17) we make use of a third field K'' isomorphic to K and algebraically independent of K and K'. For the $dw_i{}^h$ we choose the basis (2) of differentials of the first kind of KK''/K'' using the diagonal of KK'' as prime divisor $\mathfrak{q}$. This means that

$$dw_i{}^h = ((x - x'')^{\mu_i} + \gamma_{i,\mu_i+1}(x - x'')^{\mu_i+1} + \cdots)\, dx^h, \qquad \gamma_{ij} \in K''. \tag{18}$$

The exponents μ_i are those of (16). An easy computation gives

$$\operatorname{res}_{x'=x} \frac{(x' - x'')^\nu}{(x' - x)^{\mu+1}}\, d'x' = \binom{\nu}{\mu}(x - x'')^{\nu-\mu}. \tag{19}$$

Now let i be the largest index such that the coefficient $c_{-\mu_i-1} \neq 0$ in (16). Then Eqs. (16), (18), and (19) yield

$$(d'^{1-h}\, d\mathfrak{G}^h, d'w_i'^h) = (c_{-\mu_i-1} + \cdots)\, dx^h,$$

the dots indicating a function which vanishes on the diagonal of KK''. But, by (17), this is possible only if $\mu_i = 0$, that is, if $i = 1$ and $c_{-1} = 1$. This proves our assertion.

Remembering our notion of equality of principal part systems we have

$$d'^{1-h}\, d\mathfrak{G}^h = (d'G_{\mathfrak{p}'}^{1-h}\, dx^h), \qquad d'G_{\mathfrak{p}'}^{1-h}\, dx^h = \begin{cases} \dfrac{d'x'^{1-h}\, dx^h}{x' - x} & \text{for } \mathfrak{p}' = \mathfrak{D} \\ 0 & \text{for } \mathfrak{p}' \neq \mathfrak{D} \end{cases} \tag{20}$$

with any separating variable x. (20) may be considered to be an analog of Green's function.

The following theorem will show that the principal part system (20) is, up to minor alterations, uniquely determined. We state this fact in a different way, however, more adapted to an application in V, §2, distinguishing the two alternative assumptions on $\mathfrak{a}$ made in §5,1.

The exact principal part system $d'^{1-h}\, d\mathfrak{G}_1{}^h = G\, d'x'^{1-h}\, dx^h$ has the following properties as a divisor of K and K'. For every divisor $\mathfrak{p}'$ of KK'/K other than the poles of $d'^{1-h}w_i'$ and the diagonal,

$$v_{\mathfrak{p}'}(\mathfrak{a}'^{-1}(G\, d'x'^{1-h})) \geqq 0 \tag{21}$$

holds. At the diagonal,

$$G\,d'x'^{1-h}\,dx^h = \left(\frac{1}{x-x'} + \cdots\right) d'x'^{1-h}\,dx^h, \tag{22}$$

the dots indicating a function holomorphic here. Furthermore, under assumption (1)

$$v_{\mathfrak{p}}(\mathfrak{a}(G\,dx^h)) \geqq 0 \tag{23}$$

holds for all prime divisors $\mathfrak{p}$ *of* KK'/K' *other than the diagonal. Under assumption* (2) *there exists* (*up to constant multiples*) *exactly one differential* dv^{1-h} *with divisor* $(dv^{1-h}) = \mathfrak{a}$ *in* K. *Now the addition of* $dv'^{1-h}\,du^h$ *with a suitable differential* du^h *of* K *brings* $G\,d'x'^{1-h}\,dx^h$ *into a normalization for which* (23) *also holds save for one arbitrarily chosen place* $\mathfrak{p}_0$ *of* K *at which there is a pole of principal part shown by*

$$G\,d'x'^{1-h}\,dx^h = -\frac{d'v'^{1-h}}{dv^{1-h}}\frac{dp_0}{p_0} + \cdots,$$

p_0 *denoting a prime element at* $\mathfrak{p}_0$.

Proof. Equations (21) and (22) hold by our construction. The field KK'/K' has two sorts of prime divisors. The first are already prime divisors of K/k_0, while the others are generated in the algebraic extension $KK'/k_0(x, x')$ by prime polynomials $p(x, x')$ dependent upon x as well as x'. $G\,dx^h$ can have no pole of the latter sort aside from the diagonal, for it would occur in the denominator of G. The corresponding prime divisor of KK'/K would divide the denominator, contradicting (21).

To show that in general $\mathfrak{a}G\,dx^h$ can have no poles of the first sort either, we note that if such poles existed they would remain if the field of constants k_0 were extended. Thus, k_0 can be assumed algebraically closed. Let $\mathfrak{p}$ be a prime divisor of the denominator of $\mathfrak{a}G\,dx^h$. Choose some differential dw^h of K whose divisor has numerator and denominator both prime to $\mathfrak{p}$. Also, let x be some element integral at the place $\mathfrak{p}$. We now show that $G\,dx^h/dw^h$ is, in general, integral at $\mathfrak{p}$.

Assume the contrary, and let q be a prime element for $\mathfrak{p}$ and $l > 0$ the least exponent such that $q^l G\,dx^h/dw^h$ is integral at $\mathfrak{p}$. We defined G to be such that $G\,d'^{1-h}\,dx^h$ and $\sum d'^{1-h}\mathfrak{w}_i\,dw_i^h - d'^{1-h}\,d\mathfrak{G}^h$ are identical (in the sense of §5,2) as principal part systems of KK'/K. Then the difference of

$$q^l G\frac{dx^h}{dw^h}\,d'x'^{1-h} \qquad \text{and} \qquad q^l \sum \frac{dw_i^h}{dw^h}\,d'^{1-h}\mathfrak{w}_i - q^l\,d'^{1-h}\left(\frac{d\mathfrak{G}^h}{dw^h}\right)$$

multiplied by $\mathfrak{a}'^{-1}$ is everywhere holomorphic in KK'/K. Residue formation for $\mathfrak{p}$-integral elements of KK' modulo $\mathfrak{p}$ is a homomorphic map of the integral domain of those elements onto K'. The latter principal part system being considered is clearly mapped to zero. Thus the residue class of the former, namely, $q^l G\, d'^{1-h} dx^h/dw^h$, with respect to the module $\mathfrak{p}$ is a differential of degree $1-h$ in KK'/K, whose divisor is divisible by $\mathfrak{a}'$. This divisor then is $\mathfrak{b}'\mathfrak{a}'$ with some integral divisor $\mathfrak{b}'$, and it lies in the $(1-h)$th power of the canonical class W'. So $\mathfrak{b}' \in \mathfrak{a}'^{-1} W'^{1-h}$. Under the first assumption on $\mathfrak{a}$ the class $\mathfrak{a}'^{-1} W'^{1-h}$ has dimension 0 according to the Riemann–Roch theorem. So in this case we have a contradiction, and (23) is proved.

The second assumption on $\mathfrak{a}$ implies that $\mathfrak{b}'$ lies in the principal class and, being integral, $\mathfrak{b}' = (1)$. Now furthermore there exists (up to a constant multiple) exactly one differential dv^{1-h} whose divisor is $\mathfrak{a}$. With the corresponding differential $d'v'^{1-h}$ in K' our result that $\mathfrak{b}' = (1)$ means that, at $\mathfrak{p}$,

$$G\, d'x'^{1-h}\, dx^h = d'v'^{1-h}(c_l q^{-l} + \cdots)\, dq^h$$

with $c_l \in k_0$. After subtraction of $d'v'^{1-h}\, c_l q^{-l}\, dq^h$ the same reasoning can be repeated. Thus $G\, d'x'^{1-h}\, dx^h$ may possess poles at such places, but their principal parts must be of the form $d'v'^{1-h}\,(c_l q^{-l} + \cdots + c_1 q^{-1})\, dq^h$.

These poles can yet be compensated by subtracting from $G\, d'x'^{1-h}\, dx^h$ differentials $d'v'^{1-h}\, du^h$, where du^h has one pole of exactly order l at $\mathfrak{p}$, and so forth. A differential du^h of this property exists if and only if

$$\begin{aligned}\dim(\mathfrak{p}^l\mathfrak{a}W^h) &= g(\mathfrak{p}^l\mathfrak{a}W^h) + 1 - g + \dim(\mathfrak{p}^{-l}\mathfrak{a}^{-1}W^{1-h})\\ &> \dim(\mathfrak{p}^{l-1}\mathfrak{a}W^h) + g(\mathfrak{p}^l\mathfrak{a}W) + 2 - g + \dim(\mathfrak{p}^{1-l}\mathfrak{a}^{-1}W^{1-h}),\end{aligned}$$

and this is the case if and only if

$$\dim(\mathfrak{p}^{-l}\mathfrak{a}^{-1}W^{1-h}) = \dim(\mathfrak{p}^{1-l}\mathfrak{a}^{-1}W^{1-h}).$$

For $l > 1$ this is indeed so because of our assumption on $\mathfrak{a}$. For $l = 1$ the same reasoning gives a differential du^h with poles of first orders exactly at $\mathfrak{p}$ and an arbitrary further place $\mathfrak{p}_0$, and we subtract it from $G\, d'x'^{1-h}\, dx^h$. So we eventually arrive at a form of $G\, d'x'^{1-h}\, dx^h$ where this differential has two poles at $\mathfrak{D}$ and $\mathfrak{p}_0$. According to (22)

$$G \frac{d'x'^{1-h}}{d'v'^{1-h}} \frac{dv^{1-h}}{dx^{1-h}}\, dx$$

has the residue 1 in $\mathfrak{D}$. By virtue of the residue theorem, this differential then has the residue -1 in the pole $\mathfrak{p}_0$. Thus, in $\mathfrak{p}_0$,

$$G\, d'x'^{1-h}\, dx^h = -\frac{d'v'^{1-h}}{dv^{1-h}} \frac{dp}{p_0} + \cdots,$$

and the proof is complete. ▼

Notes

The principal theorem of §5,3 is sometimes called the "inhomogeneous Riemann–Roch theorem." A generalization is given in the paper [10] of §3,7, but only for the case of zero characteristic. The theorem was first stated in our form by Teichmüller† and proved in a different, somewhat quicker manner. The proof given here is by Kappus.‡ Another proof is due to Petersson,¶ the theorem there being formulated within the framework of the theory of automorphic forms.

§6*. Reduction of a Function Field with Respect to a Prime Ideal of the Constant Field

1. The Irreducibility Theorem

Let the function field K be separably generated over its exact constant field k_0, that is, let $K = k_0(x, y)$ with y satisfying the irreducible separable equation

$$f(y, x) = a_0(x)y^n + a_1(x)y^{n-1} + \cdots + a_n(x) = 0, \qquad a_0(x) \neq 0, \tag{1}$$

over $k_0(x)$. The exactness of the constant field assures the continued irreducibility of (1) in $k_1(x)$ for every extension k_1/k_0, and conversely. Of course, this need be confirmed only for finite extensions, and then

$$[Kk_1 : k_0(x)] = [Kk_1 : k_1(x)][k_1(x) : k_0(x)] = [Kk_1 : K][K : k_0(x)]$$

By §3, (12), namely

$$[k_1 : k_0] = [k_1(x) : k_0(x)] = [Kk_1 : K],$$

this leads to

$$[Kk_1 : k_1(x)] = [K : k_0(x)],$$

expressing the irreducibility of (1) in $k_1(x)$.

Let the constant field k_0 contain an integral domain $\mathfrak{o}_0$ with quotient field k_0 and satisfying the ideal theory of Chapter II. It is then no loss of generality to assume the $a_i(x)$ of (1) to be polynomials in $\mathfrak{o}_0[x]$. We now want to investigate the congruence $f(y, x) \equiv 0$ modulo prime $\mathfrak{o}_0$-ideals $\mathfrak{p}_0$ under these

† *Drei Vermutungen über algebraische Funktionenkörper*, J. Reine Angew. Math. **185**, 1–11 (1943).

‡ *Darstellungen von Korrespondenzen algebraischer Funktionenkörper und ihre Spuren*, J. Reine Angew. Math. **210**, 123–140 (1962).

¶ *Konstruktion von Modulformen* etc., S.-B. Heidelberger Akad. Wiss. Math. Nat. Kl., 417–494 (1950).

assumptions. Residue class formation mod $\mathfrak{p}_0$ will be indicated throughout by overlining. Thus, in the following we are to consider the equation

$$\overline{f(y, x)} = \bar{f}(\bar{y}, \bar{x}) = 0 \qquad \text{in} \quad \mathfrak{o}_0/\mathfrak{p}_0 = \bar{k}.$$

As a first example, let $k_0 = Q$, $\mathfrak{o}_0 = Z$. We then have an equation in the finite field Z/p which, if it is irreducible, defines a function field over that finite field. Second, let $k_0 = k_{00}(z)$, where z is some variable (independent of x) and $\mathfrak{o}_0 = k_{00}[z]$. Taking residues modulo $z - \zeta$ with $\zeta \in k_{00}$ becomes the familiar operation of substituting $z = \zeta$ into polynomials. The theorems to be developed thus include the definition of such substitution in algebraic functions.

Irreducibility Theorem. *For only a finite number of $\mathfrak{p}_0$ in k_0 does $f(y, x)$ become reducible or even inseparable in the residue class field $\bar{k}_0$ modulo $\mathfrak{p}_0$ or in an extension of that field $\bar{k}_0{}'$.*

Proof (Deuring†). The finiteness of the number of inseparable cases is obvious. If, after a finite extension $k_0{}'/k_0$ of the constant field, the polynomial is irreducible modulo a prime divisor $\mathfrak{p}_0{}'$ of $\mathfrak{p}_0$ in $k_0{}'$, then it is also irreducible modulo $\mathfrak{p}_0$. This permits finite extensions $k_0{}'/k_0$ in the course of the proof.

As in §1,4, replacing y by $a_0(x)y$ leads to an equation (1) in which $a_0(x) = 1$ and the other $a_i(x)$ remain polynomials of $\mathfrak{o}_0[x]$. Except for the finitely many $\mathfrak{p}_0$ which divide all the coefficients of the original $a_0(x)$, irreducibility of the new and old polynomials mod $\mathfrak{p}_0$ coincide.

If k_0 contains only a finite number of elements the theorem has no substance, and we eliminate that case. Thus k_0 and therefore $\mathfrak{o}_0$ contain infinitely many elements. In §1,4 we saw that all γ algebraic over k_0 for which $f(y, \gamma) = 0$ has multiple zeros satisfy an equation $D(x) = 0$ where $D(x) \in \mathfrak{o}_0[x]$. Now take some $\gamma \in \mathfrak{o}_0$ with $D(\gamma) \neq 0$. As we could substitute $x \to x - \gamma$ we can immediately assume $\gamma = 0$.

Further, let η_{i0} $(i = 1, \ldots, n)$ be the solutions of $f(y, 0) = 0$. Finite algebraic extension of k_0 is permissible, so that we may assume the η_{i0} to lie in $\mathfrak{o}_0$. Then, by §1,4 there exist n power series

$$y_i = \eta_{i0} + \eta_{i1}x + \cdots$$

and an $N \in \mathfrak{o}_0$ such that $\eta_{i\mu}N^{\mu} \in \mathfrak{o}_0$, satisfying

$$f(y, x) = \prod_{i=1}^{n} (y - y_i). \tag{2}$$

† *Reduktion algebraischer Funktionenkörper nach Primdivisoren des Konstantenkörpers*, Math. Z. **47**, 643–654 (1941).

After the finitely many prime divisors of N are eliminated, residue class formation modulo the remaining $\mathfrak{p}_0$ yields power series

$$\bar{y}_i = \bar{\eta}_{i0} + \bar{\eta}_{i1}\bar{x} + \cdots$$

with coefficients in $\bar{k}_0$ which then satisfy

$$\bar{f}(\bar{y}, \bar{x}) = \prod_{i=1}^{n} (\bar{y} - \bar{y}_i). \tag{3}$$

Now, if (3) were reducible then one of its partial products would have to be a polynomial in $\bar{y}$, $\bar{x}$, whose degree in $\bar{x}$ is at most the degree n_x of $f(y, x)$ in x. The corresponding partial product of (2) would be a polynomial in y with power series in x as coefficients which are infinite because, otherwise, $f(y, x)$ would be reducible in $k_0(x)$. So these series would have coefficients which are divisible by $\mathfrak{p}_0$ with but finitely many exceptions. Therefore form all the possible proper partial products of (2) and in each take the first non-vanishing coefficient of a power of x which is $> n_x$, and multiply all these coefficients. The product is divisible by $\mathfrak{p}_0$ while by hypothesis on $\mathfrak{o}_0$ only a finite number of $\mathfrak{p}_0$ can go into such a product. Hence $f(y, x)$ is irreducible mod $\mathfrak{p}_0$ save for finitely many exceptional $\mathfrak{p}_0$.

Any finite extension $\bar{k}_0'/\bar{k}_0$ arises from a finite extension k_0'/k_0 by taking residues of the principal order $\mathfrak{o}_0'$ of k_0' with respect to $\mathfrak{o}_0$ modulo a prime divisor $\mathfrak{p}_0'$ of $\mathfrak{p}_0$. Replace k_0 in the above argument by all possible k_0'; this gives the last part of the theorem. The set of $\mathfrak{p}_0$ to be eliminated is finite, for these are always divisors of a fixed element. ▼

One consequence of the irreducibility of $f(y, x)$ mod $\mathfrak{p}_0$ is the following.

Theorem of Inertia. *Let the polynomial $f(y, x)$ be irreducible and separable mod $\mathfrak{p}_0$, $\mathfrak{o}_{\mathfrak{p}_0}$ denote the integral domain of those $a_0 \in k_0$ with denominators not divisible by $\mathfrak{p}_0$, and $\mathfrak{J}_{\mathfrak{p}_0}$ be the principal order of K with respect to $\mathfrak{o}_{\mathfrak{p}_0}[x]$. Then $\mathfrak{J}_{\mathfrak{p}_0}\mathfrak{p}_0$ is a prime ideal and the quotient field $\bar{K}$ of the residue class ring $\mathfrak{J}_{\mathfrak{p}_0}/\mathfrak{J}_{\mathfrak{p}_0}\mathfrak{p}_0$ can be formed.*

This means that every $\kappa \in K$ can be written as

$$\kappa = p_0^{\,h}(\alpha/\beta) \tag{4}$$

with $\alpha, \beta \in \mathfrak{J}_{\mathfrak{p}_0}$ and $p_0^{-1}\alpha, p_0^{-1}\beta \in \mathfrak{J}_{\mathfrak{p}_0}$, where p_0 is some prime element for $\mathfrak{p}_0$ in k_0. If $h = 0$ the element κ is called $\mathfrak{p}_0$-*normed.* Thus any principal divisor (κ) can be represented by a $\mathfrak{p}_0$-normed element.

Proof. It suffices to show that if $\alpha \in \mathfrak{J}_{\mathfrak{p}_0}$ has a norm $n_{K/k_0(x)}(\alpha)$ divisible by $\mathfrak{p}_0$, then $p_0^{-1}\alpha \in \mathfrak{J}_{\mathfrak{p}_0}$. We use the generating equation (1) and, as before, may assume that $a_0(x) = 1$ and that the other $a_i(x) \in \mathfrak{o}_0[x]$. Then $y \in \mathfrak{J}_{\mathfrak{p}_0}$. As we assume $\bar{f}(\bar{y}, \bar{x})$ to be separable, the discriminant $D = D(1, y, \ldots, y^{n-1}) \not\equiv 0 \bmod \mathfrak{p}_0$, so that α can be represented as (cf. II,§1,2)

$$\alpha = D^{-1} \sum_{i=0}^{n-1} b_i y^i, \qquad b_i = s_{K/k_0(x)}(\alpha y^i) \in \mathfrak{o}_{\mathfrak{p}_0}[x].$$

As $D \not\equiv 0 \bmod \mathfrak{p}_0$ we may replace α by αD in our verification. So now let $n(\sum b_i y^i) \equiv 0 \bmod \mathfrak{p}_0$. As the order of norm and residue class formation may be interchanged, we have $n(\sum \bar{b}_i \bar{y}^i) = 0$. Further, the order of residue class and discriminant formation is interchangeable, so that $D(\bar{1}, \bar{y}_1, \ldots, \bar{y}^{n-1}) \neq 0$. The vanishing of the norm then implies all $\bar{b}_i = 0$. Thus all the b_i are divisible by $\mathfrak{p}_0$, completing the proof. ▼

2. Regular Prime Ideals

From now on we always omit the prime ideals $\mathfrak{p}_0$ excluded in the proof of the irreducibility theorem of the last section from our considerations. Let $\bar{K}$ be the field found in the inertia theorem. It is defined over $\bar{k}_0$ by the equation $(\bar{x}, \bar{y}) = 0$, and is a separable extension of degree n over $\bar{k}_0(\bar{x})$, the exact field of constants being $\bar{k}_0$.

Further limitations are desirable. Consider the following example. Let k_{00} be an algebraically closed field of prime characteristic $p > 2$ and set $k_0 = k_{00}(\bar{u})$ with an indeterminate u. Finally, let $K = k_0(x, \sqrt{x^p - u})$. The genus of K is $g = (p-1)/2$. Modulo any prime ideal $\mathfrak{p}_0$ of $\mathfrak{o}_0 = k_{00}[u]$, the residue class $\bar{u}$ is a pth power in $\bar{k}_0 = k_{00}$. Thus $\bar{K} = k_{00}(\sqrt{\bar{x} - \sqrt[p]{\bar{u}}})$ is a field of genus $\bar{g} = 0$. We should like to exclude exceptional cases of such a nature.

A field K has been called *conservative* if its genus never changes under extension of constants. The K of this example is certainly not conservative, as adjunction of $\sqrt[p]{u}$ to k_0 leads to a field with $g = 0$.

A prime ideal $\mathfrak{p}_0$ of $\mathfrak{o}_0$ is called *regular* if $\bar{K}$ is a separable function field of degree $[\bar{K} : \bar{k}_0(\bar{x})] = [K : k_0(x)] = n$ over $\bar{k}_0(\bar{x})$ with exact constant field $\bar{k}_0$ and with the same genus $\bar{g} = g$ as K.

Let K be separable over $k_0(x)$ and conservative. Then, except for a finite number of exceptions, all prime $\mathfrak{o}_0$-ideals of k_0 are regular.

For the proof† we again form the integral domain $\mathfrak{o}_{\mathfrak{p}_0}$ of all $a \in k_0$ with denominators not divisible by $\mathfrak{p}_0$. Let $\mathfrak{J}_{\mathfrak{p}_0}, \mathfrak{J}, \bar{\mathfrak{J}}$ be the principal orders of K with respect to $\mathfrak{o}_{\mathfrak{p}_0}[x]$, $\mathfrak{i} = k_0[x]$, and of $\bar{K}$ with respect to $\bar{\mathfrak{i}} = \bar{k}_0[\bar{x}]$. Clearly

$$\bar{\mathfrak{J}} \supseteq \mathfrak{J}_{\mathfrak{p}_0}/\mathfrak{J}_{\mathfrak{p}_0}\mathfrak{p}_0. \tag{5}$$

Let ω_i be a basis of $\mathfrak{J}$ with respect to $\mathfrak{i}$. In II,§4,2 we saw that the discriminant

† Another proof is found in M. Deuring, *Die Zetafunktion einer algebraischen Kurve vom Geschlecht* 1, Nachr. Akad. Wiss. Göttingen Math.-Phys.-Chem. Kl., 13–42 (1955).

$D(\omega_i)$ is the norm of the different ideal of $\mathfrak{J}$. We now exclude the finite number of $\mathfrak{p}_0$ for which $\omega_i \in \mathfrak{J}_{\mathfrak{p}_0}$ is not true, and then the discriminant of the residue classes of the ω_i mod $\mathfrak{J}_{\mathfrak{p}_0}\mathfrak{p}_0$, namely $D(\bar{\omega}_i)$, is a polynomial in $\bar{x}$ having the same degree as $D(\omega_i)$. Using (5) then, the degree in $\bar{x}$ of the discriminant of $\bar{\mathfrak{J}}$ is not greater than the degree in x of the discriminant of $\mathfrak{J}$.

The same argument is then repeated with $x^{-1} = x'$ in place of x, but only insofar as to see what power of x' divides the discriminant of $\mathfrak{J}'$. Together with the degree in x of $D(\omega_i)$ this gives the degree of the discriminant divisor: almost always

$$g(\mathfrak{D}_{K/\bar{k}_0(\bar{x})}) \leqq g(\mathfrak{D}_{K/k_0(x)}). \tag{6}$$

Thus, by §3,(8) the genera of $\bar{K}$ and K satisfy $\bar{g} \leqq g$.

To prove the equality for almost all $\mathfrak{p}_0$ we must show that equality actually holds almost always in (5), and that the same is true when x is replaced by x'. Then equality holds for (6) as well, and the assertion $\bar{g} = g$ is verified. Due to the symmetry of x and x^{-1} we must actually only do one.

Let the polynomials $q_j = q_j(x)$ $(j = 1, \ldots, h)$ be the prime divisors of $D(\omega_i)$. Immediately exclude all $\mathfrak{p}_0$ by which the denominators or highest coefficients of the q_j are divisible. Now choose some finite algebraic extension k_0'/k_0 so that the q_j decompose into linear factors (which we again call q_j), and so that the prime divisors of the q_j in $K' = Kk_0'$ are of the first degree. The lemma in §4,4 makes this possible. Simultaneously, replace $\mathfrak{p}_0$ by one of its prime divisors $\mathfrak{p}_0'$ in k_0'. K was assumed conservative, so the genus g remains unchanged by this extension. By §3,4 the genus of $\bar{K}$ is at most reduced. If we can show that $\bar{g} = g$ after the extension, this was also so before.

For simplicity we omit the primes, thus again writing k_0, $\mathfrak{p}_0$, etc., for k_0', $\mathfrak{p}_0'$, etc. We also now exclude those $\mathfrak{p}_0$ which lead to the same residue class for different q_j. Now let q be one of the q_j, and the numerator divisor $\mathfrak{q}$ of q decomposes in the manner

$$\mathfrak{q} = \mathfrak{q}_1^{e_1} \cdots \mathfrak{q}_r^{e_r}. \tag{7}$$

By assumption the $\mathfrak{q}_\varrho$ are prime divisors of the first degree and II,§3,(14) then shows the ramification indices have the sum

$$e_1 + \cdots + e_r = n. \tag{8}$$

We now show that, in general, the numerator divisor of $\bar{q}$ decomposes in the same manner. To this end we apply II,§2,2. The prime divisors $\mathfrak{q}_i$ are represented by prime $\mathfrak{J}_q$-ideals where $\mathfrak{J}_q$ denotes as usual the principal order of K with respect to $\mathfrak{i}_q$. No confusion is to be feared if we denote these prime ideals by the same symbols $\mathfrak{q}_i$. As $\mathfrak{J}_q$ is a principal ideal domain, $\mathfrak{q}_i = \mathfrak{J}_q\kappa_i$. The elements κ_i may be assumed to be $\mathfrak{p}_0$-normed. We now form the domains $\mathfrak{J}_{q,\mathfrak{p}_0}$ consisting of all $\kappa \in \mathfrak{J}_q$ of the form (4) with $h \geqq 0$ and furthermore the domains of their residues mod $\mathfrak{p}_0$. Then the $\bar{\mathfrak{q}}_i = \bar{\mathfrak{I}}_{\bar{q}}\bar{\kappa}_i$ are integral $\bar{\mathfrak{I}}_{\bar{q}}$-ideals satisfying

$$\bar{\mathfrak{I}}_q\bar{q} = \bar{\mathfrak{q}}_1^{e_1} \cdots \bar{\mathfrak{q}}_r^{e_r}. \tag{9}$$

We maintain that the $\bar{\mathfrak{q}}_\varrho$ are prime ideals. For, if

$$\bar{\mathfrak{q}}_\varrho = \prod_\nu \bar{\mathfrak{p}}_{\varrho\nu}^{\varepsilon_{\varrho\nu}}$$

were their decomposition and $f_{\varrho\nu}$ the degrees of the $\bar{\mathfrak{p}}_{\varrho\nu}$ with respect to $\bar{k}_0(\bar{x})$, then II, §3,(14) would give

$$\sum_{\varrho,\nu} f_{\varrho\nu}\varepsilon_{\varrho\nu}e_\varrho = n.$$

But this equation is only compatible with (8) if

$$\sum_\nu f_{\varrho\nu}\varepsilon_{\varrho\nu} = 1$$

for every ϱ, that is, if the $\bar{\mathfrak{q}}_\varrho$ were prime ideals.

If K has characteristic 0 then the discriminant $D(\mathfrak{I}) = D(\omega_i)$ is divisible by q to the power $\sum(e_\varrho - 1)$, by §2,(36). But the Dedekind discriminant theorem assures us that, assuming (9), q divides the discriminant $D(\bar{\mathfrak{I}})$ to at least that degree. If several of the $\bar{\mathfrak{q}}_\varrho$ were equal, an even higher power of $\bar{q}$ would divide $D(\bar{\mathfrak{I}})$. But as the $\bar{q} = \bar{q}_i$ were assumed to be different and because of (6), it is impossible for $D(\bar{\mathfrak{I}})$ to be divisible by a single $\bar{q}_i$ to a higher power than $D(\mathfrak{I})$ by the corresponding q_i. Thus the discriminants $D(\bar{\mathfrak{I}})$ and $D(\mathfrak{I})$ are of the same degree, so that equality holds in (5).

This argument fails, though, for K of prime characteristic, for then it is possible for $D(\bar{\mathfrak{I}})$ to be divisible by a higher power of $\bar{q}$. But now, construct the κ_ϱ in order, making them pairwise prime. Then for any two of them there exist elements $\alpha_{\varrho\sigma}, \beta_{\varrho\sigma} \in \mathfrak{I}_q$, such that

$$\alpha_{\varrho\sigma}\kappa_\varrho + \beta_{\varrho\sigma}\kappa_\sigma = 1, \qquad \varrho \neq \sigma. \tag{10}$$

Again, $\alpha_{\varrho\sigma}, \beta_{\varrho\sigma} \in \mathfrak{I}_{q,\mathfrak{p}_0}$ except for finitely many $\mathfrak{p}_0$. Residues mod $\mathfrak{p}_0$ can now be taken for (10), showing the $\bar{\mathfrak{q}}_\varrho$ to be differing prime ideals.

Finally, for each $\bar{q} = \bar{q}_i$ consider the element

$$\bar{\lambda}_{\varrho j} = \bar{\kappa}_1^{e_1} \cdots \bar{\kappa}_r^{e_r}\bar{\kappa}_\varrho^{j-e_\varrho}, \qquad j = 0, 1, \ldots, e_\varrho - 1. \tag{11}$$

They all lie in $\mathfrak{I}_{q,\mathfrak{p}_0}/\mathfrak{I}_{q,\mathfrak{p}_0}\mathfrak{p}_0 = \bar{\mathfrak{I}}_{\bar{q}}$ and have the orders

$$v_{\bar{\mathfrak{p}}_\sigma}(\bar{\lambda}_{\varrho j}) = \begin{cases} j & \text{for } \varrho = \sigma \\ e_\sigma & \text{for } \varrho \neq \sigma. \end{cases} \tag{12}$$

The method of §2,4, using (12), can now verify that the $\bar{\lambda}_{\varrho j}$ are a basis of $\bar{\mathfrak{I}}_{\bar{q}}$ with respect to $\bar{\mathfrak{i}}_{\bar{q}}$. Therefore, if both of the $\bar{\mathfrak{i}}$-modules of (5) are extended to $\bar{\mathfrak{i}}_{\bar{q}}$-modules by permitting denominators prime to $\bar{q}$, they coincide. But this is so for all q that divide the residue class of $D(\omega_i)$ mod $\mathfrak{p}_0$, which proves (5).

3. Behavior of Ideals under Residue Formation

From now on let $\mathfrak{p}_0$ always be a regular prime ideal. As above, let $\mathfrak{o}_{\mathfrak{p}_0}$ be the integral domain of $a \in k_0$ with denominators indivisible by $\mathfrak{p}_0$, a principal ideal domain, by II, §2,2. Finally, let $\mathfrak{K}_{\mathfrak{p}_0}$ be the integral domain of elements (4) with $h \geqq 0$. Note that the definition of $\mathfrak{K}_{\mathfrak{p}_0}$ depends upon the element x, so that it is sometimes practical to denote it by $\mathfrak{K}_{\mathfrak{p}_0} = \mathfrak{K}_{x,\mathfrak{p}_0}$.

This section will be devoted to the study of $\mathfrak{J}$-ideals and their residue classes mod $\mathfrak{p}_0$. The latter have yet to be defined; to do so consider, to any ideal $\mathfrak{a}$,

$$\mathfrak{a}_{\mathfrak{p}_0} = \mathfrak{a} \cup \mathfrak{K}_{\mathfrak{p}_0}, \qquad \text{in particular} \qquad \mathfrak{J}_{\mathfrak{p}_0} = \mathfrak{J} \cap \mathfrak{K}_{\mathfrak{p}_0}. \tag{13}$$

These $\mathfrak{a}_{\mathfrak{p}_0}$ are $\mathfrak{o}_{\mathfrak{p}_0}[x]$-modules of rank n, and they satisfy the conditions of the basis theorem of I, §1,5. In fact, let $a, b \in \mathfrak{o}_{\mathfrak{p}_0}[x]$ be two mutually prime elements and $\alpha a, \alpha b \in \mathfrak{a}_{\mathfrak{p}_0}$. Then $\alpha \in \mathfrak{a}$. As a and b are not both divisible by $\mathfrak{p}_0$ we see that $\alpha \in \mathfrak{K}_{\mathfrak{p}_0}$ and therefore also $\alpha \in \mathfrak{a}_{\mathfrak{p}_0}$. Hence the $\mathfrak{a}_{\mathfrak{p}_0}$ have bases of n elements, α_i, with respect to $\mathfrak{o}_{\mathfrak{p}_0}[x]$, which are clearly also bases of $\mathfrak{a}$ with respect to $\mathfrak{i}$.

Taking residues mod $\mathfrak{p}_0$ gives

$$\bar{\mathfrak{a}} = \mathfrak{a}_{\mathfrak{p}_0}/\mathfrak{a}_{\mathfrak{p}_0}\mathfrak{p}_0, \qquad \text{in particular} \qquad \bar{\mathfrak{J}} = \mathfrak{J}_{\mathfrak{p}_0}/\mathfrak{J}_{\mathfrak{p}_0}\,\mathfrak{p}_0. \tag{14}$$

These $\bar{\mathfrak{a}}$ are $\bar{\mathfrak{J}}$-ideals

$\bar{\mathfrak{J}}$ is the principal order of $\bar{K}$ with respect to $\bar{\mathfrak{i}} = \bar{k}_0[\bar{x}]$.

The proof uses the inverse of an argument of the last section. Were the principal order $\bar{\mathfrak{J}}' \supset \bar{\mathfrak{J}}$ (the notation is different here), then the discriminant of $\bar{\mathfrak{J}}'$ with respect to $\bar{\mathfrak{i}}$ would be of lower degree in $\bar{x}$ than that of $\bar{\mathfrak{J}} = \mathfrak{J}_{\mathfrak{p}_0}/\mathfrak{J}_{\mathfrak{p}_0}\mathfrak{p}_0$, while the degree of the latter is at most the same as the degree of the discriminant of $\mathfrak{J}$ with respect to $\mathfrak{i}$. But then we should have $\bar{g} < g$ by §6,2, contrary to hypothesis. ▼

The residue classes of the bases α_i of $\mathfrak{a}_{\mathfrak{p}_0}$ are bases of the $\bar{\mathfrak{a}}$ with respect to $\bar{\mathfrak{i}}$. From the definition of the norm in II, §3,2 we immediately see

$$n(\bar{\mathfrak{a}}) = \overline{n(\mathfrak{a})}; \tag{15}$$

that is, *norm and residue formation for ideals are commutable operations.*

Any two ideals satisfy

$$\overline{\mathfrak{a}\mathfrak{b}} = \bar{\mathfrak{a}}\,\bar{\mathfrak{b}}. \tag{16}$$

Proof. Clearly $(\mathfrak{a}\mathfrak{b})_{\mathfrak{p}_0} \supseteq \mathfrak{a}_{\mathfrak{p}_0}\mathfrak{b}_{\mathfrak{p}_0}$ and therefore

$$\overline{\mathfrak{a}\mathfrak{b}} \supseteq \bar{\mathfrak{a}}\,\bar{\mathfrak{b}}. \tag{17}$$

By setting $\mathfrak{b} = \mathfrak{a}^{-1}$ we have

$$\bar{\mathfrak{J}} \supseteq \bar{\mathfrak{a}}\,\overline{\mathfrak{a}^{-1}}.$$

This implies

$$\overline{\mathfrak{a}^{-1}} \subseteq \bar{\mathfrak{a}}^{-1} \tag{18}$$

because the ideal theory of II §,1 holds for the $\bar{\mathfrak{J}}$-ideals.

As the residues of the bases of $\mathfrak{I}_{\mathfrak{p}_0}$ and $(\mathfrak{a}^{-1})_{\mathfrak{p}_0}$ are bases of $\bar{\mathfrak{I}}$ and $\overline{\mathfrak{a}^{-1}}$, a linear substitution with determinant $n(\mathfrak{a}^{-1})$ transforms a basis of $\bar{\mathfrak{I}}$ to one of $\overline{\mathfrak{a}^{-1}}$. By (15) we have

$$n(\overline{\mathfrak{a}^{-1}}) = \overline{n(\mathfrak{a}^{-1})} = \overline{n(\mathfrak{a})^{-1}} = \overline{n(\mathfrak{a})}^{-1} = n(\bar{\mathfrak{a}})^{-1} = n(\bar{\mathfrak{a}}^{-1}).$$

On the other hand, a linear substitution with determinant $n(\bar{\mathfrak{a}}^{-1})$ transforms a basis of $\bar{\mathfrak{I}}$ to a basis of $\bar{\mathfrak{a}}^{-1}$. The equality of the norms, together with (18), yields

$$\overline{\mathfrak{a}^{-1}} = \bar{\mathfrak{a}}^{-1} \qquad \text{or} \qquad \bar{\mathfrak{a}}\overline{\mathfrak{a}^{-1}} = \bar{\mathfrak{I}}. \tag{19}$$

Multiplication of (17) with $\overline{\mathfrak{a}^{-1}}$ then yields

$$\overline{\mathfrak{a}^{-1}}\,\overline{\mathfrak{a}\mathfrak{b}} \supseteq \bar{\mathfrak{b}}. \tag{20}$$

But (17) can also be applied in the form $\bar{\mathfrak{b}} = \overline{\mathfrak{a}^{-1}\cdot\mathfrak{a}\mathfrak{b}} \supseteq \overline{\mathfrak{a}^{-1}}\,\overline{\mathfrak{a}\mathfrak{b}}$, and then equality prevails in (20). But now multiplication by $\bar{\mathfrak{a}}$ gives the asserted equation (16). ▼

4. Behavior of Divisors under Residue Formation

We first review, in a slightly revised form, the introduction of the notion of divisor. The group of divisors of K, of II, §5,3, can be described as follows. Separate the prime divisors $\mathfrak{p}$ into two sets, F and I, the "finite" and "infinite" prime divisors, on the basis of whether $\mathfrak{p}$ divides a prime divisor $\mathfrak{z}_{q(x)}$ of the rational function field $k = k_0(x)$ which is the numerator of a prime polynomial $q(x) \in k_0[x]$, or whether it divides the denominator divisor $\mathfrak{z}_\infty$ of x. Correspondingly, split an arbitrary divisor up into a "finite" and an "infinite" part,

$$\mathfrak{a} = \mathfrak{a}_F \mathfrak{a}_I = \prod_{\mathfrak{p}\in F} \mathfrak{p}^{v_\mathfrak{p}(\mathfrak{a})} \prod_{\mathfrak{p}\in I} \mathfrak{p}^{v_\mathfrak{p}(\mathfrak{a})}. \tag{21}$$

Clearly

$$(\mathfrak{a}\mathfrak{b})_F = \mathfrak{a}_F \mathfrak{b}_F, \qquad (\mathfrak{a}\mathfrak{b})_I = \mathfrak{a}_I \mathfrak{b}_I. \tag{22}$$

The group of "finite" divisors $\mathfrak{a} = \mathfrak{a}_F$ is isomorphic to the group of $\mathfrak{I}$-ideals, where $\mathfrak{I}$ is the principal order of K with respect to $\mathfrak{i} = k_0[x]$.

This isomorphism can be materialized as follows. For any divisor $\mathfrak{a}$, take the totality $\mathfrak{a}_\mathfrak{I}$ of all $\alpha \in K$ with $v_\mathfrak{p}(\alpha) \geqq v_\mathfrak{p}(\mathfrak{a})$ for all $\mathfrak{p} \in F$. Certainly $(1)_\mathfrak{I} = \mathfrak{I}$, and the $\mathfrak{a}_\mathfrak{I}$ are $\mathfrak{I}$-ideals. We call them the $\mathfrak{I}$-multiple ideals to the divisors. Our mapping $\mathfrak{a} \to \mathfrak{a}_\mathfrak{I}$ is a homomorphism of the divisor group onto the group of $\mathfrak{I}$-ideals which maps precisely the "infinite" divisors $\mathfrak{a}$, those for which $\mathfrak{a} = \mathfrak{a}_\mathfrak{I}$, onto the unit element.

In the same vein the mapping of the divisors onto the $\mathfrak{I}_\infty$-ideals $\mathfrak{a}_\infty$ is a homomorphism which maps the "finite" divisors $\mathfrak{a}$, those for which $\mathfrak{a} = \mathfrak{a}_F$, onto the unit element. Thus the groups of finite components $\mathfrak{a}_F$ and infinite components $\mathfrak{a}_I$ of the divisors are isomorphic to the groups of $\mathfrak{I}$-ideals and $\mathfrak{I}_\infty$-ideals, and the group of all divisors is the direct product.

All this can be used to define a homomorphic mapping of the group of divisors of K into that of $\bar{K}$. First, (14) defines a homomorphic [because of (16)] mapping of the group of $\mathfrak{J}$-ideals into that of $\bar{\mathfrak{J}}$-ideals. To distinguish ideals from divisors we now write

$$\overline{\mathfrak{a}_{\mathfrak{J}}\mathfrak{b}_{\mathfrak{J}}} = \overline{\mathfrak{a}_{\mathfrak{J}}}\,\overline{\mathfrak{b}_{\mathfrak{J}}}$$

though. The "finite" divisors $\bar{\mathfrak{a}}_F$ are associated, as described, with the $\bar{\mathfrak{J}}$-ideals so that

$$\mathfrak{a} \to \mathfrak{a}_{\mathfrak{J}} \to \bar{\mathfrak{a}}_{\mathfrak{J}} \to \bar{\mathfrak{a}}_F \tag{23}$$

produces a homomorphic mapping of the divisor group of K into the group of finite divisors of $\bar{K}$.

A second homomorphism $\mathfrak{a} \to \bar{\mathfrak{a}}_I$ must now be defined. Just as the principal order $\mathfrak{J}$ of K with respect to $\mathfrak{i} = k_0[x]$ was formed, now form the principal order $\mathfrak{J}'$ of K with respect to $\mathfrak{i}' = k_0[x']$, $x' = x^{-1}$, and, as before, $\bar{\mathfrak{J}}'$ as well as the sets F', I'. We have $I \subset F'$ and $\bar{I} \subset \bar{F}'$. The homomorphism $\mathfrak{a} \to \bar{\mathfrak{a}}_{F'}$ corresponds to (23). The divisors $\bar{\mathfrak{a}}_{F'}$ can be decomposed by (21) to $\bar{\mathfrak{a}}_{F'} = (\bar{\mathfrak{a}}_{F'})_F \cdot (\bar{\mathfrak{a}}_{F'})_I$, so that

$$\mathfrak{a} \to \mathfrak{a}_{\mathfrak{J}'} \to \overline{\mathfrak{a}_{\mathfrak{J}'}} \to \bar{\mathfrak{a}}_{F'} \to (\bar{\mathfrak{a}}_{F'})_I = \bar{\mathfrak{a}}_I \tag{24}$$

is a homomorphism of the divisor group of K into the group of "infinite" divisors of $\bar{K}$.

Finally, by composing (23) and (24) we find the mapping

$$\mathfrak{a} \to \bar{\mathfrak{a}}_F\bar{\mathfrak{a}}_I = \bar{\mathfrak{a}} \tag{25}$$

which is a homomorphism of the divisor group of K into that of $\bar{K}$, that is,

$$\overline{\mathfrak{a}\mathfrak{b}} = \bar{\mathfrak{a}}\,\bar{\mathfrak{b}}. \tag{26}$$

We call $\bar{\mathfrak{a}}$ the *residue of* $\mathfrak{a}$ mod $\mathfrak{p}_0$. This mapping has the following properties.

Integral divisors $\mathfrak{a}$ *are mapped onto integral divisors* $\bar{\mathfrak{a}}$ *so that* $\bar{\mathfrak{a}}$ *divides* $\bar{\mathfrak{b}}$ *if* $\mathfrak{a}$ *divides* $\mathfrak{b}$.

It leaves degrees invariant:

$$g(\bar{\mathfrak{a}}) = g(\mathfrak{a}). \tag{27}$$

Norm and residue formation commute:

$$\overline{n_{K/k}(\mathfrak{a})} = n_{\bar{K}/\bar{k}}(\bar{\mathfrak{a}}). \tag{28}$$

The principal divisor of an element $\alpha \in K$ *is mapped onto the principal divisor of the residue class of* α mod $\mathfrak{p}_0$ *if only* α *is assumed to be* $\mathfrak{p}_0$*-normed*:

$$\overline{(\alpha)} = (\bar{\alpha}). \tag{29}$$

In view of (29) *the mapping* $\mathfrak{a} \to \bar{\mathfrak{a}}$ *is also a mapping of the divisor class group of* K *into that of* $\bar{K}$.

Proof. The assertions concerning integral divisors are immediate consequences of the corresponding and obvious statements for ideals.

Equation (28) follows from the analogous equation (15) for $\mathfrak{I}$-ideals and $\mathfrak{I}'$-ideals, the latter computation using only the "infinite" components.

Remembering (28) and II, §5,(16), it now suffices to verify (27) for rational function fields. Thus, let $K = k_0(x)$. Further, it suffices to use only prime ideals. The denominator divisor $\mathfrak{z}_\infty$ of x is mapped onto the denominator divisor $\bar{\mathfrak{z}}_\infty$ of $\bar{x}$; here (27) is self-evident. Now let $\mathfrak{a} = \mathfrak{z}_{q(x)}$ be the numerator divisor of the prime polynomial

$$q(x) = a_0 x^m + \cdots + a_r x^{m-r} + \cdots + a_m, \qquad a_i \in \mathfrak{o}_{\mathfrak{p}_0},$$

and assume a_r to be the first coefficient not divisible by $\mathfrak{p}_0$. Then, the "finite" part of $\bar{\mathfrak{a}}$ is the numerator divisor of the polynomial $\bar{a}_r\bar{x}^{m-r} + \cdots + \bar{a}_m$ and is of degree $m - r$. Further,

$$\mathfrak{a}_{\mathfrak{I}'} = \mathfrak{I}'(a_0 + \cdots + a_r x'^r + \cdots + a_m x'^m),$$

$$\overline{\mathfrak{a}_{\mathfrak{I}'}} = \bar{\mathfrak{I}}'(\bar{a}_r + \cdots + \bar{a}_m \bar{x}'^{m-r})\bar{x}'^r.$$

Thus, the "infinite" part of $\bar{\mathfrak{a}}$ is $\bar{\mathfrak{a}}_I = \bar{\mathfrak{z}}_\infty{}^r$ and of degree r. We have shown that m is the degree of $\bar{\mathfrak{a}}$ as well as of $\mathfrak{a}$, proving (27).

Equation (29) can be found directly from the equations

$$\overline{\mathfrak{I}\mathfrak{a}} = \bar{\mathfrak{I}}\bar{\alpha}, \qquad \overline{\mathfrak{I}'\mathfrak{a}} = \bar{\mathfrak{I}}'\bar{\alpha}$$

for principal ideals, by assuming α to be $\mathfrak{p}_0$-normed (cf. §6,1). ▼

The homomorphic mapping $\mathfrak{a} \to \bar{\mathfrak{a}}$ *of divisor groups described coincides for the two models of the field,* $K = k_0(x, y)$ *with* $f(y, x) = 0$ *and* $K = k_0(x', y')$ *with* $f'(y', x') = 0$ *if the rings* $\mathfrak{R}_{x,\mathfrak{p}_0}$ *and* $\mathfrak{R}_{x',\mathfrak{p}_0}$ *associated with these models in* §6,3 *coincide.*

Remark. The hypothesis of the theorem is not satisfied by the two models $k_0(x)$ and $k_0(x')$ with $x' = p_0 x$. For, the numerator divisor of $p_0 x + 1$ is mapped onto the denominator of $\bar{x}$ by the former but not by the latter.

Proof. A divisor $\mathfrak{a}$ can be uniquely characterized by the totality of principal divisors (α) by which $\mathfrak{a}$ is *exactly* divisible in the following sense: the product of the numerator and denominator of $\mathfrak{a}$ is prime to $\mathfrak{a}^{-1}\alpha$. This characterization is an immediate consequence of the ideal-theoretical introduction of divisors in II, §5. Now, under the hypothesis, residue formation $\alpha \to \bar{\alpha}$ leads to the same result in the two models. This is then also the case for principal ideals, by (29), and therefore even for arbitrary ideals. ▼

It is worthwhile mentioning that it is not always known whether every divisor $\bar{\mathfrak{a}}$ of $\bar{K}$ has an inverse image in K, as would be the case for residue formation in rings.

5. Continuation; Behavior of Differentials under Residue Formation

For the residue class $\bar{\mathfrak{a}}$ *of a divisor* $\mathfrak{a}$ *of* K,

$$\dim(\bar{\mathfrak{a}}) \geqq \dim(\mathfrak{a}). \tag{30}$$

Of course, the equality sign holds for all divisors whose degrees are $g(\bar{\mathfrak{a}}) = g(\mathfrak{a}) \geqq 2g-2$, *due to the theorem of Riemann–Roch. If* (30) *is an equality the multiples of* $\bar{\mathfrak{a}}^{-1}$ *are residues of multiples of* $\mathfrak{a}^{-1}$ *mod* $\mathfrak{p}_0$.

Proof. The $\mathfrak{p}_0$-normed multiples of $\mathfrak{a}^{-1}$ generate a finite $\mathfrak{o}_{\mathfrak{p}_0}$-module. Let α_i be a basis of it. The $\bar{\alpha}_i$ are then multiples of $\bar{\mathfrak{a}}^{-1}$. They are linearly independent, because a nontrivial linear relation $\sum c_i\alpha_i \equiv 0 \bmod \mathfrak{p}_0$ with $c_i \in \mathfrak{o}_{\mathfrak{p}_0}$ would contradict the assumption that the α_i form a basis. ▼

The commutability of trace and residue formation applied to the discussion of §6,3 implies that the different ideals satisfy

$$\mathfrak{d}_{\bar{\mathfrak{J}}} = \bar{\mathfrak{d}}_{\mathfrak{J}}, \qquad \mathfrak{d}_{\bar{\mathfrak{J}}'} = \overline{\mathfrak{d}_{\mathfrak{J}'}},$$

the second equation implying that

$$\mathfrak{d}_{\bar{\infty}} = \overline{\mathfrak{d}_{\infty}}.$$

Together they give the different divisor:

$$\mathfrak{d}_{K/\bar{k}} = \overline{\mathfrak{d}_{K/k}}. \tag{31}$$

A consequence of (31) is: *the residue of the canonical class* W *of* K *is the canonical class* $\overline{W}$ *of* $\overline{K}$.

Let the residue class $\bar{u}$ of the element u be a separating element of $\overline{K}$. Then certainly u is a separating element of K. According to §4,2 there exists an irreducible equation in x and u, $g(u, x) = 0$, whose partial derivatives with respect to u and x do not vanish. Now, $\bar{u}$ generates a subfield of $\overline{K}$ over $\bar{k}_0(\bar{x})$. Because $[\overline{K} : \bar{k}_0(\bar{x})] = [K : k_0(x)]$ we also have $[\bar{k}_0(\bar{x}, \bar{u}) : \bar{k}_0(\bar{x})] = [k_0(x, u) : k_0(x)]$. This means that $\bar{g}(\bar{u}, \bar{x})$ is also irreducible over $\bar{k}_0$, so that it is the minimal equation between $\bar{u}$ and $\bar{x}$. We can now again cite §4,2 in that the partial derivatives $\bar{g}_{\bar{x}}$ and $\bar{g}_{\bar{u}}$ do not vanish. §4,(1) then states that

$$\overline{du/dx} = -\bar{g}_{\bar{x}}/\bar{g}_{\bar{u}} = -\overline{g_u/g_x} = \overline{du/dx}.$$

This equation also holds for the derivative of any other separating element $\bar{v}$ with respect to $\bar{x}$, so that application of the chain rule in K and $\overline{K}$ finally gives

$$d\bar{u}/d\bar{v} = \overline{du/dv}; \tag{32}$$

residue formation and differentiation commute.

Our treatment of differentials (of the first or arbitrary degree h) is based on

(32). Associate with the differential $u\,dv^h$, where u and v are $\mathfrak{p}_0$-normed and $\bar{v}$ is a separating element, the differential

$$\overline{u\,dv^h} = \bar{u}\,d\bar{v}^h \tag{33}$$

as its residue mod $\mathfrak{p}_0$. The independence of this definition of the particular choice of the variable v follows from (32) as in §4,3.

If the different divisors $\mathfrak{d}_{K/k}$ and $\bar{\mathfrak{d}}_{\bar{K}/\bar{k}}$ are divisible by separable prime divisors only (*this is particularly so if the constant fields k_0 and $\bar{k}_0$ are perfect*), *or if the fields K and $\bar{K}$ are conservative, then the divisors of differentials satisfy*

$$\overline{(u\,dv^h)} = (\bar{u}\,d\bar{v}^h). \tag{34}$$

Remark. The assumption is essential. As an exercise the reader can construct the example in which k_0 is of characteristic 0 while $\bar{k}_0$ is of characteristic $p > 2$, $\bar{K}$ being the field in the example of §3,4. Then, not even $\overline{(dx)} = (d\bar{x})$ holds.

Proof. Using §4,8, the assumptions imply that

$$\overline{(dx)} = \overline{\mathfrak{d}_{K/k}\mathfrak{z}_\infty^{-2}} = \bar{\mathfrak{d}}_{\bar{K}/\bar{k}}\bar{\mathfrak{z}}_\infty^{-2} = (d\bar{x}). \tag{35}$$

With this and (32) we can now compute

$$\overline{(u\,dv^h)} = \bar{u}\overline{(dv/dx)^h}\overline{(dx^h)} = \bar{u}(d\bar{v}/d\bar{x})^h\overline{(dx^h)} = (\bar{u}\,d\bar{v}^h),$$

which proves (34). ▼

Let K and $\bar{K}$ be conservative (cf. §6,2). *Let the residue class of a prime divisor $\mathfrak{p}$ of K decompose in $\bar{K}$ into*

$$\bar{\mathfrak{p}} = \bar{\mathfrak{p}}_1^{e_1} \cdots \bar{\mathfrak{p}}_r^{e_r},$$

where these prime divisors are separable. Then, if the residue of $u\,dv$ at $\mathfrak{p}$ lies in $\mathfrak{o}_{\mathfrak{p}_0}$, we have

$$\overline{\operatorname{res}_{\mathfrak{p}}(u\,dv)} = \sum_{i=1}^{r} \operatorname{res}_{\bar{\mathfrak{p}}_i}(\bar{u}\,d\bar{v}). \tag{36}$$

Proof. If $\mathfrak{p}$ is a first degree prime divisor then so is $\bar{\mathfrak{p}}$, and (36) is trivial. In the next section we will show that extension of the constant field and residue class formation mod $\mathfrak{p}_0$ are commutable operations. By choosing an extension in which $\mathfrak{p}$ decomposes to first degree prime divisors we derive (36) from the sum formula §4,(31). ▼

The notion of residue of a differential leads naturally to the residue of a principal part system. For these (36) shows that

$$\overline{(d^{1-h}\mathfrak{w}, dw^h)} = (d^{1-h}\bar{\mathfrak{w}}, d\bar{w}^h). \tag{37}$$

6. Behavior of the Field under Residue Formation and Extension

In this section we assume $\mathfrak{p}_0$ to be regular and both K and $\bar{K}$ to be conservative. Apply some extension of constants k_0'/k_0 and let $\mathfrak{o}_0'$ be the principal order of k_0' and $\mathfrak{p}_0'$ be a prime ideal of k_0' dividing $\mathfrak{p}_0$. The prime will be applied correspondingly elsewhere too, e.g., $K' = Kk_0'$. The following is obvious.

If K and $\bar{K}$ are conservative fields then the regularity of $\mathfrak{p}_0$ carries down to all prime divisors $\mathfrak{p}_0'$ of $\mathfrak{p}_0$ in k_0'.

The ideals in K and $\bar{K}$ occurring in §6,3 can be extended to ideals in K' and $\bar{K}'$, residues being taken mod $\mathfrak{p}_0'$ after the extension. Extension and residue formation thus commute for ideals. The results of §4,4 carry this over to the divisors.

Let K and $\bar{K}$ be conservative fields. The homomorphism $\mathfrak{a} \to \bar{\mathfrak{a}}$ of the divisor group of K into that of $\bar{K}$ commutes with the embedding of these groups into those of K' and $\bar{K}'$. The same is also true for the groups of divisor classes.

We finally take up the question whether all divisors of $\bar{K}$ are really residues of divisors in K. In general the answer is negative. For example, let $k_0 = Q$, $\mathfrak{o}_0 = Z$, $\mathfrak{p}_0 = Z\,3$, and $K = k_0(x, y)$ with $x^2 + y^2 + 1 = 0$. This equation has no solutions in k_0, so that by §3,5 all divisors are of even degree. On the other hand, the equation is solvable in $\bar{k}_0 = Z/3$, so that divisors exist in $\bar{K}$ with arbitrary degrees.

Let K and $\bar{K}$ be conservative fields and $\bar{\mathfrak{a}}$ be a divisor in $\bar{K}$. Then for some finite extension of constants k_0'/k_0 there exists a divisor $\mathfrak{a}$ in $K' = Kk_0'$ which is carried into $\bar{\mathfrak{a}}$ by residue formation modulo a prime divisor $\mathfrak{p}_0'$ of $\mathfrak{p}_0$†.

Proof. First choose some extension such that $\bar{\mathfrak{a}}$ decomposes completely into prime divisors of the first degree. Then we need only prove the theorem for such a prime divisor. Choose an element $\bar{\alpha} \in \bar{K}$ with numerator divisible by $\bar{\mathfrak{a}}$. $\bar{\alpha}$ has an inverse image $\alpha \in K$. The numerator divisor $\mathfrak{b}$ of α has an image $\bar{\mathfrak{b}}$ divisible by $\bar{\mathfrak{a}}$ in $\bar{K}$. Now extend k_0 so that $\mathfrak{b}$ decomposes into prime divisors of the first degree; $\bar{\mathfrak{a}}$ will keep this property. One of the prime divisors $\mathfrak{a}$ of $\mathfrak{b}$ must have the image $\bar{\mathfrak{a}}$. ▼

Notes

Deuring uses a different definition of the residue class mapping $\mathfrak{a} \to \bar{\mathfrak{a}}$ for divisors in the paper cited on page 172, which is continued and supplemented by Lamprecht [4]. The problems we have touched have been extended in varying directions. If the residue class $\bar{f}(\bar{y}, \bar{x})$ of the polynomial in (1) is

† Such an extension is not even necessary if k_0 is perfect with respect to a discrete valuation with the same prime ideal $\mathfrak{p}$ (cf. Ref. [7]).

reducible, then $\mathfrak{p}_0$ decomposes to $\mathfrak{p}_0 = \mathfrak{p}_1^{e_1}\mathfrak{p}_2^{e_2} \cdots$ in K. The irreducible factors $\bar{f}_i(\bar{y}, \bar{x})$ of $\bar{f}(\bar{y}, \bar{x})$ define fields of functions $\bar{K}_i/\bar{k}_0(\bar{x})$, the genera of which are limited by the inequality $\sum e_i(\bar{g}_i - 1) \leqq g - 1$. (For this, in particular, cf. Lamprecht [5].) Lamprecht [6] considers to what extent the field $\bar{K}$ is dependent upon the choice of the variable x used in its definition.

Roquette [9] investigates reduction modulo $\mathfrak{p}_0$ in function fields of arbitrarily many variables. The notion of the function field is subordinated, however, to that of the algebraic-geometric variety. The definition given there of the homomorphism $\mathfrak{a} \to \bar{\mathfrak{a}}$ does not require $\bar{K}$ to have the same genus as K. Numerous further references to the literature are given.

In several cases it has been possible to gain insight into all regular prime ideals $\mathfrak{p}_0$ of $\mathfrak{o}_0$ as well as to completely characterize the behavior of the irregular $\mathfrak{p}_0$ in K (cf. Deuring [1], Hasse [2], and Igusa [3]).

References

[1] M. Deuring, *Die Zetafunktion einer algebraischen Kurve vom Geschlecht* 1, Nachr. Akad. Wiss. Göttingen Math.-Phys.-Chem. Kl., *Part I*, 85–94 (1953), *Part II*, 13–42 (1955), *Part III*, 37–76 (1956), *Part IV*, 55–80 (1957).

[2] H. Hasse, *Zetafunktionen und L-Funktionen zu zeinem arithmetischen Funktionenkörper vom Fermatschen Typus*, Abh. Deutsch. Akad. Wiss. Berlin, Kl. Math.-Naturwiss. No. 4 (1954).

[3] J.-I. Igusa, *Kroneckerian model of fields of modular functions*, Amer. J. Math. **81**, 561–577 (1959).

[4] E. Lamprecht, *Restabbildung von Divisoren I*, Arch. Math. **8**, 255–264 (1957). *Part II*, ibid. **10**, 428–437 (1959).

[5] E. Lamprecht, *Bewertungssysteme und Zetafunktionen algebraischer Funktionenkörper I*, Math. Ann. **131**, 313–335 (1956). *Part II*, Arch. Math. **7**, 225–234 (1956).

[6] E. Lamprecht, *Zur Eindeutigkeit von Funktionalprimdivisoren*, Arch. Math. **8**, 30–38 (1957).

[7] A. Mattuck, *Reduction* mod *p of p-adic divisor classes*, J. Reine Angew. Math. **200**, 45–51 (1958).

[8] E. D. Nering, *Reduction of an algebraic function field modulo a prime in the constant field*, Ann. of Math. (2) **67**, 590–606 (1958).

[9] P. Roquette, *Zur Theorie der Konstantenreduktion algebraischer Mannigfaltigkeiten*, J. Reine Angew. Math. **200**, 1–44 (1958).

CHAPTER IV

Algebraic Functions over the Complex Number Field

Now that the methodical foundations of the theory have been developed in Chapter III, their applications to specific cases and the problems involved therein take over the foreground. Here we encounter the oldest parts of the theory. The impulse for the creation of a new mathematical discipline is, in the course of history, always given by the desire to solve certain problems, a minimum of methodical apparatus being ready at hand. It is only at a later stage in the development that a broader conceptual foundation is laid. Chapter IV primarily introduces two special classes of functions, the elliptic functions (§2) and the modular functions (§4), to which the deepest applications of our theory are bound. This serves, on the one hand, to illustrate and establish firmly the general concepts, and on the other to lay the groundwork for the last chapter in which the theory is again developed further, but which also introduces applications to number-theoretical problems. §1 deals with the notion of Riemann surfaces, and is not more than a brief account. There is no shortage of detailed expositions of this theory in the usual texts. The famous Abel theorem is dealt with in §3, but the modern theory of abelian varieties, to which it leads, can no longer be subject matter for this book.

§1. Riemann Surfaces

1. The Riemann Surface of an Algebraic Function

We assume throughout §1 that the constant field $k_0 = C$, the field of all complex numbers. Let an algebraic function field $K = k_0(x, y)$ be given by Eq. III, §1,(12), which is assumed to be irreducible. Let its degree in y be n. We will first consider y as a function of x. According to Puiseux's theorem (III, §1,6) one can, for every place $x = \xi$ ($\xi \in k_0 = C$ or $\xi = \infty$), find $r = r(\xi) \leq n$ power series III, §1,(21) in the variable

$$z_\xi = x - \xi \qquad \text{or} \qquad z_\xi - x^{-1},$$

each satisfying Eq. III, §1,(12). The "ramification indices" $e_\varrho = e_\varrho(\xi)$ are limited by Eq. III, §1,(19), and all $e_\varrho(\xi) = 1$ with at most a finite number of exceptions. These power series all have positive radii of convergence, the minimum of which we denote $R(\xi)$.

If $n = 1$, so that y is a rational function of x, the Riemann number sphere $\mathfrak{R}_0$ is associated with the function y as its Riemann surface. On it, the totality of Euclidean circles about the points ξ with radii less than some now arbitrary limits $R(\xi)$ represents a system of neighborhoods, giving it the structure of a Hausdorff space. (The reader should consider the full meaning of this statement, which includes the point $\xi = \infty$ along with all others.) According to the covering theorem of Heine–Borel, a finite number of these circles already forms a covering of $\mathfrak{R}_0$.

For $n > 1$ the Riemann surface $\mathfrak{R}$ of y is defined as a covering space of $\mathfrak{R}_0$ as follows: to every circular disk $U(\xi)$ of the number sphere $\mathfrak{R}_0$ associate $r = r(\xi)$ disks $U(\xi_1), \ldots, U(\xi_r)$, some of them "wound" or "ramified," with ramification indices $e_1(\xi), \ldots, e_r(\xi)$. The radii of the $U(\xi_\varrho)$ are the same as those of the $U(\xi)$, and all lie under the limits $R(\xi)$. The $U(\xi_\varrho)$ are associated with the r separate places ξ_ϱ defined in III, §2,2 to lie over the place ξ of $\mathfrak{R}_0$, and are neighborhoods of them. Still lacking is a determination of the connectedness of this structure, which is now given as follows: a point $\xi'_{\varrho'}$ over ξ' belongs to a neighborhood $U(\xi_\varrho)$ if, first, ξ' belongs to the corresponding neighborhood $U(\xi)$ of $\mathfrak{R}_0$ and, second, the power series III, §1,(21) corresponding to ξ'_ϱ coincides as an analytic function in a sufficiently small circle about ξ' with the power series associated with ξ_ϱ, under a suitable choice for the e_ϱth root $\sqrt[e_\varrho]{z_\xi}$. The neighborhoods of the places ξ_ϱ lying over the points ξ of $\mathfrak{R}_0$ thus form a Hausdorff space $\mathfrak{R}$; it is a covering space of $\mathfrak{R}_0$ with n sheets. $\mathfrak{R}$ also can be covered by a finite number of the given neighborhoods.

$\mathfrak{R}$ is called the *Riemann surface of the function y*. $\mathfrak{R}$ is connected, and y is univalent everywhere on $\mathfrak{R}$ and, except for poles, continuous everywhere. The same is naturally true for all functions of the field K. Conversely, every function univalent on $\mathfrak{R}$ which permits an expansion III, §1,(21) with non-vanishing convergence radius at every point of $\mathfrak{R}$ belongs to the field K. (These functions are said to be meromorphic.) This last statement does not belong to algebra in the strict sense, as the notion of the Riemann surface is of essentially analytic nature.

2. The Riemann Surface as a Complex Manifold

Just as a distinction is made between the field K and its particular model $K = k_0(x, y)$ with $f(y, x) = 0$, we also distinguish an (abstract) Riemann surface $\mathfrak{R}$ from its particular description as a covering space of $\mathfrak{R}_0$. A Hausdorff space is called a (topological) manifold if the neighborhoods $U(\xi)$ of the points ξ are homeomorphic to an open Euclidean circular disk.

The Riemann surfaces defined in the last section are then manifolds.

A Hausdorff space is called a complex manifold if to every neighborhood $U(\xi)$ there is given a homeomorphism $\varphi_{U(\xi)}$ which maps it onto the interior of a circle, this system of mappings having the following property: whenever two neighborhoods $U(\xi)$ and $V(\eta)$ have a nonvanishing intersection, the mapping $\varphi_{V(\eta)}\varphi_{U(\xi)}^{-1}$ defined in the region $W = \varphi_{U(\xi)}(U(\xi) \cap V(\eta))$ is a conformal, orientation-preserving mapping of W onto another region of the circle $\varphi_{V(\eta)}(V(\eta))$.

By choosing the mapping $\varphi_{U(\xi)}(x) = \sqrt[e_\varrho]{z_\xi}$, the Riemann surface given in the last section is made into a complex manifold.

The complex structure of a Riemann surface carries with it the concept of a conformal mapping of one Riemann surface onto another. A local homeomorphism $f(x') = x$ of $\mathfrak{R}'$ onto $\mathfrak{R}$ is said to be conformal if in sufficiently small neighborhoods $U(\xi) \subset \mathfrak{R}$ and $U'(\xi') \subset \mathfrak{R}'$ the mappings $\varphi_{U(\xi)} f \varphi'^{-1}_{U'(\xi')}$ are conformal in their regions of definition [that is, in the regions $\varphi_{U(\xi)}(f(U'(\xi')) \cap U(\xi))$].

Two fields of functions, $K = k_0(x, y)$ with $f(x, y) = 0$ and $K' = k_0(x', y')$ with $f'(x', y')$, are isomorphic if and only if their Riemann surfaces $\mathfrak{R}$ and $\mathfrak{R}'$ are conformally equivalent.

Proof. If $\mathfrak{R}$ and $\mathfrak{R}'$ are conformally equivalent, then every univalent analytic function on $\mathfrak{R}$ can also be considered as a univalent analytic function on $\mathfrak{R}'$, and conversely. Thus the univalent analytic functions of $\mathfrak{R}$ and $\mathfrak{R}'$ coincide, so that the fields which these functions form, according to §1,1 coincide.

If, conversely, K and K' are isomorphic, it is no loss of generality to assume $K = K'$. Then x and y are rational functions of x' and y', and conversely. Each point of $\mathfrak{R}$ is given by a value $x = \xi$ and a power series $y = \sum a_\mu (\sqrt[e_\varrho]{z_\xi})^\mu$. But this also defines a value x' and a power series $y' = \sum a_\mu' (\sqrt[e'_{\varrho'}]{z'_{\xi'}})^\mu$, that is, a univalent mapping of $\mathfrak{R}$ onto $\mathfrak{R}'$. The inverse mapping is clearly also univalent.

It remains to be shown that this mapping is conformal. A prime element for a place ξ_ϱ (cf. III, §1,2) is called a *local uniformizing variable* here. It is a function $q \in K$, vanishing at ξ_ϱ, with respect to which any function in K can be expanded in a power series. A prime element is a convergent power series

$$q = \sum_{\mu=1}^{\infty} c_\mu (\sqrt[e_\varrho]{z_\xi})^\mu, \qquad c_1 \neq 0.$$

The lemma of III, §1,3 showed that it has an inverse power series

$$\sqrt[e_\varrho]{z_\xi} = \sum_{\mu=1}^{\infty} c_\mu' q^\mu, \qquad c_1' \neq 0,$$

which is also convergent. Equations of the same form can be written for $\sqrt[e'_{\varrho'}]{z_{\xi}'}$, and then substitution yields the convergent power series

$$\sqrt[e_\varrho]{z_\xi} = \sum_{\mu=1}^{\infty} d_\mu (\sqrt[e'_{\varrho'}]{z'_{\xi'}})^\mu, \qquad d_1 \neq 0.$$

Now, $\sqrt[e_\varrho]{z_\xi}$ and $\sqrt[e'_{\varrho'}]{z'_{\xi'}}$ are the mappings of neighborhoods of ξ_ϱ and $\xi'_{\varrho'}$ onto the interior of a circle, introduced above. We thus see that the last equation, taken in the above context, yields a conformal mapping of $\mathfrak{R}$ onto $\mathfrak{R}'$ in a neighborhood of ξ_ϱ and $\xi'_{\varrho'}$. ▼

The theorem just proved shows that a Riemann surface is associated with each function field invariantly, that is, independently of the particular model of the field. By the definition in III, §2,2 the prime divisors $\mathfrak{z}_{\xi_\varrho}$ of K correspond one-to-one with the points of $\mathfrak{R}$. The invariance of $\mathfrak{R}$ thus implies the invariance of the prime divisors.

3. The Riemann Surface as a Topological Manifold

It was seen in §1,1 that a finite number of the disks $U(\xi)$ suffice to cover the Riemann surface $\mathfrak{R}$. The bounding circles themselves, any two of which have at most two points in common, divide $\mathfrak{R}$ into a finite number e^2 of simply connected open surface sections, a finite number e^1 of curve segments, and a finite number e^0 of points. The alternating sum

$$-e^0 + e^1 - e^2 = \chi(\mathfrak{R})$$

is called the Euler characteristic of $\mathfrak{R}$. According to the Euler polyhedron theorem it is the same for all subdivisions of $\mathfrak{R}$ and thus a topological invariant of $\mathfrak{R}$. It is always even, and

$$g = g(\mathfrak{R}) = \tfrac{1}{2}\chi(\mathfrak{R}) + 1 \tag{1}$$

is called the *topological genus* of $\mathfrak{R}$.

The genus g of a function field K defined in III, §3 *coincides with the topological genus of the Riemann surface $\mathfrak{R}$ of K.*

To prove this, consider $\mathfrak{R}$ as the n sheet covering space of the sphere $\mathfrak{R}_0$, as in §1,1, and use a special subdivision. Say the ramification points lie over the points $x = \xi^{(\sigma)} (\sigma = 1, \ldots, h)$ of the sphere $\mathfrak{R}_0$. From some other point $\xi^{(0)}$ make nonintersecting cuts to all the points $\xi^{(\sigma)}$, cutting through all the sheets. $\mathfrak{R}$ decomposes into $n = e^2$ copies of a cutup sphere; these are simply connected. Altogether, $hn = e^1$ cuts were made. $\mathfrak{R}$ contains n points over $\xi^{(0)}$, but only $r(\xi^{(\sigma)}) < n$ points over each $\xi^{(\sigma)}$. Altogether this gives $n + \sum r(\xi^{(\sigma)}) = e^0$ points for this subdivision.

For the ramification indices $e_\varrho(\xi^{(\sigma)})$, $\varrho = 1, \ldots, r(\xi^{(\sigma)})$, III, §1,(19) states $\sum_\varrho e_\varrho(\xi^{(\sigma)}) = n$. We can also write

$$\sum_{\varrho=1}^{r} (e_\varrho(\xi^{(\sigma)}) - 1) = n - r, \qquad r = r(\xi^{(\sigma)}).$$

This is then used in the computation of (1):

$$g(\mathfrak{R}) = \tfrac{1}{2}(-e^0 + e^1 - e^2) + 1 = \tfrac{1}{2}\sum_{\sigma=1}^{h} (n - r(\xi^{(\sigma)})) - n + 1,$$

$$g(\mathfrak{R}) = \tfrac{1}{2}\sum_{\sigma,\varrho} (e_\varrho(\xi^{(\sigma)}) - 1) - n + 1. \tag{2}$$

In III, §2,(36) the sum $\sum(e_\varrho(\xi^{(\sigma)}) - 1)$ taken over σ and ϱ was seen to be equal to the degree of the different $\mathfrak{d}_{K/k}$, and then, by III, §3,(8), $g(\mathfrak{R})$ turns out to be equal to the genus of K. ▼

It is a theorem of the topology of two-dimensional manifolds that, if its genus is $g > 0$, $\mathfrak{R}$ can be cut open to a simply connected surface section by $2g$ closed incisions. The boundary then consists of $4g$ curves, which we denote $a_1, b_1, a_1', b_1', a_2, b_2, a_2', \ldots, b_g, a_g', b_g'$ in the sense of positive traversal. To reconstruct $\mathfrak{R}$ one must attach the a_i with the a_i' and the b_i with the b_i', always in the opposing sense. (All corners then meet at a single point.) These $2g$ incisions a_i, b_i which leave $\mathfrak{R}$ simply connected are called a *canonical system of incisions*. The a_i, b_i form a homology basis. Any two canonical incision systems satisfy the homologies

$$a_i' \sim \sum a_{ij}a_j + \sum b_{ij}b_j, \qquad b_i' \sim \sum c_{ij}a_j + d_{ij}b_j,$$

where

$$G = \begin{pmatrix} a_{ij} & b_{ij} \\ c_{ij} & d_{ij} \end{pmatrix}$$

is a symplectic modular matrix in the sense of §1,4 of the Appendix to Chapter I, that is, $G \in \Gamma^g$. Conversely, a canonical system of incisions is associated with every such G (cf. §3,1).

Any closed path on $\mathfrak{R}$ is homotopic to a combination of the paths a_i, a_i', b_i, b_i' (that is, it can be deformed to such a combination). The a_i differ from the a_i' (and the b_i from the b_i') only in their orientation. The combination $a_1b_1a_1'b_1' \cdots a_gb_ga_g'b_g'$ is homotopic to 0.

For $g = 1$ this last statement implies that ab and ba are homotopic. This means that any path is homotopic to μ traversals of a and then ν traversals of b, where μ and ν are rational integers which are taken negative for traversals in the opposite direction of a and b.

§2. Fields of Elliptic Functions

1. Introduction

By a field of elliptic functions K we mean a separably generated and conservative function field of genus $g = 1$. Throughout §2 we assume K to have at least one prime divisor $\mathfrak{o}$ of degree 1, which would not necessarily be true.† For the start we make no assumptions concerning the base field k_0.

The canonical class W of an elliptic function field must always be the principal class, for being of dimension 1 and degree 0, it must contain an integral divisor of degree 0.

The class of the divisor $\mathfrak{o}^2$ is of dimension 2, according to the Riemann–Roch theorem. Then let $\mathfrak{x}$ be an integral divisor of this class linearly independent of $\mathfrak{o}^2$, and let x be an element such that

$$(x) = \mathfrak{x}/\mathfrak{o}^2. \tag{1}$$

The numerator divisor $\mathfrak{x}$ of x is of degree 1 in $k_0(x)$ and of degree 2 in K. Using the concluding remark of II, §5,4 we see that K is then an extension of $k = k_0(x)$ of the second degree; the same could be seen from Eq. III, §2,(20) by letting $\mathfrak{a}$ there be the denominator divisor $\mathfrak{o}^2$ of x. All multiples of $\mathfrak{o}^{-2}$ are now linear combinations of 1 and x.

Again using the Riemann–Roch theorem, the dimension of the class of $\mathfrak{o}^3$ is 3. Thus there exists a $y \in K$ with irreducible divisor representation

$$(y) = \mathfrak{y}/\mathfrak{o}^3. \tag{2}$$

1, x, y are now a basis of the multiples of $\mathfrak{o}^{-3}$. The element y is not contained in $k_0(x)$ so that, as $[K : k_0(x)] = 2$, necessarily $K = k_0(x, y)$. An irreducible equation $f(x, y) = 0$ over k_0 must hold between x and y, naturally, but its exact form is of no interest now.

In the case of an algebraically closed constant field k_0 of characteristic 0, all prime divisors $\mathfrak{p}$ correspond to pairs of solutions (x_1, y_1) of the equation $f(x, y) = 0$ in k_0, in the manner of III, §2,2, with the one exception of the prime divisor $\mathfrak{o}$. We now want to investigate the general case of first degree prime divisors $\mathfrak{p} \neq \mathfrak{o}$ where the constant field is arbitrary. By II, §5,5 such

---

† I. R. Šafarevič [*Exponents of elliptic curves*, Dokl. Akad. Nauk SSSR **114**, 714–716 (1957)] shows that there exists an elliptic function field over $k_0 = Q$ with an arbitrary natural m as the minimum degree of its prime divisors.

Exercise. Show: (a) $K = k_0(x, y^{1/2})$ with Eq. III, §3,(14), unsolvable in k_0, holding between x and y, is an elliptic function field with 2 as least degree of a prime divisor; (b) $K = Q(x, y)$ where $x^3 + y^3 + 1 = 0$ is of genus $g = 1$, with minimal divisor degree 3. Use the fact that the Fermat equation with exponent 3 has no integral rational solutions.

a $\mathfrak{p}$ produces a homomorphic mapping of every ring $\mathfrak{J}$ of functions of K, whose denominators are not divisible by $\mathfrak{p}$, onto k_0. If further $\mathfrak{J}$ has K as its quotient field, this homomorphism can be extended to the ring $\mathfrak{J}_\mathfrak{p}$ of all elements of K with denominators not divisible by $\mathfrak{p}$. One such ring $\mathfrak{J}$ is that generated by x and y from above. The residues x_1, y_1 in k_0 of x, y mod $\mathfrak{p}$ are often denoted

$$x_1 = x\mathfrak{p}, \qquad y_1 = y\mathfrak{p} \tag{3}$$

As $f(x, y)\mathfrak{p} = 0\mathfrak{p} = 0$ they are a solution of the equation $f(x, y) = 0$ in k_0. Conversely, any mapping of the pair x, y onto a solution x_1, y_1 defines a homomorphism $\mathfrak{J} \to k_0$ whose kernel is a prime $\mathfrak{J}$-ideal, which in turn defines a prime divisor $\mathfrak{p} \neq \mathfrak{o}$.

Hence, the prime divisors $\mathfrak{p} \neq \mathfrak{o}$ *of degree* 1 *are in one-to-one correspondence with the pairs of solutions* (x_1, y_1) *in* k_0 *of the equation* $f(x, y) = 0$.

Finding all the first degree prime divisors is a plausible task for the theory. The following theorem relates it with a deeper notion, which clearly goes to the core of the theory.

Every divisor class P *of degree* 0 *contains exactly one divisor* $\mathfrak{p}\mathfrak{o}^{-1}$, *where* $\mathfrak{p}$ *is a prime divisor of degree* 1. *This correspondence* $P \leftrightarrow \mathfrak{p}$ *is a one-to-one mapping of the divisor classes of degree* 0 *onto the prime divisors of degree* 1.

Proof. If $\mathfrak{p}$ is such a divisor, then $\mathfrak{p}\mathfrak{o}^{-1}$ lies in such a class P. P is the unit class if and only if $\mathfrak{p} = \mathfrak{o}$. For, there can exist no element $z \notin k_0$ with the divisor decomposition $(z) = \mathfrak{p}\mathfrak{o}^{-1}$; the argument used right after Eq. (1) would then lead to $K = k_0(z)$. Conversely, let P be given. By the Riemann–Roch theorem, the dimension of $P\mathfrak{o}$ is 1, so that there is exactly one divisor $\mathfrak{p}$ of degree 1 in the class $P\mathfrak{o}$, and, of course, it is a prime divisor. ▼

2. The Addition Theorem

Let two prime divisors of the first degree, $\mathfrak{p}$ and $\mathfrak{q}$, be given; by the last theorem they define another such divisor $\mathfrak{r}$, with

$$\frac{\mathfrak{p}}{\mathfrak{o}}\frac{\mathfrak{q}}{\mathfrak{o}} \sim \frac{\mathfrak{r}}{\mathfrak{o}}.$$

Writing this relation as

$$\mathfrak{p} + \mathfrak{q} = \mathfrak{r} \tag{4}$$

we define an addition of prime divisors of degree 1. Any such divisor could play the part of $\mathfrak{o}$, but it must then be held fixed. The prime divisors of the first degree are thus made into an additive abelian group, isomorphic to the multiplicative group of divisor classes of degree 0. The neutral element (zero element) is $\mathfrak{o}$.

Addition Theorem. *Let $\mathfrak{p}_1, \mathfrak{p}_2, \mathfrak{p}_3$ be three prime divisors of the first degree, not coinciding with each other or with $\mathfrak{o}$, and satisfying*

$$\mathfrak{p}_1 + \mathfrak{p}_2 + \mathfrak{p}_3 = \mathfrak{o} \qquad \text{or} \qquad \mathfrak{p}_1 + \mathfrak{p}_2 = -\mathfrak{p}_3. \tag{5}$$

Then,

$$\begin{vmatrix} 1 & x\mathfrak{p}_1 & y\mathfrak{p}_1 \\ 1 & x\mathfrak{p}_2 & y\mathfrak{p}_2 \\ 1 & x\mathfrak{p}_3 & y\mathfrak{p}_3 \end{vmatrix} = 0. \tag{6}$$

If some of the $\mathfrak{p}_i$ coincide with each other or with $\mathfrak{o}$, we have the following supplements. When

$$\mathfrak{p}_1 = -\mathfrak{p}_2 \neq \mathfrak{o} \tag{5'}$$

then

$$\begin{vmatrix} 1 & x\mathfrak{p}_1 \\ 1 & x\mathfrak{p}_2 \end{vmatrix} = 0. \tag{6'}$$

When

$$\mathfrak{p}_1 = -\mathfrak{p}_2 - \mathfrak{p}_2 = -2\mathfrak{p}_2 \neq \mathfrak{o}, \qquad \mathfrak{p}_2 \neq \mathfrak{o}, \tag{5''}$$

then

$$\begin{vmatrix} 1 & x\mathfrak{p}_1 & y\mathfrak{p}_1 \\ 1 & x\mathfrak{p}_2 & y\mathfrak{p}_2 \\ 0 & \dfrac{dx}{dp_2}\mathfrak{p}_2 & \dfrac{dy}{dp_2}\mathfrak{p}_2 \end{vmatrix} = 0, \tag{6''}$$

with some prime element p_2 for $\mathfrak{p}_2$. For

$$\mathfrak{p} + \mathfrak{p} + \mathfrak{p} = 3\mathfrak{p} = \mathfrak{o}, \qquad \mathfrak{p} \neq \mathfrak{o}, \tag{5'''}$$

we have

$$\begin{vmatrix} 1 & x\mathfrak{p} & y\mathfrak{p} \\ 0 & \dfrac{dx}{dp}\mathfrak{p} & \dfrac{dy}{dp}\mathfrak{p} \\ 0 & \dfrac{d^2x}{dp^2}\mathfrak{p} & \dfrac{d^2y}{dp^2}\mathfrak{p} \end{vmatrix} = 0 \tag{6'''}$$

Finally, if

$$\mathfrak{p} + \mathfrak{p} = \mathfrak{o}, \qquad \mathfrak{p} \neq \mathfrak{o}, \tag{5''''}$$

then

$$\begin{vmatrix} 1 & x\mathfrak{p} \\ 0 & \dfrac{dx}{dp}\mathfrak{p} \end{vmatrix} = 0. \tag{6''''}$$

Conversely, Eqs. (6), *etc., imply the truth of Eqs.* (5), *etc.*

Proof. In the general case, Eq. (5) states that $\mathfrak{p}_1\mathfrak{p}_2\mathfrak{p}_3\mathfrak{o}^{-3} = (z)$ is a principal divisor. z is a multiple of $\mathfrak{o}^{-3}$, so that it is of the form $z = \alpha + \beta x + \gamma y$ with $\alpha, \beta, \gamma \in k_0$. This means that we can now replace one column of (6) by $z\mathfrak{p}_i$.

But these are all 0-residues, which proves (6). An analogous argument leads from (5′) to (6′).

Given (5″), there exists a $z = \alpha + \beta x + \gamma y$ with numerator divisible by $\mathfrak{p}_1\mathfrak{p}_2^2$. Then dz/dp_2 is still divisible by $\mathfrak{p}_2$, and as above we can replace a column by $z\mathfrak{p}_1, z\mathfrak{p}_2, dz/dp_2\ \mathfrak{p}_2$. The vanishing of these residues proves the assertion. A similar argument holds in the cases (5‴) and (5⁗). ▼

The addition theorem can be visualized geometrically. We must assume k_0 to be algebraically closed and not to have characteristic 2 or 3. In §2,6 we shall show that x and y can be so chosen that (21) there is the defining equation; it represents the curve of Fig. 2. Write x_i, y_i $(i = 1, 2, 3)$ for short instead of $x\mathfrak{p}_i, y\mathfrak{p}_i$. These element pairs lie on the curve. Equation (6) states that the three points (x_i, y_i) lie on a single line or, better, that (x_3, y_3) is the third point of intersection of the curve with the line given by

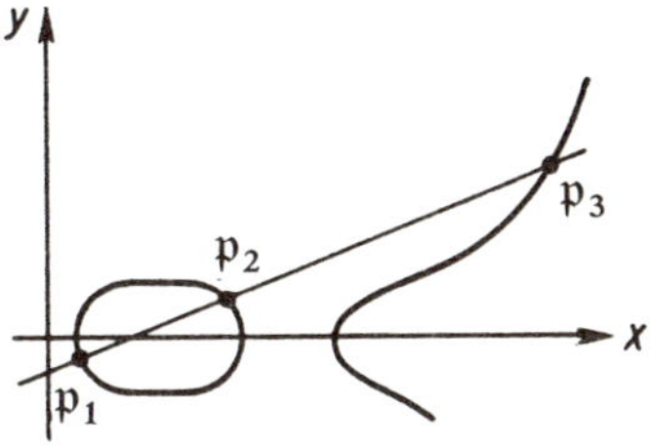

FIG. 2

$$\begin{vmatrix} 1 & x_1 & y_1 \\ 1 & x_2 & y_2 \\ 1 & x & y \end{vmatrix} = 0.$$

Substitution into (21) of the value this gives for y yields

$$x^3 - \frac{1}{4}\left(\frac{y_2 - y_1}{x_2 - x_1}\right)^2 x^2 - \left(\frac{1}{4} g_2 - \frac{1}{2}\frac{(x_1 y_2 - x_2 y_1)(y_2 - y_1)}{(x_1 - x_2)^2}\right)x - \frac{1}{4}\left(g_3 + \left(\frac{x_1 y_2 - x_2 y_1}{x_2 - x_1}\right)^2\right) = 0.$$

As we have two zeros, $x = x_1$ and x_2, the third is

$$x_3 = \frac{1}{4}\left(\frac{y_2 - y_1}{x_2 - x_1}\right)^2 - x_1 - x_2. \tag{7}$$

It corresponds to

$$y_3 = \frac{1}{4}\left(\frac{y_2 - y_1}{x_2 - x_1}\right)^3 - 2\,\frac{x_1 y_2 - x_2 y_1}{x_2 - x_1} - \frac{x_2 y_2 - x_1 y_1}{x_2 - x_1}. \tag{8}$$

3. AUTOMORPHISMS

In this section we will discuss the automorphisms σ of K which leave the elements of k_0 fixed. Application of such a σ to the element z and divisor $\mathfrak{p}$ yields the element z^σ and the divisor $\mathfrak{p}^\sigma$.

As preparation we need a lemma: An automorphism is uniquely characterized by its action on the prime divisors. In other words, we must prove that if $\mathfrak{p}^\sigma = \mathfrak{p}$ for all prime divisors, then σ is the identity. The hypothesis implies (cf. II, §5,3 or III, §2,2) that σ transforms all elements to multiples of themselves:

$$z^\sigma = c_z z, \qquad c_z \in k_0 .$$

If z_1, z_2 are linearly independent over k_0 we then immediately see that $c_{z_1+z_2} = c_{z_1} = c_{z_2}$, and, since $c_1 = 1$ must hold, $c_z = 1$ for all z. Thus σ is the identical automorphism.

Let $\mathfrak{o}$, $\mathfrak{o}'$ be two prime divisors, not necessarily differing, of the first degree, and let $\mathfrak{z}$ be the integral divisor linearly independent of but equivalent to $\mathfrak{o}\mathfrak{o}'$, which we know to exist by the Riemann–Roch theorem. Let $\mathfrak{z}(\mathfrak{o}\mathfrak{o}')^{-1} = (z)$. The argument of §2,1 then shows K to be an extension of the second degree over $k_0(z)$. This extension is always separable, even if the characteristic is 2. For, if this were not the case, the composition $\bar{K} = K\bar{k}_0$ of K and the algebraic closure $\bar{k}_0$ of k_0 would be an elliptic function field on the one hand and an inseparable extension of $\bar{k}_0(z)$ of the second degree on the other. We should then have $\bar{K} = \bar{k}_0(z, \sqrt{a(z)})$ with a polynomial $a(z) = a_0 z^n + \cdots + a_n$. Now, $\bar{k}_0$ is algebraically closed, so that $\sqrt{a(z)} = \sqrt{a_0}(\sqrt{z})^n + \cdots + \sqrt{a_n} = b(\sqrt{z})$ is a polynomial in $\sqrt{z}$ over $\bar{k}_0$, which would mean $\bar{K} \subseteq \bar{k}_0(\sqrt{z})$. Comparison of the degrees would lead to $[\bar{k}_0(\sqrt{z}) : \bar{k}_0(z)] = [\bar{K} : \bar{k}_0(z)]$, so that $\bar{K} = \bar{k}_0(\sqrt{z})$. But this would give $\bar{K}$ the genus 0.

As $K/k_0(z)$ is now a second degree separable extension, there exists an automorphism $\sigma = \sigma_{\mathfrak{o},\mathfrak{o}'}$ not the identity, of order 2 of K, which leaves $k_0(z)$ and therefore $(z) = \mathfrak{z}(\mathfrak{o}\mathfrak{o}')^{-1}$ fixed. We call it the *reflection automorphism* of $\mathfrak{o}$ and $\mathfrak{o}'$ in K. It interchanges $\mathfrak{o}$ and $\mathfrak{o}'$ if $\mathfrak{o} \neq \mathfrak{o}'$ was chosen, as is usually done in the decomposition of a prime divisor of $k_0(z)$ to two prime divisors in a quadratic extension K.

Now let $\mathfrak{p}$ be some divisor of degree 1. Then $(\mathfrak{p}/\mathfrak{o})(\mathfrak{p}/\mathfrak{o})^\sigma$ is a divisor of degree 0 in the rational function field $k_0(z)$, and thus a principal divisor. In terms of the convention (4) this means

$$\mathfrak{p}^\sigma = \mathfrak{o}' - \mathfrak{p} \qquad \text{for} \quad \sigma = \sigma_{\mathfrak{o},\mathfrak{o}'} \tag{9}$$

and, in particular, $\mathfrak{p}^\sigma = -\mathfrak{p}$ for $\sigma = \sigma_{\mathfrak{o},\mathfrak{o}}$. Setting

$$\tau_{\mathfrak{o},\mathfrak{o}'} = \sigma_{\mathfrak{o},\mathfrak{o}}\sigma_{\mathfrak{o},\mathfrak{o}'}, \tag{10}$$

this yields, for every prime divisor of degree 1,

$$\mathfrak{p}^\tau = \mathfrak{p} + \mathfrak{o}' \qquad \text{for} \quad \tau = \tau_{\mathfrak{o},\mathfrak{o}'} . \tag{11}$$

Because of (11) $\tau_{\mathfrak{o},\mathfrak{o}'}$ is called the *translation automorphism* to $\mathfrak{o}$, $\mathfrak{o}'$.

If $\tau' = \tau_{\mathfrak{o},\mathfrak{o}''}$ is another translation automorphism, then by (11) we have†

$$\mathfrak{p}^{\tau\tau'} = \mathfrak{p} + \mathfrak{o}' + \mathfrak{o}''.$$

But the translation automorphism $\tau_{\mathfrak{o},\mathfrak{o}'+\mathfrak{o}''}$ has the same effect on the prime divisors of first degree. This fact will enable us to prove

$$\tau_{\mathfrak{o},\mathfrak{o}'}\tau_{\mathfrak{o},\mathfrak{o}''} = \tau_{\mathfrak{o},\mathfrak{o}'+\mathfrak{o}''}. \tag{12}$$

Indeed, in the algebraic closure of k_0 the reflection and translation automorphisms remain such, but all prime divisors are of the first degree. We have seen, though, that the automorphisms $\tau_{\mathfrak{o},\mathfrak{o}'}\,\tau_{\mathfrak{o},\mathfrak{o}''}$ and $\tau_{\mathfrak{o},\mathfrak{o}'+\mathfrak{o}''}$ have the same effect on these divisors, and this remains valid if k_0 is left unextended. By the lemma above, then, they are identical. The purpose of Eqs. (11) and (12) is the following theorem, which they clearly imply.

The translation automorphisms form a group isomorphic to the additive group of prime divisors of the first degree and to the multiplicative group of divisor classes of degree 0.

For the full group of automorphisms we have: *the translations are a normal subgroup of the group of all automorphisms, and the factor group is finite.*

Proof. Let σ be any automorphism and $\tau = \tau_{\mathfrak{o},\mathfrak{o}'}$ be a translation. Because of (11), application of τ to a class of degree 0 amounts to multiplication by the class of $\mathfrak{o}'/\mathfrak{o}$. Hence

$$\left(\frac{\mathfrak{p}}{\mathfrak{o}}\right)^{\sigma\tau\sigma^{-1}} \sim \left(\frac{\mathfrak{p}^{\sigma}}{\mathfrak{o}^{\sigma}}\,\frac{\mathfrak{o}'}{\mathfrak{o}}\right)^{\sigma^{-1}} = \frac{\mathfrak{p}}{\mathfrak{o}}\,\frac{\mathfrak{o}'^{\sigma^{-1}}}{\mathfrak{o}^{\sigma^{-1}}} \sim \frac{\mathfrak{p}}{\mathfrak{o}}\,\frac{\mathfrak{o}''}{\mathfrak{o}} \tag{13}$$

with a certain $\mathfrak{o}''$, and then the same argument that verified (12) now assures that

$$\sigma\tau_{\mathfrak{o},\mathfrak{o}'}\sigma^{-1} = \tau_{\mathfrak{o},\mathfrak{o}''}, \qquad \mathfrak{o}'' = \mathfrak{o}'^{\sigma^{-1}} - \mathfrak{o}^{\sigma^{-1}}. \tag{14}$$

This is exactly the first assertion.

Now let $\mathfrak{o}^{\sigma} = \mathfrak{o}'$, where σ is still an arbitrary automorphism, so that σ and $\tau_{\mathfrak{o},\mathfrak{o}'}$ both map $\mathfrak{o}$ onto $\mathfrak{o}'$. Thus $\tau_{\mathfrak{o},\mathfrak{o}'}\sigma^{-1} = \sigma^{-1}\tau_{\mathfrak{o},\mathfrak{o}''}$ leaves $\mathfrak{o}$ fixed.

We see that the cosets with respect to the translation group each have a single representative which leaves $\mathfrak{o}$ *fixed.*

All that remains to be shown is that the group of automorphisms leaving $\mathfrak{o}$ fixed is of finite order. As this proof is more closely connected with the subject of §2,6, we postpone it until then. No further use shall be made of this result. ▼

† One might now incorrectly argue that $\mathfrak{p}^{\tau\tau'} = (\mathfrak{p} + \mathfrak{o}')^{\tau'} = \mathfrak{p} + \mathfrak{o}' + 2\mathfrak{o}''$. Instead, it should be noted that, in the sense of (4), $\mathfrak{p} + \mathfrak{o}'$ must be represented by a prime divisor $\mathfrak{q}$ of the first degree. The formula is then clear.

4. The Integral of the First Kind

From now on the field C of all complex numbers is taken for the constant field k_0. By III, §4 the divisor of every differential then (and even more generally, if the constant field is perfect) belongs to the canonical class. According to the Riemann–Roch theorem there exists, up to multiples with constants, a single differential du with an integral divisor, and that divisor is then the unit divisor.

We will now denote the points of the Riemann surface and the corresponding prime divisors with the same symbol. In particular, $\mathfrak{o}$ will be the infinitely distant point over the x-plane. We introduce the *integral of the first kind*

$$u = u(\mathfrak{o}, \mathfrak{p}) = \int_{\mathfrak{o}}^{\mathfrak{p}} du, \tag{15}$$

and first compute it in a canonically incised copy $\mathfrak{F}$ of the Riemann surface $\mathfrak{R}$ (cf. §1,3). $\mathfrak{F}$ is a direct connected surface section, and the integral (15) depends only upon the starting and end points. The fact that the divisor of the differential du is the unit divisor has the effect that, for a local uniformizing variable p at the place $\mathfrak{p}$, the differential quotient $du/dp \neq 0$ and $\neq \infty$. This is so for any place $\mathfrak{p}$. The image of $\mathfrak{F}$ under the function u in a plane (the u-plane) is nonoverlapping and finite, that is, it is a finite, simply connected surface section $\mathfrak{F}_u$. $\mathfrak{F}_u$ is a quadrangle bounded by four curves (a, b, a', b') which are, in general, curved.

Now let $\mathfrak{p}$ be a point on the side a of $\mathfrak{F}$, and $\mathfrak{p}'$ the other copy of the point $\mathfrak{p}$ lying on a'. We compute the integral $\int du$ from $\mathfrak{p}$ to $\mathfrak{p}'$ in $\mathfrak{F}$. We may choose to do so by first following a path of integration along a until we meet the curve b. We then integrate along b, and finally along a' until we reach $\mathfrak{p}'$. The integral consists of three parts, the first and last of which are equal but opposite, because du is the same on a and a'. Hence,

$$\int_{\mathfrak{p}}^{\mathfrak{p}'} du = \int_b du = \omega_b,$$

independent of the location of $\mathfrak{p}$ on a. This equation shows that the images of $\mathfrak{p}$ and $\mathfrak{p}'$ in the u-plane differ by the translations of a fixed complex number ω_b whenever $\mathfrak{p}$ and $\mathfrak{p}'$ are corresponding points on a and a'.

Of course the analog holds for $\mathfrak{p}$ and $\mathfrak{p}'$ on the boundaries b and b' of $\mathfrak{F}$. Translations of the amounts

$$\int_a du = \omega_a, \qquad \int_b du = \omega_b$$

thus displace the image $\mathfrak{F}_u$ of $\mathfrak{F}$ in the u-plane in such a manner that $\mathfrak{F}_u$ and the new images meet exactly along the boundaries b and a'.

The translations by $\mu\omega_a + \nu\omega_b$ with arbitrary rational integers μ and ν

fully pave (i.e., cover without omissions or overlappings) the u-plane with copies of $\mathfrak{F}_u$.

Proof. We give a concrete realization of the paving. The basic figure $\mathfrak{F}_u$ is found by computing (15) for all points of $\mathfrak{F}$. To the "basic path" from $\mathfrak{o}$ to $\mathfrak{p}$ lying inside $\mathfrak{F}_u$ now add a closed path from $\mathfrak{p}$ to $\mathfrak{p}$, homotopic to the path a. This yields the translated figure $\mathfrak{F}_u + \omega_a$. According to §1,3, every closed path on $\mathfrak{R}$ is homotopic to μ traversals of a followed by ν traversals of b. Thus the addition of an arbitrary closed path to the path of integration increases the integral (15) by the number $\mu\omega_a + \nu\omega_b$.

Now, say the regions $\mathfrak{F}_u$ translated by all $\mu\omega_a + \nu\omega_b$ cover some part U of the u-plane. There exists a bound ε such that all u of distance not more than ε from some arbitrary $u_0 \in \mathfrak{F}_u$ must lie in $\mathfrak{F}_u$ or one of the eight neighboring regions $\mathfrak{F}_u + \mu\omega_a + \nu\omega_b$ with $\mu, \nu = 0, \pm 1$. As the translations leave the metric invariant, the fact that some $u_0 \in U$ implies that all u not further than ε from u_0 also lie in U. This entails that U is the entire u-plane.

We must yet eliminate the possibility that our regions (partially) cover the u-plane multiply. This would imply the existence of two different points $\mathfrak{p}$ and $\mathfrak{q}$ and some integration path in $\mathfrak{R}$ such that $u(\mathfrak{o}, \mathfrak{p}) = u(\mathfrak{o}, \mathfrak{q})$ or $u(\mathfrak{p}, \mathfrak{x}) = 0$. Let the point $\mathfrak{x}$ traverse this path W from $\mathfrak{p}$ to $\mathfrak{q}$. Then $u(\mathfrak{p}, \mathfrak{x})$ follows a closed path W_u in the u-plane. Such a path can be continuously deformed into a point. But then W is also deformed into a point which is impossible, because $\mathfrak{p} \neq \mathfrak{q}$. The theorem is proved. ▼

These $\mu\omega_a + \nu\omega_b$ which we have found are the *periods* of the integral u. They form an abelian group generated by the elements ω_a and ω_b, called the *period module* $\omega = [\omega_a, \omega_b]$.

The points $\mathfrak{p}$ *of* $\mathfrak{R}$ *and the residue classes of the complex numbers* mod ω *are placed in one-to-one correspondence by* (15). This is already proved once it is shown that one point $\mathfrak{p}$ is associated with every residue class u mod ω. Choose some representative of this class in $\mathfrak{F}_u$. Then, as $\mathfrak{F}$ is mapped onto $\mathfrak{F}_u$ homeomorphically, exactly one $\mathfrak{p}$ corresponds to that representative.

Note the incidental result: *the ratio*

$$\tau = \omega_a/\omega_b \tag{16}$$

is not real. For otherwise, the translations by ω_b and ω_a would be parallel, and then, $\mathfrak{F}_u$ being a proper surface section, only a strip of the u-plane could be paved.

5. The Addition Theorem and the Abel Theorem

As the differential du is determined up to a constant factor by its divisor, we have, for any automorphism σ,

$$du^\sigma = \mu_\sigma\, du, \qquad \mu_\sigma \neq 0 \in k_0. \tag{17}$$

For a reflection automorphism σ we have $\mu_\sigma = -1$, while if τ is a translation automorphism then $\mu_\tau = 1$.

Proof. Let $\mathfrak{o}$, $\mathfrak{o}'$ be two prime divisors and z an element with the divisor $(z) = \mathfrak{z}(\mathfrak{o}\mathfrak{o}')^{-1}$. Our differential of the first kind is put in the form $du = dz/w$, so that, by III, §4, its divisor takes the form

$$(1) = (du) = \frac{\mathfrak{d}_{K/k_0(z)}}{\mathfrak{o}^2\mathfrak{o}'^2(w)},$$

with the different divisor $\mathfrak{d}_{K/k_0(z)}$. This gives

$$(w^2) = \mathfrak{d}^2_{K/k_0(z)}\mathfrak{o}^{-4}\mathfrak{o}'^{-4}.$$

From III, §2,(35) and $e_1 = \cdots = 2$ we see that $\mathfrak{d}^2_{K/k_0(z)}$ is a divisor in $k_0(z)$. All divisors of degree 0 in $k_0(z)$ are principal, so that $w^2 \in k_0(z)$. Now, w itself cannot lie in $k_0(z)$, as this would make du a differential of $k_0(z)$ so that the corresponding divisors in the two fields would satisfy

$$(du)_K = (du)_{k_0(z)}\, \mathfrak{d}_{K/k_0(z)}.$$

But $\mathfrak{d}_{K/k_0(z)}$, as the product of ramified divisors, could not compensate for the poles of $(du)_{k_0(z)}$. According to §2,3, K is an extension of degree 2 of $k_0(z)$. So we see that K is obtained from $k_0(z)$ by the adjunction of the square root of w^2.

Furthermore, the reflection automorphism $\sigma = \sigma_{\mathfrak{o},\mathfrak{o}'}$ is that of $K/k_0(z)$, and thus it replaces w by $-w$. This means that $du^\sigma = -du$, as asserted. Definition (10) then immediately gives the result for the translation automorphism. ▼

Application of the translation $\tau = \tau_{\mathfrak{o},\mathfrak{o}'}$ leads to the formula

$$\int_{\mathfrak{o}}^{\mathfrak{p}} du = \int_{\mathfrak{o}}^{\mathfrak{p}} du^{\tau^{-1}} = \int_{\mathfrak{o}^\tau}^{\mathfrak{p}^\tau} du \equiv \int_{\mathfrak{o}'}^{\mathfrak{p}+\mathfrak{o}'} du \mod \omega \tag{18}$$

in which $\mathfrak{p} + \mathfrak{o}'$ must be considered a point (or prime divisor) in the sense of §2,2. A result is the following Addition theorem for integrals of the first kind:

$$\int_{\mathfrak{o}}^{\mathfrak{p}_1} du + \int_{\mathfrak{o}}^{\mathfrak{p}_2} du \equiv \int_{\mathfrak{o}}^{\mathfrak{p}_1+\mathfrak{p}_2} du \mod \omega, \tag{19}$$

where the coordinates $x\mathfrak{p}$, $y\mathfrak{p}$ of the points $\mathfrak{p} = \mathfrak{p}_1, \mathfrak{p}_2, \mathfrak{p}_1 + \mathfrak{p}_2$ necessary to compute the integrals are found by the process of §2,2.

Indeed, because of (18) with $\mathfrak{p} = \mathfrak{p}_2$, $\mathfrak{o}' = \mathfrak{p}_1$, the left side is

$$\equiv \int_{\mathfrak{o}}^{\mathfrak{p}_1} du + \int_{\mathfrak{p}_1}^{\mathfrak{p}_1+\mathfrak{p}_2} du \mod \omega. \quad ▼$$

This surprising interplay between the results of algebra and analysis undoubtedly represents a highlight of the theory of algebraic functions. More surprising, though, is the fact that it was discovered by Euler long before any systematic theory of algebraic functions existed. A special case (the so-called case of the lemniscate, with $\mathfrak{p}_1 = \mathfrak{p}_2$) was even known earlier to Fagnano.

Because the additive group of prime divisors is isomorphic to the multiplicative group of divisor classes of degree 0, one can express (19) as: *By representing the divisor classes of degree* 0 *as quotients* $\mathfrak{p}\mathfrak{o}^{-1}$ *of prime divisors and assigning to them the integrals* (15), *the multiplicative group of divisor classes of degree* 0 *is mapped homomorphically onto the additive group of residue classes of complex numbers with respect to the period module* ω.

Abel Theorem. *This mapping is even an isomorphism. In other words, a divisor of degree* 0 *and with the prime decomposition* $\mathfrak{p}_1 \cdots \mathfrak{p}_h(\mathfrak{q}_1 \cdots \mathfrak{q}_h)^{-1}$ *is a principal divisor if and only if the congruence*

$$\sum_{i=1}^{h} \int_{\mathfrak{q}_i}^{\mathfrak{p}_i} du \equiv 0 \qquad \text{mod } \omega$$

holds. The irrelevance of the numbering of the $\mathfrak{p}_i$ and $\mathfrak{q}_i$ can be seen, for example, from the difference of the integrals

$$\int_{\mathfrak{q}_1}^{\mathfrak{p}_1} du + \int_{\mathfrak{q}_2}^{\mathfrak{p}_2} du, \qquad \int_{\mathfrak{q}_1}^{\mathfrak{p}_2} du + \int_{\mathfrak{q}_2}^{\mathfrak{p}_1} du$$

being the same integral over the closed path $\mathfrak{q}_1, \mathfrak{p}_1, \mathfrak{q}_2, \mathfrak{p}_2, \mathfrak{q}_1$ which is $\equiv 0$ mod ω, as all closed paths can be combined from the paths a and b.

Proof. Equivalent to the assertion is the statement: $\mathfrak{p}_1 \ldots \mathfrak{p}_h\mathfrak{o}^{-h} \sim \mathfrak{q}_1 \ldots \mathfrak{q}_h\mathfrak{o}^{-h}$ if and only if

$$\sum_{i=1}^{h} \int_{\mathfrak{o}}^{\mathfrak{p}_i} du \equiv \sum_{i=1}^{h} \int_{\mathfrak{o}}^{\mathfrak{q}_i} du \qquad \text{mod } \omega$$

holds. By introducing two further points in the sense of (4),

$$\sum_{i=1}^{h} \mathfrak{p}_i = \mathfrak{p}, \qquad \sum_{i=1}^{h} \mathfrak{q}_i = \mathfrak{q}$$

and remembering (19) we arrive at another equivalent form of the assertion: $\mathfrak{p}\mathfrak{o}^{-1} \sim \mathfrak{q}\mathfrak{o}^{-1}$ if and only if

$$\int_{\mathfrak{o}}^{\mathfrak{p}} du \equiv \int_{\mathfrak{o}}^{\mathfrak{q}} du \qquad \text{mod } \omega$$

holds. But by §2,1, $\mathfrak{p}\mathfrak{o}^{-1} \sim \mathfrak{q}\mathfrak{o}^{-1}$ is equivalent to $\mathfrak{p} = \mathfrak{q}$. Now, if the path of integration is chosen completely within a copy of $\mathfrak{R}$ with canonical incisions, then the last congruence is equivalent to the equation

$$\int_{\mathfrak{o}}^{\mathfrak{p}} du = \int_{\mathfrak{o}}^{\mathfrak{q}} du.$$

As the mapping of $\mathfrak{R}$ onto $\mathfrak{F}_u$ in the u-plane given by u is one to one, this equation can only hold if $\mathfrak{p} = \mathfrak{q}$, so that the theorem is proved. ▼

At this point we break off our line of thought. In §3, we will carry these theorems proved for elliptic functions over to fields of higher genus. To close §2 we undertake some considerations, a part of whose importance lies in their applications in §4.

6. The Weierstrass Normal Form

In this section we again permit arbitrary k_0. For simplicity's sake, though, we assume the characteristic not to be 2 or 3.† Let x and y be the elements of §2,1 which generate K, and consider the divisor class of $\mathfrak{o}^6$. Its dimension, by the Riemann–Roch theorem, is 6. But, $1, x, x^2, x^3, y, y^2, xy$ are seven multiples of $\mathfrak{o}^{-6}$, which means that there must be a linear relation in k_0 between them. The element y^2 must occur in it, for otherwise $K = k_0(x)$; and x^3 must also occur, if K is not to have genus 0 by III, §3,5. Hence, K is defined by an equation

$$\begin{gathered} y^2 + (ax + b)y + cx^3 + dx^2 + ex + f = 0, \\ a, \ldots, f \in k_0, \qquad c \neq 0. \end{gathered} \tag{20}$$

For characteristic unequal to 2 and 3 we can set

$$x + (d - a^2/4)/3c = cx', \qquad 2y + ax + b = c^2 y'$$

and then again write x and y for x' and y'. The equation becomes

$$y^2 = 4x^3 - g_2 x - g_3, \qquad g_2, g_3 \in k_0, \tag{21}$$

which is called the *Weierstrass normal form* of the generating equation of an elliptic field. Of course the polynomial on the right cannot have multiple zeros, that is, its discriminant does not vanish. Up to the factor 16 it is given by $g_2^3 - 27g_3^2$.‡

† Cf. M. Deuring, *Invarianten und Normalformen elliptischer Funktionenkörper*, Math. Z. **47**, 47–56 (1941), continued and corrected in M. Deuring, *Zur Theorie der elliptischen Funktionenkörper*, Abh. Math. Sem. Univ. Hamburg **15**, 211–261 (1945). The cases excluded here are investigated there.

‡ B. L. van der Waerden, *Algebra*, 2nd ed., Vol. I, §58. Berlin, 1937. (This edition reprinted New York, 1943.)

The substitution

$$x \to t^2 x, \qquad y \to t^3 y, \qquad t \neq 0, \tag{22}$$

leaves (21) in the same form, but with the coefficient change

$$g_2 \to t^{-4} g_2, \qquad g_3 \to t^{-6} g_3. \tag{23}$$

Let the characteristic be $\neq 2, 3$. *The Weierstrass normal form then corresponds uniquely, up to the substitution* (22), (23), *to the field K. The invariant*

$$j(K) = 12^3 g_2{}^3/(g_2{}^3 - 27 g_3{}^2) \tag{24}$$

remains unchanged by such substitutions.

Proof. Let

$$y'^2 = 4x'^3 - g_2' x' - g_3' \tag{25}$$

be some other equation generating K. Then $(x') = \mathfrak{x}'\mathfrak{o}'^{-2}$ with a first degree prime divisor $\mathfrak{o}'$. The reflection automorphism $\sigma_{\mathfrak{o},\mathfrak{o}'}$ maps x' into an element with denominator $\mathfrak{o}^2$, that is, into an element $\alpha + \beta x$ with $\alpha, \beta \in k_0$, $\beta \neq 0$. Thus there is no loss of generality in assuming that $x' = \alpha + \beta x$. Equation (25) then shows that y' is of the form $y' = \gamma + \delta x + \varepsilon y$ with $\gamma, \delta, \varepsilon \in k_0$, $\varepsilon \neq 0$. These quantities substituted into (25) must yield (21), which is only possible if $\alpha = \gamma = \delta = 0$ and $\beta^3 = \varepsilon^2$. Hence $\beta = t^2$, $\varepsilon = t^3$ with some nonzero $t \in k_0$. This is the substitution (22) which leads to (23). ▼

Let the characteristic be $\neq 2, 3$. *For every* $c \in k_0$ *there exists an elliptic function field K with the invariant* $j(K) = c$.

Proof. For $c = 12^3$ and $c = 0$ the fields in question have the defining equations

$$y^2 = 4(x^3 - x), \qquad y^2 = 4(x^3 - 1).$$

The older literature calls the first of these lemniscatic, as the field is found in the computation of the arc of the lemniscate, while the second field is called equianharmonic. For any c other than 12^3 or 0, the field defined by

$$y^2 = 4x^3 - 3c(c - 12^3)x - c(c - 12^3)^2$$

has that c as its invariant. ▼

Let the characteristic be $\neq 2, 3$. *If two fields, K and K′, defined by Eqs.* (21) *and* (25), *have the same invariants,* $j(K) = j(K')$, *then there exists an extension* k_1 *of the constant field* k_0 *such that the extensions* $K_1 = Kk_1$ *and* $K_1' = K'k_1$ *are isomorphic. This* k_1 *can even be chosen such that* $[k_1 : k_0] \leqq 2, 4,$ *or* 6, *depending upon whether* $12^3 \neq j \neq 0$, $j = 12^3$, *or* $j = 0$.

Proof. A substitution (22) with t in a suitable extension k_1/k_0 can be found yielding $g_2' = g_2$. The equality of the invariant then assures that $g_3' = \pm g_3$.

Should the minus sign hold, a further substitution (22) with $t = (-1)^{1/2}$ is still necessary, but then (25) has been transformed to (21). This shows the equations

$$g_2' = t^{-4} g_2, \qquad g_3' = t^{-6} g^3$$

to be solvable in a suitable k_1. In the three respective cases, a second, fourth, or sixth root must be adjoined to k_0. ▼

These same considerations also show that the automorphisms of K which leave $\mathfrak{o}$ fixed must be of the form (22) and even satisfy the equations $g_2 = t^{-4} g_2, g_3 = t^{-6} g_3$. *Let the characteristic* $\neq 2, 3$. *Then for* $12^3 \neq j(K) \neq 0$, $j(K) = 12^3, j(K) = 0$, *respectively, the group of automorphisms leaving a first degree divisor of K invariant is the cyclic group of order* 2, 4, *or* 6 *elements.* This proves an assertion made in §2,3. But the cases with characteristic 2 and 3 are also needed there. Such an automorphism, though, satisfies

$$x^\sigma = \alpha + \beta x, \qquad y^\sigma = \gamma + \delta x + \varepsilon y, \qquad \beta\varepsilon \neq 0.$$

By substitution into (20) it is seen that only a finite number of values are possible for α, ... The exact procedure is left to the reader.†

7. Elementary Elliptic Functions

We conclude with several formulas, referring to other texts for the proofs.‡ Consider the classical case, and let K be defined by the Weierstrass normal form (21). Then, the Weierstrass $\wp$*-function*

$$x = \wp(u) = \frac{1}{u^2} + \sum_{\mu,\nu=-\infty}^{+\infty}{}' \left(\frac{1}{(u - \mu\omega_a - \nu\omega_b)^2} - \frac{1}{(\mu\omega_a + \nu\omega_b)^2} \right) \tag{26}$$

relates x to the integral u of the first kind. The prime over the summation symbol indicates that the pair $\mu = \nu = 0$ is to be left out. Up to transforms under $\wp(u) \to \alpha\wp(u) + \beta$ with constant $\alpha \neq 0, \beta$, this $\wp(u)$ is the only analytic function of the residue classes of the complex numbers mod ω with only one pole, of the second order at $u \equiv 0 \bmod \omega$, corresponding to the denominator divisor $\mathfrak{o}^2$ of x. The coefficients of (21) are then

$$\begin{aligned} g_2(\omega_a, \omega_b) &= 60 \sum_{\mu,\nu=-\infty}^{+\infty}{}' \frac{1}{(\mu\omega_a + \nu\omega_b)^4}, \\ g_3(\omega_a, \omega_b) &= 140 \sum_{\mu,\nu=-\infty}^{+\infty}{}' \frac{1}{(\mu\omega_a + \nu\omega_b)^6}. \end{aligned} \tag{27}$$

† It can be found explicitly in the second paper by Deuring mentioned on page 200.

‡ K. Knopp, *Funktionentheorie*, Vol. II. Göschen, Leipzig, und Berlin, 1926. English translation, Dover Press, New York, 1945.

A. Hurwitz and R. Courant, *Funktionentheorie*, 3rd ed., Berlin, 1929. Revised edition, Springer-Verlag, Berlin, 1964.

The series (26) and (27) converge absolutely and uniformly in u, ω_a, ω_b, if u remains in a compact region of the u-plane not containing any lattice points $\mu\omega_a + \nu\omega_b$ and the quotient $\tau = \omega_a/\omega_b$ remains in a compact region of the τ-plane not containing any real points.

The verification that $du = dx/y$ is the differential of the first kind is an easy application of (21). But this means that $y = dx/du = \wp'(u)$, so that $\wp(u)$ satisfies the differential equation

$$\wp'(u)^2 = 4\wp(u)^3 - g_2\wp(u) - g_3.$$

The series (27) are now seen to be the coefficients g_2 and g_3 by substitution of (26) in this equation. ▼

This implies the following.

For any pair ω_a, ω_b with nonreal quotient, there exists an elliptic function field whose integral of the first kind has that pair as its periods.

The integral

$$\zeta = \int_{\mathfrak{o}}^{\mathfrak{p}} x\,du = \int_{\mathfrak{o}}^{\mathfrak{p}} \frac{x\,dx}{(4x^3 - g_2x - g_3)^{1/2}} \tag{28}$$

is of the second kind. Its only pole is at the point $\mathfrak{o}$, and it is of the first order with residue 1.

Any integral of the second kind is of the form $\alpha\zeta + \beta u + z$, with $\alpha, \beta \in k_0$, $z \in K$. The proof is contained in the final theorem of III, §5,4.

Exercise. Express the reflection and translation automorphisms explicitly in terms of u and the functions $x = \wp(u)$, $y = \wp'(u)$ [use Eqs. (7) and (8)].

NOTES

In §2,2 and 3 we essentially followed a paper [2] by Hasse. The finiteness of the group of automorphisms leaving a point fixed is proved differently there, though. The paper cited is only one of a series, whose full contents will come up later. We shall then also discuss the extensive literature related to it, as well as that of other topics concerning elliptic functions (V, §6).

In this book we omit discussion of two problems in the theory of elliptic functions. The first problem is the connection between elliptic functions whose periods are related by equations

$$\omega_a' = \alpha\omega_a + \beta\omega_b, \qquad \omega_b' = \gamma\omega_a + \delta\omega_b, \qquad \alpha\delta - \beta\gamma \neq 0$$

with rational $\alpha, \beta, \gamma, \delta$. An introduction to this question can be found in Tricomi's text [7], while Igusa [5] treats it from a modern point of view.

The second problem is the description of all prime divisors of the first degree for the case that k_0 is the rational, or, more generally, an algebraic number field. A theorem of Mordell [6] shows that they form a group, in the sense of §2,2, with a finite system of generators (see also Weil [8]). A new proof was given by Hasse [3], while A. Weil generalized the theorem to fields of higher genus. Newer results as well as extensive reference to other literature are found in Cassels [1]. Hofmann [4] surveys Euler's work in this field.

References

[1] J. W. S. Cassels, *Proc. Internat. Congr. Mathematicians 1962*, Stockholm, pp. 234–246.

[2] H. Hasse, *Zur Theorie der abstrakten elliptischen Funktionenkörper II*, J. Reine Angew. Math. **175**, 69–88 (1936).

[3] H. Hasse, *Der n-Teilungskörper eines abstrakten elliptischen Funktionenkörpers als Klassenkörper, nebst Anwendung auf den Mordell–Weilschen Endlichkeitssatz*, Math. Z. **48**, 48–66 (1942).

[4] J. E. Hofmann, *Über zahlentheoretische Methoden Fermats und Eulers, ihre Zusammenhänge und ihre Bedeutung*, Arch. History Exact Sci. **1**, 122–159 (1961).

[5] J.-I. Igusa, *On the transformation theory of elliptic functions*, Amer. J. Math. **81**, 436–452 (1959).

[6] L. J. Mordell, *On the rational solutions of the indeterminate equations of the third and fourth degrees*, Proc. Cambridge Philos. Soc. **21** (1922).

[7] F. Tricomi, *Funzioni ellittiche*, Bologna, 1937. [German revised edition, edited and translated by M. Krafft, *Elliptische Funktionen.* Leipzig, 1948].

[8] A. Weil, *Sur un théorème de Mordell*, Bull. Sci. Math. **54**, 1–10 (1930).

§3. The Group of Divisor Classes of Degree 0

In fields of genus $g = 0$ this group consists only of the unit element, according to II, §5,1 and III, §2. For fields with $g = 1$ we at least determined this group for the classical case in §2,4 and 5. The method used there is not applicable to higher genus, though, for the translation automorphism (§2,3) has no analog. Such can only be found again in certain associated function fields of g variables.† We attain our goal by a different method, for which we must first make several preparations.

1. The Riemann Period Matrix

Our considerations concern an algebraic function field K with constant field C, the trivial case in which the genus $g = 0$ being excluded. Canonical incisions are made in the Riemann surface $\mathfrak{R}$, as in §1,3. The differentials are considered divided into the three kinds, as in III, §5,1. Let du_i $(i = 1, \ldots, g)$ be a basis of differentials of the first kind, and dv_i be differentials of the second kind satisfying Eqs. III, §5,(12).

† For the extensive theory of abelian function fields or abelian varieties see S. Lang *Abelian Varieties*. New York, 1958. A complete bibliography is found there. Classical function-theoretical methods are not used there, as opposed to the book by F. Conforto, *Abelsche Funktionen und algebraische Geometrie*. Berlin-Göttingen-Heidelberg, 1956.

For every differential of the first kind du, form the *integral of the first kind*

$$u = \int_{\mathfrak{p}_0}^{\mathfrak{p}} du, \tag{1}$$

with the lower limit $\mathfrak{p}_0$ held fixed while the upper limit $\mathfrak{p}$ is allowed to vary over $\mathfrak{R}$. Integrals of the second and third kinds are introduced correspondingly. The path of integration must be given before these integrals are uniquely defined, a continuous deformation of that path being permitted though, so long as all residues are 0, that is, for differentials of the first and second kind. Given only the limits of (1), two such integrals may differ by an integral over a closed path w. Now, by §1 any closed path w is made up of the paths a_i, b_i. By calling the integral over w *the period of u corresponding to w*, we can say that all the periods are linear combinations of the base periods

$$\int_{a_i} du_j = \omega_{ij}^a, \qquad \int_{b_i} du_j = \omega_{ij}^b, \qquad \int_{a_i} dv_j = \eta_{ij}^a, \qquad \int_{b_i} dv_j = \eta_{ij}^b \tag{2}$$

with rational integers as coefficients. The matrices

$$\Omega = \begin{pmatrix} \omega_{ij}^a \\ \omega_{ij}^b \end{pmatrix}, \qquad \mathrm{H} = \begin{pmatrix} \eta_{ij}^a \\ \eta_{ij}^b \end{pmatrix}, \qquad \Pi = (\Omega \mathrm{H}), \tag{3}$$

the first two having $2g$ rows and g columns and the last with $2g$ rows and columns, compile these periods. Ω is called the *Riemann period matrix.*

The periods η_{ij}^a, η_{ij}^b do not change when differentials of functions of K are added to the dv_j, as these have periods 0. And, the dv_j are determined only up to these by the du_i.

The scalar products of differentials of the first or second kind du, dv, introduced by III, §5,(10), can, with the help of the Cauchy integral theorem, now be written as

$$(du, dv) = \sum_{\mathfrak{p}} \operatorname{res}_{\mathfrak{p}}(u\, dv) = (2\pi i)^{-1} \oint u\, dv, \tag{4}$$

taken over the canonical system of incisions a_i, b_i. The relationship of the function u on a_i and a_i', and on b_i and b_i', is expressed by

$$u(a_i') = u(a_i) + \int_{b_i} du = u(a_i) + \omega_i^b,$$

$$u(b_i') = u(b_i) + \int_{a_i'} du = u(b_i) - \int_{a_i} du = u(b_i) - \omega_i^a.$$

Hence the integral on the right in (4), taken over the paths a_i, b_i, a_i', b_i', is

$$\int u\, dv = -\omega_i^b \int_{ai} dv + \omega_i^a \int_{bi} dv = \omega_i^a \eta_i^b - \omega_i^b \eta_i^a.$$

All that is left to compute (4) is summation over the i. Hence we can express the relations III, §5,(12) in the form

$$\sum_{l=1}^{g}(\omega_{li}^{a}\omega_{lj}^{b}-\omega_{li}^{b}\omega_{lj}^{a})=\sum_{l=1}^{g}(\eta_{li}^{a}\eta_{lj}^{b}-\eta_{li}^{b}\eta_{lj}^{a})=0,$$

$$\sum_{l=1}^{g}(\omega_{li}^{a}\eta_{lj}^{b}-\omega_{li}^{b}\eta_{lj}^{a})=-\sum_{l=1}^{g}(\eta_{li}^{a}\omega_{lj}^{b}-\eta_{li}^{b}\omega_{lj}^{a})=\begin{cases}2\pi i & \text{for } i=j\\ 0 & \text{for } i\neq j.\end{cases}$$

The single matrix equation

$$\Pi^{t}J\Pi=2\pi iJ, \tag{5}$$

where J is as defined in the appendix to Chapter I, §1,(1), compiles these results. This means that $(2\pi i)^{1/2}\Pi$ is a complex symplectic matrix. Equation (5) expresses the *Riemann period relations*, also called the *Legendre relations* in the elliptic case.

Replacing the canonical incision system used by another such system leads to periods that are linear combinations of the old ones, with rational integers as coefficients. Indeed, all closed paths can be combined from the a_i, b_i. This implies that the new period matrix is $\Pi_1=S\Pi$, with an integral $2g$ by $2g$ matrix S. S must also be symplectic, as $(2\pi i)^{1/2}\Pi_1$ is symplectic. *Under a change of canonical incisions the period matrix transforms by a matrix S in the symplectic modular group Γ^g.*†

2. A Hermitian Metric for Differentials of the First Kind

Let z be some variable of K. Then $\mathfrak{R}$ can be seen as a covering space of the complex z-plane, and the decomposition into real and imaginary parts, $z=x+iy$, has meaning. Now, for two differentials du and dv of K consider the surface differential

$$du\,d\bar{v}=\frac{du}{dz}\overline{\frac{dv}{dz}}\,dx\,dy; \tag{6}$$

it represents a surface differential on $\mathfrak{R}$, which is independent of the variable z used. For, if z' is some other variable, then

$$\frac{du}{dz'}\overline{\frac{dv}{dz'}}\,dx'\,dy'=\frac{du}{dz}\overline{\frac{dv}{dz}}\left|\frac{dz}{dz'}\right|^{2}\left|\frac{\partial(x',y')}{\partial(x,y)}\right|dx\,dy,$$

the Cauchy–Riemann differential equations giving the functional determinant

$$\left|\frac{\partial(x',y')}{\partial(x,y)}\right|=\left|\frac{dz'}{dz}\right|^{2}.$$

† Conversely, to every symplectic modular matrix S there exists a canonical system of incisions. Cf. C. L. Siegel, *Ausgewählte Fragen der Funktionentheorie*, *Lecture Notes*, Vol. II, p. 108 ff. Göttingen, 1954.

The differential $du\, d\bar{v}$ is finite everywhere on $\mathfrak{R}$ for two differentials of the first kind, and its integral taken over the entire Riemann surface,

$$\{du, dv\} = \iint_{\mathfrak{R}} du\, d\bar{v}, \tag{7}$$

exists. *This defines a hermitian metric in the space of differentials of the first kind*, that is, it has the following properties:

$$\{\alpha_1\, du_1 + \alpha_2\, du_2, du_3\} = \alpha_1\{du_1, du_3\} + \alpha_2\{du_2, du_3\},$$

$$\{du, dv\} = \overline{\{dv, du\}}, \qquad \{du, du\} > 0 \quad \text{if} \quad du \neq 0.$$

The surface differential can also be expressed in terms of divergence. The indefinite integrals $\int du = u$ and $\int dv = v$ are introduced, and then

$$\operatorname{div}(u\bar{v}_x, u\bar{v}_y) = u_x\bar{v}_x + u_y\bar{v}_y = u_x(\bar{v}_x + i\bar{v}_y) = 2u_x\bar{v}_x = 2\,\frac{du}{dz}\,\overline{\frac{dv}{dz}}\,.$$

By the Gauss integral theorem the surface integral (7) is equal to the integral

$$\frac{1}{2}\oint u(\bar{v}_x\, dy - \bar{v}_y\, dx) = \frac{i}{2}\oint u\, d\bar{v},$$

taken along the path of the canonical system of incisions. This line integral can be expressed in terms of the periods exactly as the integral $\oint u\, dv$ of the last section. This yields

$$\{du_i, du_j\} = \frac{i}{2}\sum_{l=1}^{g}(\omega_{li}^a\bar{\omega}_{lj}^b - \omega_{li}^b\bar{\omega}_{lj}^a). \tag{8}$$

The submatrix (ω_{ij}^a) of the Riemann period matrix is nonsingular, for otherwise there would exist a first kind differential $du \neq 0$ with periods 0 on the incisions a_i. This would imply $\{du, du\} = 0$ in (8), contradicting the property of the metric.

By applying the linear transformation with the matrix $(\omega_{ij}^a)^{-1}$ to the du_i, the matrices (ω_{ij}^a) and (ω_{ij}^b) pass to E and

$$\mathrm{T} = (\omega_{ij}^b)(\omega_{ij}^a)^{-1}. \tag{9}$$

The matrix T *is symmetric and its imaginary component is the matrix of a positive definite quadratic form* (or, for short, its imaginary component is positive definite).

Proof. By the above it is no loss of generality to assume that $(\omega_{ij}^a) = \mathrm{E}$ and $(\omega_{ij}^b) = \mathrm{T}$. The symmetry of T follows from the fact that $(2\pi i)^{1/2}\Pi$ is a symplectic matrix, by the appendix to Chapter I, §1,(3). Furthermore, (8) becomes

$$\{du_i, du_j\} = -\frac{i}{2}(\omega_{ij}^b - \bar{\omega}_{ij}^b) = -i\,\mathrm{Im}(\mathrm{T}).$$

But this is a real symmetric matrix. The metric (7) being definite, our second assertion is also proved. ▼

Let $\begin{pmatrix} A & B \\ C & D \end{pmatrix}$ be an element of the symplectic modular group Γ^g. In §1,(3) of the appendix to Chapter I it was seen that then the same is true of $\begin{pmatrix} D & C \\ B & A \end{pmatrix}$. But it is by such a matrix that the periods of the du_i are transformed, according to the last section, under a change of incisions. Equation (9) then becomes

$$\mathrm{T}' = (A\mathrm{T} + B)(C\mathrm{T} + D)^{-1}.$$

This lends concrete importance to the substitutions that were investigated in §1 of the appendix to Chapter I.

3. Abelian Integrals of the Third Kind

Our next task is to construct special differentials of the third kind, by which all others can be represented.

Let $\mathfrak{p}$, $\mathfrak{q}$ be arbitrary points on $\mathfrak{R}$, not coinciding. We need a differential $dw'(\mathfrak{p}, \mathfrak{q})$ whose divisor $(dw'(\mathfrak{p}, \mathfrak{q})) = \mathfrak{b}\mathfrak{p}^{-1}\mathfrak{q}^{-1}$ with integral $\mathfrak{b}$ prime to $\mathfrak{p}$ and $\mathfrak{q}$. The class of $\mathfrak{b}$ is $W\mathfrak{p}\mathfrak{q}$ which, by the Riemann–Roch theorem, has a dimension $\dim(W\mathfrak{p}\mathfrak{q}) > \dim(W\mathfrak{q})$. Thus, such a divisor $\mathfrak{b}$, prime at least to $\mathfrak{p}$, exists. The corresponding differential has a pole of order 1 at $\mathfrak{p}$, and can have, at the most, another such pole at $\mathfrak{q}$. The residue theorem of III, §4,7 assures that this pole really does occur, and that the residues at $\mathfrak{p}$ and $\mathfrak{q}$ are equal but opposite. Multiplication by a constant sets the residues of $dw'(\mathfrak{p}, \mathfrak{q})$ at $\mathfrak{p}$ and $\mathfrak{q}$ to 1 and -1.

This property of $dw'(\mathfrak{p}, \mathfrak{q})$ is not lost when it is replaced by $dw(\mathfrak{p}, \mathfrak{q}) = dw'(\mathfrak{p}, \mathfrak{q}) - du$ with some differential du of the first kind. We try to choose the latter in such a manner that the periods of the indefinite integral

$$w(\mathfrak{p}, \mathfrak{q}) = \int dw(\mathfrak{p}, \mathfrak{q})$$

along the incisions a_i, b_i of a canonical system for $\mathfrak{R}$, as in §1,3, become pure imaginary. This will turn out to be possible in a unique way, and then $dw(\mathfrak{p}, \mathfrak{q})$ is called the *normal integral of the third kind for the points* $\mathfrak{p}$, $\mathfrak{q}$.

In the search for such a du we start with the normal integrals u_j of the first kind, whose period matrix Ω has the two parts $(\omega_{ij}^a) = \mathrm{E}$, $(\omega_{ij}^b) = \mathrm{T}$. Let the periods of $w'(\mathfrak{p}, \mathfrak{q})$ be η_i^a, η_i^b. Express the desired differential, or rather its integral, formally as $u = \sum \alpha_j u_j$, so that the α_j can be determined from the equations

$$\operatorname{Re}(\eta_l^a) = \operatorname{Re}\left(\sum_j \alpha_j \omega_{lj}^a\right), \qquad \operatorname{Re}(\eta_l^b) = \operatorname{Re}\left(\sum_j \alpha_j \omega_{lj}^b\right).$$

The first system of equations has the matrix $(\omega_{lj}^a) = \mathrm{E}$, and thus the solution $\operatorname{Re}(\alpha_j) = \operatorname{Re}(\eta_j{}^a)$. The imaginary components remain without influence, and are then computed by

$$-\operatorname{Re}(\eta_l^b) + \sum_j \operatorname{Re}(\alpha_j)\operatorname{Re}(\omega_{lj}^b) = \sum_j \operatorname{Im}(\alpha_j)\operatorname{Im}(\omega_{lj}^b).$$

The matrix on the right is $(-i/2)(\mathrm{T} - \overline{\mathrm{T}})$ and, being the coefficient matrix of a definite quadratic form by the last section, it has a nonvanishing determinant. ▼

Let some arbitrary differential dw have the poles $\mathfrak{p}_\nu$ with the residues ϱ_ν, where $\sum \varrho_\nu = 0$ because of the residue theorem. The differential $dw_1 = dw - \sum \varrho_\nu \, dw(\mathfrak{p}_\nu, \mathfrak{q})$ has poles of residue 0 only (where we take $dw\,(\mathfrak{q}, \mathfrak{q}) = 0$), and is thus of the second kind. By III, §5,4 then, dw_1 is a linear combination of g differentials of both the first and second kinds as well as an exact differential (that is, the differential of a function of K).

4. Abel's Theorem

A divisor of degree 0,

$$\mathfrak{x} = \prod_\nu \mathfrak{p}_\nu / \mathfrak{q}_\nu, \tag{10}$$

where $\mathfrak{p}_\nu$ and $\mathfrak{q}_\nu$ are an equal number of prime divisors, is a principal divisor if and only if there exist $2g$ rational integers $e_l{}^a$, $e_l{}^b$, such that for $j = 1, \ldots, g$ the equation

$$\sum_\nu \int_{\mathfrak{p}_\nu}^{\mathfrak{q}_\nu} du_j = \sum_{l=1}^{g} (e_l{}^b \omega_{lj}^a - e_l{}^a \omega_{lj}^b) \tag{11}$$

holds. The du_j here are an arbitrary basis of differentials of the first kind and the ω_{lj}^a, ω_{lj}^b their periods, computed in §3,1. The particular pairing of the $\mathfrak{p}_\nu$, $\mathfrak{q}_\nu$ is irrelevant, as already seen in §2,5.

Proof. (a) First let $\mathfrak{x} = (x)$ be a principal divisor and let all the $\mathfrak{p}_\nu$ and $\mathfrak{q}_\nu$ differ from another. Then $\log x = \int dx/x$ is an integral of the third kind. The Riemann surface $\mathfrak{R}$ is assumed to have a system of canonical incisions, as in §1,3, that do not cut the poles of dx/x. Further incisions, from some border point $\mathfrak{r}$ to the points $\mathfrak{p}_\nu$, $\mathfrak{q}_\nu$, are made without intersections. Along the border of the cut-up surface take the integrals

$$(2\pi i)^{-1} \int \log x \, du_j, \tag{12}$$

which must vanish according to the Cauchy integral theorem.

The integrals (12) consist of two parts. The first of these, over the incisions a_l, b_l, can be expressed as in §3,1 by the periods η_l^a, η_l^b corresponding to $\log x$ and by the periods of the u_j as

$$(2\pi i)^{-1}\sum_{l=1}^{g}(\eta_l^a\omega_{lj}^b - \eta_l^b\omega_{lj}^a).$$

Now, $\log x$ is determined up to integral multiples of $2\pi i$. Hence $\eta_l^a = -2\pi i e_l^a$, $\eta_l^b = -2\pi i e_l^b$ with integral e_l^a, e_l^b. This gives the right side of (11).

The other parts of the integrals (12) are

$$(2\pi i)^{-1}\sum_{\nu}\left(\int_{\mathfrak{r}}^{\mathfrak{p}_\nu}\log x\,du_j + \int_{\mathfrak{p}_\nu}^{\mathfrak{r}}\log x\,du_j + \int_{\mathfrak{r}}^{\mathfrak{q}_\nu}\log x\,du_j + \int_{\mathfrak{q}_\nu}^{\mathfrak{r}}\log x\,du_j\right),$$

where the paths $\mathfrak{r}\mathfrak{p}_\nu$ and $\mathfrak{p}_\nu\mathfrak{r}$ lie on the different edges of the same incisions. On the second of these, $2\pi i$ must be subtracted from $\log x$. The integrals over $\mathfrak{r}\mathfrak{q}_\nu$ and $\mathfrak{q}_\nu\mathfrak{r}$ correspond, but here $2\pi i$ must be added. Thus these integrals contribute the sum

$$-\sum_{\nu}\left(\int_{\mathfrak{r}}^{\mathfrak{p}_\nu}du_j - \int_{\mathfrak{r}}^{\mathfrak{q}_\nu}du_j\right) = -\sum_{\nu}\int_{\mathfrak{q}_\nu}^{\mathfrak{p}_\nu}du_j$$

to (12). This proves (11) for the case that the $\mathfrak{p}_\nu$ and $\mathfrak{q}_\nu$ all differ.

If, say, $\mathfrak{p}_1 = \cdots = \mathfrak{p}_h$, and the other $\mathfrak{p}$ differ from $\mathfrak{p}_1$, then dx/x has the residue h at $\mathfrak{p}_1$, so that the second part of (12) contributes $-h\int_{\mathfrak{r}}^{\mathfrak{p}_1}du_j$. This is in concurrence with the asserted equation (11), where $\mathfrak{p}_1$ occurs h times as upper limit.

(b) Now let (11) hold with certain rational integers e_l^a, e_l^b. Using the normal integral $w(\mathfrak{p}_\nu, \mathfrak{q}_\nu)$ of the third kind, set

$$w = \sum_{\nu} w(\mathfrak{p}_\nu, \mathfrak{q}_\nu).$$

This is an integral of the third kind with logarithmic singularities at the $\mathfrak{p}_\nu, \mathfrak{q}_\nu$, but no others. It is diminished by $2\pi i h_\nu$ every time $\mathfrak{p}_\nu$ is encircled in the negative sense, where h_ν is the number of points $\mathfrak{p}_\mu$ equal to $\mathfrak{p}_\nu$; and it is augmented by $2\pi i h_\nu'$, every time $\mathfrak{q}_\nu$ is encircled in the negative sense, where h_ν' has the analog meaning. If the periods of w on the incisions a_l, b_l are also multiples of $2\pi i$, then $x = e^w$ is a univalent function on $\mathfrak{R}$ with the zeros $\mathfrak{p}_\nu$ and poles $\mathfrak{q}_\nu$ of correct multiplicity. A remark made in §1 then assures that $x \in K$ so that $\mathfrak{x}$ is the principal divisor (x). The theorem would then be proved.

Hence we need now only show that the periods of w at the incisions a_l, b_l are multiples of $2\pi i$. The first part of the proof can be used literally, replacing log x by w to yield the equation

$$\sum_{\nu} \int_{\mathfrak{q}_\nu}^{\mathfrak{p}_\nu} du_j = (2\pi i)^{-1} \sum_{l=1}^{g} (\eta_l^{\,a}\, \omega_{lj}^{b} - \eta_l^{\,b}\, \omega_{lj}^{a}).$$

Introducing the abbreviations

$$(2\pi i)^{-1}\eta_l^{\,a} - e_l^{\,a} = c_l^{\,a}, \qquad (2\pi i)^{-1}\eta_l^{\,b} - e_l^{\,b} = c_l^{\,b},$$

we may use (11) to write

$$\sum_{l=1}^{g} (c_l^{\,a}\omega_{lj}^{b} - c_l^{\,b}\omega_{lj}^{a}) = 0. \tag{13}$$

Now clearly our theorem does not depend upon the use of a particular basis for integrals of the first kind. According to §3,2 we can choose a basis which has $(\omega_{ij}^{a}) = \text{E}$. Equation (13) becomes

$$\sum_{l=1}^{g} c_l^{\,a}\omega_{lj}^{b} = c_j^{\,b}. \tag{14}$$

We finally apply the fact that the $\eta_l^{\,a}$, $\eta_l^{\,b}$ were constructed so as to be pure imaginary. Then, the $c_l^{\,a}$, $c_l^{\,b}$ are real, and the equations conjugate to (14) also hold. They can be subtracted from (14), yielding

$$-i \sum_{l} c_l^{\,a}(\omega_{lj}^{b} - \bar{\omega}_{lj}^{b}) = 0.$$

But the determinant of this system of equations cannot vanish, as its matrix represents a positive definite form, by §3,2. Thus, the $c_l^{\,a} = 0$, and by (13) the $c_l^{\,b} = 0$ as well. The proof is finished. ▼

5. The Jacobian Variety

With any divisor class A of degree 0, represented, say, by the divisor $\prod \mathfrak{p}_\nu\, \mathfrak{q}_\nu^{-1}$, associate the vector

$$\mathfrak{c}(A) = \left(\sum_{\nu} \int_{\mathfrak{q}_\nu}^{\mathfrak{p}_\nu} du_1, \ldots, \sum_{\nu} \int_{\mathfrak{q}_\nu}^{\mathfrak{p}_\nu} du_g\right) \tag{15}$$

in the g-dimensional vector space C^g, some path of integration being arbitrarily fixed for the present.

Further, introduce the $2g$ *period vectors*

$$\mathfrak{o}_i^{\,a} = (\omega_{i1}^{a}, \ldots, \omega_{ig}^{a}), \qquad \mathfrak{o}_i^{\,b} = (\omega_{i1}^{b}, \ldots, \omega_{ig}^{b}). \tag{16}$$

They are linearly independent over the real numbers, a consequence of our demonstration that the system of Eqs. (13) can have no nontrivial real solution. These period vectors span a point lattice $\mathfrak{O}$ in C^g, which forms an additive group. The residue classes of the points of C^g modulo $\mathfrak{O}$ form a finite variety $\mathfrak{J}$, the so-called $2g$-dimensional (real) torus. $\mathfrak{J}$ is called the *jacobian variety* of K. As the factor group $C^g/\mathfrak{O}$ of the additive vector group C^g with respect to $\mathfrak{O}$, $\mathfrak{J}$ itself is an additive abelian group.

The Abel theorem states that (15) associates every divisor class of degree 0 with a unique (i.e., independent of the representative $\prod \mathfrak{p}_\nu \mathfrak{q}_\nu^{-1}$ of A) point of $C^g/\mathfrak{O} = \mathfrak{J}$. Moreover, we can prove the following.

Equation (15) *defines a one-to-one mapping of the divisor classes of K of degree* 0 *onto the entire jacobian variety. Furthermore, this mapping satisfies*

$$\mathfrak{c}(AB) = \mathfrak{c}(A) + \mathfrak{c}(B). \tag{17}$$

This theorem gives a complete mastery of the divisor class group of degree 0, for the classical case. In formula (17) we also have the *addition theorem for abelian integrals*, which took the form §2,(19) for $g = 1$. For, in that case multiplication of classes and addition of points coincide by the convention of §2,2.

Proof. Let $\prod \mathfrak{p}_\nu \mathfrak{q}_\nu^{-1}$, $\prod \mathfrak{s}_\nu \mathfrak{r}_\nu^{-1}$ be representatives of the classes A and B. Then

$$\mathfrak{c}(AB) = \left(\sum_\nu \int_{\mathfrak{q}_\nu}^{\mathfrak{p}_\nu} du_j + \sum_\nu \int_{\mathfrak{r}_\nu}^{\mathfrak{s}_\nu} du_j\right) \equiv \mathfrak{c}(A) + \mathfrak{c}(B) \bmod \mathfrak{O}.$$

Taking the vectors as points in $\mathfrak{J}$, this is (17).

Now let g arbitrary points $\mathfrak{p}_1, \ldots, \mathfrak{p}_g$ be given on $\mathfrak{R}$, and let z_ν be local uniformizing variables at the $\mathfrak{p}_\nu$. There then exist constants $\varepsilon_\nu > 0$ such that to every value of $z_\nu \in C$ with $|z_\nu| < \varepsilon_\nu$ there corresponds only one point $\mathfrak{p}_\nu'$ in a suitable neighborhood of $\mathfrak{p}_\nu$, and conversely.

Construct g algebraically independent copies $K_1, \ldots, K_g$ of K, with fixed isomorphisms $K_\nu \to K$. Compute the integrals

$$u_j(z_\nu) = \int_{\mathfrak{q}_\nu}^{\mathfrak{p}_\nu'} du_j$$

in K_ν. They are analytic functions of the z_ν when $|z_\nu| < \varepsilon_\nu$, and their derivatives with respect to z_ν are the functions $du_j/dz_\nu \in K_\nu$. The du_j being linearly independent and the columns of the matrix (du_j/dz_ν) lying in algebraically independent fields, the determinant of that matrix is a nonvanishing function in the composite of those fields. Setting this determinant to 0 thus defines an algebraic variety, which certainly can cover no complete neighborhood of the point $z_1 = \cdots = z_g = 0$.

Thus in some sufficiently small neighborhood of that point a complex g-dimensional domain (an open connected point set) of the z-space exists on which $|\partial u_j/\partial z_\nu| \neq 0$. It is mapped onto a domain $\mathfrak{K}_0$ of $\mathfrak{J}$ by the $u_j(z_\nu)$. But each point of the domain also corresponds to a system of g points $\mathfrak{p}_1', \ldots, \mathfrak{p}_g'$ on $\mathfrak{R}$, by the isomorphism $K_\nu \to K$. This means that a point of $\mathfrak{K}_0$ corresponds to every divisor class. Abel's theorem assures that it can be only one, of course.

Now let $\mathfrak{K}$ be the set of all points of $\mathfrak{J}$ to which divisor classes correspond. By (17), $\mathfrak{K}$ is a subgroup of $\mathfrak{J}$. Furthermore, $\mathfrak{K}$ contains a domain $\mathfrak{K}_0$. If $\mathfrak{k}_0 \in \mathfrak{K}_0$, then $\mathfrak{K}_0 - \mathfrak{k}_0$ is a domain of $\mathfrak{K}$ containing the neutral element, and for simplicity we can say $\mathfrak{K}_0$ is such. Were $\mathfrak{K} \neq \mathfrak{J}$ there would exist a frontier point $\mathfrak{k}$ of $\mathfrak{K}$. Were $\mathfrak{k} \in \mathfrak{K}$, then so would $\mathfrak{k} + \mathfrak{K}_0 \subset \mathfrak{K}$, and $\mathfrak{k} + \mathfrak{K}_0$ would be a domain containing $\mathfrak{k}$, contrary to the assumption that $\mathfrak{k}$ is a frontier point. But, were $\mathfrak{k} \notin \mathfrak{K}_0$ then also $\mathfrak{k} + \mathfrak{K}_0 \not\subset \mathfrak{K}$, leading to the same contradiction. Thus $\mathfrak{K} \neq \mathfrak{J}$ is refuted and the theorem proved. ▼

NOTES

As we already mentioned in §2,8 for the case of elliptic functions, the divisor classes of degree 0 form a finitely generated group if k_0 is a finite algebraic number field. The newest proof of this fact is found in [2]. It is essential for the theorem that the group of divisor classes of degree 0 is abelian and also represents an algebraic variety defined over k_0. Such a group is called an *abelian variety*. Abel's theorem gives an analytic model for this group for the case $k_0 = C$ in the $2g$-dimensional period torus. Orientation into the theory of abelian varieties is given in [1].

The theory of abelian integrals has been generalized as follows. Let $D\begin{pmatrix} a & b \\ c & d \end{pmatrix}$ be a representation of a group $\mathfrak{G}$ of fractional linear substitutions §4,(1) to which a field of automorphic functions belongs. Now, consider one-column matrices of differentials that transform in the manner

$$\mathfrak{u}\left(\frac{a\tau + b}{c\tau + d}\right) d\,\frac{a\tau + b}{c\tau + d} = D\begin{pmatrix} a & b \\ c & d \end{pmatrix} \mathfrak{u}(\tau)\, d\tau$$

under elements $\begin{pmatrix} a & b \\ c & d \end{pmatrix}$ of $\mathfrak{G}$. Indefinite integrals

$$\mathfrak{U}(\tau) = \int_{\tau_0}^{\tau} \mathfrak{u}(\tau)\, d\tau$$

undergo substitutions of $\mathfrak{G}$ as follows:

$$\mathfrak{U}\left(\frac{a\tau + b}{c\tau + d}\right) = D\begin{pmatrix} a & b \\ c & d \end{pmatrix} \mathfrak{U}(\tau) + \mathfrak{O}\begin{pmatrix} a & b \\ c & d \end{pmatrix},$$

where the $\mathfrak{O}\left(\begin{smallmatrix} a & b \\ c & d \end{smallmatrix}\right)$ are constant one-column matrices, depending upon the element involved, called the periods. The system of these periods represents a generalization of the Riemann period matrix (3). Information concerning the present state of this theory is found in a paper by Shimura [4] (cf. also [3]). Shimura remarks that under certain assumptions these periods can be associated with analytic models of abelian varieties. To date nothing is known about the meaning of these varieties for the function fields.

References

[1] S. Lang, *Abelian Varieties*, Interscience tracts in pure and applied mathematics, No. 7. New York-London, 1959.

[2] S. Lang and A. Néron, *Rational points of abelian varieties over function fields*, Amer. J. Math. **81**, 95–118 (1959).

[3] M. Kuga and G. Shimura, *On vector differential forms attached to automorphic functions*, J. Math. Soc. Japan **12**, 258–270 (1960).

[4] G. Shimura, *Sur les integrales attachées aux formes automorphes*, J. Math. Soc. Japan **11**, 291–311 (1959).

§4. Modular Functions

The field of modular functions is among the few examples for which the uniformization is completely known. The arithmetic definition of the modular group $\Gamma = \Gamma^1$, and with it of the Riemann surface of modular functions, permits the deep application of this theory in arithmetic.

1. The Modular Surface

In the following we build up on §§1 and 2 of the appendix to Chapter I and on IV, §1. The modular group consists of the fractional linear mappings

$$\tau \to \tau' = \frac{a\tau + b}{c\tau + d} = \begin{pmatrix} a & b \\ c & d \end{pmatrix}\tau, \qquad a, b, c, d \in Z, \ ad - bc = 1, \tag{1}$$

of the upper half-plane $\mathfrak{H}$ onto itself. For (1) we introduce the abbreviation $L\tau$, where L can be any two-row matrix. We also remark that the mapping (1) and the determinant $ad - bc$ define the matrix $\left(\begin{smallmatrix} a & b \\ c & d \end{smallmatrix}\right)$ up to the sign, thus exactly bivalently.

The mappings (1) represent motions of the Poincaré model $\mathfrak{H}$ of the hyperbolic plane. A fundamental domain $\mathfrak{F}$ exists and was found in the appendix to Chapter I, §1.

Identification of the frontier points of $\mathfrak{F}$ equivalent under Γ yields a surface $\mathfrak{R}$. To it is added the point $i\infty$. This makes $\mathfrak{R}$ a closed surface of genus $g = 0$, the *modular surface*. $\mathfrak{H}$ is an infinite-sheeted covering space of $\mathfrak{R}$, where a point $\tau \in \mathfrak{H}$ is said to lie over a point $\tau_0 \in \mathfrak{R}$, that is, over $\tau_0 \in \mathfrak{F}$, whenever τ and τ_0 are equivalent under Γ. The neighborhood of a point $\tau \in \mathfrak{H}$ lies smoothly over a neighborhood $\tau_0 \in \mathfrak{F}$ except for the points over

$$\mathfrak{v}_1 = i\infty, \qquad \mathfrak{v}_2 = \zeta = \tfrac{1}{2}(-1 + i\sqrt{3}), \qquad \mathfrak{v}_3 = i$$

for which the number of sheets of $\mathfrak{H}$ cyclically connected are $v_1 = \infty$, $v_2 = 3$, $v_3 = 2$, respectively.

We proceed to make $\mathfrak{R}$ into a Riemann surface. For any inner point τ_0 of $\mathfrak{F}$ call $q_{\tau_0} = \tau - \tau_0$ a local uniformizing variable. This can even be done for almost all frontier points. For $\mathfrak{F}$ in the form we defined it, equivalent parts of the boundary are mapped onto each other by the substitutions (1) with the matrices

$$C_1 = \begin{pmatrix} 1 & 1 \\ 0 & 1 \end{pmatrix}, \qquad C_2 = \begin{pmatrix} 1 & 1 \\ -1 & 0 \end{pmatrix}, \qquad C_3 = \begin{pmatrix} 0 & 1 \\ -1 & 0 \end{pmatrix}.$$

They leave the points $\mathfrak{v}_\varrho$ $(\rho = 1, 2, 3)$, respectively, fixed. The fundamental domain can be changed in such a manner as to cut off part of a boundary and to attach an equivalent part to an equivalent boundary. This is a hyperbolic motion, and thus a conformal mapping, so that $q_{\tau_0} = \tau - \tau_0$ passes into a power series $q'_{\tau_0} = c_1 q_{\tau_0} + \cdots$ with $c_1 \neq 0$. Thus at all points other than the $\mathfrak{v}_\varrho$ the local uniformizing variables may be taken the same.

For these exceptional points, though, set

$$q_{i\infty} = e^{2\pi i\tau}, \qquad q_\zeta = \tau + C_2\tau + C_2{}^2\tau - 3\zeta = (\tau - \zeta)^3/\tau(\tau + 1),$$
$$q_i = \tau + C_3\tau - 2i = (\tau - i)^2/\tau.$$

As τ traverses the part of a neighborhood of $\tau_0 = i\infty, \zeta, i$ lying in the interior of $\mathfrak{F}$, $q = q_{i\infty}, q_\zeta, q_i$ traverse a full neighborhood of $q = 0$ in the q-plane. As these functions are holomorphic for $\tau \neq i\infty, \zeta, i$ and have nonvanishing derivatives, our convention gives $\mathfrak{R}$ a conformal structure.

The definition of $\mathfrak{R}$ can also be formulated otherwise. The points of $\mathfrak{R}$ are the totalities $\Gamma\tau_0$ of equivalent points $L\tau_0$ for all $L \in \Gamma$, in addition to $\Gamma i\infty$. This formulation has the advantage of being free from an arbitrary choice of a fundamental domain of Γ. On the other hand, a fundamental domain is a welcome aid to the intuition. In effect there is no difference between the definitions.

2. Covering Spaces of the Modular Surface

Now let H be a subgroup of Γ with finite index and

$$\Gamma = \bigcup_{j=1}^{n} \mathrm{H}L_j, \qquad n = [\Gamma:\mathrm{H}], \tag{2}$$

a decomposition into right cosets. If, as before, $\mathfrak{F}$ is taken as a fundamental domain of Γ, then

$$\mathfrak{F}_{\mathrm{H}} = \bigcup_{j=1}^{n} L_j \mathfrak{F} \tag{3}$$

is a fundamental domain of H. Here $L_j\mathfrak{F}$ stands for the set of points $L_j\tau_0$, τ_0 traversing all points of $\mathfrak{F}$. To verify this, simply multiply both sides of (3) by H noting (2).

A surface $\mathfrak{R}_{\mathrm{H}}$ is now defined by the "points" $\mathrm{H}\tau_0$, once the $\mathrm{H}L_j i\infty$ are added as above ($j = 1,\ldots, n$). More than one point must generally be added, namely, all $L_j i\infty$ not equivalent by H. The points of $\mathfrak{R}$ consist of the totalities $\Gamma\tau_0$, and these decompose by (2), each into n totalities $\mathrm{H}L_j\tau_0$, that is, into n points of $\mathfrak{R}_{\mathrm{H}}$. We speak of each point of $\mathfrak{R}_{\mathrm{H}}$ lying over some point of $\mathfrak{R}$. This makes $\mathfrak{R}_{\mathrm{H}}$ into an n-sheeted covering space of $\mathfrak{R}$ and, as such, $\mathfrak{R}_{\mathrm{H}}$ is also a Riemann surface.

We must investigate the nature of this covering space more closely. It will turn out, in particular, that $\mathfrak{R}_{\mathrm{H}}$ has only a finite number of ramification points over $\mathfrak{R}$. Furthermore, $\mathfrak{R}_{\mathrm{H}}$ is connected, as two points $\mathrm{H}\tau_0$ and $\mathrm{H}\tau_1$ can be connected by a curve in $\mathfrak{R}_{\mathrm{H}}$, by so connecting τ_0 and τ_1 in $\mathfrak{H}$.

$\mathfrak{R}_{\mathrm{H}}$ is covered by $\mathfrak{H}$ just as $\mathfrak{R}$ is. As the only ramifications of $\mathfrak{H}$ lie over the points $\Gamma\mathfrak{v}_\varrho$, all ramifications of $\mathfrak{R}_{\mathrm{H}}$ with respect to $\mathfrak{R}$ must lie over these same points. In order to discuss them more closely, assume further that H is a normal subgroup of Γ.

Let N_1 denote the smallest natural number such that

$$C_1^{N_1} = \begin{pmatrix} 1 & N_1 \\ 0 & 1 \end{pmatrix} \in \mathrm{H}.$$

Then the cosets $\mathrm{H}C_1{}^{\mu}M$ and $\mathrm{H}C_1{}^{\nu}M$ are disjoint for $0 \leqq \nu \neq \mu < N_1$ and any $M \in \Gamma$, and if H is a normal subgroup, the same is true for the cosets $\mathrm{H}MC_1{}^{\mu}$ and $\mathrm{H}MC_1{}^{\nu}$. This implies the existence of a decomposition into cosets

$$\Gamma = \bigcup_{j=1}^{n_1} \bigcup_{\nu=0}^{N_1-1} \mathrm{H}M_jC_1{}^{\nu}, \qquad n_1 = nN_1^{-1}.$$

This means, incidentally, that n is divisible by N_1. As C_1 leaves the point $\mathfrak{v}_1 = i\infty$ fixed, only the nN_1^{-1} points $\mathrm{H}M_j i\infty$ of $\mathfrak{R}_{\mathrm{H}}$ lie over the point $\Gamma i\infty$ of $\mathfrak{R}$, and these are ramification points of order N_1. This means that we may choose $q_{\mathrm{H},i\infty} = \exp(2\pi i N_1^{-1}\tau)$ as a local uniformizing variable here.

Further, let N_σ ($\sigma = 2, 3$) be the smallest natural numbers such that $C_\sigma^{N_\sigma} \in \mathrm{H}$. The same argument shows that nN_σ^{-1} points of $\mathfrak{R}_{\mathrm{H}}$ lie over the points $\Gamma\zeta$ and Γi of $\mathfrak{R}$, and that they are ramification points of orders N_σ. Local uniformizing variables are $q_{\mathrm{H},\zeta} = \tau - \zeta$ and $q_{\mathrm{H},i} = \tau - i$ or $q_{\mathrm{H},\zeta} = q_\zeta$ and $q_{\mathrm{H},i} = q_i$, depending upon whether $N_\sigma = v_\sigma$ or 1.

$\mathfrak{R}$ is homeomorphic to a sphere and thus of genus $g = 0$. $\mathfrak{R}_{\mathrm{H}}$ is an $n = [\Gamma:\mathrm{H}]$ sheet covering of $\mathfrak{R}$, with ramification points over, and only over, $\mathfrak{v}_\sigma$ ($\sigma = 1, 2, 3$). They number exactly n/N_σ and are of order N_σ. The genus g_{H} is given by §1,(2) where now the $e_\sigma(\xi^{(\varrho)}) = N_\sigma$ ($\varrho = 1, \ldots, n/N_\sigma$). This gives

$$g_{\mathrm{H}} = \tfrac{1}{2}(nN_1^{-1}(N_1 - 1) + nN_2^{-1}(N_2 - 1) + nN_3^{-1}(N_3 - 1)) - n + 1,$$

$$= 1 + \frac{n}{2}\left(1 - \frac{1}{N_1} - \frac{1}{N_2} - \frac{1}{N_3}\right). \tag{4}$$

We further show that, in general, $N_2 = 3$, $N_3 = 2$. Clearly the N_σ are divisors of $v_\sigma = 3$ or 2. Now, were $N_2 = N_3 = 1$, we should have $g_{\mathrm{H}} < 0$. Were $N_2 = 3$, $N_3 = 1$, we should have

$$g_{\mathrm{H}} = 1 - (n/2N_1) - (n/6) \geqq 0,$$

so that $n < 6$. But n is divisible by $N_2 = 3$. This leaves $n = 3$. Furthermore, as N_1 also divides n, the possibilities $N_1 = 1$ or 3 remain. The first would make g_{H} negative. Finally, $N_2 = 1$, $N_3 = 2$ would imply

$$g_{\mathrm{H}} = 1 - (n/2N_1) - (n/4) \geqq 0$$

so that $n < 4$. The same argument then shows $n = 2$, and $N_1 = 1$ or 2, the first case again making g_{H} negative. Summing up we can state the following.

The genus of $\mathfrak{R}_{\mathrm{H}}$, *for normal* $\mathrm{H} \subset \Gamma$, *is generally given by*

$$g_{\mathrm{H}} = 1 + \frac{([\Gamma:\mathrm{H}]}{12N}(N - 6), \tag{5}$$

where $N = N_1$ *is the smallest natural number such that* $\begin{pmatrix} 1 & N \\ 0 & 1 \end{pmatrix} \in H$, *the exceptions being* $[\Gamma:\mathrm{H}] = N_1 = N_2 = 3$, $N_3 = 1$, $g_{\mathrm{H}} = 0$, *and* $[\Gamma:\mathrm{H}] = N_1 = N_3 = 2$, $N_2 = 1$, $g_{\mathrm{H}} = 0$.

The points $\mathrm{H}M_j i\infty$ of $\mathfrak{R}_{\mathrm{H}}$ are called the *cusps* of $\mathfrak{R}_{\mathrm{H}}$. The points $HM_j i\infty$ ($H \in \mathrm{H}$) belonging to a fundamental domain $\mathfrak{F}_{\mathrm{H}}$ of H are its cusps. *The number of cusps is, in the case of a normal subgroup, given by*

$$\sigma_{\mathrm{H}} = [\Gamma:\mathrm{H}]/N. \tag{6}$$

3. Congruence Subgroups

A particular normal subgroup is the *principal congruence subgroup* $\mathrm{H} = \Gamma(N)$ *of level* N, defined for every natural N by the congruence

$$H = \begin{pmatrix} a & b \\ c & d \end{pmatrix} \equiv \pm\begin{pmatrix} 1 & 0 \\ 0 & 1 \end{pmatrix} \bmod N; \tag{7}$$

that is, H consists of all matrices $H \in \Gamma$ satisfying this congruence. In general, any subgroup is called a congruence subgroup if it contains some

$\Gamma(N)$. To investigate these $\Gamma(N)$ as well as for later purposes we need the following lemma.

Lemma. *For every solution of the congruence*

$$\alpha\delta - \beta\gamma \equiv 1 \bmod N \tag{8}$$

there exists an integral matrix $\left(\begin{smallmatrix} a & b \\ c & d \end{smallmatrix}\right)$ *of determinant* 1 *such that*

$$\begin{pmatrix} a & b \\ c & d \end{pmatrix} \equiv \begin{pmatrix} \alpha & \beta \\ \gamma & \delta \end{pmatrix} \bmod N. \tag{9}$$

Proof. Because of (8) the *g.c.d.* of α, β, N is 1. Thus there exists a β' such that α and $\beta + \beta' N$ have *g.c.d.* 1. Hence, we may assume that $(\alpha, \beta) = 1$. There then exist γ', δ' such that $\alpha\delta' - \beta\gamma' = 1$. Now set

$$a = \alpha, \qquad b = \beta, \qquad c = \gamma + \gamma'(1 - \alpha\delta + \beta\gamma), \qquad d = \delta + \delta'(1 - \alpha\delta + \beta\gamma)$$

which gives $ad - bc = 1$ and the congruence (9).

As a consequence of the lemma we have the following. *The factor group*

$$\mathfrak{M}(N) = \Gamma/\Gamma(N) \tag{10}$$

is isomorphic to the group of two-row matrices $\left(\begin{smallmatrix} \alpha & \beta \\ \gamma & \delta \end{smallmatrix}\right)$ *of determinant* 1, *whose entries are residue classes* mod N, *where matrices are to be identified if they are the same up to multiplication by* -1. For $N = 2$ this last reservation is meaningless, as $-1 \equiv 1 \bmod 2$.

The order of this factor group is equal to the index $[\Gamma : \Gamma(N)]$. In particular

$$\begin{aligned} [\Gamma:\Gamma(N)] &= \tfrac{1}{2} N^3 \prod_{q/N} (1 - q^{-2}) \qquad &&\text{for} \quad N \geqq 3 \\ &= 6 &&\text{for} \quad N = 2, \end{aligned} \tag{11}$$

where q traverses all prime divisors of N. We find the number on the right, without the factor $\frac{1}{2}$, in three steps. (*a*) Let $N = q$ be prime. Then $\alpha\delta - \beta\gamma \equiv 0$, mod q has exactly $q^3 + q^2 - q$ solutions, so that $\alpha\delta - \beta\gamma \not\equiv 0 \bmod q$ has $q^4 - q^3 - q^2 + q$ solutions. The number of solutions of (8) is $1/(q-1)$ as many, *i.e.*, $q^3 - q$. (*b*) Let $N = q^h$ be a power of a prime. A solution $\alpha_h, \ldots, \delta_h$ is found by setting $\alpha_h = \alpha_{h-1} + q^{h-1}\alpha, \ldots$ in a solution $\alpha_{h-1}, \ldots, \delta_{h-1}$ of (8) with q^{h-1} in place of q^h and finding $\alpha, \ldots$ which solve

$$\alpha\delta_{h-1} + \delta\alpha_{h-1} - \beta\gamma_{h-1} - \gamma\beta_{h-1} \equiv (1 - \alpha_{h-1}\delta_{h-1} + \beta_{h-1}\gamma_{h-1})q^{1-h} \bmod q.$$

The number of solutions of this linear congruence is q^3. The number of solutions of this case can now be found by induction on h. (*c*) If N is the product of powers q^h of primes, then the number of solutions of (8) is equal to the product of the numbers of solutions for the factors q^h. Finally, due to the ambiguity of the sign in (7), we must divide by 2 unless $N = 2$.

As the exceptions to formula (5) cannot occur here, since neither C_2 nor C_3 belong to any $\Gamma(N)$, $N > 1$, we may compute the genus by that formula.

By a similar computation we find that the number of cusps of $\mathfrak{R}_{\Gamma(N)}$ as given by (6), namely,

$$\sigma_N = [\Gamma:\Gamma(N)]/N, \tag{12}$$

is equal to the number of pairs (a, c) of residue classes mod N having g.c.d. 1 with N, where (a, c) is identified with $(-a, -c)$ for $N > 2$ (by calculating the latter number along with $[\Gamma:\Gamma(N)]$). *Thus, there exists a one-to-one correspondence between these pairs and the cusps* $\Gamma(N)\binom{a\ \ b}{c\ \ d}i\infty = \Gamma(N)a/c$ *of* $\mathfrak{R}_{\Gamma(N)}$.†

From the equation

$$\begin{pmatrix} \alpha & \beta \\ \gamma & \delta \end{pmatrix}\Gamma(N)\begin{pmatrix} a & b \\ c & d \end{pmatrix} i\infty = \Gamma(N)\frac{\alpha a + \beta c}{\gamma a + \delta c}$$

it also follows that the factor group $\mathfrak{M}(N) = \Gamma/\Gamma(N)$ permutes these cusps $\Gamma(N)a/c$ and the pairs (a, c) in the same manner. In particular, we clearly have: $\mathfrak{M}(N)$ *is faithfully represented by the permutations suffered by the pairs* (a, c).

A local uniformizing variable at the cusp $\Gamma(N)a/c$ is

$$q = q_{\Gamma(N),\Gamma(N)ac^{-1}} = \exp[2\pi i N^{-1}(d\tau - b)(-c\tau + a)^{-1}] = \exp[2\pi i N^{-1}H^{-1}(\tau)]$$

if $ac^{-1} \neq i\infty$ or

$$q = q_{\Gamma(N),\Gamma(N)i\infty} = \exp[2\pi i N^{-1}\tau], \tag{13}$$

where b and d are to be determined for given a, c by $ad - bc = 1$. The ambiguity of this computation shows up in a constant factor of q, and is thus without significance. Indeed, the substitution $\binom{a\ \ b}{c\ \ d}$ of τ which maps $\Gamma(N)i\infty$ onto $\Gamma(N)a/c$ also transforms q into $q_{\Gamma(N),\Gamma(N)i\infty}$.

4. Modular Forms

By a modular form of the dimension $-k$ for a subgroup $\mathrm{H} \subseteq \Gamma$ we mean a function $M(\tau)$, defined in the complex upper half-plane $\mathfrak{H}$ and analytic there except for singular points, satisfying the functional equations

$$M(\tau)\begin{bmatrix} a & b \\ c & d \end{bmatrix}^{-k} = M\left(\frac{a\tau + b}{c\tau + d}\right)(c\tau + d)^{-k} = \chi\begin{pmatrix} a & b \\ c & d \end{pmatrix}M(\tau), \quad \begin{pmatrix} a & b \\ c & d \end{pmatrix} \in \mathrm{H}, \tag{14}$$

where $\chi\binom{a\ \ b}{c\ \ d}$ is a root of unity depending upon $\binom{a\ \ b}{c\ \ d}$. It is called the *character* of $M(\tau)$. Of particular interest are modular forms with character 1, but they

† The notation $\Gamma(N)a/c$ must not mislead us into believing that this cusp is uniquely determined by the ratio $a : c$ mod N.

cannot occur if k is odd. Indeed, substitution of $\left(\begin{smallmatrix} a & b \\ c & d \end{smallmatrix}\right) = \left(\begin{smallmatrix} -1 & 0 \\ 0 & -1 \end{smallmatrix}\right)$ into (14) gives $\chi\left(\begin{smallmatrix} -1 & 0 \\ 0 & -1 \end{smallmatrix}\right) = (-1)^k = -1$.

In (14) we defined an operator $\left[\begin{smallmatrix} a & b \\ c & d \end{smallmatrix}\right]^{-k}$ on the space of all functions in τ for every substitution (1). Its easily verifiable properties are

$$(M_1(\tau) + M_2(\tau))\begin{bmatrix} a & b \\ c & d \end{bmatrix}^{-k} = M_1(\tau)\begin{bmatrix} a & b \\ c & d \end{bmatrix}^{-k} + M_2(\tau)\begin{bmatrix} a & b \\ c & d \end{bmatrix}^{-k},$$
$$\left(M(\tau)\begin{bmatrix} a & b \\ c & d \end{bmatrix}^{-k}\right)\begin{bmatrix} \alpha & \beta \\ \gamma & \delta \end{bmatrix}^{-k} = M(\tau)\left[\begin{pmatrix} a & b \\ c & d \end{pmatrix}\begin{pmatrix} \alpha & \beta \\ \gamma & \delta \end{pmatrix}\right]^{-k}. \tag{15}$$

We will choose either the group Γ itself or one of its congruence subgroups $\Gamma(N)$ as H from now on. A modular form is said to be *meromorphic* at the cusp $\Gamma(N)ac^{-1}$ or $\Gamma(N)i\infty$ if it permits an expansion

$$M(\tau) = (-c\tau + a)^{-k} \sum_{n=n_0}^{\infty} c_n q^n \qquad \text{or} \qquad = \sum_{n=n_0}^{\infty} c_n q^n \tag{16}$$

in $q = q_{\Gamma(N),\, \Gamma(N)a/c}$ or $q = q_{\Gamma(N),\, \Gamma(N)i\infty}$ convergent for sufficiently small q. Modular forms are called *meromorphic* if they are meromorphic in the interior of the upper half-plane as well as at all cusps. It suffices to check this at all the cusps of a fundamental domain of $\Gamma(N)$, as the behavior (16) at the other cusps then follows from the functional equation (14). A modular form is called *holomorphic* or *integral* if it is holomorphic in the interior of $\mathfrak{H}$ and only terms with $n \geqq 0$ occur in the expansions (16). In what follows we will consider only integral modular forms, so that we may omit the specification "integral."

The modular forms of a fixed dimension $-k$ for the same subgroup $\mathrm{H} \subseteq \Gamma$ form a vector space whose dimension will be calculated for certain cases in §4,6.

As special modular forms of dimension $-k$ for $\Gamma(N)$ we define for all natural $k > 2$ the *Eisenstein series*

$$G_k(\tau; a_1, a_2, N) = \sum_{m_i \equiv a_i(N)}' \frac{1}{(m_1\tau + m_2)^k} = (-1)^k G_k(\tau; -a_1, -a_2, N), \tag{17}$$

the prime over the summation indicating that the pair $m_1 = m_2 = 0$ is omitted even if it satisfies the conditions $m_i \equiv a_i \bmod N$. This series converges absolutely and uniformly in every compact set of $\mathfrak{H}$ for $k > 2$. Its expansions (16) will be given explicitly in §4,7, as their calculation here would require much computation and interrupt our line of thought. The series vanish identically for $2a_1 \equiv 2a_2 \equiv 0 \bmod N$ and k odd, for then each term is cancelled by its opposite.

It is easy to verify that

$$G_k(\tau; a_1, a_2, N)\begin{bmatrix} a & b \\ c & d \end{bmatrix}^{-k} = G_k\left(\frac{a\tau + b}{c\tau + d}; a_1, a_2, N\right)(c\tau + d)^{-k}$$

$$= G_k(\tau; a_1 a + a_2 c, a_1 b + a_2 d, N) \tag{18}$$

for every $\begin{pmatrix} a & b \\ c & d \end{pmatrix} \in \Gamma$. Equation (18) carries with it the functional equation (14) for every $\begin{pmatrix} a & b \\ c & d \end{pmatrix} \in \Gamma(N)$, with $\chi\begin{pmatrix} a & b \\ c & d \end{pmatrix} = (\pm 1)^k$, depending upon whether $d \equiv \pm 1 \bmod N$. ▼

We say that $G_k(\tau; a_1, a_2, N)$ is *primitive* if the g.c.d. (a_1, a_2, N) is 1. Primitive Eisenstein series can vanish identically only if $N \leqq 2$ and k is odd. Call two Eisenstein series $G_k(\tau; a_1, a_2, N)$ and $G_k(\tau; a_1', a_2', N)$ *formally equal* if $a_1 \equiv a_1' \bmod N$ and $a_2 \equiv a_2' \bmod N$ or $a_1 \equiv -a_1' \bmod N$ and $a_2 \equiv -a_2' \bmod N$, otherwise *formally different*. Formally different Eisenstein series then correspond one to one with the pairs (a_1, a_2) of residue classes mod N such that $(a_1, a_2, N) = 1$, where two pairs (a_1, a_2) and $(-a_1, -a_2)$ are counted as one. In §4,7 we shall undertake a computation to show the formally different Eisenstein series for $k > 2$ to be linearly independent. Anticipating this result, a remark in the last section permits us to observe the following.

The number of linearly independent primitive Eisenstein series of an even dimension k is the same as the number of cusps. If $N > 2$, the same holds true for odd k.

Equation (18) shows that the operator $\begin{bmatrix} a & b \\ c & d \end{bmatrix}^{-k}$ associated with an element $\begin{pmatrix} a & b \\ c & d \end{pmatrix} \in \Gamma$ permutes the Eisenstein series in the same manner as $\begin{pmatrix} a & c \\ b & d \end{pmatrix}$ itself permutes the pairs (a_1, a_2) or $\begin{pmatrix} a & c \\ b & d \end{pmatrix}$ the cusps. *By* §4,3 *then, the operators $\begin{bmatrix} a & b \\ c & d \end{bmatrix}^{-k}$ are a group of permutations of the $G_k(\tau; a_1, a_2, N)$ faithfully representing the group $\mathfrak{M}(N)$.*

5. The Field of Modular Functions

Meromorphic modular forms of dimension 0 are *modular functions.* One speaks of such with respect to a subgroup $\mathrm{H} \subseteq \Gamma$, and in particular of modular functions of level N if $\mathrm{H} \subseteq \Gamma(N)$. For $N = 1$ we even omit mention of the level. The modular functions of a group H are clearly a field K_{H}. We will now investigate the fields K_Γ and $K_{\Gamma(N)}$.

As early as §2,(27) we considered the special Eisenstein series

$$\begin{aligned} g_2(\omega_1, \omega_2) &= 60\omega_2^{-4} G_4(\omega_1/\omega_2; 0, 0, 1), \\ g_3(\omega_1, \omega_2) &= 140\omega_2^{-6} G_6(\omega_1/\omega_2; 0, 0, 1). \end{aligned} \tag{19}$$

There we wrote ω_a, ω_b instead of ω_1, ω_2. These expressions are homogeneous of degrees -4 and -6 in both variables ω_1, ω_2. The function §2,(24), in particular,

$$j(\tau) = 12^3 \frac{g_2(\tau, 1)^3}{g_2(\tau, 1)^3 - 27g_3(\tau, 1)^2}, \qquad \tau = \omega_1/\omega_2, \tag{20}$$

is a modular function.

The function (20) *maps the modular surface* $\mathfrak{R}$ *onto the complex number sphere, conformally and one to one. Every modular function is a rational function of* $j(\tau)$.

Proof. By §2,6 there exists an elliptic field of functions L over the constant field C with any given $\gamma \in C$ as its invariant $j(L) = \gamma$. Any two such fields are related by a transformation §2,(22). The two base periods $\omega_a = \omega_1$, $\omega_b = \omega_2$ are multiplied, under it, by t^{-1}, because $du = dx/y$. But their ratio remains unchanged. If $\omega_1' = a\omega_1 + b\omega_2$, $\omega_2 = c\omega_1 + d\omega_2$ is another basis of the period module, then $\tau' = \omega_1'/\omega_2' = \left(\begin{smallmatrix} a & b \\ c & d \end{smallmatrix}\right)\tau$. This means that exactly one point $\Gamma\tau$, with $j(\tau) = \gamma$, corresponds to every value of γ.

The mapping of $\mathfrak{R}$ onto the number sphere given by $j(\tau)$ is conformal everywhere. Indeed, it is certainly conformal almost everywhere, that is, for those $\tau = \tau_0$ where $j(\tau)$ is regular, $dj(\tau)/d\tau \neq 0$ and $\tau - \tau_0$ is a local uniformization. If the mapping were not conformal at some point, then in a neighborhood of that point the equation $j(\tau) = \gamma$ would have to have several solutions in the fundamental domain $\mathfrak{F}$. This is not the case. The denominator of (20), which is the discriminant of the polynomial in §2,(21), does not vanish in $\mathfrak{H}$, according to §2,6, so that the point $\tau = i\infty$ of $\mathfrak{R}$ is mapped onto the infinitely distant point of the number sphere.

An arbitrary (meromorphic) modular function $f(\tau)$ assumes the value 0 only finitely often on $\mathfrak{R}$. For, as $\mathfrak{R}$ is a compact point set, an infinite number of zeros would have a cluster point, which is not permissible for a meromorphic function. The same argument shows that a meromorphic function has only finitely many poles.

Let the number of zeros and poles of the modular function $f(\tau)$ be n_0 and n_∞. We maintain these numbers to be equal. The proof is found by taking the integral

$$(2\pi i)^{-1} \int f'(\tau)\, d\tau/f(\tau) = 0$$

over the boundary of a small circle $\mathfrak{k}$ in $\mathfrak{R}$ not containing zeros or poles of $f(\tau)$. It is, of course, simultaneously the boundary of the complementary region $\mathfrak{R} - \mathfrak{k}$. Hence the value of the integral is known, from the familiar theorems of function theory, to be $\pm(n_0 - n_\infty)$, where poles and zeros are counted with their respective multiplicities. Hence, $n_0 = n_\infty$.

Now let ν_i and π_i be the zeros and poles of $f(\tau)$. The function

$$g(\tau) = f(\tau) \prod_{i=1}^{n_0} \frac{j(\tau) - j(\pi_i)}{j(\tau) - j(\nu_i)}$$

is then a modular function without poles or zeros, if $j(\tau) - j(i\infty)$ is given meaning by setting it to $j(\tau)^{-1}$. This function must be a constant, for otherwise we could find a $\gamma \in C$ such that $g(\tau) - \gamma$ has zeros, but does not vanish identically. But this function would have no poles, and thus contradict what was shown above. This proves the theorem. ▼

Thus the modular functions form a rational function field

$$K_\Gamma = C(j(\tau)). \tag{21}$$

Now assume $N > 1$. Adjoin to K_Γ the quotients of the primitive Eisenstein series of level N and some fixed dimension k, excluding all $k \equiv 1 \bmod 2$ if $N = 2$. *The field extension $K_{\Gamma(N)}$ thus found coincides with the field of all modular functions of level N. It is a Galois extension of K_Γ with the group $\mathfrak{M}(N)$.*

Proof. It is clear from the last section that these quotients are invariant under the substitutions (1) for $\left(\begin{smallmatrix} a & b \\ c & d \end{smallmatrix}\right) \in \Gamma(N)$, but that they permute under arbitrary substitutions (1) of Γ. It was seen at the end of the last section that these permutations even represent $\mathfrak{M}(N)$ faithfully. This means that $K_{\Gamma(N)}/K_\Gamma$ is normal with that group.

Let $f(\tau)$ be an arbitrary modular function of level N and let $L_1 = E, L_2, \ldots, L_n$ with $n = [\mathfrak{M}(N):1] = [K_{\Gamma(N)} : K_\Gamma]$ represent the cosets of $\Gamma(N)$ in Γ. The elementary symmetric functions of the $f(L_i(\tau))$ remain invariant under Γ and thus produce an equation of degree n for $f(\tau)$ over K_Γ. Were $f(\tau) \notin K_{\Gamma(N)}$, there would exist a function generating the field $K_{\Gamma(N)}(f(\tau))/K_\Gamma$. No generality is lost if we assume $f(\tau)$ to be that function. Then $f(\tau)$ satisfies an irreducible equation of degree higher than $[K_{\Gamma(N)}:K_\Gamma]$ over K_Γ, a contradiction.

6. Modular Forms and Differentials

$M(\tau)\, d\tau^h$, where h is a natural number and $M(\tau)$ is a modular form with character $\chi = 1$ and dimension $-2h$ for some group $\mathrm{H} \subseteq \Gamma$, is invariant under H. For, the functional equation (14) can be written

$$M\left(\frac{a\tau + b}{c\iota + d}\right)\left(\frac{d}{d\tau}\frac{a\tau + b}{c\tau + d}\right)^h = M(\tau).$$

If conversely $u(\tau)$ is a modular function and

$$u(\tau)\, dj(\tau)^h = u(\tau)(dj(\tau)/d\tau)^h\, d\tau^h = M(\tau)\, d\tau^h$$

a differential of degree h, then $M(\tau)$ is such a modular form. Thus there is a one-to-one correspondence between modular forms with character 1 and differentials.

We want to determine the denominator of the divisor that corresponds, by III, §5,1, to such a modular form. First let $N > 1$. Then, except for the cusps, $\tau - \tau_0$ is a local uniformizing variable everywhere. Thus at most the prime divisors $\mathfrak{s}_\nu$ of the cusps can occur in the denominator of the divisor of $M(\tau)\, d\tau^h$. By (13) and (16) we have, for the cusps,

$$M(\tau)\, d\tau^h = M(\tau)\,(d\tau/dq)^h\, dq^h = (N/2\pi i)^h \sum_{\nu=0}^{\infty} c_\nu q^{\nu-h}\, dq^h.$$

This puts the divisor of $M(\tau)\, d\tau^h$ in the form $\mathfrak{m}(\mathfrak{s}_1, \ldots, \mathfrak{s}_{\sigma_N})^{-h}$ with an integral divisor $\mathfrak{m}$.

On the other hand, the divisor of $M(\tau)\, d\tau^h$ must lie in the hth power of the canonical class W, which places $\mathfrak{m}$ in the class $W^h(\mathfrak{s}_1, \ldots, \mathfrak{s}_{\sigma_N})^h$. By the Riemann–Roch theorem this class contains $(2h-1)(g-1) + h\sigma_N$ linearly independent integral $\mathfrak{m}$. This then is exactly the dimension of the linear space of all integral modular forms $M(\tau)$ of this sort. An explicit expression for this dimension is gotten from (5), (11), and (12). *There exist*

$$\frac{(2h-1)N+6}{24} N^2 \prod_{q/N} (1 - q^{-2}) \quad \text{for } N > 2 \quad \text{and} \quad h+1 \quad \text{for } N = 2$$

linearly independent integral modular forms of level $N > 1$ *and of dimension* $-2h$ *with the character* 1.

It is somewhat more complicated to determine this number for $N = 1$, as the local uniformization at the points ζ and i deviate, as seen in §4,1. We have

$$\frac{dq_\zeta}{d\tau} = 3\,\frac{(\tau-\zeta)^2}{\tau(\tau+1)} - (2\tau+1)\,\frac{(\tau-\zeta)^3}{\tau^2(\tau+1)^2}, \qquad \frac{dq_i}{d\tau} = 2\,\frac{(\tau-i)}{\tau} - \frac{(\tau-i)^2}{\tau^2}.$$

Thus the divisor of a holomorphic modular form $M(\tau)$ of this sort generally contains $\mathfrak{q}_\zeta$ to a power $\leqq 2h/3$ and $\mathfrak{q}_i$ to a power $\leqq h/2$ in its denominator, the notation being obvious. The rest of the argument is as above, yielding the following. *There exist*

$$\begin{aligned} [2h/3] + [h/2] - h + 1 &= [h/6] && \text{for} \quad h \equiv 1 \bmod 6 \\ &= [h/6] + 1 && \text{for} \quad h \not\equiv 1 \bmod 6 \end{aligned}$$

linearly independent integral modular forms of level 1, *dimension* $-2h$, *and character* $\chi = 1$.

We finally remark: *for an odd dimension* $-k$ *no integral modular forms of level* 1 *exist having characters with values* ± 1 *exclusively*. For if $M(\tau)$ were such, then $M(\tau)^2\,d\tau^k$ would be a differential. Then $M(\tau)^2\,(d\tau/dq_i)^k$ would be a power series in q_i at $\tau = i$. But then it can be read directly from q_i and $dq_i/d\tau$ that $M(\tau)$ cannot be expanded in a power series in $\tau - i$; therefore $M(\tau)$ cannot be holomorphic at $\tau = i$.

7. Fourier Expansions of Eisenstein Series

In this section we assume $(a_1, a_2, N) = 1$. Starting point is the equation

$$\sum_{m=-\infty}^{+\infty} \frac{1}{(\tau+m)^k} = \frac{(-2\pi i)^k}{(k-1)!} \sum_{n=1}^{\infty} n^{k-1} e^{2\pi i n\tau} \tag{22}$$

which holds for $k \geqq 2$. On the left is a function periodic in τ, whose Fourier coefficients are correctly given, as is easily verified. The sum (17), which we want to compute for $k > 2$, contains terms with $m_1 = 0$ if and only if $a_1 \equiv 0$ mod N. By setting $\delta(a_1/N) = 1$ or 0, depending upon whether a_1/N is integral or not, we have

$$G_k(\tau; a_1, a_2, N) = \delta\left(\frac{a_1}{N}\right) \sum_{m_2 \equiv a_2(N)}' \frac{1}{m_2^k} + \sum_{\substack{m_1 \equiv a_1(N) \\ m_1 \neq 0}} \sum\nolimits_2 \sum_{m=-\infty}^{+\infty} \frac{1}{(m_1\tau + m_2 + m m_1 N)^k},$$

where the sum $\sum_2$ is taken over all m_2 running through a residue system mod $m_1 N$ with $m_2 \equiv a_2$ mod N. Now, the inner sum over m is $(m_1 N)^{-k}$ times the sum on the left of (22), with $N^{-1}(\tau + m_2/m_1)$ in place of τ. By applying Eq. (22) we find that

$$G_k(\tau; a_1, a_2, N) = \delta\left(\frac{a_1}{N}\right) \sum_{m_2 \equiv a_2(N)}' \frac{1}{m_2^k} + \frac{(-2\pi i)^k}{(k-1)!N^k} \sum_{n=1}^{\infty} \sum_{\substack{m_1 \equiv a_1(N) \\ m_1 \neq 0}} \sum\nolimits_2 \frac{n^{k-1}}{m_1^k} \exp[2\pi i n N^{-1}(\tau + m_2 m_1^{-1})].$$

The inner sum over m_2 can be realized by setting $m_2 = a_2 + N m_2'$ and then letting m_2' take on all values from 0 to $m_1 - 1$. It is $|m_1|$ times $n^{k-1} m_1^{-k}$ $\exp[2\pi i n N^{-1}(\tau + a_2 m_1^{-1})]$ or 0, depending upon whether n is divisible by m_1 or not. Set $n = t m_1$, so that, for $k > 2$,

$$\begin{aligned} G_k(\tau; a_1, a_2, N) &= \delta\left(\frac{a_1}{N}\right) \sum_{m_2 \equiv a_2(N)}' \frac{1}{m_2^k} + \sum_{n=1}^{\infty} c_k(n; a_1, a_2, N) \exp[2\pi i n N^{-1}\tau], \\ c_k(n; a_1, a_2, N) &= \frac{(-2\pi i)^k}{(k-1)!N^k} \sum_{\substack{t/n \\ nt^{-1} \equiv a_1(N)}} (\operatorname{sign} t) t^{k-1} \exp[2\pi i t a_2 N^{-1}]. \end{aligned} \tag{23}$$

In particular

$$G_{2k}(\tau; 0, 0, 1) = 2\zeta(2k) + \sum_{n=1}^{\infty} c_{2k}(n) e^{2\pi i n \tau},$$
$$c_{2k}(n) = 2\,\frac{(2\pi i)^{2k}}{(2k-1)!} \sum_{\substack{t/n \\ t>0}} t^{2k-1}. \tag{24}$$

Here $\zeta(s)$ denotes the Riemann zeta function.†

The constant term of the Fourier series (23) is $\neq 0$ if and only if $a_1 \equiv 0 \bmod N$. Assume $G_k(\tau; a_1, a_2, N)$ to be primitive and apply to it the substitution (18). Now let τ approach $i\infty$, but by using the expansion (16) for the cusp $\Gamma(N)a/c$ on the left and the Fourier series (23) on the right. This yields

$$\sum_{n=0}^{\infty} c_n \exp[2\pi i N^{-1}\tau] = G_k(\tau; a_1 a + a_2 c, a_1 b + a_2 d, N)$$

and *the constant term of the expansion* (16) *of a primitive* $G_k(\tau; a_1, a_2, N)$ *at the cusp* $\Gamma(N)a/c$ *is* $c_0 \neq 0$ *if and only if* $a_1 a + a_2 c \equiv 0 \bmod N$.

Now we can prove that all formally different Eisenstein series are linearly independent if $k > 2$. By the last statement it suffices to show that these series with $a_1 a + a_2 c \equiv 0 \bmod N$ are linearly independent. Such Eisenstein series are uniquely determined by the residue of $a_1 b + a_2 d = a_2' \bmod N$, and among them there are as many formally different as a_2' takes values mod N with $(a_2', N) = 1$ and $0 < a_2' \leqq N/2$; the latter inequality takes account of the fact that $G_k(\tau; a_1, a_2, N)$ and $G_k(\tau; -a_1, -a_2, N)$ do not differ formally (in the sense of §4,4). It amounts to the same to prove that all primitive $G_k(\tau; 0, a, N)$ with $(a, N) = 1$ and $0 < a \leqq N/2$ are linearly independent. This we derive from their Fourier expansions (23) [where we neglect the common factor $(-2\pi i)^k/(k-1)!N^k$]. In the Fourier expansions we only consider the terms $c_k(mN; 0, a, N)\exp[2\pi i m\tau]$ with $m = 1$ and further prime values of m by which N is not divisible. According to (23) the coefficients are

$$c_k(N; 0, a, N) = \exp[2\pi i a N^{-1}] + (-1)^k \exp[-2\pi i a N^{-1}] = \gamma(a)$$

$$\begin{aligned} c_k(mN; 0, a, N) &= \exp[2\pi i a N^{-1}] + (-1)^k \exp[-2\pi i a N^{-1}] \\ &\quad + m^{k-1}[\exp[2\pi i m a N^{-1}] + (-1)^k \exp[-2\pi i m a N^{-1}]], \\ &= \gamma(a) + m^{k-1}\gamma(ma) \end{aligned}$$

† The numbers $2\zeta(2k)\,(2k)!\,(2\pi)^{-2k} = B_k$ are the so-called *Bernoulli numbers*. They are rational. In particular, $2\zeta(4) = (2\pi)^4/720$, $2\zeta(6) = (2\pi)^6/42\cdot 720$.

the latter for a prime m; $\gamma(a)$ is an abbreviation. If our Fourier series were linearly dependent, so would the series

$$\sum \gamma(ma) \exp[2\pi i m\tau].$$

At first let k be even. Then the $\gamma(a)$ form a basis with respect to Q of the largest real subfield Q_N^{real} of the Nth cyclotomic field Q_N. To each residue m mod N prime to N there corresponds an automorphism σ_m of Q_N taking $\gamma(a)$ into $\gamma(a)^{\sigma_m} = \gamma(ma)$. Therefore the determinant $|\gamma(ma)|$ with m and a running through a system as described above is $\neq 0$ which proves our contention. If k is odd the $\gamma(a)$ form one-half of a basis of Q_N/Q, the other half given by $\exp[2\pi i a N^{-1}] + \exp[-2\pi i a N^{-1}]$. Now the same reasoning as before can be repeated. ▼

Integral modular forms of dimension -2 can, by §4,6, exist only for $N > 1$. Then

$$N^{-2}\wp\left(\frac{a_1\tau + a_2}{N}; \tau, 1\right) = \sum_{m_i \equiv a_i(N)}' \left(\frac{1}{(m_1\tau + m_2)^2} - \frac{1}{((m_1 - a_1)\tau + (m_2 - a_2))^2}\right) \tag{25}$$

are such modular forms. These are the *partial values* of the Weierstrass $\wp$ function of §2,(26), of periods τ and 1. The case $a_1 \equiv a_2 \equiv 0$ mod N is now excluded by assumption; the prime over the summation indicates that whenever $m_1 - a_1 = m_2 - a_2 = 0$ that subtrahend is to be omitted. The easy verification of the functional equations (18) is left to the reader as an exercise.

The Fourier expansion of (25) is found by summing first over m_2 and then over m_1, preliminarily limiting m_1 to a finite range:

$$N^{-2}\wp\left(\frac{a_1\tau + a_2}{N}; \tau, 1\right) = \delta\left(\frac{a_1}{N}\right) \sum_{m_2 \equiv a_2(N)}' \frac{1}{m_2{}^2} - \frac{2}{N^2}\zeta(2)$$

$$+ \sum_{n=1}^{\infty} c_2(n; a_1, a_2, N) \exp[2\pi i n N^{-1}\tau], \tag{26}$$

$$c_2(n; a_1, a_2, N) = \frac{(-2\pi i)^2}{N^2}\left(\sum_{\substack{t/n \\ nt^{-1} \equiv a_1(N)}} |t| \exp[2\pi i t a_2 N^{-1}] - \sum_{t/n} |t|\right),\dagger$$

the sums in the last equation having $nt^{-1} = m_1$ or $m_1 - a_1$, respectively, limited to that finite range. This range is then increased until all integers $m_1 \equiv a_1$ mod N are included. The result is a sequence of Fourier series, whose nth coefficients c_n remain unchanged following some member depending upon n, and each of which converges absolutely and uniformly in $\text{Im}(\tau) \geqq \varepsilon > 0$. The same is true of their limit, which therefore represents the function (25).

† t takes on values of the positive and negative divisors of n.

The demonstration of the linear independence of the primitive $G_k(\tau; a_1, a_2, N)$ cannot be taken over to the partial $\wp$-values because of the extra terms $2N^{-2}\zeta(2)$ in their Fourier series (26). But the argument, applied to their differences, still shows that $\sigma_N - 1$ linearly independent modular forms (25) exist. Actually, the only linear relation between them is the vanishing of their sum. For this sum, being an integral modular form of level 1, must vanish according to the last section.

To every integral modular form of dimension $-k$ *there exists a linear combination of primitive Eisenstein series* ($k > 2$) *or of primitive partial* $\wp$*-values* ($k = 2$) *such that their difference vanishes in all the cusps.* These differences are modular forms, called *cusp forms.*

Proof. The assertion immediately follows the fact, for $k > 2$, that there are just as many linearly independent Eisenstein series as cusps. For $k = 2$ a linear combination of the partial $\wp$-values can still be found, such that the difference $f(\tau)$ vanishes at all except at most one cusp. Now, by the last section, $f(\tau)\,d\tau$ is a differential having at most a pole in that cusp, which at most is of the first order. By the residue theorem of III, §4,7 the residue is then 0. Thus no pole occurs. ▼

A cusp form $f(\tau)$ *of dimension* -2 *determines a differential* $f(\tau)\,d\tau$ *of the first kind, and conversely,* as follows immediately from the definition of the divisor of $f(\tau)\,d\tau$ of the last section.

Another theorem of importance is found from the series (24): *the discriminant* $\Delta(\tau) = g_2(\tau, 1)^3 - 27g_3(\tau, 1)^2$ *can be expanded in the Fourier series*

$$\Delta(\tau) = (2\pi)^{12} \sum_{n=1}^{\infty} \tau(n) e^{2\pi i n\tau}$$

with integral coefficients $\tau(n)$ *and with* $\tau(1) = 1$. *Furthermore,*

$$j(\tau) = \frac{12^3 g_2(\tau, 1)^3}{\Delta(\tau)} = e^{-2\pi i\tau} + \sum_{n=0}^{\infty} c_n e^{2\pi i n\tau}$$

with rational integers c_n. Incidentally, the surprising formula

$$\Delta(\tau) = (2\pi)^{12} e^{2\pi i\tau} \prod_{n=1}^{\infty} (1 - e^{2\pi i n\tau})^{24}$$

can be proved, but it is difficult to interpret it in the framework of our considerations.†

† A simple proof of this is found in R. Fueter, *Vorlesungen uber die singulären Moduln und die komplexe Multiplikation der elliptischen Funktionen*, p.24. Leipzig-Berlin, 1924.

Proof. By (19) and (24) we have

$$g_2(\tau, 1) = (2\pi)^4\left(\frac{1}{12} + 20\sum_{n=1}^{\infty} \gamma_n e^{2\pi i n\tau}\right)$$

$$g_3(\tau, 1) = (2\pi)^6\left(\frac{1}{9\cdot 24} - \frac{7}{3}\sum_{n=1}^{\infty} \gamma_n' e^{2\pi i n\tau}\right)$$

with rational integers γ_n, γ_n', and further,

$$\Delta(\tau) = (2\pi)^{12}\sum_{n=1}^{\infty}\sum_{\substack{t/n\\ t>0}} \frac{5+7t^2}{12}\, t^3 e^{2\pi i n\tau} + \cdots,$$

where the $\cdots$ indicate a Fourier series with integral coefficients and starting with $e^{4\pi i\tau}$. But, by a theorem of elementary number theory, $(5+7t^2)t^3 \equiv 0$ mod 12 must always hold, so that the $\tau(n)$ are integers. As $\tau(1) = 1$, the c_n are also integers. ▼

8. Theta Functions

Special modular forms of dimension $-k$ and level N were already derived in the appendix to Chapter I, §2,3, as the theta functions

$$\vartheta(\tau, F) = \sum_m e^{\pi i\tau F[m]} = \sum_{n=0}^{\infty} c(n, F)e^{2\pi i n\tau} \tag{27}$$

to definite quadratic forms $F[m]$ in $2k$ variables and of level N. The $c(n, F)$ were the numbers of ways the integer n could be represented by the quadratic form $\frac{1}{2}F[m]$. It turns out that in certain cases these functions can be represented as linear combinations of Eisenstein series. The expansion coefficients of the latter were computed in the last section, and are functions of elementary number theory. In such cases, then, the $c(n, F)$ can be expressed by such functions. Otherwise one can at least find approximations of the $c(n, F)$ by elementary expressions.

As an example we consider the form $F[m] = 2(m_1{}^2 + \cdots + m_4{}^2)$ in detail. In this case it is best to verify directly from (27) and the appendix to Chapter I, §1,(13), that $\vartheta(\tau/2, F)$ is a modular form of dimension -2 for the subgroup Θ of Γ generated by $\left(\begin{smallmatrix}1 & 2\\ 0 & 1\end{smallmatrix}\right)$ and $\left(\begin{smallmatrix}0 & 1\\ -1 & 0\end{smallmatrix}\right)$. §1,4 of that appendix shows that $\Theta \supset \Gamma(2)$. Thus, not only is $\vartheta(\tau, F)$ a modular form of level 4 as seen in the appendix to Chapter I, §2,3, but $\vartheta(\tau/2, F)$ is one of level 2. This simplifies our investigations.

$\vartheta(\tau/2, F)$ is a linear combination of the two linearly independent partial $\wp$-values

$$\wp\left(\frac{\tau}{4}; \frac{\tau}{2}, 1\right) = -\frac{\pi^2}{3} - 8\pi^2 \sum_{n=1}^{\infty}\left(\sum_{\substack{t/n \\ nt^{-1}\equiv 1(2)}} t - \sum_{t/n} t\right)e^{\pi i n(\tau/2)}$$

$$\wp\left(\frac{\tau}{2} + \frac{1}{2}; \frac{\tau}{2}, 1\right) = \frac{2\pi^2}{3} - 8\pi^2 \sum_{n=1}^{\infty}\left(\sum_{\substack{t/n \\ nt^{-1}\equiv 0(2)}} (-1)^t t - \sum_{t/n} t\right)e^{\pi i n(\tau/2)}$$

(the sums being taken only over the positive divisors t of n, here), and of a cusp form. Further, the cusp forms correspond to differentials of the first kind. Since, by (11) and (5), the genus of the field $K_{\Gamma(2)}$ is zero, there can be no cusp forms. For the two coefficients $c(0, F) = 1$ and $c(1, F) = 8$ to agree, we must have

$$\vartheta(\tau/2, F) = \pi^{-2}(\wp(\tau/2 + \tfrac{1}{2}; \tau/2, 1) - \wp(\tau/2; \tau, 1)).$$

For the Fourier coefficients this implies

$$c(n, F) = 8\left(\sum_{\substack{t/n \\ nt^{-1}\equiv 1(2)}} t - \sum_{\substack{t/n \\ nt^{-1}\equiv 0(2)}} (-1)^t t\right).$$

An argument from elementary number theory improves this: $c(n, F)$ is 8 or 24 times the sum of the odd divisors of n, depending upon whether n is odd or even. This result was attained in essentially the same way by Jacobi.

The quadratic forms with matrices

$$\begin{pmatrix} 2 & 1 & 1 & 1 \\ 1 & 2 & 0 & 0 \\ 1 & 0 & 2 & 0 \\ 1 & 0 & 0 & 2 \end{pmatrix}, \quad \begin{pmatrix} 2 & 1 & 0 & 0 \\ 1 & 2 & 0 & 0 \\ 0 & 0 & 2 & 1 \\ 0 & 0 & 1 & 2 \end{pmatrix}, \quad \begin{pmatrix} 2 & 1 & 0 & 0 \\ 1 & 2 & 0 & 1 \\ 0 & 0 & 10 & 5 \\ 0 & 1 & 5 & 4 \end{pmatrix}$$

have the determinants $|F| = 4, 9, 25$ and levels $N = 2, 3, 5$. The corresponding $\vartheta(\tau, F)$ are modular forms of that level and of dimension -2. By (11) and (5) the genera of these $K_{\Gamma(N)}$ are still 0, so that no cusp forms occur as yet. Hence the functions $\vartheta(\tau, F)$ can be linearly combined out of the partial $\wp$-values. An elementary computation shows that the number $c(n, F)$ of representations of n by these forms F is equal to the sum of the divisors of n prime to 2, 3, 5, multiplied by 24, 12, 6, respectively

Quaternary forms F with larger determinants and forms in more variables lead to modular forms really including cusp forms. The functions $\vartheta(\tau, F)$ are then sums of partial $\wp$-values or Eisenstein series and cusp forms. Denoting the Fourier coefficients of the latter by $\gamma_k(n)$ we have, for the general case,

$$c(n, F) = \sum_{a_1, a_2} \alpha(a_1, a_2)c_k(n, a_1, a_2, N) + \gamma_k(n). \tag{28}$$

A simple approximation given by Hecke† shows that the $\gamma_k(n)$ are of the order

$$\gamma_k(n) = O(n^{k/2}), \tag{29}$$

while the $c_k(n; a_1, a_2, N)$ are, by (23) and (26), of the order $O(n^{k-1+\varepsilon})$ with arbitrarily small ε. Thus the $c_k(n; a_1, a_2, N)$ prevail if $k > 2$. More advantageous approximations of $\gamma_k(n)$ exist, but lie far deeper (see Chapter V).

Notes

The theory can be carried over to an essentially more general class of modular functions, which satisfy the functional equation

$$M\left(\frac{a\tau + b}{c\tau + d}\right)(c\tau + d)^{-k} = v\begin{pmatrix} a & b \\ c & d \end{pmatrix} M(\tau) \tag{30}$$

instead of (14). An arbitrary complex number can even be taken for k. The no longer uniquely defined function $(c\tau + d)^k$ must be fixed if k is not an integer. The numbers $v\left(\begin{smallmatrix} a & b \\ c & d \end{smallmatrix}\right)$ are called the *multipliers* of $M(\tau)$. Divisors in a generalized sense can be associated with these modular forms, and a suitable generalization of the Riemann–Roch theorem permits the determination of the number of linearly independent modular forms satisfying certain integrity conditions, in many cases. The theory was built up by Petersson in a series of papers [4] which, incidentally, touch on the paper by Weil [10] mentioned on page 142. In [8] Petersson discusses the possibility of a generalization beyond (30).

Petersson [5], using an idea which will occur in V, §3,4, proved that all modular forms of level 1 and of real dimension $-k < -2$ can be represented by special Poincaré series

$$G_k(\tau, m) = \sum (c\tau + d)^{-k} \exp[2\pi i m(a\tau + d)(c\tau + d)^{-1}], \tag{31}$$

where c, d run through a full system of mutually prime numbers and a, b are chosen correspondingly so that $ad - bc = 1$. The corresponding generalization holds true for arbitrary subgroups of finite index in Γ, and even for the most general discontinuous groups of motions of the hyperbolic plane [6]. In another direction, Eichler [1] shows that the cusp forms of dimension $-2k$ can be represented by theta series $\vartheta(\tau, F)$ whenever N is a prime number and the functional equation (14) is assumed to hold with $\chi\left(\begin{smallmatrix} a & b \\ c & d \end{smallmatrix}\right) = 1$. A similar statement continues to hold for the generalized theta series (32) limited to quaternary forms [2]. The proof is of a number-theoretic nature, but enters upon algebraic properties of modular forms (see Chapter V).

The method of the last section for constructing modular forms using quadratic forms has been generalized as follows. Let $F = VV^T$ with a real matrix V and $f(x)$ be a homogeneous spherical function of degree $2r$ in $2k$

---

† E. Hecke, *Mathematische Werke*, Paper 24, pp.461–486. Göttingen, 1959.

variables. Then the function

$$\vartheta(\tau; F, f) = \sum_{m} f(V^T(m)) e^{\pi i \tau F[m]} \tag{32}$$

satisfies the functional equation (14) if $k + r$ replaces k. The proof, given by Schoeneberg [10], uses a method essentially dating back to Hermite.

It has finally become clear that the proof of Eq. §2,(15) of the appendix to Chapter I also holds for quadratic forms in odd numbers of variables. The determination of the sign, now less simple, is carried out by Pfetzer [3], immediately for the generalized series (32).

Modular forms of positive dimensions and their Fourier series can also be applied in number theory [9]. Finally, we cite a paper by Petersson [7], which conceptually characterizes the Eisenstein series $G_k(\tau; a_1, a_2, N)$ as against other integral modular forms by one of its properties.

Completely ignored in this survey of the literature are the various treatments of the question as to Weierstrass points on the modular surface $\mathfrak{R}_{\Gamma(N)}$. Furthermore, all papers on the analytic number theory of quadratic forms were passed over, insofar as they offer no substantial contributions to the theory of modular forms as an essentially algebraic notion. No sharp boundary can be drawn, but were one to neglect any sort of division, the literature to be cited would swell substantially. The boundary between modular functions and automorphic functions is also vague, but literature on the latter has been suppressed as far as possible.

Further literature on modular forms must still be discussed in Chapter V.

References

[1] M. Eichler, *Über die Darstellbarkeit der Modulformen durch Thetareihen*, J. Reine Angew. Math. **195**, 156–171 (1956).

[2] M. Eichler, *Quadratische Formen und Modulfunktionen*, Acta Arith. **4**, 217–239 (1958).

[3] W. Pfetzer, *Die Wirkung der Modulsubstitutionen auf Thetareihen zu quadratischen Formen ungerader Variablenzahl*, Arch. Math. **4**, 448–454 (1953).

[4] H. Petersson, *Zur analytischen Theorie der Grenzkreisgruppen*, Part I, Math. Ann. **115**, 23–67 (1938); Part II, 175–204; Part III, 518–572; Part IV, 670–709.

[5] H. Petersson, *Über die Metrisierung der ganzen Modulformen*, Jahrb. Deutsch. Math. Verein. **49**, 49–75 (1939).

[6] H. Petersson, *Über eine Metrisierung der automorphen Formen und die Theorie der Poincaréschen Reihen*, Math. Ann. **117**, 453–537 (1939).

[7] H. Petersson, *Über die systematische Bedeutung der Eisensteinschen Reihen*, Abh. Math. Sem. Univ. Hamburg **16**, 104–126 (1945).

[8] H. Petersson, *Über die Transformationsfaktoren der relativen Invarianten linearer Substitutionsgruppen*, Monatsh. Math. Phys. **53**, 17–41 (1949).

[9] H. Petersson, *Konstruktion der Modulformen und der zu gewissen Grenzkreisgruppen gehörigen automorphen Formen von positiver reeller Dimension und die vollständige Bestimmung ihrer Fourier-Koeffizienten*, S.-B. Heidelberger Akad. Wiss. pp. 3–80 (1950).

[10] B. Schoeneberg, *Das Verhalten von mehrfachen Thetareihen bei Modulsubstitutionen* Math. Ann. **116**, 511–523 (1939).

CHAPTER V

Correspondences between Fields of Algebraic Functions

By a correspondence between two Riemann surfaces $\mathfrak{R}_1$ and $\mathfrak{R}_2$ we mean a mapping of $\mathfrak{R}_1$ onto $\mathfrak{R}_2$ which is, along with its inverse, finitely multivalued and conformal, the latter condition being permitted to fail in finitely many exceptional points. §1 parallels this geometric notion with that of an algebraically defined correspondence, the Riemann surfaces being replaced by arbitrary algebraic function fields. The correspondences of an algebraic function field with itself form an associative ring. In §2 the representations of that ring by matrices are considered. A number of important applications of algebraic function theory to number theory are based on the notion of correspondences. §3 will deal with the correspondences of the field of modular functions with itself, in particular, the results including deep-lying facts concerning the number theory of quadratic forms. §4 returns to the main line of the general theory and includes the proof of an important principal theorem. Its meaning becomes particularly clear by its almost surprising implications in §5. A short survey of correspondences between elliptic function fields is finally given in §6.

§1. The Correspondences

1. Basic Concepts

Let K and K' be two fields of algebraic functions of one variable each, both having k_0 as exact constant field. K and K' are assumed to be *separably generated*, *conservative*, and *algebraically independent* of each other, the latter meaning that no two nonconstant elements $x \in K$ and $y' \in K'$ can satisfy an algebraic equation $f(x, y') = 0$ with coefficients in k_0. It is obvious that the algebraic independence of K and K' is preserved under extension of the constant field k_0.

We agree to always distinguish elements of K and K' by denoting the former without, the latter with primes. We follow the same convention for ideals and divisors.

If the constant field k_0 of K is extended to K', then the algebraic independence assures us that the function field KK' has the exact constant field K'. The field $K'K$ can be formed analagously. Then

$$KK' = K'K. \tag{1}$$

All this equation states is that the order of adjunction of the primed and unprimed quantities has no effect on the final field. This is also a consequence of the algebraic independence.

The fields K/k_0 and KK'/K' have the same genera, as do also K'/k_0 and KK'/K, because K and K' were assumed conservative.

According to III, §3,4 this implies that *a divisor of K/k_0 keeps its same degree as a divisor of KK'/K', and then so does a divisor of K'/k_0 extended to KK'/K.*

We shall denote elements of KK' by upper-case Greek letters, such as $\mathrm{A}, \mathrm{B}, \Gamma$.

For two integral domains $\mathfrak{J}, \mathfrak{J}'$ in K, K', let the *ring product*

$$\mathfrak{J} \times \mathfrak{J}' = \{\textstyle\sum \iota_i \iota_i'; \quad \iota_i \in \mathfrak{J}, \quad \iota_i' \in \mathfrak{J}'\} \tag{2}$$

be the integral domain of all finite sums $\sum \iota_i \iota_i'$.

In particular now, let $\mathfrak{J}$ be the principal order of K with respect to $k_0[x]$. *Then $\mathfrak{J} \times K'$ is the principal order of KK' with respect to $K'[x]$.* For the proof denote this order by $\mathfrak{K}$; clearly it contains $\mathfrak{J} \times K'$. If $\mathfrak{K}$ were larger than $\mathfrak{J} \times K'$, the determinant of a linear transformation (c_{ij}) transforming a basis K_i of $\mathfrak{K}/K'[x]$ into a basis ι_i of $\mathfrak{J}/k_0[x]$ which is also a basis of $\mathfrak{J} \times K'/K'[x]$ would be a nonconstant polynomial in x. Thus the extension $k_0 \to K'$ of the constant field would decrease the degree of the discriminant divisor of K and consequently of the genus of K. This is a contradiction.

We shall apply the ideal theory of II, §1 to the principal order of KK' with respect to $K'[x]$. The $\mathfrak{J} \times K'$-ideals will be denoted $A_{\mathfrak{J} \times K'}, B_{\mathfrak{J} \times K'}$, etc. By

$$A_{\mathfrak{J} \times K'} \to A_{K \times K'} = K A_{\mathfrak{J} \times K'} \tag{3}$$

these ideals are mapped onto $K \times K'$-ideals.

The mapping (3) *is a homomorphism. The kernel of this homomorphism is the group consisting of all ideals $\mathfrak{a} \times K'$ where $\mathfrak{a}$ is any $\mathfrak{J}$-ideal in K.*

An $\mathfrak{J} \times K'$-ideal $A_{\mathfrak{J} \times K'}$ has a finite system of generators. But a system of generators of $A_{\mathfrak{J} \times K'}$ is also a system of generators of $A_{K \times K'}$ because of (3). By calling a $K \times K'$-ideal *finite* whenever it has a finite system of generators we may state: any finite $K \times K'$-ideal is the image of an $\mathfrak{J} \times K'$-ideal. The principal theorem of II, §1 may then be carried over as follows.

The finite $K \times K'$-ideals form a group, each of whose elements can be uniquely decomposed to prime ideals.

With the elements $A_{K \times K'}$ of this group we associate elements of an isomorphic abstract group, but now writing the group operation as addition we call the elements *correspondences between K and K′*. The sum of two correspondences A and B is thus defined by stipulating

$$(A + B)_{K\times K'} = A_{K\times K'}B_{K\times K'}. \tag{4}$$

Correspondences belonging to prime ideals are called *prime correspondences.*

There is an isomorphic image of the group of correspondences in the divisor group of the field KK′ of functions of one variable over the constant field K′. The same holds if K and K′ are interchanged. Indeed, we have already associated every correspondence A with an ideal $A_{\mathfrak{J}\times K'}$ to the principal order $\mathfrak{J} \times K'$ of KK' with respect to $K'[x]$, and actually with the set of such ideals $\mathfrak{a}A_{\mathfrak{J}\times K'}$, where $\mathfrak{a}$ is any $\mathfrak{J}$-ideal. The $\mathfrak{J} \times K'$-ideals factor uniquely to prime ideals. Now, the set of $\mathfrak{J} \times K'$-ideals $A_{\mathfrak{J}\times K'}$ not containing any prime $\mathfrak{J}$-ideal in their factorization form a group which is isomorphic to the group of correspondences, by definition. But this group is also isomorphic to a subgroup of the divisor group of KK'/K' from the definition of the latter in II, §5,3.

The theorem just proved permits us to identify correspondences with divisors and to associate properties of those divisors with the correspondences. Thus we can first speak of *equivalent* correspondences A and B:

$$A \sim B \qquad \text{if} \quad A_{K\times K'} = \mathrm{A}B_{K\times K'} \qquad \text{with an} \quad \mathrm{A} \in KK'$$

(moreover $A = B$ if $\mathrm{A} \in K$ or K'). The classes again form a group, and we can write $A \sim 0$ for $A_{K\times K'} = \mathrm{A}K \times K'$.†

We can also associate two *degrees*, $g(A)$ and $g'(A)$, with each correspondence. We mean the degrees of the associated divisors in KK' with K' and then with K as constant field. KK' is first considered an extension of constants of K and then of K'. Addition of correspondences clearly leads to

$$g(A + B) = g(A) + g(B), \qquad g'(A + B) = g'(A) + g'(B). \tag{5}$$

One misunderstanding should be avoided. The degree of the principal divisor (A) of KK'/K' is $g((\mathrm{A})) = 0$. However, the degree of the correspondence represented by the ideal $\mathrm{A}K \times K' = A_{K\times K'}$ may differ from $g((\mathrm{A})) = 0$, as the ideal $\mathrm{A}I \times K' = A_{I\times K'}$ may have prime factors of the sort $\mathfrak{a} \times K'$.

We frequently want to extend one of the fields K or K' algebraically to K_1 or $K_1{}'$. A $K \times K'$-ideal $A_{K\times K'}$ then passes to

$$A^1_{K_1\times K'} = K_1 A_{K\times K'}, \qquad A^{1'}_{K\times K_1'} = K_1{}' A_{K\times K'}. \tag{6}$$

The associated correspondences are denoted A^1 and $A^{1'}$. Equation (6) indicates isomorphic mappings $A \to A^1$ and $A \to A^{1'}$ of the groups of correspondences of K and K' into those of K_1 and K' and of K and $K_1{}'$. An

† We have seen that a divisor of KK' over the constant field K and a divisor of KK' over K' is associated with each correspondence. Thus, the equivalence of two correspondences A, B really means: the divisor associated with A–B in KK'/K' is a principal divisor up to a factor of a divisor of K/k_0, and that associated with A–B in KK'/K is a principal divisor up to a divisor of K'/k_0.

extension of K or of K' may imply an extension k_1/k_0 of the constant field. In this case A^1 or $A^{1'}$ clearly are correspondences between K_1 and $K'k_1$ or between Kk_1 and K_1'.

Under an extension of the constant field the degrees of the correspondences remain unchanged, due to our assumptions. For extensions which leave the constant fields invariant we have

$$\begin{aligned} g(A^1) &= [K_1 : K]g(A), \qquad g(A^{1'}) = g(A), \\ g'(A^1) &= g'(A), \qquad g'(A^{1'}) = [K_1' : K']g'(A). \end{aligned} \tag{7}$$

The first and last of these equations result from the last theorem of II, §5,4. Taking into account that K' and K_1' are the exact fields of constants of KK' and KK_1', and that K has been assumed separably generated and conservative, we see by III, §3,4 that $g(A^{1'}) = g(A)$. The other equation follows symmetrically.

Examples of prime correspondences can be given. *Let an isomorphism $a \to a'$ of K onto a subfield K_0' of K' leave the field of constants, k_0, fixed. Then the differences $a - a'$ for all $a \in K$, together with their products with $K \times K'$, generate a $K \times K'$-ideal $P_{K \times K'}$ which is prime and of degrees $g(P) = 1$, $g'(P) = [K' : K_0']$.*

Proof. Clearly $P_{K \times K'}$ is an ideal. To see that it is finite remember that all we need are the differences $a_\nu - a_\nu'$ with a system a_ν of generators of K/k_0. For, if $a = f(a_\nu)g(a_\nu)^{-1}$ with polynomials $f(a_\nu)$, $g(a_\nu)$ in those generators, then clearly $f(a_\nu)\, g(a_\nu)^{-1} - f(a_\nu')\, g(a_\nu')^{-1}$ lies in $K \times K'(a_1 - a_1', a_2 - a_2', \ldots)$. The residue classes of $K \times K'$ mod $P_{K \times K'}$ can be represented by elements of K'. The field of these residue classes is therefore equal to the constant field K' of KK'/K', so that $g(P) = 1$. But at the same time the residue classes of the subfield $K \times K_0'$ mod $P_{K \times K'}$ can be represented by elements of K, all residue classes of $K \times K'$ mod $P_{K \times K'}$ generating an extension K_1 of K isomorphic to K'. This means that the latter residue field is an extension of degree $g'(P) = [K_1 : K] = [K' : K_0']$ of the constant field K of KK'/K. ▼

2. Multiplication of Correspondences

Let three algebraically independent fields K, K', K'' of functions with common constant field k_0 be given. Let A and B be correspondences between K and K', and K' and K''. The *product* $AB = C$ will now be defined as a certain correspondence between K and K'', this being done in several steps. The degrees of B will be denoted $g'(B)$ and $g''(B)$ in an obvious manner.

(*a*) Let B be a prime correspondence of degree $g'(B) = 1$. By II, §5,5 then, the residue class ring of $K' \times K''$ mod $B_{K' \times K''}$ is a field isomorphic to K''. Moreover, every residue class has exactly one representative in K''. By taking the elements of K' in particular, we have an isomorphic mapping of K' onto a subfield of K''. Thus, B has precisely the property of the example at the close of the last section. We also have an isomorphic map of $K \times K'$ onto a subring of $K \times K''$.

For this case we denote the $K \times K''$-ideal arising from $A_{K \times K'}$ under the isomorphism by $(AB)_{K \times K''}$. Clearly

$$(A_1 + A_2)B = A_1 B + A_2 B. \tag{8}$$

(*b*) Let B be a *purely inseparable* prime correspondence with respect to K'', that is, let the residue class field of $K' \times K''$ mod $B_{K' \times K''}$ be a purely inseparable extension of K'' of degree p^h, where p is the characteristic. Then every $a' \in K'$ satisfies some congruence $a'^{p^h} \equiv a''$ mod $B_{K \times K''}$ with $a'' \in K''$. Thus there exists an isomorphic mapping $a' \to a'^{p^h} \to a''$ of K' onto a subfield of K'', and $B_{K' \times K''}$ is generated by the differences $a'^{p^h} - a''$ as in the above example, again with only a finite system a_ν' of generators of K' being necessary.

Extend K'' to K_1'' by adjoining all p^hth roots.† The extended ideal of $B_{K' \times K''}$ then becomes

$$B^{1''}_{K' \times K''} = (G_{K' \times K_1''})^{p^h}$$

and the corresponding divisors are

$$B^{1''} = p^h G,$$

$G_{K' \times K_1''}$ denoting the ideal generated by the differences $a' - \sqrt[p^h]{a''}$. $G_{K' \times K_1''}$ maps $K' \times K_1''$ onto K_1'' and is therefore of degree 1. The residue classes of elements of $A_{K \times K'}$ in KK_1'' mod $G_{K' \times K_1''}$ form a $K \times K_1''$-ideal $V_{K \times K_1''}$; let Γ_ν be a system of generators. The $\Gamma_\nu^{p^h}$ then generate a $K \times K''$-ideal $C_{K \times K''}$. The correspondence C between K and K'' associated with $C_{K \times K''}$ is called the product AB. Again (8) holds. Later we will use a property of the product in this case. Clearly

$$C_{K \times K''} K_1'' \subseteq (V_{K \times K_1''})^{p^h}.$$

The opposite inclusion holds, too. To prove it, let $\mathfrak{P}$ be a prime $\mathfrak{J} \times K_1''$-ideal. The $\mathfrak{P}$-components (in the sense of II, §2) are principal ideals:

$$(V_{\mathfrak{J} \times K_1''})_{\mathfrak{P}} = \Gamma_{\mathfrak{P}}(\mathfrak{J} \times K_1'')_{\mathfrak{P}}, \quad \Gamma_{\mathfrak{P}} = \sum \Delta_\nu \Gamma_\nu, \quad \Delta_\nu \in (\mathfrak{J} \times K_1'')_{\mathfrak{P}}.$$

But then,

$$(V_{K \times K_1''})_{\mathfrak{P}}^{p^h} = \Gamma_{\mathfrak{P}}^{p^h}(K \times K_1'')_{\mathfrak{P}}, \quad \Gamma_{\mathfrak{P}}^{p^h} = \sum \Delta_\nu^{p^h} \Gamma_\nu^{p^h}$$

so that

$$(V_{K \times K_1''})_{\mathfrak{P}}^{p^h} \subseteq ((C_{K \times K''})K_1'')_{\mathfrak{P}}$$

for all $\mathfrak{P}$, of course. Thus we also have

$$(V_{K \times K_1''})^{p^h} \subseteq C_{K \times K''} K_1''$$

† For this extension, as for similar cases which follow, the exact constant field is generally also extended. But, this in no way impairs the arguments which follow. When two, or more such extensions are carried out, however, the resulting constant fields should be checked to see if they are still identical.

and therefore

$$C_{K\times K''}K_1'' = (V_{K\times K_1''})^{p^h}.$$

Remembering the definitions of C and V, we can write this equation as

$$(AB)^{1''} = p^h AG.$$

(c) Let B be a separable prime correspondence with respect to K''. By the lemma of III, §4,4 there exists a finite separable extension K_1''/K'' such that $B^{1''}$ decomposes to prime correspondences all of degree 1. It is no loss of generality to assume K_1''/K'' to be Galois. Then, if

$$B^{1''} = \sum B_\varrho,$$

we may use Step (a) and set

$$AB^{1''} = \sum AB_\varrho.$$

Under the automorphisms of K_1''/K'' the ideals $(AB_\varrho)_{K\times K_1''}$ permute just as the $(B_\varrho)_{K\times K_1''}$. This leaves their product invariant, which, we maintain, is generated by a $K\times K''$-ideal. For, if that product is decomposed to its prime factors over KK_1'', the hypothesis assures that, along with any factor, its conjugates over KK'' occur. Now, with a suitable basis the discriminant of $K\times K_1''$ with respect to $K\times K''$ is an element of K'' and thus a unit of $K\times K''$. Hence, by the Dedekind discriminant theorem there can be no ramified prime $K\times K_1''$-ideals, so that the product of all the conjugate prime ideals of KK_1'' is actually the corresponding ideal of KK''. But this implies the assertion. Hence $AB^{1''} = C^{1''}$, where C is a correspondence between K and K'', which we define to be the product of A and B.

This definition of the product is independent of the choice of K_1''. For, if K_2'' is a larger Galois extension the B_ϱ can nevertheless be decomposed no further, as their degrees $g'(B_\varrho) = 1$.

(d) Now let B be an arbitrary prime correspondence. Say K_0'' is the largest subfield in the residue class field of $K'\times K''$ mod $B_{K'\times K''}$ separable with respect to K''. Let K_1'' be the Galois closure of K_0''. Then the ideal $B^{1''}_{K'\times K_1''}$ decomposes to a product of purely inseparable prime ideals. The definition of the product AB is now derived by combining Steps (b) and (c).

(e) Finally let B be the sum of arbitrary prime correspondences B_ϱ. Use Steps (b) through (d) to define

$$A\left(\sum m_\varrho B_\varrho\right) = \sum m_\varrho AB_\varrho. \tag{9}$$

The conventions of Steps (b) through (d) assure the invariance of the product AB under an extension of K'' even if the B_ϱ decompose.

Exercise. Let $K = k_0(x)$, $K' = k_0(x')$, $K'' = k_0(x'')$. Further, set $A_{K\times K'} = \mathrm{A}(x, x')K\times K'$ and $B_{K'\times K''} = \mathrm{B}(x', x'')K'\times K''$, where neither of the polynomials $\mathrm{A}(x, x')$, $\mathrm{B}(x', x'')$ are divisible by a polynomial in a single variable. Then, $(AB)_{K\times K''} = \Gamma(x, x'')K\times K''$, where $\Gamma(x, x'')$ is the resultant of the polynomials $\mathrm{A}(x, x')$ and $\mathrm{B}(x', x'')$ with respect to the variable x'.

3. PROPERTIES OF THE PRODUCT

As it is in effect derived from the first two steps above, the product always satisfies the distributive laws (8) and (9).

Let k_1/k_0 be an extension of the constant field and A^1, B^1 the extended correspondences between Kk_1, $K'k_1$, and $K''k_1$. Then the definition of the product shows

$$A^1B^1 = (AB)^1;$$

in other words, the product is invariant under an extension of the constant field.

Lemma. Let K_1' be a finite extension of K' with the same exact constant fields and $A^{1'}$, $B^{1'}$ be the correspondences between K and K_1' and between K_1' and K'' defined by the ideals $A_{K\times K'}K_1'$ and $K_1'B_{K'\times K''}$, respectively. Then

$$A^{1'}B^{1'} = [K_1':K']AB.$$

Proof. Indeed we may extend K'' in such a way to K_1'' that $B^{1''}$ decomposes to prime factors of degree $g' = 1$. Assuming the lemma to be proved for such factors, Eq. (9) shows that

$$A^{1'}B^{1'1''} = [K_1':K']AB^{1''}.$$

The corresponding $K \times K_1''$-ideals are obtained from $(A^{1'}B^{1'})_{K\times K''}$ and $((AB)_{K\times K''})^{[K_1':K']}$ by multiplication with K_1''. Evidently this process cannot make different ideals equal. The proof is thus reduced to this case.

Now let $g'(B) = 1$. By (7), then, $g'(B^{1'}) = [K_1':K']$. To carry out the construction of $A^{1'}B^{1'}$ we must extend K'' suitably to K_1'' so that

$$B^{1'1''} = \sum B_\varrho$$

with B_ϱ again of degree $g'(B_\varrho) = 1$. Then we see by (5) that the number of B_ϱ occurring is exactly $[K_1':K']$.

Any system of generators of the ideal $A_{K\times K'}$ also generates the ideal $(A^{1'})_{K\times K_1'}$. Residues of these generators mod $K_1'B_{K'\times K''}K_1'' = (B^{1'1''})_{K_1'\times K_1''}$ in $K \times K_1''$ are the same as their residues in $K \times K''$ mod $B_{K'\times K''}$. But residues of these generators modulo the $(B_\varrho)_{K_1'\times K_1''}$ are also the same. The assertion now follows from the fact that exactly $[K_1':K']$ such B_ϱ occur. ▼

For the degrees we have the equations

$$g(AB) = g(A)g'(B), \qquad g''(AB) = g'(A)g''(B). \tag{10}$$

To prove the first of Eqs. (10) extend K' until A decomposes to prime correspondences A_ϱ of degree $g(A_\varrho) = 1$. This extension of K' may be effected in two steps. In the first, only the constant field is extended, so that both sides of (10^1) remain unchanged. Now we consider the second step which leaves the exact constant field invariant. Here $g(A)$ remains unchanged, while $g'(B)$ takes on a factor $[K_1':K']$, because of (7), and the lemma shows that the product AB behaves like the latter. Because of (5) we need only consider the A_ϱ, which amounts to assuming $g(A) = 1$. Similarly, a suitable extension of K'' decomposes B to prime correspondences B_ϱ of degree $g'(B_\varrho) = 1$, so that we may assume $g'(B) = 1$. Now, A defines an isomorphic mapping of K onto a subfield of K', and B an isomorphic mapping of K' onto a subfield of K''. From the example at the end of §1,1 we see that the composed mapping corresponds to the product AB, so that we have $g(AB) = 1$.

The second of Eqs. (10) results from the same argument, again setting $g(A) = g'(B) = 1$. The isomorphisms $A: a \to a'$ and $B: a' \to a''$ of K onto a subfield $K_0' \subseteq K'$ and of K' onto a subfield $K_0'' \subseteq K''$, of indices $[K':K_0'] = g'(A)$ and $[K'':K_0''] = g''(B)$, lead to a composed isomorphism $AB: a \to a''$ of K onto a subfield K_{00}'' of K_0'' of index $[K_0'':K_{00}''] = [K':K_0']$. But then $g''(AB) = [K'':K_{00}''] = [K'':K_0''][K_0'':K_{00}''] = g'(A)\,g''(B)$.

We can now prove the *associative law*

$$(AB)C = A(BC). \tag{11}$$

By the lemma both sides of (11) are multiplied by the same factor under finite extension of K' and K''. Extension of K''' is necessary in any case to form the product. Now extend K', K'', and K''' so that A, B, C all decompose to first degree prime correspondences with respect to their first fields. By distributivity we need only prove (11) for the resulting prime correspondences; i.e., we may assume $g(A) = g'(B) = g''(C) = 1$. As A, B, C induce isomorphisms of K, K', K'' onto subfields of K', K'', K''', the composed mapping is associative, and both products are generated by the differences $a - a'''$.

Symmetry law for the product. *Choose some extension K_1 of K such that the extended correspondence A^1 between K_1 and K' decomposes to a sum $A^1 = \sum m_\varrho A_\varrho$ of prime correspondences A_ϱ of degree $g'(A_\varrho) = 1$. Now let the product $A_\varrho B$ be that correspondence whose ideal is the residue class* mod A_ϱ *of the ideal of B in K_1K''. Further let AB be the correspondence whose extension is $A^1B = \sum m_\varrho A_\varrho B$. The product thus defined coincides with that of our definition above.* The roles of A and B have been interchanged in this new product formation. The fact that this new product is at least a correspondence between K and K'' is seen as above, in the original definition.

Proof. Extend K and K'' so that A and B both decompose to prime correspondences of degrees $g' = 1$. We need only carry out our proof for these prime correspondences. Thus, let $g'(A) = g'(B) = 1$. This means that isomorphisms of K' onto subfields of K and K'' are associated with A and B, respectively. But this implies that to every $\mathrm{M} \in K \times K' \times K''$ there exists exactly one $\mathrm{N}_B \in K \times B_{K' \times K''}$ and one $\mathrm{N}_A \in A_{K \times K'} \times K''$ such that $\mathrm{M} - \mathrm{N}_B$

and $M - N_A$ lie in $K \times K''$. We denote them $N_B = \varphi_B(M)$ and $N_A = \varphi_A(M)$; these are linear functions.

For any $A \in K \times K' \times K''$ both $A - \varphi_B(A)$ and $\varphi_B(A) - \varphi_A(\varphi_B(A))$ lie in $K \times K''$, and then so does $A - \varphi_A(\varphi_B(A)) = A - \varphi_B(A) + \varphi_B(A) - \varphi_A(\varphi_B(A))$. Thus, the last difference is a constant in $KK'K''$ taken as a field of functions over KK''. But, choosing $A \in A_{K \times K'} \times K''$, this difference also lies in the ideal $A_{K \times K'} \times K''$, which is now possible only if $A - \varphi_A(\varphi_B(A)) = 0$. Interchanging A and B leads to a similar equation. Our results, stated together, are:

$$\varphi_A(M) \in A_{K \times K'} \times K'', \qquad \varphi_B(M) \in K \times B_{K' \times K''},$$

$$\varphi_A(\varphi_B(A)) = A \qquad \text{for} \quad A \in A_{K \times K'} \times K'',$$

$$\varphi_B(\varphi_A(B)) = B \qquad \text{for} \quad B \in K \times B_{K' \times K''}.$$

The product $(AB)_{K \times K''}$ is generated by all the differences $A - \varphi_B(A)$, by definition, and the ideal formed by interchanging roles of A and B is generated by all the differences $B - \varphi_A(B)$. But, among the former, the differences $\varphi_A(B) - \varphi_B(\varphi_A(B)) = \varphi_A(B) - B$ occur, that is the latter. The converse is shown similarly. This completes the proof. ▼

4. Correspondences of a Field with Itself

A particularly important case is that in which K and K' are isomorphic extensions of the constant field. We then speak of correspondences of K with itself.

Multiplication can be defined for the correspondences of K with itself so that these, using the addition already defined, form a ring with unity. This ring also has an involutive antiautomorphism, called the antiautomorphism of Rosati.

Proof. To define the product of two correspondences A and B between K and K' consider also a third isomorphic but algebraically independent field K'', and the ideals $A_{K \times K''}$ and $B_{K'' \times K'}$ isomorphic to $A_{K \times K'}$ and $B_{K \times K'}$, along with the associated correspondences. These can now be multiplied as above. The associative and distributive laws are consequences of (8), (9), and (11).

The unit correspondence D, taken as a divisor, is the diagonal divisor $\mathfrak{D}$ introduced in III, §5,5.

By interchanging the fields K and K' in the ideals $A_{K \times K'}$ we are lead to an isomorphic involution of the additive group of correspondences onto itself. It is denoted $A \leftrightarrow A^*$, and satisfies

$$(AB)^* = B^*A^* \tag{12}$$

because of the symmetry law of multiplication. ▼

We can also immediately see that

$$g(A^*) = g'(A), \qquad g'(A^*) = g(A). \tag{13}$$

The correspondences $A \sim 0$ form a two-sided ideal in the ring of correspondences, and the residue classes with respect to it, that is, the correspondence classes, also form a ring with unity and with an involutive antiautomorphism.

Proof. If $A \sim 0$, $B \sim 0$ then clearly also $A - B \sim 0$. If $A \sim 0$, that is, $A_{K \times K'} = \mathrm{A}K \times K'$, and B is arbitrary, our definition of the product in §1,2 assures $AB \sim 0$. And, as $A \sim 0$ implies $A^* \sim 0$, we have

$$BA = (A^*B^*)^* \sim 0^* \sim 0.$$

It is the classes of correspondences, above all, which have important applications. For characteristic 0 there are generally no classes other than the integral multiples of the class generated by the unit correspondence D.† Nontrivial correspondence classes occur in special fields only; some examples will be considered in §§3 and 4.

5. Effect of Correspondences on Divisors

With a correspondence A between K and K' we now associate a mapping $\mathfrak{a} \to \mathfrak{a}A = \mathfrak{a}'$ of divisors $\mathfrak{a}$ of K onto divisors $\mathfrak{a}'$ of K'. It turns out practical to write the divisor group additively. The steps of the definition are virtually a repetition of §1,2.

(a) If A is of degree $g(A) = 1$, then $\mathfrak{a}A$ is that divisor of K' arrived at from $\mathfrak{a}$ by taking residue classes in K mod $A_{K \times K'}$. Clearly

$$(\mathfrak{a} + \mathfrak{b})A = \mathfrak{a}A + \mathfrak{b}A, \tag{14}$$

so the mapping is a homomorphism.

(b) For a prime correspondence A, purely inseparable with respect to K' of exponent p^h, let K_1' be the field extended by adjoining all p^hth roots. Then $(A^{1'})_{K \times K_1'} = (B_{K \times K_1'})^{p^h}$ with $g(B^{1'}) = 1$, and $B_{K \times K_1'}$ furnishes a mapping of K onto a subfield of K_1'. Denote the image of $\mathfrak{a}$ under this map by $\mathfrak{a}_1'$ and set $\mathfrak{a}' = \mathfrak{a}A = p^h \mathfrak{a}_1'$. This is a divisor of K'.

(c) If A is a prime correspondence separable with respect to K' and $A^{1'} = \sum A_\varrho$ in a suitable Galois extension K_1' of K', set $\mathfrak{a}' = \mathfrak{a}A = \sum \mathfrak{a}A_\varrho$. This is again a divisor of K'. As in §1,2 the image $\mathfrak{a}'$ is independent of the choice of the extension K_1'.

(d) For an arbitrary prime correspondence A choose a suitable Galois extension such that $A^{1'} = \sum A_\varrho$ with purely inseparable prime correspondences A_ϱ.

---

† A. Hurwitz, *Über algebraische Korrespondenzen und das allgemeine Korrespondenzprinzip*, Math. Werke, Vol. I pp. 163–188. Basel, 1932.

Then define $\mathfrak{a}A_\varrho$ as in Step (b) and let $\mathfrak{a}' = \mathfrak{a}A = \sum \mathfrak{a}A_\varrho$, again a divisor in K'.

(e) For a completely arbitrary correspondence A define $\mathfrak{a}' = \mathfrak{a}A$ using Steps (b) through (d) in connection with $\mathfrak{a}(\sum m_\varrho A_\varrho) = \sum m_\varrho \mathfrak{a}A_\varrho$.

As all other cases were referred to the first, the mapping $\mathfrak{a} \to \mathfrak{a}' = \mathfrak{a}A$ is always a homomorphism of the divisor group of K into that of K'. We also have

$$\mathfrak{a}(A + B) = \mathfrak{a}A + \mathfrak{a}B. \tag{15}$$

Furthermore, with another correspondence B between K' and a third field K'' we can show that

$$(\mathfrak{a}A)B = \mathfrak{a}(AB). \tag{16}$$

Proof. As in §1,3 in the proof of (11) no generality is lost in assuming that $g(A) = g'(B) = 1$. But then we are dealing with an isomorphic embedding of K in K' and K' in K'', which also implies a homomorphic mapping of divisor groups. Equation (16) simply states that AB is the correspondence associated with the composed mapping. Similarly to §1.3, the operations of extension of the constant field and formation of the product $\mathfrak{a}A$ are permutable. ▼

The most important case is again that in which K and K' are isomorphic extensions of k_0, a fixed isomorphism $K \to K'$ being given. *The image $\mathfrak{a}A = \mathfrak{a}'$ can then be carried back to K in the sense of this fixed isomorphism, leading to a ring of homomorphic mapping of the divisor group of K into itself; this ring is homomorphic with the ring of correspondences between K and itself. In particular, the image $\mathfrak{a}D$ under the unit correspondence is $\mathfrak{a}$ itself.*

Actually, it can even be shown that the ring of these mappings is isomorphic to the ring of correspondences.† This is why these mappings are often taken as the correspondences themselves. Our procedure, however, leads to certain simplifications.

The lemma of §1,3 can now be carried over. Let K_1/K be a finite extension with the same exact constant field and $\mathfrak{a}_1$ the image of $\mathfrak{a}$ under the natural embedding of the divisor group of K into that of K_1. Then we have

$$\mathfrak{a}_1 A^1 = [K_1 : K]\mathfrak{a}A. \tag{17}$$

Proof. Let K'' be a third field isomorphic to K and D the unit correspondence between K and K''. Then $\mathfrak{a}D = \mathfrak{a}''$ is the image of the divisor $\mathfrak{a}$ under the given isomorphism. Further, let K_1'' be the extension of K'' corresponding to K_1, and also rewrite A as a correspondence between K'' and K'. The lemma of §1,3 then yields

$$[K_1 : K]\mathfrak{a}A - [K_1 : K]\mathfrak{a}DA = \mathfrak{a}(D^{1''}A^{1''}) = \mathfrak{a}_1 A^1,$$

which was to be shown. ▼

† M. Deuring, *Arithmetische Theorie der Korrespondenzen algebraischer Funktionenkörper I*, J. Reine Angew. Math. **177**, 161–191 (1937).

Now that we have (17) we can carry the proof of (10) over (considering $\mathfrak{a}A$ as a divisor of K'), arriving at

$$g'(\mathfrak{a}A) = g(\mathfrak{a})\, g'(A). \tag{18}$$

Correspondences map principal divisors onto principal divisors so that, with (14), *they are mappings of divisor classes.*

Dually, equivalent correspondences map a divisor onto equivalent divisors. This only holds, though, for correspondences A and B equivalent in a stronger sense than that of our definition, in that the Γ which generates $(A - B)_{K \times K'}$ must generate ideals in the rings $\mathfrak{J} \times K'$ and $K \times \mathfrak{J}'$ not divisible by any prime ideals $\mathfrak{p} \times K'$ or $K \times \mathfrak{p}'$, respectively. ($\mathfrak{p}$ and $\mathfrak{p}'$ are any prime ideals in $\mathfrak{J}$ and $\mathfrak{J}'$). *Using equivalence in this sense* we have the theorem:

If the constant field is algebraically closed, then equivalent correspondences map divisors onto equivalent divisors. Thus the mappings of divisor classes depend only upon the class of the correspondence.

Proof. Because of (15) it suffices to prove the first assertion for prime correspondences A. To form $(a)A$ we take an extension K_1'/K' in which $A^{1'}$ decomposes to prime divisors A_ϱ of degrees $g(A_\varrho) = 1$. Then $(a)A = \sum (a)A_\varrho$. The $(a)A_\varrho = (a_\varrho')$ are principal divisors in K_1', and their sum is the principal divisor $(\Pi\, a_\varrho')$. Discussing the cases of a purely inseparable, a separable, and a mixed extension K_1'/K' as above, we see that $\Pi\, a_\varrho'$ is an element in K', remembering that the a_ϱ' are algebraically conjugate to each other since the A_ϱ are conjugate correspondences.

Some preparation is needed for the proof of the second assertion. Let $\mathfrak{p}$ be a prime divisor of K; as we now assume k_0 to be algebraically closed its degree is 1. Let $\mathfrak{J}$ be the principal order of K with respect to $k_0[x]$, x being some element whose denominator is not divisible by $\mathfrak{p}$. An $\mathfrak{J}$-ideal $\mathfrak{p}_\mathfrak{J}$ is thus associated with $\mathfrak{p}$. Also, let A be a prime correspondence of degree $g(A) = 1$ represented by the ideal $A_{\mathfrak{J} \times K'}$. The divisor $\mathfrak{p}A = \mathfrak{p}'$ in K' is found by taking the residues of $\mathfrak{p}_\mathfrak{J}$ mod $A_{\mathfrak{J} \times K'}$ in K'; they form an $\mathfrak{J}'$-ideal $\mathfrak{p}'_{\mathfrak{J}'}$ in a subfield K_0' of K'. Then $\mathfrak{p}'$ is the prime divisor in K_0' associated with $\mathfrak{p}'_{\mathfrak{J}'}$, where $\mathfrak{J}'$ is the isomorphic image of $\mathfrak{J}$ projected into K_0' by A. In general $\mathfrak{p}'$ is not a prime divisor in K'.

Our discussion of the law of symmetry in the multiplication of divisors can now be carried over. Form the $\mathfrak{J} \times \mathfrak{J}'$-ideal $\mathfrak{p}_\mathfrak{J} \times \mathfrak{J}'$. To every $\mathrm{M} \in \mathfrak{J} \times \mathfrak{J}'$ there exists a $\varphi_\mathfrak{p}(\mathrm{M}) \in \mathfrak{p}_\mathfrak{J} \times \mathfrak{J}'$ and a $\varphi_A(\mathrm{M}) \in A_{\mathfrak{J} \times K'} \cap \mathfrak{J} \times \mathfrak{J}'$ such that $\mathrm{M} - \varphi_\mathfrak{p}(\mathrm{M}) \in \mathfrak{J}'$ and $\mathrm{M} - \varphi_A(\mathrm{M}) \in \mathfrak{J}'$. As above we have $\varphi_A(\varphi_\mathfrak{p}(\mathrm{A})) = \mathrm{A}$ for any $\mathrm{A} \in A_{\mathfrak{J} \times K'} \cap \mathfrak{J} \times \mathfrak{J}'$ and $\varphi_\mathfrak{p}(\varphi_A(\pi)) = \pi$ for any $\pi \in \mathfrak{p}_\mathfrak{J} \times \mathfrak{J}'$. The ideal $\mathfrak{p}'_{\mathfrak{J}'}$ consists of all differences $\pi - \varphi_A(\pi)$, for these are the residue classes of the π in K'. Also, as before, it is seen that this ideal is generated by the differences $\mathrm{A} - \varphi_{\mathfrak{p}}(\mathrm{A})$, and these are exactly the residue classes of the $\mathrm{A} \in A_{\mathfrak{J} \times K'} \cap \mathfrak{J} \times \mathfrak{J}'$ with respect to the module $\mathfrak{p}_{\mathfrak{J}} \times \mathfrak{J}'$.

This supplies us with *another mode of calculation* of $\mathfrak{p}A$. We have shown that for prime correspondences A, the divisor $\mathfrak{p}A$ is given by the $\mathfrak{J}'$-ideal consisting of all residues of $A_{\mathfrak{J} \times K'} \cap \mathfrak{J} \times \mathfrak{J}' \bmod \mathfrak{p}_{\mathfrak{J}} \times \mathfrak{J}'$. The modified equivalence concept assures us, that this representation holds for all correspondences with which we must deal. As in III, §6, we here have a reduction of a field KK'/K modulo the prime divisors of the constant field K. We need not exclude any $\mathfrak{p}$. The divisors of KK'/K which are already divisors of K'/k_0 could cause trouble, though, as they are associated with the unit correspondence. Therefore our new mode of calculation is applicable if and only if $A_{K \times K'} \cap K \times \mathfrak{J}'$ contains no such factors in its prime decomposition.

Little remains to prove the second assertion. It suffices to treat the case of a prime divisor $\mathfrak{a} = \mathfrak{p}$. Let A be the principal correspondence generated by the element A satisfying our conditions. The computation thus permitted shows $\mathfrak{p}A$ to be the principal divisor (a') where a' is the residue class of A mod $\mathfrak{p}$ in K'. The assertion is proved. ▼

6. Prime Correspondences

In what follows the fields K and K' need not be isomorphic. Let A be a prime correspondence between them. The residue class field of $K \times K'$ mod $A_{K \times K'}$ is an extension $\overline{K}$ of K and $\overline{K}'$ of K' of degrees

$$[\overline{K}:K] = g'(A), \qquad [\overline{K}:K'] = g(A),$$

respectively. In the second of these equations we have abused the symbols and set $\overline{K} = \overline{K}'$, which actually cannot be as K and K' are algebraically independent. But, there really is a field $K'' \subseteq \overline{K}$ isomorphic to K' and of index $g(A)$ in $\overline{K}$, and occasionally writing K' for K'' can lead to no errors. In this sense a prime correspondence between two fields establishes a certain algebraic dependence between them, and it pays to investigate it from the point of view of field theory.

In this section we assume the extension $\overline{K}/K'$ to be separable. According to the last section we need a Galois extension L'/K' to form $\mathfrak{a}A$, the associated extension $A^{L'}$ of A decomposing to prime correspondences of the first degree: $A^{L'} = \sum A_\varrho$, $g(A_\varrho) = 1$. (We had previously written K_1' for L'; the symbol K_1' must now be reserved for another field.) The residue classes of $K \times L'$ modulo all the ideals $(A_\varrho)_{K \times L'}$ can be represented by elements of L'. Thus exactly one $x_\varrho' \in L'$ is associated with any $x \in K$, the mapping $x \to x_\varrho'$ being an isomorphism of K onto a subfield K_ϱ' of L'. All the x_ϱ' can never coincide for different ϱ, as this would lead to equal A_ϱ, which is impossible due to the assumed separability of $\bar{K}/K'$ and the Dedekind discriminant theorem.

The Galois group of L'/K' permutes the A_ϱ. The totalities of the x_ϱ' are permuted in the same manner, and must therefore be algebraic conjugates with respect to K'. Thus, any function symmetric in them must lie in K'. Now, in the last section we had set $\mathfrak{a}' = \mathfrak{a}A = \sum \mathfrak{a}A_\varrho$. From our present vantage point we have the following interpretation. *First make* $\mathfrak{a}$ *a divisor in the extension field* $\bar{K} = K \times K'/A_{K \times K'}$ *of* K *in the natural manner and denote it* $\bar{\mathfrak{a}}$. *Then* $\mathfrak{a}'$ *is the sum of the divisors conjugate to* $\bar{\mathfrak{a}}$ *with respect to* K'. *This is exactly the norm*: $\mathfrak{a}' = n_{\bar{K}/K'}(\bar{\mathfrak{a}})$.

In conclusion consider the classical case, $k_0 = C$. The prime divisors of K now correspond one to one with the points of the Riemann surface $\mathfrak{R}$ of K. Let $\bar{\mathfrak{R}}$ and $\mathfrak{R}'$ be the Riemann surfaces of $\bar{K}$ and K'. As a covering space of $\mathfrak{R}$ and of $\mathfrak{R}'$, the number of sheets of $\bar{\mathfrak{R}}$ is $g'(A)$ and $g(A)$. The image $\mathfrak{p}' = \mathfrak{p}A$ of a point $\mathfrak{p}$ of $\mathfrak{R}$ is found by first finding all the points $\bar{\mathfrak{p}}_i$ of $\bar{\mathfrak{R}}$ lying over $\mathfrak{p}$, counting them with correct multiplicity if $\mathfrak{p}$ is a ramification point. The $g'(A)$ points $\bar{\mathfrak{p}}_i$ thus found for $\mathfrak{p}$ satisfy the equation $\mathfrak{p} = \sum \bar{\mathfrak{p}}_i$. Each of these $\bar{\mathfrak{p}}_i$ is then mapped onto the $\mathfrak{p}_i'$ of $\mathfrak{R}'$ over which it lies. The prime divisors thus denoted satisfy $\mathfrak{p}_i' = n_{\bar{K}/K'}(\bar{\mathfrak{p}}_i)$. The image $\mathfrak{p}' = \mathfrak{p}A$ is now the group of points $\mathfrak{p}_i$. The following is also clear:

The Rosati–conjugate correspondence A^* maps a point $\mathfrak{p}'$ of $\mathfrak{R}'$ onto the totality of points $\mathfrak{p}$ of $\mathfrak{R}$ whose images under A include $\mathfrak{p}'$.

7. Inseparable Extensions

Let the field of constants k_0 be an arbitrary field of prime characteristic p. Let A be a prime correspondence between K and K' and use the convention of the last section which makes the residue class field $\bar{K} = K \times K'/A_{K \times K'}$ to a finite extension of both K and K'.

$\bar{K}$ is also a minimal field in this respect. In fact, any subfield of $\bar{K}$ containing K and K' would contain the residues mod $A_{K \times K'}$ of all finite sums $\sum x_i y_i'$ with $x_i \in K$, $y_i' \in K'$, but these form the field $\bar{K}$ by definition.

If k_0 *is perfect then* $\bar{K}$ *is separable over either* K *or* K', *or over both.*

Proof. Use the considerations of III, §4,2. As k_0 is perfect, formula (3) there goes into $K = K^p(x)$ with any p-variable x in K. This means that there can only be a single inseparable extension of K of degree p, that being $K(\sqrt[p]{x})$, and only a single subfield of index p over which K is inseparable, that being K^p. The same is true of all finite extensions of K. But now, if contrary to the assertion $\bar{K}$ were inseparable over both K and K', then $\bar{K}^p$ would contain both K and K', a property shared by no field smaller than $\bar{K}$ as we have seen before. ▼

Assume now that K' has the property that the field K'^p of its pth powers is an extension of the constant field k_0 isomorphic to K. This means that if K is generated as $k_0(x, y)$ with a polynomial equation $f(x, y) = 0$ over k_0, then there exist two elements x', y' of K' such that $K'^p = k_0(x'^p, y'^p)$ with $f(x'^p, y'^p) = 0$. This assumption is equivalent to: $K' = k_0(x', y')$ with $f^{p^{-1}}(x', y') = 0$, where $f^{p^{-1}}$ is found by extracting pth roots of the coefficients of f. Remember that taking pth powers is an isomorphism of $K' \to K'^p$. Hence the assumption implies the isomorphism of K and K'. In general, however, K and K' are not isomorphic extensions of k_0, this being true only if (in a suitable model) the coefficients of $f^{p^{-1}}$ and f coincide, that is, if they lie in the prime field.

Under our assumptions a prime correspondence P between K and K' is given by

$$P_{K \times K'} = K \times K'(x - x'^p, y - y'^p).$$

The residue class field $K \times K'/P_{K \times K'}$ is the field K'; we have $g(P) = 1$ and, with III, §4,(6), $g'(P) = p$. It is easy to see that this correspondence does not depend upon the model of K used.

Now choose another field K'' isomorphic to K', but independent of K and K'. We form the product with the Rosati adjoint

$$P^*_{K'' \times K} = K'' \times K(x''^p - x, y''^p - y)$$

and find that

$$\begin{aligned}(P^*P)_{K'' \times K'} &= K'' \times K'(x''^p - x'^p, y''^p - y'^p)\\ &\subseteq (K'' \times K'(x'' - x', y'' - y'))^p = (pD)_{K'' \times K'}\end{aligned}$$

holds. Because of (10) the degree of the left side is $g(P^*P) = p$, and $g(pD) = p$ as well. We therefore have equality throughout, so that

$$P^*P = pD. \tag{19}$$

We also prove the opposite equation

$$PP^* = pD \tag{20}$$

noting, though, that where before D was the unit correspondence of K' it is now that of K. The ideal associated with P^* in (20) is generated by the elements $x'^p - x''$, $y'^p - y''$. Following the second step in §1,2 we extend K'' by adjoining all pth roots, the resulting field K''_1 containing, in particular, $x''_1 = \sqrt[p]{x''}$, $y''_1 = \sqrt[p]{y''}$.

Our notation for $P_{K'' \times K}$ might appear arbitrary, inasmuch as it also arises from $P^*_{K \times K'}$ by the isomorphism $K'' \cong K$ without the antiautomorphism. Our product establishes an order between the fields, though, and the notation expresses the intermediate position of K. It even becomes imperative when K and K' are isomorphic extensions of k_0. We leave the verification of the relation

$$P^2_{K \times K'} = K \times K'(x - x'^{p^2}, y - y'^{p^2})$$

for the product $P_2 = PP$ which can then be formed to the reader as an exercise.

If, for a prime correspondence A, the field extension $\bar{K}/K$ defined in the last section is inseparable, then $A = BP$ with the correspondence P just defined, and some other prime correspondence B. If $\bar{K}/K'$ is inseparable, then $A = P^*B$ with a prime correspondence B.

Conversely, products of this sort always have the given inseparable extensions.

Proof. Let $K = k_0(x, y)$ and $K' = k_0(x', y')$ with $f(x, y) = 0$ and $f'(x', y') = 0$. As elements of $\bar{K}$, x' and y' satisfy certain equations $F(x, y; x') = 0$, $G(x, y; y') = 0$, and the functions F and G generate a $K \times K'$-ideal $B_{K \times K'}$. This ideal contains $A_{K \times K'}$ since x' and y' satisfy the above equations; on the other hand, $B_{K \times K'}$ is a prime ideal because its residue ring is a field. So $B_{K \times K'} = A_{K \times K'}$. Now we assumed $\bar{K}/K$ to be inseparable. Therefore F and G are inseparable with respect to x' and y', and the ideal $A_{K \times K'}$ is generated by functions of the form $F(x, y; x'^p, y'^p)$. These generate a $K \times K'^p$-ideal whose extension is exactly $A_{K \times K'}$. Choose a third independent field K'' which is an extension of k_0 isomorphic to K'^p. The correspondence P is then defined between K'' and K', and the $F(x, y; x'', y'')$ generate a $K \times K''$-ideal $B_{K \times K''}$. It is a prime ideal, and associated with a prime correspondence B between K and K''. It can now be seen that $A = BP$. If $\bar{K}/K'$ is inseparable the above result yields the second assertion by the use of the Rosati antiautomorphism.

The converse, in the form asserted, is obvious. ▼

Let K and K' be isomorphic extensions of the constant field. Prime correspondences $A = BP$ then generate a two-sided ideal in the correspondence ring. The same is true of prime correspondences $A = P^*B$.

Proof. We must show that, for another prime correspondence C, the product BPC is again in the ideal. This is obvious for C of the form $C = EP$. For C not of this form, though, the last theorem shows the residue class field $K' \times K''/C_{K' \times K''}$ to be separable over K''. Then C can be decomposed to a sum of prime correspondences C_ν of degree $g'(C_\nu) = 1$ in a separable extension K''_1/K''. If we can show that for each ν there exists a prime correspondence E_ν such that $BPC_\nu = E_\nu P$, our assertion is proved.

Thus we need only consider the case where $g'(C) = 1$. By hypothesis and the previous theorem certain functions $F(x, y; x'^p, y'^p)$ generate BP. Say C is associated with the isomorphism $a' \to a''$ of K' onto a subfield of K''; then BPC is generated by the functions $F(x, y; x''^p, y''^p)$. A final application of the same theorem completes the proof. ▼

All assertions and proofs of this section retain their validity if p is replaced by any power $q = p^h$.

8. The Frobenius Automorphism

In addition to the assumptions of the last section we now also let K and K' be generated as $K = k_0(x, y)$ and $K' = k_0(x', y')$ with the same equation, $f(x, y) = f(x', y') = 0$, the coefficients of f being elements of a finite subfield of k_0 with $q = p^h$ elements. Thus K and K' are isomorphic. Also, let k_0 be perfect.

Exponentiation with q is an automorphism of k_0; denote it by the symbol κ, so that $\alpha^\kappa = \alpha^q$. As k_0 is perfect the inverse automorphism κ^{-1} exists. By setting $x^\kappa = x^{\kappa^{-1}} = x$, $y^\kappa = y^{\kappa^{-1}} = y$ we can extend κ and κ^{-1} to automorphisms of K, and similarly for K', for our hypothesis assures the invariance of the coefficients of the defining equation $f = 0$ under κ, κ^{-1}.

The same argument shows that K'^q is generated as $K'^q = k_0(x'^q, y'^q)$ with $f(x'^q, y'^q) = 0$. By the last section then, a correspondence F can be defined by the ideal

$$F_{K \times K'} = K \times K'(x - x'^q, y - y'^q). \tag{21}$$

It is called the *Frobenius correspondence* of K with itself, is of degrees

$$g(F) = 1, \qquad g'(F) = q = p^h,$$

and satisfies the equations

$$FF^* = F^*F = qD. \tag{22}$$

An element $a \in K$, expressed as a rational function $\varphi(x, y)$, is mapped onto $a' = \varphi(x'^q, y'^q) \in K'^q$ by F. The principal divisor (a) is also mapped onto (a'). In the sense of the isomorphism $x \leftrightarrow x'$, $y \leftrightarrow y'$ between K and K', the image in K of a' is

$$a_1 = \varphi(x^q, y^q) = (\varphi^{\kappa^{-1}}(x, y))^q.$$

This equation is valid for principal divisors, and then applies for all divisors. We thus have the following. *Denote the image of* $\mathfrak{a}F$ *in* K' (cf. §1,5), *carried back into K by the above isomorphism* $K \cong K'$, *by* $\mathfrak{a}F$ *as well. Then*

$$\mathfrak{a}F = (\mathfrak{a}^{\kappa^{-1}})^q. \tag{23}$$

To be able to compute $\mathfrak{a}F^*$ we must extend K' to K_1' by adjoining all qth roots, according to our rule. The extended ideal is then

$$F_{K \times K_1'} = (K \times K_1'(x - \sqrt[q]{x'}, y - \sqrt[q]{y'}))^q = G^q_{K \times K_1'} = (qG)_{K \times K_1}.$$

An element $a = \varrho(x, y) \in K$ is mapped onto $a_1 = \varrho(\sqrt[q]{x'}, \sqrt[q]{y'})$ by the correspondence G thus defined. Hence the divisor (a) is mapped onto $(a_1)^q = (\varrho^\kappa(x', y'))$ by $F = qG$. Carried back into K this gives the principal divisor

(a^{κ}). Of course this result is valid for arbitrary divisors:

$$\mathfrak{a}F^{*} = \mathfrak{a}^{\kappa}. \tag{24}$$

9. Correspondences of a Field of Automorphic Functions with Itself

We finally consider certain examples of correspondences that will occur in §3. Let a group $\mathfrak{G}$ of fractional linear substitutions

$$\tau \to (a\tau + b)/(c\tau + d) = M(\tau) = \tau' \tag{25}$$

of a complex variable τ be given. Consider the field K of all meromorphic functions $f(\tau)$ satisfying the functional equation $f(\tau') = f(\tau)$ for all substitutions (25) of $\mathfrak{G}$, and assume that $\mathfrak{G}$ is the greatest group under which K is invariant. In brief, let K be the field of automorphic functions of the group $\mathfrak{G}$. The field of constants is C. Assume further that K is generated by two such functions $x(\tau)$, $y(\tau)$, between which an algebraic equation $f(y(\tau), x(\tau)) = 0$ holds identically in τ. We will not discuss the nature of the group and the range of the variable τ leading to the realization of this hypothesis here. It suffices to say that the modular functions represent an example, choosing $\mathfrak{G} = \Gamma(N)$, the congruence subgroup of the modular group of some level $N \geqq 1$.

Now, say R is another linear substitution not necessarily belonging to $\mathfrak{G}$, but such that the group $\overline{\mathfrak{G}} = \mathfrak{G} \cap R^{-1}\mathfrak{G}R$ has finite indices $g'(R)$ and $g(R)$ in $\mathfrak{G}$ and $\mathfrak{G}' = R^{-1}\mathfrak{G}R$. Let

$$\mathfrak{G} = \overset{g'(R)}{\underset{\nu=1}{\dot{\bigcup}}} \overline{\mathfrak{G}}M_{\nu}, \qquad \mathfrak{G}' = \overset{g(R)}{\underset{\nu=1}{\dot{\bigcup}}} \overline{\mathfrak{G}}M_{\nu}' \tag{26}$$

be decompositions into right cosets. The notation of $g'(R)$ and $g(R)$ should be noted, although it appears inconsistent.

The functions $f(R(\tau))$ form a field K' of automorphic functions with respect to the group $\mathfrak{G}'$, and K' is clearly isomorphic to K. Because of the nature of the M_{ν}, the functions $f(RM_{\nu}(\tau))$ $(\nu = 1, \ldots, g'(R))$ suffer the same permutations as the cosets $\overline{\mathfrak{G}}M_{\nu}$ under substitutions in $\mathfrak{G}$ for τ. The symmetric functions, being invariant under $\mathfrak{G}$, must lie in K. As the $f(R(\tau))$ occur among the $f(RM_{\nu}(\tau))$, every function of K' satisfies an equation of degree $g'(R)$ in K. Adjoining $x(R(\tau))$ and $y(R(\tau))$ to K gives a field $\overline{K}$ containing K' and with $[\overline{K}:K] \leqq g'(R)$.

The degree of this extension is really $[\overline{K}:K] = g'(R)$. For were it smaller, a subset of the $f(RM_{\nu}(\tau))$ would be permuted into itself under $\mathfrak{G}$. But $\mathfrak{G}$ permutes the cosets $\overline{\mathfrak{G}}M_{\nu}$ transitively. This would lead to a $G \in \mathfrak{G}$ and a pair of indices ν, μ such that for all functions $f(\tau)$, $f(RM_{\nu}G(\tau)) = f(RM_{\mu}(\tau))$ would hold, while $M_{\nu}G$ and M_{μ} would lie in different cosets. Thus all $f(\tau)$ would be invariant under the element $RM_{\nu}GM_{\mu}^{-1}R^{-1}$, not in $\mathfrak{G}$, contrary to hypothesis.

Hence the situation of §1,6 applies. It only remains to be shown that the field $\overline{K}$ can be defined by residue class formation in $K \times K'$ modulo a finite ideal $A_{K \times K'}$. To this end choose a duplicate τ' for the variable τ for the field K' and subject it to the substitutions of $\mathfrak{G}'$. Further, choose two indeterminates u and v.

$$\prod_{\nu=1}^{g'(R)} (ux(R\tau') + vy(R\tau') - ux(RM_\nu\tau) - vy(RM_\nu\tau))$$
$$= \sum_{i=0}^{g'(R)} u^i v^{g'(R)-i} \mathrm{A}_i(\tau, \tau') \tag{27}$$

is a polynomial in u, v. The coefficients $\mathrm{A}_i(\tau, \tau')$ are invariant under the groups $\mathfrak{G}$ and $\mathfrak{G}'$, and therefore belong to the composite field KK', and even to the ring product $K \times K'$. Set $\xi = ux(R\tau') + vy(R\tau')$, then (27) becomes a polynomial $A(\xi')$ with coefficients in $K[u, v]$. As $\mathfrak{G}$ permutes the cosets $\overline{\mathfrak{G}}M_\nu$ transitively $A(\xi')$ is irreducible in $K(u, v)$, and the residue ring of $KK'[u, v]$ mod $A(\xi')$ is generated over $K[u, v]$ by adjoining a root of $A(\xi')$, for instance $ux(R\tau) + vy(R\tau)$. Now the coefficients $A_i(\tau, \tau')$ in (27) generate a $K \times K'$-ideal $A_{K \times K'}$ having the property that the residue ring of $K \times K'$ mod $A_{K \times K'}$ is generated over K by adjoining $x(R\tau)$ and $y(R\tau)$. This means that the residue ring is the field $\overline{K}$, as was to be proved. ▼

We can complement our algebraic description of this correspondence A_R by a geometric description. In doing so we assume the points of the Riemann surfaces $\mathfrak{R}$, $\overline{\mathfrak{R}}$, $\mathfrak{R}'$ of K, $\overline{K}$, K' to correspond one to one with the totalities $\mathfrak{G}(\tau)$, $\overline{\mathfrak{G}}(\tau)$, $\mathfrak{G}'(\tau)$ of points (from among a certain range of τ) equivalent under $\mathfrak{G}$, $\overline{\mathfrak{G}}$, $\mathfrak{G}'$, respectively, as we saw to be the case in IV, §4,2 for the group $\mathfrak{G} = \Gamma$ and its subgroups of finite index. By (26), over the point $\mathfrak{G}(\tau)$ of $\mathfrak{R}$ there lie the points $\overline{\mathfrak{G}}M_\nu(\tau)$ of $\overline{\mathfrak{R}}$ which in turn lie over the points $\mathfrak{G}'M_\nu(\tau)$ of $\overline{\mathfrak{R}}'$. This means that

$$\mathfrak{G}(\tau) \to \bigcup_{\nu=1}^{g'(R)} \mathfrak{G}'M_\nu(\tau)$$

is the point mapping of $\mathfrak{R}$ onto $\mathfrak{R}'$ associated with A_R.

It is generally desirable to have correspondences of $\mathfrak{R}$ with itself, which can be easily multiplied. This is achieved by following through with the mapping of $\mathfrak{R}'$ onto $\mathfrak{R}$ given by $\mathfrak{G}'(\tau) \to R\mathfrak{G}'(\tau) = \mathfrak{G}R(\tau)$ associated with the isomorphism $f(\tau) \leftrightarrow f(R\tau)$. The correspondence of points is now written

$$\mathfrak{G}(\tau) \to \mathfrak{G}(\tau)A_R = \bigcup_{\nu=1}^{g'(R)} \mathfrak{G}RM_\nu(\tau). \tag{28}$$

Conversely, map the point $\mathfrak{G}(\tau)$ of $\mathfrak{R}$ onto the point $R^{-1}\mathfrak{G}(\tau) = \mathfrak{G}'R^{-1}(\tau)$ of $\mathfrak{R}'$. Over it there lie the points $\overline{\mathfrak{G}}M_\nu'R^{-1}(\tau)$ of $\overline{\mathfrak{R}}$ which lie over the points $\mathfrak{G}M_\nu'R^{-1}(\tau)$ of $\mathfrak{R}$. Thus

$$\mathfrak{G}(\tau) \to \mathfrak{G}(\tau)A_R{}^* = \bigcup_{\nu=1}^{g(R)} \mathfrak{G}M_\nu{}'R^{-1}(\tau) \tag{29}$$

represents the Rosati adjoint correspondence of §1,4.

It is interesting to note the formal analogy between formulas (27) and (28). The formation of a system of generators for the ideal associated with A_R can be read directly from (28).

The same notation can be applied to represent composed correspondences $A = \sum m_\varrho A_\varrho$ whose prime components A_ϱ arise from substitutions R_ϱ as above. The cosets occurring in the right side of (28) must now be taken by their proper multiplicities m_ϱ:

$$\mathfrak{G}(\tau)A = \bigcup_{\varrho,\nu} m_\varrho \times \mathfrak{G}R_\varrho M_{\varrho\nu}(\tau), \qquad \text{with} \quad \mathfrak{G}(\tau)A_\varrho = \bigcup_\nu R_\varrho M_{\varrho\nu}(\tau).$$

§2. Representations of Correspondences in the Space of Differentials

1. Definitions

As in §1, K, K', etc., will always denote algebraic function fields with exact constant field k_0. *In* §2 *the latter will always be assumed perfect.* Then by III, §4 the divisors of differentials of K, etc., are always divisors of the fields and belong to the canonical class.

To a correspondence A between K and K' we now define a linear mapping of the differentials of degree h ($h = 0, \pm 1, \pm 2, \ldots$) of K onto differentials of K'. As in §1,3 we again use several steps, starting with the simplest case in which $g(A) = 1$. The residue class mapping $K \times K'/A_{K\times K'}$ then leads to an isomorphism of K onto a subfield K_0' of K'. By $dw'^h = dw^h A$ we denote, in this case, the image of dw^h in K_0', but take it as a differential in the entire field K'. This is achieved as follows: let $u_0' \, dv_0'^h$ be that image in K_0' and and v' a separating element in K'; then $u_0' \, dv_0'^h = u_0' \, (dv_0'/dv')^h \, dv'^h$. If K' is inseparable over K_0', $dv_0'/dv' = 0$. Therefore we must assume throughout §2: *if A contains inseparable prime correspondences, h must be* $\geqq 0$.

The rest of the steps follow those of §1,3 exactly, and in the general case they yield a linear mapping $dw^h \to dw'^h = dw^h A$ with the properties

$$dw^h\,(A + B) = dw^h\,A + dw^h\,B, \qquad (dw^h\,A)B = dw^h\,(AB). \tag{1}$$

The ring R of correspondences of K with itself is thus represented by linear mappings of the space of differentials dw^h of K. If the characteristic is prime and $h > 0$ the representation is of the residue class ring of R modulo the ideal $RP \cup P^*R$ generated by the correspondence P of §1,7.

The second of Eqs. (1) shows that we need only prove that $dw^hP = dw^hP^* = 0$. For the first term, dw^h must first be mapped isomorphically into a subfield $K_0' \subset K'$ and taken as a differential of K'. But K'/K_0' is purely inseparable so that the differentials in K' of all functions of K_0' vanish. Thus $dw^h\,P = 0$.

To compute $dw^h P^*$ we start from $P_{K\times K'} = K \times K'(x^p - x', y^p - y')$. Adjoining the pth roots of x' and y' we get an extension K_1'/K', and P^* becomes $P^* = pQ$ with $Q_{K\times K'} = K \times K'(x - \sqrt[p]{x'}, y - \sqrt[p]{y'})$. Now the first equation of (1) gives $dw^h P^* = p\, dw^h Q = 0$. ▼

As we wish a representation of finite degree, we limit ourselves to the effect of correspondences on differentials of the first kind. It will turn out that these differentials keep that property if only certain limitations are made in the case of $h \neq 1$.

To this end we distinguish certain prime divisors $\mathfrak{p}_1, \ldots, \mathfrak{p}_m$ of K arbitrarily, and, given a prime correspondence A between K and a field K' isomorphic to K for which $\bar{K} = K \times K'/A_{K\times K'}$ is separable over K and K', we say that correspondence is of *bounded ramification* if it satisfies the following three conditions.

(a) The images $\mathfrak{p}_\mu A$ in K' are composed of the $\mathfrak{p}_\mu'$ which are associated with the $\mathfrak{p}_\mu$ under the isomorphism $K' \cong K$.

(b) Any ramifications of $\bar{K}/K$ and $\bar{K}/K'$ that exist lie over the places $\mathfrak{p}_\mu$ and $\mathfrak{p}_\mu'$, respectively.

(c) The differents are of the form

$$\mathfrak{d}_{\bar{K}/K} = \prod_{\mathfrak{p}} \prod_i \bar{\mathfrak{p}}_i^{e_i - 1}, \qquad \mathfrak{d}_{\bar{K}/K'} = \prod_{\mathfrak{p}'} \prod_i \bar{\mathfrak{p}}_i^{e_i' - 1}, \tag{2}$$

where

$$\mathfrak{p} = \prod_i \bar{\mathfrak{p}}_i^{e_i'}, \qquad \mathfrak{p}' = \prod_i \bar{\mathfrak{p}}_i^{e_i}$$

are the decompositions of the prime divisors $\mathfrak{p}$ of K and $\mathfrak{p}'$ of K' in $\bar{K}$.

If $A = BP^l$ or $A = P^{*l}B$ with maximal l and a prime correspondence B, then the field extensions $\bar{K}/K$ and $\bar{K}/K'$ are separable for B instead of A, as was shown in §1,7. We then say A is of bounded ramification if this is so of B. Finally an arbitrary correspondence $A = \sum m_\varrho A_\varrho$ is said to be of bounded ramification if all prime correspondences A_ϱ occurring satisfy the criterion.

The Rosati adjoint A^* of a correspondence A of bounded ramification is also of bounded ramification.

The special form (2) of the different is often said to be *regular*, or $\bar{K}/K$ and $\bar{K}/K'$ are said to *ramify regularly* at all places. According to III, §2 this is the case whenever k_0 is of characteristic 0 and algebraically closed. But the latter condition is not even necessary, for the ramification indices e_i and the different remain invariant under the algebraic closure of k_0 carried out as an extension of constants of K. The concept of regular ramification was introduced in II, §6,2 for Galois field extensions in a different manner.

We now hold the power product

$$\mathfrak{a} = (\mathfrak{p}_1 \cdots \mathfrak{p}_m)^{h-1} \tag{3}$$

of the prime divisors distinguished above fixed, in order to operate with the notion of a differential dw^h of degree h and of the first kind with respect to $\mathfrak{a}$, as introduced in III, §5,1. We prove the following theorem.

The correspondences of K with itself map all differentials of the first kind of K among themselves. The correspondences of bounded ramification map all differentials dw^h of the first kind with respect to $\mathfrak{a}$ among themselves. Remember that, in the case of inseparable correspondences, $h > 0$ is required.

It naturally suffices to prove this for a prime correspondence A, where the separability of $\bar{K}/K$ and $\bar{K}/K'$ may even be assumed, for otherwise the assertion is a trivial consequence of the vanishing of $dw^h\,A$. Now, in this special case the considerations of §1,6 must be applied in the sense of differentials to determine $dw^h\,A$. That is, dw^h is taken as a differential of $\bar{K}$ and the sum of the differentials conjugate to dw^h with respect to K' is formed. Letting x' denote a separating nonconstant element of K' this sum is

$$dw'^h = s_{\bar{K}/K'}(dw^h/dx'^h)\,dx'^h, \tag{4}$$

where $s_{\bar{K}/K'}$ is the trace. Thus we must study the behavior of dw^h taken in the field $\bar{K}$ and then its trace.

According to III, §4,(16) the differential dw'^h is associated with the divisor

$$(dw'^h)_{K'} = (dw'/dx')^h\,\mathfrak{d}^h_{K'/k_0(x')}\mathfrak{z}'^{-2h}_\infty \tag{5}$$

in K', where $\mathfrak{z}_\infty'$ is the denominator divisor of x'. An analogous formula holds also for the divisor $(dw^h)_{\bar{K}}$ in $\bar{K}$, namely,

$$(dw^h)_{\bar{K}} = (dw/dx')^h\,\mathfrak{d}^h_{K'/k_0(x')}\mathfrak{z}'^{-2h}_\infty\mathfrak{d}^h_{\bar{K}/K'}, \tag{6}$$

where the transitivity formula II, §4,(9) has been applied. In the same sense we have the formula

$$(dw^h)_{\bar{K}} = (dw^h)_K\,\mathfrak{d}^h_{\bar{K}/K}. \tag{7}$$

The fact that $dw^h = (dw^h)_K$ is a differential of the first kind with respect to $\mathfrak{a}$ means $v_{\mathfrak{p}_\mu}((dw^h)_K) \geqq 1-h$ for all prime divisors $\mathfrak{p}_\mu$ of $\mathfrak{a}$ and $v_\mathfrak{p}((dw^h)_K) \geqq 0$ for all other prime divisors. Now we consider $dw^h = (dw^h)_{\bar{K}}$ as a differential of $\bar{K}$ and use (7) as well as the assumption of bounded ramification. The above conditions on the local orders of $(dw^h)_K$ are then translated into the following conditions on the local orders of $(dw^h)_{\bar{K}}$:

$$v_{\mathfrak{p}_{\mu i}}((dw^h)_{\bar{K}}) \geqq e_{\mu i}(1-h) + h(e_{\mu i}-1) = e_{\mu i} - h \geqq 1-h$$

for all prime divisors $\mathfrak{p}_{\mu i}$ of the $\bar{\mathfrak{p}}_\mu$ in $\bar{K}$ and $v_{\bar{\mathfrak{p}}}((dw^h)_{\bar{K}}) \geqq 0$ for all other prime divisors of $\bar{K}$.

To extract our assertion from this statement, we choose a prime element x' for one of the $\mathfrak{p}_\mu'$. (It is a separating element of K' since we assume the constant field to be perfect; therefore there are no inseparable prime divisors.) Let l_μ be the exact power of $\mathfrak{p}_\mu'$ dividing the different $\mathfrak{d}_{K'/k_0(x')}$, and $e'_{\mu i}$ the ramification index of a prime divisor $\bar{\mathfrak{p}}_{\mu i}$ of $\mathfrak{p}_\mu'$ in $\bar{K}$. With (6) and (2) we now have

$$v_{\bar{\mathfrak{p}}_{\mu i}}((dw/dx')^h) \geqq 1 - h - e'_{\mu i} l_\mu h - (e'_{\mu i} - 1)h = 1 - e'_{\mu i} h(l_\mu + 1)$$

which can also be written

$$v_{\bar{\mathfrak{p}}_{\mu i}}((dw/dx')^h x'^{h-1+l_\mu h}) \geqq 1 - e'_{\mu i}.$$

This means that $(dw/dx')^h x'^{1-h+l_\mu h}$ lies in the $\mathfrak{p}_\mu'$-component of the inverse of the different $\mathfrak{d}_{\bar{K}/K'}$. But by II, §4,2, this component $\mathfrak{d}^{-1}_{\bar{K}/K',\mathfrak{p}_\mu'}$ is the complement $(\mathfrak{I}'_{\mathfrak{p}_\mu'})^*_{\bar{K}/K'}$ of the integral domain of all $\mathfrak{p}_\mu'$-integral elements of $\bar{K}$. Thus the traces with respect to K' of all these elements are $\mathfrak{p}_\mu'$-integral:

$$v_{\mathfrak{p}'_\mu}(s_{\bar{K}/K'}(((dw/dx')^h)x'^{h-1+l_\mu h})) \geqq 0,$$

which is equivalent to

$$v_{\mathfrak{p}'_\mu}(s_{\bar{K}/K'}((dw/dx')^h)) \geqq 1 - h - l_\mu h.$$

Combining this with (5) we have already proved that

$$v_{\mathfrak{p}'_\mu}(dw'^h) = v_{\mathfrak{p}'_\mu}(s_{\bar{K}/K'}((dw/dx')^h)dx'^h) \geqq 1 - h$$

for all μ.

We must carry out the same considerations for the nonramified prime divisors $\mathfrak{p}'$ of K', starting out with $v_{\mathfrak{p}_i}((dw^h)_K) \geqq 0$. Let x' again be a prime element for $\mathfrak{p}'$, and $\mathfrak{p}'$ divide the different $\mathfrak{d}_{K'/k_0(x')}$ to the power l; due to (6), the assumption can be written

$$v_{\bar{\mathfrak{p}}_i}((dw/dx')^h x'^{lh}) \geqq 0$$

leading to traces satisfying

$$v_{\mathfrak{p}'}(s_{\bar{K}/K'}((dw/dx')^h)) \geqq -lh,$$

which is equivalent to $v_{\mathfrak{p}'}(dw'^h) \geqq 0$, because of (5). The proof is complete. ▼

For a later application we need the following corollary to the theorem just proved: *the correspondences of bounded ramification of K with itself map all differentials dw^{1-h} of the first kind with respect to $\mathfrak{a}^{-1}$ among themselves.*

Although replacing h by $h-1$ is not generally permissible here, this is possible in (4) through (7). The proof now proceeds through the inequalities

$$v_{\mathfrak{p}_\mu}((dw^{1-h})_K) \geqq h - 1,$$

$$v_{\bar{\mathfrak{p}}_{\mu i}}((dw^{1-h})_{\bar{K}}) \geqq (h-1)e_{\mu i} + (e_{\mu i} - 1)(1 - h) = h - 1,$$

$$v_{\bar{\mathfrak{p}}_{\mu i}}\left(\frac{dw^{1-h}}{dx'^{1-h}}\right) \geqq h - 1 + e'_{\mu i} l_\mu (h-1) + (e'_{\mu i} - 1)(h-1),$$

$$- (h-1)e'_{\mu i}(l_\mu + 1),$$

$$v_{\bar{\mathfrak{p}}_{\mu i}}\left(\frac{dw^{1-h}}{dx'^{1-h}} x'^{-(h-1)(l_\mu+1)}\right) \geqq 0.$$

Because the traces of integral elements are integral,

$$v_{\mathfrak{p}_\mu'}(s_{\bar{K}/K'}(dw^{1-h}/dx'^{1-h})) \geqq (h-1)(l_\mu+1),$$

$$v_{\mathfrak{p}_\mu'}(s_{\bar{K}/K'}(dw^{1-h}/dx'^{1-h})\, dx'^{1-h}) \geqq (h-1)(l_\mu+1)+(1-h)l_\mu = h-1.$$

The considerations concerning the other prime divisors can be repeated with $1-h$ instead of h. ▼

The correspondences of bounded ramification generate a subring $R_\mathfrak{a}$ of the full ring R of correspondences, this subring clearly depending upon the distinguished divisor $\mathfrak{a}$. The correspondences of bounded ramification include, as special cases, the unramified correspondences, that is, those composed of prime correspondences with unramified extensions $\bar{K}/K$ and $\bar{K}/K'$. Examples will be found in the next two sections.

Exercise. If the requirement for bounded ramification is reduced to the divisibility of the differents by at most $\bar{\mathfrak{p}}_i^{r(e_i-1)}$ or $\bar{\mathfrak{p}}_i^{r(e_i'-1)}$, respectively, with an arbitrarily given natural integer $r \geqq 1$, then setting $\mathfrak{a} = (\mathfrak{p}_1 \cdots)^{rh-1}$ retains the validity of the theorems.

2. The Classical Case

Throughout this section we take $k_0 = C$ and $h = 1$. Let du_i be a basis of differentials of the first kind. For an arbitrary correspondence A of K with itself the $du_i\, A$ are differentials of the first kind and satisfy a system of equations

$$du_i\, A = \sum_{j=1}^{g} a_{ij}\, du_j \tag{8}$$

with coefficients $a_{ij} \in C$. We have implicitly identified the fields K and K', which is quite natural here. Equations (1) assure us that this gives a representation of the ring of correspondences by g row matrices in C.

This representation is related to another, based on a topological property of correspondences. First let A be a prime correspondence. In §1,6 we saw that such an A furnishes a conformal mapping of the Riemann surface $\mathfrak{R}$ of K onto itself, ramification points being allowed. As this mapping is continuous it maps closed paths onto closed paths. Also, the relation between a region and its boundary is retained. Both of these statements need only be verified locally, the global statement following. (The arguments in detail are left to the reader.) This shows that A maps homology classes of closed curves onto such classes. Now by IV, §1,3 the canonical system of incisions a_i, b_i is a basis of these homology classes. We can apply A to it, so that a linear system of equations

$$a_i A = \sum_{l=1}^{g} m_{il}^a a_l + \sum_{l=1}^{g} n_{il}^a b_l, \qquad b_i A = \sum_{l=1}^{g} m_{il}^b a_l + \sum_{l=1}^{g} n_{il}^b b_l \tag{9}$$

with rational integers m_{il}^a, etc., results. It is not difficult to see how these equations behave for arbitrary correspondences $A = \sum m_\varrho A_\varrho$ taken as sums of prime correspondences. Equations (9) thus yield a representation of the ring of correspondences in $2g$ row matrices with rational integers as coefficients.

To understand the relation of (8) to (9) it is better to think of (8) in its exact defined sense, writing

$$du_j\, A = \sum_{l=1}^{g} a_{jl}\, du_l',$$

the du_l' being the basis associated with du_l of a field K' isomorphic to, and corresponding with, K. Once again let A be a prime correspondence. Let the variables of K' traverse one of the period paths a_i or b_i and form the integrals IV, §3,(2) thus traced for the right side:

$$\int_{a_i} \sum_l a_{jl}\, du_l' = \sum_{l=1}^{g} \omega_{il}^a a_{jl}, \qquad \int_{b_i} \sum_l a_{jl}\, du_l' = \sum_{l=1}^{g} \omega_{il}^b a_{jl}.$$

For the left side, remember the definition of $du_j\, A$. Let z' be a local uniformizing variable in a neighborhood $\mathfrak{Z}'$ of a place of the Riemann surface $\mathfrak{R}'$. By the last section, then,

$$du_j\, A = \sum_\nu (du_j/dz')^{(\nu)}\, dz' \tag{10}$$

is a certain sum taken over $g(A)$ functions $(du_j/dz')^{(\nu)}$ algebraically conjugate with respect to K'. The neighborhood $\mathfrak{Z}'$ is the image of $g(A)$ inverse images $\mathfrak{Z}_\nu$ in $\mathfrak{R}$, and these are the images of $\mathfrak{Z}'$ by the Rosati-adjoint correspondence A^*.

Now, form the integral of the differential (10) over a path c' lying in $\mathfrak{Z}'$. Each summand on the right could just as well be taken in $\mathfrak{R}$ by replacing c' with a path c_ν lying in one of the neighborhoods $\mathfrak{Z}_\nu$. Suitable numbering of the differentials $(du_j/dz')^{(\nu)}$ in $\mathfrak{Z}_\nu$ makes them equal to du_j. Thus the sum on the right is the integral $\int du_j$ extended over the image of c' under the Rosati conjugate correspondence A^*. Letting c' exhaust one of the cycles a_i, b_i, this yields

$$\int_{a_i} du_j\, A = \int_{a_i A^*} du_j, \qquad \int_{b_i} du_j\, A = \int_{b_i A^*} du_j.$$

The homology classes $a_i A^*$, $b_i A^*$ can here be expressed in the a_i, b_i as in (9). We simplify the notation by writing the matrices which occur as m_{il}^a, etc. The equations

$$\sum_l \omega_{il}^a a_{jl} = \sum_l m_{il}^a \omega_{lj}^a + \sum_l n_{il}^a \omega_{lj}^b, \qquad \sum_l \omega_{il}^b a_{jl} = \sum_l m_{il}^b \omega_{lj}^a + \sum_l n_{il}^b \omega_{lj}^b$$

result. The complex conjugate equations must also hold, the rational integers of the matrices m_{il}^a being self-conjugate. All the equations thus found can be compiled to a single matrix equation

$$\begin{pmatrix} \omega_{il}^{a} & \overline{\omega}_{il}^{a} \\ \omega_{il}^{b} & \overline{\omega}_{il}^{b} \end{pmatrix} \begin{pmatrix} a_{lj} & 0 \\ 0 & \bar{a}_{lj} \end{pmatrix}^{t} = \begin{pmatrix} m_{il}^{a} & n_{il}^{a} \\ m_{il}^{b} & n_{il}^{b} \end{pmatrix} \begin{pmatrix} \omega_{lj}^{a} & \overline{\omega}_{lj}^{a} \\ \omega_{lj}^{b} & \overline{\omega}_{lj}^{b} \end{pmatrix}$$

or, in shorter notation,

$$P_1 \begin{pmatrix} D_1(A)^t & 0 \\ 0 & \overline{D_1(A)^t} \end{pmatrix} = H(A^*)P_1,$$

(transposition being indicated by the superscript t). We want to use this The equation with A^* in place of A. The matrix

$$\begin{pmatrix} \omega_{lj}^{a} & \overline{\omega}_{lj}^{a} \\ \omega_{lj}^{b} & \overline{\omega}_{lj}^{b} \end{pmatrix} = P_1$$

is nonsingular. For we know ω_{lj}^a has an inverse by IV, §3,2, and that

$$P_1 \begin{pmatrix} (\omega_{lj}^{a})^{-1} & 0 \\ 0 & (\omega_{lj}^{a})^{-1} \end{pmatrix} \begin{pmatrix} E & -E \\ 0 & E \end{pmatrix} = \begin{pmatrix} E & 0 \\ \mathrm{T} & \overline{\mathrm{T}}-\mathrm{T} \end{pmatrix},$$

where T was introduced in IV, §3,(9). As the imaginary component of the matrix T is positive definite, $\overline{\mathrm{T}} - \mathrm{T}$ is nonsingular. We can now compile results.

Let $D_1(A)$ be the representation of the ring of correspondences defined by (8) *and $H(A)$ be that defined by* (9). *A nonsingular matrix P_1 (composed of the Riemann period matrix and its complex conjugates) exists such that*

$$H(A)P_1 = P_1 \begin{pmatrix} D_1(A^*)^t & 0 \\ 0 & \overline{D_1(A^*)^t} \end{pmatrix}$$

holds; that is to say, the representation $H(A)$ is equivalent to the sum of the representation $D_1(A^*)^t$ and its complex conjugate.

3. Continuation; Representations of Rosati-Adjoint Correspondences

We must now prove the important formula

$$\{du\, A, dv\} = \{du, dv\, A^*\}, \tag{11}$$

where $\{\cdots\}$ is the scalar product of differentials of the first kind defined in IV, §3,(7). We may limit ourselves to prime correspondences, and use the notation of §1,6.

The scalar product on the left of (11) is computed as an integral over the Riemann surface $\mathfrak{R}'$ by carrying dv to K' by the isomorphism. The differential $du\, A$ is computed by (10). The index ν occurring in the following indicates the various sheets of $\overline{\mathfrak{R}}$ lying over $\mathfrak{R}'$. In all these sheets dv has the same value, and we can take

$$\sum_{\nu}\left(\frac{du}{dz'}\right)^{(\nu)}\overline{\frac{dv}{dz'}}\,dx'\,dy'$$

as the sum of the differentials $du\,d\bar{v}$ in $\overline{\mathfrak{R}}$, over all the points of $\overline{\mathfrak{R}}$ lying over a single point of $\mathfrak{R}'$. This leads to

$$\{du\,A,\,dv\} = \iint_{\overline{\mathfrak{R}}} du\,d\bar{v}.$$

The same transformation applies to the right side of (11). There $\mathfrak{R}$ and $\mathfrak{R}'$ are interchanged so that A must be replaced by A^*. ▼

If a basis du_i such that $\{du_i,\,du_j\} = \delta_{ij}$ ($= 1$ or 0 for $i = j$, $i \neq j$, respectively) and the abbreviation

$$du_i\,A^* = \sum_j a_{ij}^*\,du_j$$

are used, then (8) and (11) yield

$$\{du_i\,A,\,du_j\} = a_{ij} = \{du_i,\,du_j\,A^*\} = a_{ji}^*.$$

Thus in a suitable basis the Rosati-adjoint correspondence can be represented by the Hermite-adjoint matrix.

4. The Trace

Let the divisor $\mathfrak{a}$ be fixed as in (3). A basis of the differentials of the first kind with respect to $\mathfrak{a}$ is given by dw_i^h. In §1,2 we associated with any correspondence A of bounded ramification a system of equations

$$dw_i^h\,A = \sum_j a_{ij}\,dw_j^h \tag{12}$$

with $a_{ij} \in k_0$. The rest of this paragraph will be concerned with computing the trace of the representation of the ring $R_{\mathfrak{a}}$ of correspondences with bounded ramification. To this end we shall apply the results of III, §5 and therefore make the assumptions on $\mathfrak{a}$ formulated there which can now be expressed as:

(i) *either the number m of distinct prime factors of $\mathfrak{a}$ satisfies the inequality $(m + 2(g-1))(h-1) \geqq 0$ and $\mathfrak{a}$ does not lie in the* $(1-h)$th *power of the canonical class, or*

(ii) $\mathfrak{a}$ *lies in the* $(1-h)$th *power of the canonical class.*

It suffices to compute the trace of prime correspondences because of (1), nonvanishing values being found under this hypothesis only if the corresponding field extensions $\overline{K}/K$ and $\overline{K}/K'$ are separable. We may also eliminate the unit correspondence $A = D$ which maps every differential onto itself;

its trace is therefore equal to the number $g_{h,\mathfrak{a}}$ of linearly independent dw_i^h which was calculated in III, §5,1.

The representation (12) suffers no change under an extension of the constant field k_0. We may therefore assume throughout that k_0 is algebraically closed.

We use the basis $d^{1-h}\mathfrak{w}_i$ of principal part systems complementary to the dw_i^h to write the trace as

$$s_h(A) = \sum_i a_{ii} = \sum_i (d^{1-h}\mathfrak{w}_i, dw_i^h A). \tag{13}$$

which follows from III, §5,(13). The key to our computation is the application of III, §5,6, where the principal part systems

$$dd'^{1-h}\mathfrak{G}_0{}^h = \sum_i d'^{1-h}\mathfrak{w}_i' \, dw_i^h, \qquad dd'^{1-h}\mathfrak{G}_1{}^h = G\, d'x'^{1-h}\, dx^h$$

were introduced. The second of these is exact and compensates the poles of the first, as well as having a pole at the diagonal. Otherwise it has no further pole under assumption (i) or exactly one further pole of first order with a given residue under assumption (ii). Apply the correspondence A with respect to (the variables of) K to both of these principal part systems. On the one hand we have, from (13),

$$s_h(A) = \sum \operatorname{res}_{\mathfrak{p}}(dd'^{1-h}\mathfrak{G}_0{}^h A)_{K=K'}.$$

On the other, $dd'^{1-h}\mathfrak{G}_1{}^h A$ can be formed, confusion being best avoided by interpreting A for the time being as a correspondence between K and an algebraically independent K''. Our assumption that $A \neq D$ assures that the resulting exact principal part system in $K'K''$ has no pole at the diagonal. Thus we may now identify K' and K'', so that we have a differential du' of degree 1 in K':

$$du' = (dd'^{1-h}\mathfrak{G}_1{}^h A_{(K \to K'')})_{K''=K'}. \tag{14}$$

By construction du' has all the poles of $(dd'^{1-h}\mathfrak{G}_0{}^h A)_{K=K'}$ and then others. Because the sum of all residues of du' is 0, the trace is the sum of the residues, taken negatively, of those other poles. We have assumed implicitly that none of these other poles coincide with those of $d'^{1-h}\mathfrak{w}_i'$. But this we may, in fact, assume, for by III, §5 this principal part system can be chosen with poles at arbitrarily given places.

These further poles of du' under discussion will turn out (with a single exception) to be the *fixed points* of the correspondence A. They are precisely those prime divisors $\mathfrak{p}$ of K whose images $\mathfrak{p}'$ by the isomorphism $K \cong K'$ occur in the prime decomposition of $\mathfrak{p}A$.

The differential (14) is computed by formula (4), which requires that $dd'^{1-h}\mathfrak{G}_1{}^h = G\, d'x'^{1-h}\, dx^h$ be taken as a differential of degree h in the common extension field $\bar{K}$ of K and K' determined by A, and that then its trace be formed with respect to K'. It seems useful to write dx' in place of $d'x'$. This trace $s_{K/K'}(G\,(dx/dx')^h)\,dx'^h$ is a differential of degree h in K', and

$$du' = s_{\bar{K}/K'}(G\,(dx/dx')^h)\,dx'. \tag{15}$$

To investigate the behavior of du' at any place $\mathfrak{p}'$ not a pole of a $d'^{1-h}\mathfrak{w}_i'$, set

$$G = (a'/a)(x - x')^{-1} + H.$$

The $a' \in K'$ here is such that $v_{\mathfrak{p}'}(a') = v_{\mathfrak{p}'}(\mathfrak{a}')$, and x' is a prime element for $\mathfrak{p}'$, while a and x are the corresponding elements in K by the isomorphism. It can be seen that the first summand on the right yields singularities of du' at the fixed points of A. We assert that the second summand yields no contribution to the singularities of du' outside of the singular places of the $d^{1-h}\mathfrak{w}_i$.

Certainly H is integral for all divisors of KK'/K not already divisors of K', by III, §5,6. Thus $H \in K \times K'$. This means that in the residue class formation mod $A_{K \times K'}$, which is equivalent to the embedding of K and K' in a common extension $\bar{K}$, only the prime divisors $\mathfrak{p}'$ and $\mathfrak{p}$ of KK'/K and KK'/K' that are already prime divisors of K' and K must be considered. As for these, we see from III, §5,6 that

$$v_{\mathfrak{p}'}(a'^{-1}H\,dx'^{1-h}) \geqq 0, \qquad v_{\mathfrak{p}}(aH\,dx^h) \geqq 0 \tag{16}$$

also holds with the exception of the poles of the $d'^{1-h}\mathfrak{w}_i'$ and a single exceptional $\mathfrak{p} = \mathfrak{p}_0$, the latter only under assumption (ii).

Represent H in the form

$$H = \sum_i H_i H_i' \qquad \text{with} \quad H_i \in K, \quad H_i' \in K'. \tag{17}$$

We maintain: there exists a representation whose summands satisfy the inequalities

$$v_{\mathfrak{p}'}(a'^{-1}H_i'\,dx'^{1-h}) \geqq 0, \qquad v_{\mathfrak{p}}(aH_i\,dx^h) \geqq 0. \tag{18}$$

To prove this, expand

$$H_i' = \sum_\nu c_{i\nu} x'^\nu, \qquad c_{i\nu} \in k_0,$$

in power series in the prime element x'. Were not all of the first inequalities (18) true, i.e., if smaller powers of x' occurred in these series than (18) permits, these would have to cancel out in the sum (17) because of (16). This would imply that certain H_i differ but by a factor in k_0, so that H could be represented as a sum of the sort (17) with fewer terms. By assuming that (17) can no longer be simplified in this sense, we have proved the first inequalities of (18). The same argument proves the other inequalities of (18).

The second inequalities of (18) state that $aH_i\,dx^h$ satisfies the conditions for a differential of the first kind at the place $\mathfrak{p}$. Then, by the proof of the theorem in §2,1 we also have

$$v_{\mathfrak{p}'}(a's_{\bar{K}/K'}(H_i\,(dx/dx')^h)\,dx'^h) \geqq 0.$$

By applying (17) and the first inequalities of (18) we see that

$$v_{\mathfrak{p}'}(s_{K/K'}(H\,(dx/dx')^h)\,dx') \geqq 0,$$

so that H contributes nothing to the trace.

Finally we must discuss the exceptional place $\mathfrak{p}_0$ under assumption (ii). Now $G\,dx'^{1-h}\,dx^h$ has a further pole of order 1 at a place $\mathfrak{p}_0$ which can be chosen arbitrarily. We choose $\mathfrak{p}_0$ in such a way that its images under A fall neither on the singular places of the $d'^{1-h}\mathfrak{w}_i$, on the fixed places, nor on the prime factors $\mathfrak{p}_\mu'$ of $\mathfrak{a}'$. There is (up to a constant multiple) exactly one differential dv^{1-h} with the divisor $(dv^{1-h}) = \mathfrak{a}$. According to III,§5,6 the residue of $G\,dx'^{1-h}\,dx^h$ at the pole $\mathfrak{p}_0$ is made up of the contributions of the fixed points to the trace and the sum of the residues of the differential

$$s_{K/K'}\left(\frac{dv'^{1-h}}{dv^{1-h}}\,\frac{dp_0}{p_0\,dx'}\right)dx'$$

at the images $\mathfrak{p}'_{0i}$ $(i = 1,\ldots, g'(A))$ of $\mathfrak{p}_0$ in K'. Because the $\mathfrak{p}'_{0i}$ differ from the $\mathfrak{p}_\mu'$, each place $\mathfrak{p}'_{0i}$ is the image of exactly one place $\bar{\mathfrak{p}}_{0i}$ of $\bar{K}$ lying over $\mathfrak{p}_0$. We can apply Eq. (31) in III,§4 and compute the sum of residues in $\bar{K}$. Now this sum is equal to

$$\sum_i \operatorname{res}_{\mathfrak{p}_0} \frac{dv'^{1-h}}{dv^{1-h}}\,\frac{d\bar{p}_{0i}}{\bar{p}_{0i}}$$

with a prime element $\bar{p}_{0i}$ at $\bar{\mathfrak{p}}_{0i}$. But, the $\bar{\mathfrak{p}}_{0i}$ are all the places lying over $\mathfrak{p}_0$, so that the values of dv'/dv at these places are all the conjugates of dv'/dv with respect to K. Thus we get

$$\sum_i \operatorname{res}_{\mathfrak{p}_0}\left(\frac{dv'^{1-h}}{dx^{1-h}}\right)_i \frac{dx^{1-h}}{dv^{1-h}}\,\frac{dp_0}{p_0},$$

where the subscript i numbers the different conjugates of $dv'^{1-h}/dx^{1-h} \in \bar{K}$ with respect to K. But in this form it is seen to be

$$s_{K/K}\left(\frac{dv'^{1-h}}{dx^{1-h}}\right)\frac{dx^{1-h}}{dv^{1-h}} = (dv'^{1-h}\,A^*)(dv^{1-h})^{-1}.$$

The corollary in §2,1 states that $dv'^{1-h}\,A^*$ is of first kind with respect to $\mathfrak{a}^{-1}$, so $\mathfrak{a}^{-1}(dv'^{1-h}\,A^*)$ is an integral divisor. But by assumption (ii) $\mathfrak{a}$ is contained in the $(1-h)$th power of the canonical class. Therefore $\mathfrak{a}^{-1}(dv'^{1-h}\,A^*)$ is the unit divisor. This, in turn, identifies the differential $dv'^{1-h}A^*$ up to a constant multiple as dv^{1-h}, so that

$$dv'^{1-h}\,A^* = s_{1-h}(A^*)\,dv^{1-h}$$

with a constant $s_{1-h}(A^*)$.

Thus we have a first degree representation $A \to s_{1-h}(A)$ of the correspondences by the differentials of degree $1-h$ and of first kind with respect to $\mathfrak{a}^{-1}$. In the case $h = 1$ it is reduced to $s_0(A) = g(A)$ or $s_0(A^*) = g(A^*) = g'(A)$.

We can now compile our results as follows

$$s_h(A) = \sum_{\mathfrak{p}'} \operatorname{res}_{\mathfrak{p}'}\left(s_{K/K'}\left(\frac{1}{x'-x}\frac{a'}{a}\left(\frac{dx}{dx'}\right)^h\right)dx'\right) + \begin{cases} 0 & \text{in case (i)} \\ s_{1-h}(A^*) & \text{in case (ii)} \end{cases} \tag{19}$$

The summation is over all fixed places of A, which are thus incidentally shown to be of finite number. As we stipulated, x' is a prime element for $\mathfrak{p}'$ and a' an element such that $v_{\mathfrak{p}'}(a') = v_{\mathfrak{p}'}(\mathfrak{a}')$, while x and a are the isomorphically corresponding elements of K.

5. Evaluation of the Trace Formula

It becomes useful to remember III, §4,(31) and write (19) as a sum of residues in $\bar{K}$:

$$s_h(A) = \sum_{\bar{\mathfrak{p}}} \operatorname{res}_{\bar{\mathfrak{p}}}\left(\frac{1}{x'-x}\frac{a'}{a}\left(\frac{dx}{dx'}\right)^h dx'\right) + \begin{cases} 0 & \text{in case (i)} \\ s_{1-h}(A^*) & \text{in case (ii)} \end{cases} \tag{20}$$

We now define fixed points of A in $\bar{K}$ as those places $\bar{\mathfrak{p}}$ of $\bar{K}$ whose prime divisors simultaneously divide $\mathfrak{p}$ and $\mathfrak{p}'$, where $\mathfrak{p}$ and $\mathfrak{p}'$ correspond by the isomorphism. This convention permits the decomposition of a fixed point in K or K' into several fixed points in $\bar{K}$.

Let $\bar{x}$ be a prime element for a $\bar{\mathfrak{p}}$, and let

$$\begin{aligned} x &= \bar{x}^{e_{\bar{\mathfrak{p}}}}(c_{0\bar{\mathfrak{p}}} + \cdots + c_{m\bar{\mathfrak{p}}}\bar{x}^m + \cdots), \\ x' &= \bar{x}^{e'_{\bar{\mathfrak{p}}}}(c'_{0\bar{\mathfrak{p}}} + \cdots + c'_{m\bar{\mathfrak{p}}}\bar{x}^m + \cdots) \end{aligned} \tag{21}$$

with $c_{0\bar{\mathfrak{p}}}, c'_{0\bar{\mathfrak{p}}} \neq 0$. Our $m = m_{\mathfrak{p}}$ is the smallest exponent such that $c'_{m\bar{\mathfrak{p}}} \neq c_{m\bar{\mathfrak{p}}}$. We introduce the *distortion ratio*

$$\gamma_{\bar{\mathfrak{p}}} = c_{0\bar{\mathfrak{p}}}/c'_{0\bar{\mathfrak{p}}} \tag{22}$$

of A at a fixed point $\bar{\mathfrak{p}}$. Choose 1 or x^{h-1} for a, depending upon whether $\mathfrak{p}$ divides $\mathfrak{a}$ or not. We must compute the residues of

$$\left(\frac{\dfrac{1}{x}\dfrac{dx}{d\bar{x}}}{\dfrac{1}{x'}\dfrac{dx'}{d\bar{x}}}\right)^{h-1}\left(\frac{x}{x'}\right)^{h-1}\frac{1}{x}\frac{dx}{d\bar{x}}\frac{d\bar{x}}{x'x^{-1}-1}, \qquad \left(\frac{\dfrac{1}{x}\dfrac{dx}{d\bar{x}}}{\dfrac{1}{x'}\dfrac{dx'}{d\bar{x}}}\right)^{h-1}\frac{1}{x}\frac{dx}{d\bar{x}}\frac{d\bar{x}}{x'x^{-1}-1}$$

at a place $\bar{\mathfrak{p}}$. Two points must be noted. (a) For $h \neq 1$ the assumption of bounded ramification includes regular ramification in the sense of II, §6,2. The ramification indices $e_{\bar{\mathfrak{p}}}$ and $e_{\bar{\mathfrak{p}}}'$ cannot be divisible by the characteristic. Thus for $h \neq 1$ we have

$$\frac{\dfrac{1}{x}\dfrac{dx}{d\bar{x}}}{\dfrac{1}{x'}\dfrac{dx'}{d\bar{x}}} = \frac{e_{\bar{\mathfrak{p}}}}{e_{\bar{\mathfrak{p}}}'} + \cdots,$$

the dots indicating a power series in $\bar{x}$, starting with the first power. For

$h = 1$, though, irregular ramification can occur. But in that case these quotients, as well as $(xx'^{-1})^{h-1}$, contribute nothing to the residue. (b) The same reasons show that h must equal 1 in the cases 1,2 and 1,3 below. For the assumption of bounded variation would assure $e_{\bar{\mathfrak{p}}} = e_{\bar{\mathfrak{p}}}' = 1$ for $h \neq 1$.

Table I. $s_h(A) = \sum R_{\bar{\mathfrak{p}}} + s_{1-h}(A^*)$ *or* 0 *for a separable prime correspondence* $A \neq D$, *taken over all fixed points* $\bar{\mathfrak{p}}$ *of* A *in* $\bar{K}$. *We exclude the cases* $e_{\bar{\mathfrak{p}}} = e_{\bar{\mathfrak{p}}}'$, $\gamma_{\bar{\mathfrak{p}}} = 1$, *while* $e_{\bar{\mathfrak{p}}} = e_{\bar{\mathfrak{p}}}' = 1$ *must hold whenever* $\mathfrak{p} \nmid \mathfrak{a}$ *and* $h \neq 1$. *We see that*

(1,1)	$e_{\bar{\mathfrak{p}}} = e_{\bar{\mathfrak{p}}}'$	$\gamma_{\bar{\mathfrak{p}}} \neq 1$	$\mathfrak{p} \nmid \mathfrak{a}$:	$R_{\bar{\mathfrak{p}}} = \dfrac{\gamma_{\bar{\mathfrak{p}}}^h}{1-\gamma_{\bar{\mathfrak{p}}}} e_{\bar{\mathfrak{p}}}$
(1,2)	$e_{\bar{\mathfrak{p}}} > e_{\bar{\mathfrak{p}}}'$		$\mathfrak{p} \nmid \mathfrak{a}$:	$R_{\bar{\mathfrak{p}}} = 0$
(1,3)	$e_{\bar{\mathfrak{p}}} < e_{\bar{\mathfrak{p}}}'$		$\mathfrak{p} \nmid \mathfrak{a}$:	$R_{\bar{\mathfrak{p}}} = -e_{\bar{\mathfrak{p}}}$
(2,1)	$e_{\bar{\mathfrak{p}}} = e_{\bar{\mathfrak{p}}}'$	$\gamma_{\bar{\mathfrak{p}}} \neq 1$	$\mathfrak{p} \mid \mathfrak{a}$:	$R_{\bar{\mathfrak{p}}} = \dfrac{\gamma_{\bar{\mathfrak{p}}}}{1-\gamma_{\bar{\mathfrak{p}}}} e_{\bar{\mathfrak{p}}}$
(2,2)	$e_{\bar{\mathfrak{p}}} > e_{\bar{\mathfrak{p}}}'$		$\mathfrak{p} \mid \mathfrak{a}$:	$R_{\bar{\mathfrak{p}}} = 0$
(2,3)	$e_{\bar{\mathfrak{p}}} < e_{\bar{\mathfrak{p}}}'$		$\mathfrak{p} \mid \mathfrak{a}$:	$R_{\bar{\mathfrak{p}}} = -(e_{\bar{\mathfrak{p}}}/e_{\bar{\mathfrak{p}}}')^{h-1} e_{\bar{\mathfrak{p}}}$ or $-e_{\bar{\mathfrak{p}}}$ if $h = 1$

Additional coefficients of the power series (21) go into the residues of the excluded cases $e_{\bar{\mathfrak{p}}} = e_{\bar{\mathfrak{p}}}'$, $\gamma_{\bar{\mathfrak{p}}} = 1$. But then $s_h(A + A^*)$ can be found; for, the terms still needed are the residues, at a place $\bar{\mathfrak{p}}$, of

$$\frac{1}{x'-x}\frac{a'}{a}\,dx^h\,dx'^{1-h} - \frac{1}{x'-x}\frac{a}{a'}\,dx'^h\,dx^{1-h}$$

$$= -\frac{d(x'-x)}{x'-x}\frac{a'}{a}\left(\frac{dx}{dx'}\right)^{h-1} + \frac{1}{x'-x}\left(\frac{a'}{a}\left(\frac{dx}{dx'}\right)^{h-1} - \frac{a}{a'}\left(\frac{dx'}{dx}\right)^{h-1}\right)\frac{dx'}{d\bar{x}}\,d\bar{x}.$$

For $e_{\bar{\mathfrak{p}}} = e_{\bar{\mathfrak{p}}}'$, $\gamma_{\bar{\mathfrak{p}}} = 1$ this is

$$-(2h-1)(m_{\bar{\mathfrak{p}}} + e_{\bar{\mathfrak{p}}}) \qquad \text{or} \qquad -(2h-1)m_{\bar{\mathfrak{p}}} - e_{\bar{\mathfrak{p}}},$$

depending on whether $\mathfrak{p} \nmid \mathfrak{a}$ or $\mathfrak{p} \mid \mathfrak{a}$.

Table II. $s_h(A + A^*) = \sum R_{\bar{\mathfrak{p}}} + g(A) + s_{1-h}(A + A^*)$ *or* 0 *for a separable prime correspondence* $A \neq D$, *taken over the fixed points* $\bar{\mathfrak{p}}$ *of* A *in* $\bar{K}$. *Whenever* $\mathfrak{p} \nmid \mathfrak{a}$, $h \neq 1$, *then necessarily* $e_{\bar{\mathfrak{p}}} = e_{\bar{\mathfrak{p}}}' = 1$. *We see that*

(1,1)	$e_{\bar{\mathfrak{p}}} = e_{\bar{\mathfrak{p}}}'$	$\gamma_{\bar{\mathfrak{p}}} \neq 1$	$\mathfrak{p} \nmid \mathfrak{a}$:	$R_{\bar{\mathfrak{p}}} = \dfrac{\gamma_{\bar{\mathfrak{p}}}^h - \gamma_{\bar{\mathfrak{p}}}^{1-h}}{1-\gamma_{\bar{\mathfrak{p}}}} e_{\bar{\mathfrak{p}}}$
(1,2)	$e_{\bar{\mathfrak{p}}} = e_{\bar{\mathfrak{p}}}'$	$\gamma_{\bar{\mathfrak{p}}} = 1$	$\mathfrak{p} \nmid \mathfrak{a}$:	$R_{\bar{\mathfrak{p}}} = -(2h-1)(m_{\bar{\mathfrak{p}}} + e_{\bar{\mathfrak{p}}})$
(1,3)	$e_{\bar{\mathfrak{p}}} \neq e_{\bar{\mathfrak{p}}}'$		$\mathfrak{p} \nmid \mathfrak{a}$:	$R_{\bar{\mathfrak{p}}} = -\min(e_{\bar{\mathfrak{p}}}, e_{\bar{\mathfrak{p}}}')$
(2,1)	$e_{\bar{\mathfrak{p}}} = e_{\bar{\mathfrak{p}}}'$	$\gamma_{\bar{\mathfrak{p}}} \neq 1$	$\mathfrak{p} \mid \mathfrak{a}$:	$R_{\bar{\mathfrak{p}}} = -e_{\bar{\mathfrak{p}}}$
(2,2)	$e_{\bar{\mathfrak{p}}} = e_{\bar{\mathfrak{p}}}'$	$\gamma_{\bar{\mathfrak{p}}} = 1$	$\mathfrak{p} \mid \mathfrak{a}$:	$R_{\bar{\mathfrak{p}}} = -(2h-1)m_{\bar{\mathfrak{p}}} - e_{\bar{\mathfrak{p}}}$
(2,3)[a]	$e_{\bar{\mathfrak{p}}} \neq e_{\bar{\mathfrak{p}}}'$		$\mathfrak{p} \mid \mathfrak{a}$:	$R_{\bar{\mathfrak{p}}} = -\min((e_{\bar{\mathfrak{p}}}/e_{\bar{\mathfrak{p}}}')^{h-1} e_{\bar{\mathfrak{p}}},\ (e_{\bar{\mathfrak{p}}}'/e_{\bar{\mathfrak{p}}})^{h-1} e_{\bar{\mathfrak{p}}}')$

[a] In this case $(e_{\bar{\mathfrak{p}}}/e_{\bar{\mathfrak{p}}}')^{h-1}$ must be replaced by 1 if $h = 1$.

Consider the case $h = 1$ in more detail. Let us say that a fixed point has the *order*

$$f_{\bar{\mathfrak{p}}} = \begin{cases} \min(e_{\bar{\mathfrak{p}}}, e_{\bar{\mathfrak{p}}}') & \text{for} \quad \gamma_{\bar{\mathfrak{p}}} \neq 1 \quad \text{or} \quad e_{\bar{\mathfrak{p}}} \neq e_{\bar{\mathfrak{p}}}' \\ m_{\bar{\mathfrak{p}}} + e_{\bar{\mathfrak{p}}} & \text{for} \quad \gamma_{\bar{\mathfrak{p}}} = 1 \quad \text{and} \quad e_{\bar{\mathfrak{p}}} = e_{\bar{\mathfrak{p}}}'. \end{cases} \tag{23}$$

By the trace formula

$$s_1(A + A^*) = -\sum f_{\bar{\mathfrak{p}}} + g(A) + g'(A), \tag{24}$$

for $g(A)$ and $g'(A)$ interchange when A is replaced by A^*.

The quantities appearing on the right of (24) were defined as natural numbers. Because of the left side, though, they actually represent elements of the prime field of the characteristic in question. It seems plausible, nevertheless, to consider them to be natural numbers in any case. Then (24) associates a natural number with the correspondence A. In §4 this possibility will have important consequences.

Exercise. Let K be the field of elliptic functions over $k_0 = C$, generated by the Weierstrass $\wp$-function $\wp(u)$ and $\wp'(u)$, and A the correspondence given by the "natural multiplication" $u \to nu = v$, where n is a natural integer. A and A^* are unramified correspondences. Set the divisor $\mathfrak{a} = (1)$. The du^h $(h = 0, \pm 1, \pm 2, \ldots)$ are all differentials of the first kind; they lie in the hth power of the canonical class which is the principal class. So we are always in case (ii). The number of fixed points is $(n-1)^2$ both for A and A^*. Prove $s_h(A) = n^{2-h}$, $s_h(A^*) = n^h$ by using the definition of $du\,A$ and $du\,A^*$ and by applying the trace formula (Table I).

Notes

The calculation of the trace essentially reproduces Kappus's paper cited in III, §5,7. The present author had previously† determined the trace, limiting himself to correspondences of the sort in §1,9. It was incorrectly asserted, though, that the traces $s_h(A)$ and $s_h(A^*)$ were always equal. Also, the contribution of a fixed point of the type $\mathfrak{p}|\mathfrak{a}$, $e_{\bar{\mathfrak{p}}} = e_{\bar{\mathfrak{p}}}'$, $\gamma_{\bar{\mathfrak{p}}} = 1$ was incorrectly computed.

For the classical case $(k_0 = C)$ §2,3 gives us

$$s_1(A^*) = \overline{s_1(A)}. \tag{25}$$

The left side of (24) then becomes $s_1(A^*) + \overline{s_1(A^*)}$. By §2,2 this is the trace of the topologically defined representation $H(A)$. Thus, its trace is

$$s(H(A)) = -\sum f_{\bar{\mathfrak{p}}} + g(A) + g'(A). \tag{26}$$

† M. Eichler, *Eine Verallgemeinerung der Abelschen Integrale*, Math. Z. **67**, 267–289 (1957).

This formula was first proven by Hurwitz.† It is actually a special case of a formula of far greater generality, valid for finitely multivalued mappings of arbitrary manifolds onto themselves. Any textbook of topology gives further information concerning this formula, the *Lefschetz fixed point theorem.*

Also noteworthy is a theorem of Deuring's:‡ *for a correspondence* $A \sim 0$ (*that is, generated by a principal ideal*) *we have* $dw\,A = 0$ *for all first kind differentials* dw. As Deuring points out, this theorem has a close relationship to the Abel theorem. It no longer holds true for $h \neq 1$. Thus, a completely different grouping of correspondences into classes is found by calling them equivalent if they have the same effect upon first kind differentials dw^h with $h \neq 1$.

Exercise. Prove Deuring's theorem as follows. Use Green's function $d\mathfrak{G} = dd'^0\mathfrak{G}$ and show that

$$(du', d\mathfrak{G}) = du$$

for all first kind differentials. The correspondence A applied to both sides yields

$$(du', d\mathfrak{G}\,A) = du\,A.$$

For A generated by a principal ideal (A), it must now be shown that

$$d\mathfrak{G}\,A = \mathrm{A}^{-1}\,d\mathrm{A}$$

(K' being held constant in differentiation). This implies

$$(du', \mathrm{A}^{-1}\,d\mathrm{A}) = 0.$$

§3. Modular Functions

1. The Modular Correspondences

The notation of IV, §4 will be used. According to the lemma of IV, §4,3, every residue class $M = \left(\begin{smallmatrix}\alpha & \beta \\ \gamma & \delta\end{smallmatrix}\right)$ of determinant 1 modulo N contains a matrix $\left(\begin{smallmatrix}a & b \\ c & d\end{smallmatrix}\right) \in \Gamma$. It can be uniquely determined by M as the coset $\Gamma(N)\left(\begin{smallmatrix}a & b \\ c & d\end{smallmatrix}\right)$. By

$$f(\tau)\,A_M = f\left(\begin{pmatrix}a & b \\ c & d\end{pmatrix}\tau\right) \tag{1}$$

it defines a correspondence A_M of $K_{\Gamma(N)}$ with itself. Moreover, A_M is an automorphism of $K_{\Gamma(N)}/K_\Gamma$, so that both degrees $g(A_M) = g'(A_M) = 1$. This method gives us all the elements of the Galois group of $K_{\Gamma(N)}/K_\Gamma$, which is then isomorphic to the factor group $\mathfrak{M}(N) = \Gamma/\Gamma(N)$.

For each residue class r prime to N we need a matrix

$$U_r \equiv \begin{pmatrix}r & 0 \\ 0 & r^{-1}\end{pmatrix} \bmod N, \qquad U_r \in \Gamma, \tag{2}$$

which we choose arbitrarily in the coset $\Gamma(N)U_r$.

---

† Loc. cit., p. 242, Eq. (29).

‡ M. Deuring, *Arithmetische Theorie der Korrespondenzen algebraischer Funktionenkörper II*, J. Reine Angew. Math. **183**, 25–36 (1940).

Throughout §3, n will be a natural number prime to the level N and

$$R_n = \begin{pmatrix} 1 & 0 \\ 0 & n \end{pmatrix}. \tag{3}$$

By §1,9 a prime correspondence

$$T_n = A_{R_n} \tag{4}$$

is associated with the substitution $R = R_n$; it is called the *modular correspondence* for n. In a wider sense, the notion of modular correspondence is extended to include all correspondences of the field $K_{\Gamma(N)}$ with itself composed of the prime correspondences of the type given in §1,9. These then also include the automorphisms A_M given above. An outstanding role in analytic number theory is played by the T_n; the presentation of the foundation of their theory which we will give now is due to Hecke.

Calculations with the T_n must first be simplified. Several arguments using elementary number theory are needed. Set $\mathfrak{G} = \Gamma(N)$ into §1,9. To represent T_n as §1,(28) [or to generate the associated ideal as §1,(27)], the M_ν of the decomposition §1,(26) need not be explicitly known, the cosets $\Gamma(N)R_nM_\nu$ sufficing. We now maintain

$$\Gamma(N)(\tau)T_n = \bigcup_{a,b,d} \Gamma(N)U_a^{-1}\begin{pmatrix} a & b \\ 0 & d \end{pmatrix}(\tau), \tag{5}$$

with a, b, d taking on all values satisfying

$$\begin{gathered} ad = n, \qquad a > 0, \qquad b \equiv 0 \bmod N \\ b \textit{ traversing a complete residue system} \bmod d. \end{gathered} \tag{6}$$

Thus the number of cosets in (5) is the sum of divisors

$$g'(T_n) = \sum_{d/n} d. \tag{7}$$

Proof. We have $\Gamma(N)(\tau)T_n = \bigcup \Gamma(N)R_nM_\nu(\tau)$ from §1, (28), the M_ν coming from $\Gamma(N) = \bigcup (\Gamma(N) \cap R_n^{-1}\Gamma(N)R_n)M_\nu$. The $B_\nu = R_nM_\nu$ thus all satisfy the congruences $B \equiv R_n \bmod N$ and $|B_\nu| = n$. Furthermore, no $V \in \Gamma(N)$ can exist for $\mu \neq \nu$ with $R_nM_\mu = VR_nM_\nu$. For such an equation would imply $R_nM_\mu M_\nu^{-1}R_n^{-1} \in \Gamma(N)$, contradicting the nature of the M_ν.

Now, for a $B = U_a^{-1}\left(\begin{smallmatrix} a & b \\ 0 & b \end{smallmatrix}\right)$ on the right in (5) we have $B \equiv R_n \bmod N$ and $|B| = n$. The quotient of any different two of these B cannot lie in $\Gamma(N)$. For, $U_a^{-1}\left(\begin{smallmatrix} a & b \\ 0 & d \end{smallmatrix}\right)\left(\begin{smallmatrix} a' & b' \\ 0 & d' \end{smallmatrix}\right)^{-1}U_{a'} \in \Gamma(N)$ would imply $\left(\begin{smallmatrix} a & b \\ 0 & d \end{smallmatrix}\right)\left(\begin{smallmatrix} a' & b' \\ 0 & d' \end{smallmatrix}\right)^{-1} \in \Gamma$ and then $ad = a'd' = n$, $a > 0$, $a' > 0$ would lead to $a = a'$, $d = d'$. $U_a\Gamma(N)U_a^{-1} = \Gamma(N)$ along with the restriction of b to $0 \leqq b < d$ would also give $b = b'$. Thus, the right side of (5) represents at least a part of the right side of §1,(28).

On the other hand, let $B = \left(\begin{smallmatrix} \alpha & \beta \\ \gamma & \delta \end{smallmatrix}\right)$ be given with $|B| = n$ and $B \equiv R_n \bmod N$. Then find an $\left(\begin{smallmatrix} \alpha' & \gamma' \\ \beta' & \delta' \end{smallmatrix}\right) \in \Gamma$ such that

$$\begin{pmatrix}\alpha' & \beta' \\ \gamma' & \delta'\end{pmatrix}\begin{pmatrix}\alpha & \beta \\ \gamma & \delta\end{pmatrix} = \begin{pmatrix}a & b \\ 0 & d\end{pmatrix},$$

that is, such that the product has the lower left coefficient $c = 0$. Let the g.c.d. $(\alpha, \gamma) = a > 0$. Set $\delta' = \alpha a^{-1}$ and $\gamma' = -\gamma a^{-1}$, then determine α'', β'' such that $\alpha''\delta' - \beta''\gamma' = 1$. Now setting $\alpha' = \alpha'' + t\gamma'$ and $\beta' = \beta'' + t\delta'$ gives a solution of the equation dependent upon the parameter t.

Our assumption that $\alpha \equiv 1$ and $\gamma \equiv 0 \bmod N$ assures that $\delta' \equiv a^{-1}$ and $\gamma' \equiv 0 \bmod N$. Choose t such that $\beta' \equiv 0 \bmod N$. Further, since $\alpha'\delta' - \beta'\gamma' = 1$ we have $\alpha' \equiv a \bmod N$. Thus

$$\begin{pmatrix}\alpha' & \beta' \\ \gamma' & \delta'\end{pmatrix} \equiv \begin{pmatrix}a & 0 \\ 0 & a^{-1}\end{pmatrix} \equiv U_a \bmod N.$$

This means that we can write $\left(\begin{smallmatrix}\alpha' & \beta' \\ \gamma' & \delta'\end{smallmatrix}\right) = U_\alpha \left(\begin{smallmatrix}\alpha'' & \beta'' \\ \gamma'' & \delta''\end{smallmatrix}\right)$ with $\left(\begin{smallmatrix}\alpha'' & \beta'' \\ \gamma'' & \delta''\end{smallmatrix}\right) \in \Gamma(N)$, and that

$$\begin{pmatrix}\alpha'' & \beta'' \\ \gamma'' & \delta''\end{pmatrix} B = U_a^{-1}\begin{pmatrix}a & b \\ 0 & d\end{pmatrix}.$$

By now choosing a suitable integer u and multiplying from the left by $U_a^{-1}\left(\begin{smallmatrix}1 & uN \\ 0 & 1\end{smallmatrix}\right)U_a$, necessarily contained in $\Gamma(N)$, we can have b match any given residue class system modulo d. This shows that the sums in §1,(28) can be placed in the form of (5), so that (5) is proved. ▼

Similar to (5) we also have

$$\Gamma(N)(\tau)T_n^* = \bigcup_{a,b,d} \Gamma(N)U_d\begin{pmatrix}a & b \\ 0 & d\end{pmatrix}(\tau), \tag{8}$$

with a, b, d taking on the same values as in (6).

Proof. By §1,(29) we see that

$$\Gamma(N)(\tau)T_n^* = \bigcup_{\nu} \Gamma(N)M_\nu'\bar{R}_n^{-1}(\tau),$$

where the M_ν' are such that

$$R_n^{-1}\Gamma(N)R_n = \bigcup_{\nu} (\Gamma(N) \cap R_n^{-1}\Gamma(N)R_n)M_\nu'.$$

This implies that

$$\Gamma(N) = \bigcup_{\nu} (\Gamma(N) \cap R_n\Gamma(N)R_n^{-1})R_nM_\nu'R_n^{-1}.$$

As the same substitutions of the τ-plane are given by R_n^{-1} and $\bar{R}_n = nR_n^{-1}$ we may write

$$\Gamma(N)(\tau)T_n{}^* = \bigcup_{\nu} \Gamma(N)\bar{R}_nR_nM_\nu'\bar{R}_n^{-1}(\tau).$$

The same argument as above now shows that the cosets on the right are represented by a system of integral matrices $B \equiv \bar{R}_n \bmod N$ and with $|B| = n$, which cannot be transformed into each other by left multiplication with an element of $\Gamma(N)$. Such a system can be put in the form $B = U_d\begin{pmatrix} a & b \\ 0 & d \end{pmatrix}$ with a, b, d again as in (6). ▼

In particular, (8) gives us

$$g'(T_n^*) = g(T_n) = g(T_n^*) = g'(T_n). \tag{9}$$

2. Products of Modular Correspondences

The rules

$$A_{U_n} A_{U_m} = A_{U_{nm}}, \qquad A_{U_n} T_m = T_m A_{U_n} \tag{10}$$

hold, as well as

$$T_{n_1} T_{n_2} = \sum_{t/n_1,n_2} t A_{U_t}^{-1} T_{n_1 n_2 t^{-2}}, \tag{11}$$

so that the T_n and the A_{U_n} generate a commutative ring.

Proof. The first of Eqs. (10) is obvious; for the rest we remember the convention at the end of §1,9. For example, the sum $\sum t\Gamma(N)(\tau)A_{U_t}^{-1}T_{n_1n_2t^{-2}}$ which will occur on the right in (11) is taken as the union of the $\Gamma(N)B(\tau)$ associated with the individual summands, these being given the proper multiplicity. Leaving out the variable τ simplifies the notation. The right side of the other equation of (10) becomes

$$T_m A_{U_n} = \bigcup \Gamma(N) U_a^{-1} \begin{pmatrix} a & b \\ 0 & d \end{pmatrix} U_n, \qquad ad = m, \quad \text{etc.}$$

The matrices

$$B_1 = U_n^{-1} U_a^{-1} \begin{pmatrix} a & b \\ 0 & d \end{pmatrix} U_n$$

again traverse a full system with the properties: $B \equiv R_n \bmod N$, $|B_1| = n$. such that no two differ only by left multiplication by an element in $\Gamma(N)$. Thus

$$T_m A_{U_n} = \bigcup \Gamma(N) U_n U_a^{-1} \begin{pmatrix} a & b \\ 0 & d \end{pmatrix} = U_n \bigcup \Gamma(N) U_a^{-1} \begin{pmatrix} a & b \\ 0 & d \end{pmatrix},$$

which is exactly the assertion of the second equation (10).

The same argument places the left side of (11) in the form

$$T_{n_1} T_{n_2} = \bigcup \Gamma(N) U_{a_2}^{-1} U_{a_1}^{-1} \begin{pmatrix} a_1 & b_1 \\ 0 & d_1 \end{pmatrix} \begin{pmatrix} a_2 & b_2 \\ 0 & d_2 \end{pmatrix},$$

the $a_1, \ldots,$ and $a_2, \ldots,$ traversing solutions of (6) for n_1 and n_2. Ce $\Gamma(N) U_{a_2}^{-1} U_{a_1}^{-1} = \Gamma(N) U_{a_1 a_2}^{-1}$, so that

$$T_{n_1}T_{n_2} = \bigcup \Gamma(N)U_{a_1a_2}^{-1}\begin{pmatrix} a_1 & b_1 \\ 0 & d_1 \end{pmatrix}\begin{pmatrix} a_2 & b_2 \\ 0 & d_2 \end{pmatrix}. \tag{12}$$

If $(n_1, n_2) = 1$, then the product on the right,

$$\begin{pmatrix} a_1 & b_1 \\ 0 & d_1 \end{pmatrix}\begin{pmatrix} a_2 & b_2 \\ 0 & d_2 \end{pmatrix} = \begin{pmatrix} a & b \\ 0 & d \end{pmatrix},$$

traverses a system of solutions of (6) for $n = n_1 n_2$. Thus (11) is proved for this case.

We have reduced (11) to the special case of powers of primes, which remains to be proved:

$$T_{p^r}T_{p^s} = \sum_{\sigma=0}^{\min(r,s)} p^\sigma A_{U_{p^\sigma}}^{-1} T_{p^{r+s-2\sigma}}. \tag{13}$$

First show, more special yet, that

$$T_{p^r}T_p = T_{p^{r+1}} + pA_{U_p}^{-1}T_{p^{r-1}}. \tag{14}$$

According to (12) the product on the left is found by multiplying the matrices

$$\begin{pmatrix} a_1 & b_1 \\ 0 & d_1 \end{pmatrix} = \begin{pmatrix} p^u & b_1 \\ 0 & p^v \end{pmatrix} \quad \text{with} \quad u + v = r,$$

$$\text{by} \quad \begin{pmatrix} a_2 & b_2 \\ 0 & d_2 \end{pmatrix} = \begin{pmatrix} p & 0 \\ 0 & 1 \end{pmatrix} \quad \text{and} \quad \begin{pmatrix} 1 & b_2 \\ 0 & p \end{pmatrix},$$

where b_1 traverses a residue class system mod p^v, and b_2 a residue class system mod p. Thus the matrices

$$U_{p^{u+1}}^{-1}\begin{pmatrix} p^{u+1} & b_1 \\ 0 & p^v \end{pmatrix}, \qquad \begin{pmatrix} 1 & pb_1 + b_2 \\ 0 & p^{r+1} \end{pmatrix},$$

$$U_{p^u}^{-1}\begin{pmatrix} p^u & pb_1 + p^u b_2 \\ 0 & p^{v+1} \end{pmatrix} = pU_{p^u}^{-1}\begin{pmatrix} p^{u-1} & b_1 + p^{u-1}b_2 \\ 0 & p^v \end{pmatrix}$$

(where $u > 0$ has been assumed in the latter) occur, each of them once. Those of the first two kinds yield precisely $T^{p^{r+1}}$. The matrices of the third type may be divided by p, this factor being without meaning in their application as substitutions in the τ-plane. Note that

$$\Gamma(N)U_{p^u}^{-1} = U_p^{-1}\Gamma(N)U_{p^{u-1}}^{-1},$$

showing the contribution of that type matrix to be

$$\bigcup \Gamma(N)U_{p^u}^{-1}\begin{pmatrix} p^{u-1} & b_1 + p^{u-1}b_2 \\ 0 & p^v \end{pmatrix} = U_p^{-1}\bigcup \Gamma(N)U_{p^{u-1}}^{-1}\begin{pmatrix} p^{u-1} & b_1 + p^{u-1}b_2 \\ 0 & p^v \end{pmatrix}.$$

The $b_1 + p^{u-1}b_2$ each traverse the residue classes mod p^v exactly p times. Thus (14) holds.

It is now seen that the T_{p^r} $(r = 1, 2, \ldots)$ can be written as polynomials in T_p, meaning that they commute.

Equation (13) is finally demonstrated by induction on the number s. No generality is lost in assuming $s \leqq r$. The correctness of (13) for some s implies, by multiplication with T_p and application of (14), that

$$T_{p^r} T_{p^s} T_p = T_{p^r}(T_{p^{s+1}} + pA_{U_p}^{-1} T_{p^{s-1}})$$
$$= \sum_{\sigma=0}^{s} (p^\sigma A_{U_p\sigma}^{-1} T_{p^{r+s+1-2\sigma}} + p^{\sigma+1} A_{U_{p}\sigma+1}^{-1} T_{p^{r+s-1-2\sigma}})$$

and then an easy calculation gives

$$T_{p^r} T_{p^{s+1}} = \sum_{\sigma=0}^{s} p^\sigma A_{U_p\sigma}^{-1} T_{p^{r+s+1-2\sigma}} + p^{s+1} A_{U_p s+1}^{-1} T_{p^{r+s-1-2s}},$$

the last summand being omitted in both equations if $s = r$. This, however, is the desired equation (14) with s replaced by $s + 1$. ▼

3. Representations of Modular Correspondences by Differentials

The applications of modular correspondences in analytic number theory depend upon their representation §2,(12) by first kind differentials.

For the case of modular functions of level $N > 1$ the A_M and T_n are correspondences of bounded ramification in the sense of §2,1, *the exceptional set permitted there being the prime divisors $\mathfrak{s}_\nu$ associated with the cusps.*

Proof. This is clear for the A_M, as they are automorphisms. The T_n, and as a matter of fact all correspondences of type §1,9, map the neighborhood of a point $\tau = \tau_0$ onto neighborhoods of a finite number of points by

$$\tau \to \frac{a_\nu \tau + b_\nu}{c_\nu \tau + d_\nu} = \frac{a_\nu \tau_0 + b_\nu}{c_\nu \tau_0 + d_\nu} + \frac{a_\nu d_\nu - b_\nu c_\nu}{(c_\nu \tau_0 + d_\nu)^2} (\tau - \tau_0) + \cdots$$

according to §1,(28). Then we know from IV, §4,2 that, under our hypothesis, $N > 1$, $\tau - \tau_0$ uniformizes locally whenever τ_0 lies in the interior of the upper half-plane. This means that the mappings are locally single valued everywhere, ramification points being possible only at the cusps. The A_M and the T_n map each cusp into (generally several) cusps. The requirements for regular ramification are automatically satisfied for characteristic 0. ▼

The divisor $\mathfrak{a}$ used in the definition of differentials of degree h of the first kind is now set

$$\mathfrak{a} = (\mathfrak{s}_1 \cdots \mathfrak{s}_{\sigma_N})^{h-1}. \tag{15}$$

Then for $N > 1$, according to IV, §4,6, every differential of the first kind with respect to $\mathfrak{a}$ can be represented as $du^h = u(\tau)\, d\tau^h$ with a cusp form $u(\tau)$ of dimension $-2h$, while every such cusp form leads to a first kind differential.

For $N = 1$, though, we must replace our divisor by

$$\mathfrak{a} = \mathfrak{s}^{h-1} \mathfrak{q}_{\Gamma,\zeta}^{h_2} \mathfrak{q}_{\Gamma,i}^{h_3}, \qquad h_2 = [2h/3], \quad h_3 = [h/2], \tag{16}$$

to have cusp forms and differentials of first kind with respect to $\mathfrak{a}$ coincide. Furthermore, the T_n are no longer of bounded ramification, so that the trace formula of §2,3 and 4 perhaps no longer holds for $N = 1$. But this case need not be excluded, as long as the trace formula is not needed.

Now let a prime correspondence A_R of the type §1,(28) be given. By §2,1 it operates on a differential $u(\tau)\, d\tau^h$ as

$$u(\tau)\, d\tau^h\, A_R = \sum_\nu u\left(\frac{a_\nu\tau + b_\nu}{c_\nu\tau + d_\nu}\right)\left(\frac{a_\nu d_\nu - b_\nu c_\nu}{(c_\nu\tau + d_\nu)^2}\right)^h d\tau^h = \sum_\nu u(\tau)[RM_\nu]^{-2h}\, d\tau^h,$$

where $\begin{pmatrix} a_\nu & b_\nu \\ c_\nu & d_\nu \end{pmatrix} = RM_\nu$. Note that now in the operator $[RM_\nu]^{-h}$ we also have the determinant $a_\nu d_\nu - b_\nu c_\nu$, which would seem to contradict IV, §4,(14); but there that determinant is 1. As our sole interest is in modular forms, we may write this as

$$u(\tau)A_R = \sum_\nu u(\tau)[RM_\nu]^{-2h} = |R|^h \sum_\nu u\left(\frac{a_\nu\tau + b_\nu}{c_\nu\tau + d_\nu}\right)(c_\nu\tau + d_\nu)^{-2h}. \qquad (17)$$

It is clear that this sum depends upon the cosets $\Gamma(N)RM_\nu$ alone.

If $u(\tau)$ is an integral modular form, or even a cusp form, then so is $u(\tau)A_R$.

Our proof does not need application of the theorem in §2,1, which would not apply for $N = 1$. Instead we argue as follows. To any $G \in \Gamma(N)$ and index ν there exists a $G' \in \Gamma(N)$ and μ, such that $RM_\nu G = G'RM_\mu$, and if G is held fixed μ takes on all possible values along with ν. Noting IV, §4,(15) we find that

$$(u(\tau)A_R)[G]^{-2h} = \sum u(\tau)[G']^{-2h}[RM_\mu]^{-2h} = u(\tau)A_R.$$

That this modular form is integral is obvious. Furthermore, A_R maps cusps into (generally several) cusps. Thus cusp forms map into cusp forms. ▼

The theorem just proved can naturally be carried over to composed correspondences $A = \sum m_\varrho A_{R_\varrho}$.

4. The Petersson Metric

Let two integral modular forms, $u(\tau)$ and $v(\tau)$, be given, one of which, say $u(\tau)$, is a cusp form. The integral

$$\{u(\tau), v(\tau)\} = \iint_{\mathfrak{F}} u(\tau)\overline{v(\tau)}y^{2h-2}\, dx\, dy \qquad (\tau = x + iy) \qquad (18)$$

taken over a fundamental domain $\mathfrak{F}$ of $\Gamma(N)$ is called the *Petersson scalar product* of $u(\tau)$ and $v(\tau)$. Its value is finite and independent of the choice of fundamental domain $\mathfrak{F}$.

Proof. By the hypothesis and IV, §4,(16) the power series expansions at the cusp $i\infty$ are

$$u(\tau) = \sum_{n=1}^{\infty} c_n \exp[2\pi i N^{-1} n\tau], \qquad v(\tau) = \sum_{n=0}^{\infty} c_n' \exp[2\pi i N^{-1} n\tau],$$

and it is obvious that (18) converges at this cusp. The other cusps are transforms of $i\infty$ under $\tau \to \tau' = \left(\begin{smallmatrix} a & b \\ c & d \end{smallmatrix}\right)\tau$. We can now apply the easily verifiable identity

$$(ad - bc)^{2h} \frac{y^{2h-2}\, dx\, dy}{|c\tau + d|^{4h}} = y'^{2h-2}\, dx'\, dy', \tag{19}$$

which shows that the integrand of (18) is invariant under the substitution $\tau \to \tau'$. A suitable $\left(\begin{smallmatrix} a & b \\ c & d \end{smallmatrix}\right) \in \Gamma$ transforms this integral (18) in the neighborhood of such a cusp into an integral in the neighborhood of $i\infty$. Thus it converges here as well.

Any arbitrary fundamental domain is arrived at from any given one by cutting off some part of it, $\mathfrak{B}$, and supplementing with an equivalent part $\mathfrak{B}' = \left(\begin{smallmatrix} a & b \\ c & d \end{smallmatrix}\right)\mathfrak{B}$. The equality of the integrals over $\mathfrak{B}$ and $\mathfrak{B}'$ is seen as above. ▼

The scalar product (18) *clearly defines a definite hermitian metric in the space of cusp form, called the Petersson metric.* As an easy corollary we have: there exist bases $u_i(\tau)$ of the cusp forms such that

$$\{u_i(\tau), v_j(\tau)\} = \begin{cases} 1 & \text{for} \quad i = j \\ 0 & \text{for} \quad i \neq j. \end{cases} \tag{20}$$

Another formula generalizing §2,(11) is also due to Petersson; it states that

$$\{u(\tau)A, v(\tau)\} = \{u(\tau), v(\tau)A^*\}, \tag{21}$$

where $A = \sum m_\varrho A_{R_\varrho}$ is a correspondence composed of prime correspondences A_{R_ϱ} of type §1,9, and A^* is its Rosati conjugate.

The proof need only be given for prime correspondences $A = A_R$. We introduce the abbreviations

$$M_\nu = \begin{pmatrix} a_\nu & b_\nu \\ c_\nu & d_\nu \end{pmatrix}, \quad R = \begin{pmatrix} a & b \\ c & d \end{pmatrix}, \quad u(\tau)[R]^{-2h} = u\left(\frac{a\tau + b}{c\tau + d}\right) \frac{(ad - bc)^h}{(c\tau + d)^{2h}} = u_R(\tau)$$

to write the left side as

$$\{u(\tau)A_R, v(\tau)\} = \sum_\nu \iint_{\mathfrak{F}} u_R\left(\frac{a_\nu\tau + b_\nu}{c_\nu\tau + d_\nu}\right) \overline{v(\tau)} \frac{y^{2h-2}\, dx\, dy}{(c_\nu\tau + d_\nu)^{2h}}.$$

By IV, §4,(3) then,

$$\overline{\mathfrak{F}} = \sum_\nu M_\nu \mathfrak{F}$$

is a fundamental domain of $\mathfrak{G} = \Gamma(N) \cap R^{-1}\Gamma(N)R$, as the M_ν can be interpreted as in §1,(26). Because $v(\tau)[M_\nu]^{-2h} = v(\tau)$, the integral above is equal to

$$\sum_{\nu} \iint_{\mathfrak{F}} u_R(M_\nu(\tau))\overline{v(M_\nu(\tau))} \frac{y^{2h-2}\,dx\,dy}{|c_\nu\tau + d_\nu|^{4h}}$$

which in turn, because of (19), is equal to

$$\iint_{\overline{\mathfrak{F}}} u_R(\tau)\overline{v(\tau)}y^{2h-2}\,dx\,dy = |R|^h \iint_{\mathfrak{F}} u\left(\frac{a\tau + b}{c\tau + d}\right)\overline{v(\tau)} \frac{y^{2h-2}\,dx\,dy}{(c\tau + d)^{2h}}.$$

The substitution $R\tau = \tau'$ or $\tau = \overline{R}\tau' = (d\tau' - b)/(-c\tau' + a)$ is now carried out. Renewed application of (19) and insertion of $(c\tau + d)(-c\tau' + a) = |R|$ yields

$$\{u(\tau)A_R, v(\tau)\} = |R|^h \iint_{R\overline{\mathfrak{F}}} u(\tau')\overline{v\left(\frac{d\tau' - b}{-c\tau' + a}\right)} \frac{y'^{2h-2}\,dx'\,dy'}{(-c\bar{\tau}' + a)^{2h}}$$

$$= \iint_{R\overline{\mathfrak{F}}} u(\tau)\overline{v_{\overline{R}}(\tau)}y^{2h-2}\,dx\,dy,$$

where we have set

$$v_{\overline{R}}(\tau) = v(\tau)\begin{bmatrix} d & -b \\ -c & a \end{bmatrix}^{-2h}.$$

The domain $R\overline{\mathfrak{F}}$ of integration now is a fundamental domain of the group $R(\Gamma(N) \cap R^{-1}\Gamma(N)R)R^{-1} = \Gamma(N) \cap \overline{R}^{-1}\Gamma(N)\overline{R}$. By noting the relation to §1,(28) and (29), the last integral can be put in the form of the right side of (21) in the same way. ▼

By using an orthonormal basis $u_i(\tau)$ of the module of cusp forms, (21) leads to the following result. If

$$u_i(\tau)A = \sum_j a_{ij}u_j(\tau), \qquad u_i(\tau)A^* = \sum_j a_{ij}^*u_j(\tau)$$

are representations by matrices (a_{ij}) and (a_{ij}^*), then these are hermitian adjoints: *Rosati-adjoint correspondences are represented by hermitian-adjoint matrices.* Indeed,

$$\{u_i(\tau)A, u_l(\tau)\} = \sum_j a_{ij}\{u_j(\tau), u_l(\tau)\} = \{u_i(\tau), u_l(\tau)A^*\} = \sum_j \{u_i(\tau), u_j(\tau)\}\bar{a}_{lj}^*$$

so that

$$a_{ij} = \bar{a}_{ji}^*.$$

By now applying the theory of hermitian and unitary matrices we have the following theorems.

A representation of a ring of commuting Rosati-self-adjoint correspondences in the space of cusp forms can be simultaneously put in principal axis form.

A representation of an abelian group of automorphisms of $K_{\Gamma(N)}$ can be simultaneously put into principal axis form.

As for the last theorem, it suffices to know that any automorphism A taken as a correspondence of $K_{\Gamma(N)}$ with itself satisfies the equation $A^{-1} = A^*$. The matrix representing A over a basis (20) is then unitary.

5. Fourier Expansions of Modular Forms

A special abelian group of automorphisms of $K_{\Gamma(N)}$ is formed by the A_{U_n}. In accordance with the last theorem let $u_i(\tau)$ be a basis of the module of cusp forms for which the A_{U_n} are represented by diagonal matrices,

$$u_i(\tau)A_{U_n}^{-1} = \chi_i(n)u_i(\tau).$$

The $\chi_i(n)$ depend upon the residue class of $n \bmod N$ only, and are thus characters of the multiplicative group of residue classes modulo N prime to N. Such a basis will be said to be *normed*. More generally, any modular form $u(\tau)$ will be called *normed* if it satisfies the equations

$$u(\tau)A_{U_n}^{-1} = \chi(n)u(\tau) \tag{22}$$

with any character $\chi(n)$ of this group. Rewording the above, the module of cusp forms is spanned by normed cusp forms.

Particular normed integral modular forms of dimension $-2h$ were found in the appendix to Chapter I, §2,3; these are the theta series for integral definite quadratic forms $F(x)$ of level N in $4h$ variables, the character being the Legendre symbol

$$\chi(n) = (|F|/n).$$

The second of Eqs. (10) testifies that the T_n map normed modular forms of character $\chi(n)$ into each other. Their action on modular forms, in view of (17) and (5), is

$$u(\tau)T_n = \sum_{a,b,d} \chi(a)\left(\frac{a}{d}\right)^h u\left(\frac{a\tau + b}{d}\right), \tag{23}$$

where a, b, d are taken as in (6).

From now on let

$$u_i(\tau) = \sum_{m=1}^{\infty} c_{i,m} \exp[2\pi i N^{-1} m\tau]$$

be a basis of all normed cusp forms of character $\chi(n)$ and dimension $-2h$. Then

$$u_i(\tau)T_n = \sum_j t_{ij}(n)u_j(\tau) \tag{24}$$

is a representation of the T_n by matrices $(t_{ij}(n))$. More precisely, these matrices generate a ring depending on h and χ, and this ring is a homomorphic image of the ring of correspondences of $K_{\Gamma(N)}$ generated by the T_n. The implications of (11) and (22) apply to the $(t_{ij}(n))$:

$$(t_{ij}(n_1))(t_{ij}(n_2)) = \sum_{t/n_1,n_2} t\,\chi(t)(t_{ij}(n_1 n_2 t^{-2})).$$

We must study the system of Fourier coefficients $c_{i,m}$ and its behavior under the transformations. By (23) we have

$$u_i(\tau)T_n = \sum_{m=1}^{\infty} \sum_{a,b,d} c_{i,m}\chi(a)\frac{a^h}{d^{h-1}}\frac{1}{d}\exp[2\pi i N^{-1} a\, d^{-1} m\, \tau + 2\pi i b N^{-1}\, d^{-1} m].$$

First, sum over b, the result vanishing whenever m is not divisible by d. A simple calculation leads to

$$u_i(\tau)T_n = \sum_{m=1}^{\infty} \sum_{a/n} c_{i,mna^{-1}}\chi(a)n^{1-h}a^{2h-1}\exp[2\pi i N^{-1} m a \tau]. \tag{25}$$

After the left side has been replaced by the right side of (24), the coefficients of some subscript m prime to n are compared. Only the summands with $a = 1$ are needed, so that

$$c_{i,mn} = n^{h-1}\sum_j t_{ij}(n)c_{j,m}, \qquad (m, n) = 1. \tag{26}$$

This system of formulas was given the following surprising formulation by Hecke.

With the Fourier series of the normed cusp forms $u_i(\tau)$ associate formally† *the Dirichlet series*

$$\zeta_i(s) = \sum_{m=1}^{\infty} c_{i,m}m^{-s}, \qquad \zeta_{i0}(s) = \sum_{m/N\cdots} c_{i,m}m^{-s}, \tag{27}$$

where, in the latter, only numbers composed of prime factors of N, the level, are taken for m. Consider also the formal infinite matrix product

$$Z(s) = \prod_{p\nmid N}(E - (t_{ij}(p))p^{h-1-s} + E\chi(p)p^{2h-1-2s})^{-1} \tag{28}$$

taken over all prime numbers p not dividing N; E is the unit matrix. Then,

$$(\zeta_i(s)) = Z(s)(\zeta_{i0}(s)), \tag{29}$$

where $(\xi_i(s))$ *and* $(\zeta_{i0}(s))$ *are column matrices with the components indicated. The product* (28) *is independent of the order of the factors.*

For level $N = 1$ the $\zeta_{i0}(s)$ consist of the numbers $c_{i,1}$ only, so that (29) reduces the $\zeta_i(s)$ and thus the series expansion of the $u_i(\tau)$ entirely to knowledge of the representation $(t_{ij}(n))$ of T_n.

Proof. The commutability of the factors of (28) follows from the commutability of the T_n and thus of the matrices $(t_{ij}(n))$. The individual factors are the infinite series

† It is not difficult to arrive at these $\zeta_i(s)$ from the $u_i(\tau)$ using an integral transformation. But this would have no advantage here, where we are concerned only with formal properties of the Fourier coefficients $c_{i,m}$.

$$(E - (t_{ij}(p))p^{h-1-s} + E\chi(p)p^{2h-1-2s})^{-1} = \sum_{r=0}^{\infty} (t_{ij}(p^r))p^{r(h-1-s)}. \tag{30}$$

Equation (30) is verified if we can show

$$\begin{aligned}\sum_{r=0}^{\infty} ((t_{ij}(p^r))p^{r(h-1-s)} &- (t_{ij}(p^r))(t_{ij}(p))p^{(r+1)(h-1-s)} \\ &+ (t_{ij}(p^r))\chi(p)p^{(r+2)(h-1-s)+1}) \\ = \sum_{r=0}^{\infty} ((t_{ij}(p^r))p^{r(h-1-s)} &- (t_{ij}(p^r))(t_{ij}(p))p^{(r+1)(h-1-s)}) \\ &+ \chi(p)\sum_{r=1}^{\infty} (t_{ij}(p^{r-1}))p^{(r+1)(h-1-s)+1} = E,\end{aligned}$$

which, indeed, we can by using (13).

Now, because of (11) the product of the series (30) is

$$Z(s) = \sum_{(n,N)=1} (t_{ij}(n))n^{h-1-s},$$

so that (29) is a consequence of (26). ▼

Conversely, knowledge of the Fourier coefficients $c_{i,m}$ brings with it that of the coefficients $t_{ij}(n)$.

The proof is accomplished by first applying to Eq. (26) the relations for the $(t_{ij}(n))$ corresponding to (11), finding

$$n'^{h-1} \sum_{j} t_{ij}(n')c_{j,mn} = \sum_{t/n,n'} \chi(t)t^{2h-1}c_{i,mnn't^{-2}}, \tag{31}$$

where n and n' are held prime to N and m is composed of prime divisors of N. The linear independence of the $u_i(\tau)$ assures that the rank of the matrices $(c_{i,l})$, $l = 1, 2, \ldots$, formed by their coefficients coincides with the dimension d of the module of the $u_i(\tau)$. Thus d values of l can be found such that the submatrix $(c_{i,l})$ formed with them is nonsingular. These l are then decomposed to $l = mn$, where m is composed of the prime divisors of N and n of the others. Now (31) can be solved for the $n'^{h-1}t_{ij}(n')$, which are thus linear forms in the $c_{i,mnn't^{-2}}$ with coefficients independent of n'.

6. Ramanujan's Conjecture

From (5) and (8) we have

$$T_n^* = A_{U_n}T_n. \tag{32}$$

In the space of normed modular forms for a real character $\chi(n)$ this implies $T_n^* = \chi(n)^{-1}T_n = \pm T_n$. The theorem of §3,4 shows that the module of these cusp forms can be transformed so that the $(t_{ij}(n))$ are all diagonal matrices. The eigenvalues $\tau_i(n)$ of these $(t_{ij}(n))$ are real or imaginary numbers depending upon whether $\chi(n) = 1$ or -1. This means that the matrix (28) can be transformed to the diagonal matrix with axis elements

$$\zeta_i(s) = \prod_{p \nmid n} (1 - \tau_i(p)p^{h-1-s} + \chi(p)p^{2h-1-2s})^{-1}. \tag{33}$$

We are particularly interested in the decomposition

$$\begin{aligned} 1 - \tau_i(p)p^{h-1-s} &+ \chi(p)p^{2h-1-2s} \\ &= (1 - \omega_i(p)p^{h-1-s})(1 - [\chi(p)p/\omega_i(p)]p^{h-1-s}) \end{aligned} \tag{34}$$

of the individual factors of (33). For the convergence abscissa of the infinite products (33) and (28) depends upon the rate of growth of the roots $\omega_i(p)$.

The easiest example is where $N = 1$, $h = 6$ (so that $k = 12$), $\chi(n) = 1$. The rank of the module of associated modular forms is one, and the modular function is

$$\Delta(\tau) = (2\pi)^{12} e^{2\pi i \tau} \prod_{n=1}^{\infty} (1 - e^{2\pi i n \tau})^{24} = (2\pi)^{12} \sum_{n=1}^{\infty} \tau(n) e^{2\pi i n \tau},$$

called the *discriminant*, and coincides with the discriminant $g_2{}^3 - 27g_3{}^2$ of the Weierstrass normal form of the elliptic functions of base period, $\omega_a = 1$, $\omega_b = \tau$ [(cf. IV, §2,(21)]. On the basis of extensive numerical computation Ramanujan conjectured that the associated Dirichlet series

$$\zeta(s) = \sum_{n=1}^{\infty} \tau(n) n^{-s}$$

has the product development (28), that is, that

$$\zeta(s) = \prod_{p} (1 - \tau(p)p^{5-s} + p^{11-2s})^{-1},$$

which was first proved by Mordell. Ramanujan also conjectured that the root $\omega(p)$ of the associated equation (34) are complex numbers of absolute value $|\omega(p)| = (p)^{1/2}$. But, despite much effort, this conjecture has not been verified to date. For the history and rich literature regarding this problem, see the references below, especially Van der Blij [1].

Ramanujan's conjecture has been extended to the general case: $|\omega_i(p)| = (p)^{1/2}$. The first result along this path is due to Petersson [8], while the deepest penetrations were made by Igusa (III, §6,7,[3]) and Shimura [9]. These papers all consider the case $h = 1$ ($k = 2$) only. A very special result of Hecke's falls into this connection, though, and holds for arbitrary h, [6]. A discussion of it would lead us too far from our subject.

The form of (34) makes it clear, incidentally, that the generalized Ramanujan conjecture represents a best possible approximation of the convergence abscissa of the product (28).

Along with an approximation of the convergence abscissa of (28) comes, because of (29), an approximation of the rate of growth of the Fourier coefficients $c_{i,m}$ of the cusp forms. In particular, assuming the conjecture to be correct, we have

$$|c_{i,m}| = O(m^{h-\frac{1}{2}+\varepsilon})$$

for every positive ε, this also implying a corresponding improvement of the approximation of IV, §4,(29).

7. Results for Modular Forms of Odd Dimension; Notes

First let k be some odd natural number. The theory developed in §3,3–5 can be carried over to all modular forms satisfying the functional equations

$$u(\tau)\begin{bmatrix} a & b \\ c & d \end{bmatrix}^{-k} = \chi\begin{pmatrix} a & b \\ c & d \end{pmatrix} u(\tau) \qquad \text{with} \quad \chi\begin{pmatrix} a & b \\ c & d \end{pmatrix} = \pm 1,$$

depending upon whether $d \equiv \pm 1 \bmod N$ for all $\left(\begin{smallmatrix} a & b \\ c & d \end{smallmatrix}\right) \in \Gamma(N)$.† Then for any residue class r prime to N and any $G \in \Gamma(N)$ we have

$$u(\tau)[U_r]^{-k}[G]^{-k} = u(\tau)[U_r G U_r^{-1}]^{-k}[U_r]^{-k} = \chi(G)u(\tau)[U_r]^{-k}$$
$$= u(\tau)[G]^{-k}[U_r]^{-k}.$$

Thus if $u(\tau)$ is such a function, then so is $u(\tau)[U_r]^{-k}$, and we have a representation of the multiplicative group of residue classes modulo N in the space of modular forms. The commutativity of this group assures the reducibility of the representation to irreducible components of the first degree. Thus every modular form can be written as a sum of eigenfunctions of the $[U_r]^{-k}$, defined by

$$u(\tau)A_{U_r}^{-1} = u(\tau)[U_r^{-1}]^{-k} = \chi(r)u(\tau),$$

where $\chi(r)$ is a character of the multiplicative residue class group modulo N.

As suggested by (23), we let T_n operate on such modular forms by defining

$$u(\tau)T_n = \sum_{a,b,d} \chi(a)\left(\frac{a}{d}\right)^{k/2} u\left(\frac{a\tau + b}{d}\right), \tag{35}$$

the square root on the right being taken positive. The arguments of §3,1 again show this operation to preserve the property of being a modular form of character $\chi(r)$, so that the considerations of §3,5 hold literally.

This is the path followed by Hecke in his two papers, I and II [5], even and odd k being treated uniformly. These investigations contain our principal theorem of §3,5 and go considerably farther. Hecke annexes the factor

---

† $N > 2$ must be assumed. For $N = 1$ we also know from IV, §4,6 that no integral modular forms exist with character $\chi \neq 1$.

$n^{k/2-1}$ on the right in (35) eliminating the root there and thus also arriving at a simplified final formula (28). But such a factor sacrifices the representation of the ring of correspondences generated by the T_n.

The possibility of such a representation for odd k, however, is an open question anyway. Indeed, it would have to be shown that a relation $\sum m_{\varrho\sigma} T_{n_\varrho} T_{n\sigma} = 0$ implies the corresponding relation for the representing matrices. Although (10) and (11) are still carried over correctly if $(a/d)^{k/2}$ is chosen positive, it could be that other relations hold outside of these.

The possibility of defining similar linear operators in the space of integral modular forms of nonintegral dimension to an arbitrary multiplier system $v(M)$ is considered by Wohlfahrdt [10].

With regard to the Petersson scalar product, cf. IV, §4,9,[5].

From now on drop the assumption of k being odd. In hopes of applying the principal theorem of §3,5 to problems of number theory, especially in the realm of quadratic forms, a connection must be sought between correspondences and the other Fourier coefficients $c_{i,m}$, composed of prime divisors of the level. Such a connection is found in Hecke [5], Part II. However, distinctions must be made in the treatment of the modular forms depending upon whether they are already invariant under certain extensions of the group $\Gamma(N)$ or not, that is, whether they belong to certain fields between K_Γ and $K_{\Gamma(N)}$. Hecke [5, 7] could decompose the space of modular forms into subspaces of characteristic behavior under the Galois group of $K_{\Gamma(N)}/K_\Gamma$. The representations of the T_n decompose simultaneously (cf. [4a]).

The applications of the theory to definite quadratic forms of a finite number of variables are compiled by Hecke in the monograph [7]. The general principle is to find all quadratic forms of a given number of variables $2k$, determinant $|F|$ and level N, and to form their theta series. The associated Eisenstein series and cusp forms of character

$$\chi\begin{pmatrix} a & b \\ c & d \end{pmatrix} = (\text{sign } d)^k \left(\frac{(-1)^k |F|}{d} \right)$$

belong to the modular forms of character $\chi\left(\begin{smallmatrix} a & b \\ c & d \end{smallmatrix}\right) = \eta^k$ for $d \equiv \eta = \pm 1 \bmod N$. As soon as it is known that the theta series generate the same function module as the Eisenstein series and cusp forms, it is also known that all statements concerning the latter apply to quadratic forms. This premise may easily be checked in special cases. It was proved for the principal character $\chi\left(\begin{smallmatrix} a & b \\ c & d \end{smallmatrix}\right) = 1$ and sufficiently large prime levels in [3], and, allowing certain limitations, also for arbitrary levels not divisible by perfect squares, in [4]. The proof depends, in part, on a modification of the trace formula of §2 which is valid for modular functions of the subgroup $\Gamma_0(N)$ of all $\left(\begin{smallmatrix} a & b \\ c & d \end{smallmatrix}\right) \in \Gamma$ with $c \equiv 0 \bmod N$ (cf. the author's paper cited on page 265). Despite the nonbounded ramification of modular correspondences of $\Gamma_0(N)$ the formula hardly differs from §2,(19). For the other part, an arithmetic analog of correspondences is constructed for quaternary quadratic forms, and its trace is computed as well. The assertion results from the fact that the traces coincide.

For the theory of an arithmetic analog of correspondences for quadratic forms in an arbitrary number of variables, see [2]. It is perhaps strange that such could be found only under the hypothesis that $\chi(n) = 1$. Thus the analytic theory yields more.

References

[1] F. VAN DER BLIJ, *The function* $\tau(n)$ of S. Ramanujan, Math. Student (New series) **18**, Nos. 3, 4 (1951).

[2] M. EICHLER, *Quadratische Formen und orthogonale Gruppen*, Chap. IV. Berlin, Göttingen, Heidelberg, 1952.

[3] M. EICHLER, *Über die Darstellbarkeit von Modulformen durch Thetareihen*, J. Reine Angew. Math. **195**, 156–171 (1956).

[4] M. EICHLER, *Quadratische Formen und Modulfunktionen*, Acta Arith. **4**, 217–239 (1958).

[4a] M. EICHLER, *Einige Anwendungen der Spurformel im Bereich der Modularkorrespondenzen*, Math. Ann. (1966). In press.

[5] E. HECKE, *Über Modulfunktionen und Dirichletsche Reihen mit Eulerscher Produktentwicklung*, Part *I*, Math. Ann. **114**, 1–28 (1937); Part *II*, Math. Ann. **114**, 316–351 (1937). *Math. Werke*, Nos. 35, 36, Göttingen, 1959.

[6] E. HECKE, *Über die Darstellung der Determinante einer positiven quadratischen Form durch die Form*, Vierteljschr. Naturforsch. Ges. Zürich **85**, 64–70 (1940). *Math. Werke* No. 40.

[7] E. HECKE, *Analytische Arithmetik der positiven quadratischen Formen*, Kgl. Danske Videnskab. Selskab, Math.-Fys. Medd. **17**, (1940). *Math. Werke*. No. 41.

[8] H. PETERSSON, *Über die Zerlegung der den ganzen Modulformen von höherer Stufe entsprechenden Dirichletreihen in vollständige Eulersche Produkte*, Acta Math. **80**, 191–221 (1948).

[9] G. SHIMURA, *Correspondences modulaires et les fonctions* ζ *de courbes algébriques*, J. Math. Soc. Japan **10**, 1–28 (1958).

[10] K. WOHLFAHRDT, *Über Operatoren Heckescher Art bei Modulformen reeller Dimension*, Math. Nachr. **16**, 233–256 (1957).

§4. Castelnuovo's Inequality

1. Introduction

In §4, K and K' will always be conservative and separably generated fields; this hypothesis is certainly satisfied whenever k_0 is perfect.

In §2 we showed that a separable prime correspondence A, not the unit correspondence, together with its Rosati-conjugate correspondence A, satisfies the equation

$$s_1(A + A^*) = g(A) + g'(A) - f(A), \tag{1}$$

where s_1 is the trace of the representation of the correspondence in the module of differentials of degree 1 and of first kind, and $f(A)$ is the number of fixed points as discussed in §2,5. By defining the number of fixed points of a composed correspondence by

$$f(\sum m_\varrho A_\varrho) = \sum m_\varrho f(A_\varrho),$$

Eq. (1) also is valid for such sums. Even the unit correspondence D may be included. Its trace is known to be $s_1(D + D^*) = 2g$, so that setting

$$f(D) = 2 - 2g$$

leads to the validity of (1).

We saw in §2,2 that in the classical case

$$s(A) = g(A) + g'(A) - f(A) \tag{2}$$

is the trace of a particular representation $H(A)$ of the ring of correspondences. $H(A)$ is the sum of the two representations $H_1(A)$, $H_2(A)$. Suitable choice of a basis of the representation module makes the matrices $H_i(A)$ of Rosati-adjoint correspondences hermitian adjoint:

$$H_i(A)^* = \overline{H_i(A)^t}, \qquad i = 1, 2,$$

Thus we have

$$H_i(AA^*) = H_i(A)\overline{H_i(A)^t}, \qquad i = 1, 2,$$

and the traces of these matrices are $\geqq 0$. This means that

$$s(AA^*) \geqq 0 \tag{3}$$

for any correspondence A, a fact called *Castelnuovo's inequality*.

Surprisingly this inequality (3) can be proved quite generally for arbitrary fields of functions. Note that, from their derivation, the right sides of (1) and (2) are rational integers. In a function field of prime characteristic p the left side of (1) is an element of k_0 so that then this equation requires that the residue class mod p be taken on the right. If the function $s(A)$ is defined as the right side of (2), though, it becomes a rational integer independently of the characteristic; then (3) makes sense. Two proofs, valid generally, will be supplied in §4 for (3). In this general case we also call $s(A)$ the trace of A, without necessarily requiring it to be the trace of a representation of the ring of correspondences. Nevertheless, we shall see that at least in certain special cases it is.

2. Reduction to the Classical Case

We will first use the already proved validity of inequality (3) in the classical case to prove its validity in the following more general case. The characteristic is taken to be 0, and the field K to be generated as $K = k_0(x, y)$ with $f(x, y) = 0$. Further, let $A_\nu(x, y; x', y')$ $(\nu = 1, 2, \ldots)$ be a system of generators of the $K \times K'$-ideal defining the correspondence A, these A_ν being rational functions of their variables. But a finite number of constants occur in $f(x, y)$ and the $A(x, y; x', y')$, and these, adjoined to the prime field Q, give a finitely generated field k_1 which can be embedded isomorphically into the field C of complex numbers. $K_1 = k_1(x, y)$ with $f(x, y) = 0$ now defines a function field over k_1 and $A_\nu(x, y; x', x')$ a correspondence A_1 between K_1 and K_1'. Computation of $A_1A_1^*$ is the same as that of AA^*, and the numbers $g(A_1A_1^*)$, $g'(A_1A_1^*)$, and $f(A_1A_1^*)$ coincide with $g(AA^*)$, $g'(AA^*)$, and $f(AA^*)$. As $k_1 \subset C$ we know (3) to hold with A_1 in place of A. Thus (3) must also hold for the situation originally at hand. Thus we have proved our inequality for all fields of characteristic 0.

The inequality of Castelnuovo holds even if K and K' are not isomorphic. Then AA^* is at least a correspondence between K and a field K'' isomorphic to K, and the trace (2) is well defined. If K and K' are not isomorphic there do exist isomorphic finite extensions K_1/K and K_1'/K'. The lemma of §1,3 gives $A^{1'}(A^*)^{1'} = [K_1' : K']AA^*$, so that K' may be replaced by K_1' in the proof of (3). If K and K'' are also replaced by K_1 and K_1'', then $g(AA^*)$ and $g'(AA^*)$ are multiplied by the factor $[K_1 : K]$. As will be shown [in Eq. (17)], the same is true for $f(AA^*)$. These substitutions make K, K', and K'' isomorphic, so that then (3) holds. These considerations, incidentally, do not depend upon the field characteristic.

The entire proof can also be made useful for a field K of prime characteristic. It must be assumed, though, that K arises from a field K_0 of the same genus but of characteristic 0 by reduction with respect to a prime ideal $\mathfrak{p}_0$ of the constant field of K_0, the ideal $\mathfrak{p}_0$ being regular in the sense of III, §6,2. We will not discuss the question of the existence of such a K_0 for arbitrary K. The configuration occurs naturally in certain applications (e.g., §5,5–7). Our notation will now depart from that of III, §6, in that fields of characteristic 0 and associated objects will carry subscript 0, which will be dropped after residue class formation modulo $\mathfrak{p}_0$.

In adapting (3) we shall use the fact, still to be demonstrated, that the number of fixed points $f(A)$ of a correspondence A is equal to the degree of a certain divisor $\mathfrak{f}(A)$ of K associated with A, called the *divisor of fixed points*. This proof could be based on §2, but it will be more convenient to relate it to later material of §4,5. Our point of departure is inequality (3) for characteristic 0, $f(AA^*)$ being taken as a divisor degree.

The argument is as follows. Given is a correspondence A between K and K'. K' can also be considered a reduction mod $\mathfrak{p}_0$ of some field K_0' of functions with characteristic 0, the regularity of $\mathfrak{p}_0$ in this second reduction not being required. III, §6,6 assures the existence of a finite extension K_{01}'/K_0' and of a correspondence $A_0^{1'}$ between K_0 and K_{01}' which, under reduction mod $\mathfrak{p}_0$, maps into that extension $A^{1'}$ of A associated with the extension K_1'/K'. (If $\mathfrak{p}_0$ splits in the extension K_{01}'/K_0' the reduction is taken modulo one of its prime divisors.) By the lemma in §1,3 we have

$$A^{1'}A^{*1'} = [K_1' : K']AA^*.$$

Here we form the trace (2) which is a linear combination of divisor degrees in

KK'' and K. But these are invariant under reduction mod $\mathfrak{p}_0$. This means that

$$s(A_0^{1'}A_0^{*1'}) = [K_1':K']s(AA^*),$$

which is $\geqq 0$, for the correspondence on the left involves fields of characteristic 0. Two points are left to be clarified in this argument: first, that the operations of multiplying correspondences and reduction mod $\mathfrak{p}_0$ are commutable, and second, that the operations of formation of a fixed point divisor, to be described in §4,5, and of reduction mod $\mathfrak{p}_0$ are commutable. The first case results from the definition of multiplication in §1,2 as a residue class formation of ideals with respect to ideals; the fixed point divisor will also be defined with such an operation. Thus both cases depend upon double residue formation, which is always commutative. That is, for a ring $\mathfrak{O}$ and $\mathfrak{O}$-ideals $\mathfrak{a}$, $\mathfrak{b}$ we have the isomorphism theorem

$$(\mathfrak{O}/\mathfrak{a})/(\mathfrak{b}/(\mathfrak{a} \cap \mathfrak{b})) \cong (\mathfrak{O}/\mathfrak{b})/(\mathfrak{a}/(\mathfrak{a} \cap \mathfrak{b})). \quad \blacktriangledown$$

Despite the simplicity of the idea behind this proof of Castelnuovo's inequality, troublesome details were necessary, not to mention the assumption that a function field of prime characteristic could be obtained from a field of characteristic 0 by reduction mod $\mathfrak{p}_0$. Among the details was the proof of the trace formula of §2. Our second proof (in §4,6–8) will turn out to be quicker, but also more sophisticated. It will be based from the start on our new interpretation of $f(A)$ as the degree of the divisor of fixed points, and it will be precisely this interpretation which will then lead to applications in number theory (§5).

3. Extension of the Notion of Correspondence

Let a prime correspondence P be given. All those $A \in KK'$ whose denominators are not divisible by the ideal $P_{K \times K'}$ form a discrete local ring $\mathfrak{O}_P$. In II, §5 we associated a prime divisor of KK' with such an $\mathfrak{O}_P$, and we will now identify P with that prime divisor. An arbitrary correspondence $A = \Sigma m_\varrho P_\varrho$ thus becomes a divisor of KK', where divisor multiplication is now written additively.

Now consider the complete group of divisors of KK', but with a slight change of definition from II, §5. There we proceeded from a rational subfield $k_0(x, x')$ of KK'. A prime divisor $\mathfrak{z}_p$ of $k_0(x, x')$ was associated with every prime polynomial p in x and x', and one more $\mathfrak{z}_\infty$ was added. In contrast we now introduce two "infinite" prime divisors $\mathfrak{z}_\infty$ and $\mathfrak{z}_{\infty'}$ whose local rings consist of those rational functions having nonpositive degrees in x and x', respectively. A prime polynomial $p(x, x')$ of degrees G and G' in x and x', respectively, is now paired with the principal divisor

$$(p(x, x')) = \mathfrak{z}_{p(x,x')}\, \mathfrak{z}_\infty^{-G}\, \mathfrak{z}_{\infty'}^{-G'}.$$

Divisors thus defined in $k_0(x, x')$ decompose to prime divisors in any finite extension and, in particular, in KK'.

This defines a group $\mathfrak{K}$ of divisors of KK'. It contains three distinguished subgroups: (a) the prime divisors of all $\mathfrak{z}_p$ with $p = p(x)$ dependent upon x alone, and with $p = \infty$, and their power products, it is the group $\mathfrak{k}$ of divisors of K; (b) the analogous group $\mathfrak{k}'$; (c) the group $\mathfrak{C}$ generated by prime divisors of all $\mathfrak{z}_p$ with $p = p(x, x')$ dependent upon both variables. Clearly $\mathfrak{K}$ is the direct sum of $\mathfrak{k}$, $\mathfrak{k}'$, and $\mathfrak{C}$. Any divisor $\mathfrak{A}$ can be decomposed uniquely as

$$\mathfrak{A} = \mathfrak{a} + \mathfrak{a}' + A, \qquad \mathfrak{a} \in \mathfrak{k}, \quad \mathfrak{a}' \in \mathfrak{k}', \quad A \in \mathfrak{C}. \tag{4}$$

We call $\mathfrak{a}$ and $\mathfrak{a}'$ the constant and A the nonconstant components of $\mathfrak{A}$. The group $\mathfrak{C}$ is isomorphic to the group of correspondences between K and K'; indeed, the two groups coincide if viewed under the convention made above.

The divisors of more general form (4) are the *correspondences in the extended sense.* Two equivalence relationships are defined for them. The stricter requires the difference between the equivalent divisors, or correspondences to be a principal divisor. Two correspondences are said to be equivalent in the extended sense if their difference is of the form $\mathfrak{a} + \mathfrak{a}' + (A)$ with a principal divisor (A).

Two degrees $g(\mathfrak{A})$ and $g'(\mathfrak{A})$ are again defined, with

$$g(\mathfrak{A} + \mathfrak{B}) = g(\mathfrak{A}) + g(\mathfrak{B}), \qquad g'(\mathfrak{A} + \mathfrak{B}) = g'(\mathfrak{A}) + g'(\mathfrak{B}). \tag{5}$$

Indeed, the components $\mathfrak{a}$ and A of (4) are divisors of the field KK' of functions in one variable over the constant field K', and thus have degrees. We set

$$g(\mathfrak{A}) = g(\mathfrak{a} + A), \qquad g'(\mathfrak{A}) = g'(\mathfrak{a}' + A).$$

This implies (5). These degrees are functions of the class of $\mathfrak{A}$ in the strict sense.

We may also define multiplication of the correspondences in the extended sense. For algebraically closed k_0 this leads, as in §1,5, to a multiplication of correspondence classes in the extended sense, for

$$(\mathfrak{a} + \mathfrak{a}' + A)(\mathfrak{b}' + \mathfrak{b}'' + B) = g'(\mathfrak{b}' + B)\mathfrak{a} + A\mathfrak{b}' + g'(\mathfrak{a}' + A)\mathfrak{b}'' + \mathfrak{a}'B + AB. \tag{6}$$

AB is defined as in §1,2, $\mathfrak{a}'B$ as in §1,5, and $A\mathfrak{b}'$ as the obvious adaptation of the process of §1,5 (residue formation for $\mathfrak{b}'$ mod A). The associative and distributive laws hold, and Eq. §1,(10) as well as the lemma of §1,3 remain valid. These facts are all just incidental, and will not be applied later.

Important later is a global representation of the divisors. The process used was already described in II, §5,5, but will be repeated here, now eliminating the introduction of homogeneous coordinates at the cost of certain advantages and gaining others. We adjoin an indeterminate z and form the Kronecker divisors of KK' with respect to the order $\mathfrak{h} = k_0[x,x']$ in the sense of II, §1,6. These divisors are the cosets of the elements $\alpha(z) \in KK'(z)$ with respect to the group of units $\mathfrak{E}_{\mathfrak{h}}(z)$ of the principal order $\mathfrak{H}(z)$ with respect to $\mathfrak{h}(z)$. *The group $\mathfrak{K}_{\mathfrak{h}}$ of Kronecker divisors of KK' with respect to $\mathfrak{h}$ is a homomorphic* *image of the divisor group $\mathfrak{K}$ defined above; the divisors of $\mathfrak{K}$ mapped onto the unit element are exactly those composed of prime divisors of the denominators of x and x'.*

4. The Fixed Points of a Correspondence

From now on we suppose K and K' to be isomorphic. Moreover, we determine a fixed isomorphic map of K onto K' leaving each element of k_0 invariant, once and for all. This isomorphism will be indicated by a prime: $x \leftrightarrow x'$, $\mathfrak{a} \leftrightarrow \mathfrak{a}'$.

To any correspondence $\mathfrak{A}$ between K and K' we define a divisor $\mathfrak{f}(\mathfrak{A})$ of K to be the *divisor of fixed points* of $\mathfrak{A}$. The definition itself depends upon whether or not $\mathfrak{A}$ contains the unit correspondence D in its prime decomposition. For D itself we set

$$\mathfrak{f}(D) = -\mathfrak{d}, \tag{7}$$

with some arbitrarily determined divisor $\mathfrak{d}$ in the class of differentials of K.

If D does not occur in the prime decomposition of $\mathfrak{A}$ we define the $\mathfrak{p}$-component $\mathfrak{f}_{\mathfrak{p}}(\mathfrak{A}) = \mathfrak{p}^{\varphi_{\mathfrak{p}}(\mathfrak{A})}$ for any place $\mathfrak{p}$ of K in another manner. The representation of correspondences by Kronecker divisors is used, a special preparation being needed first. Choose some element $x \in K$ whose denominator is not divisible by $\mathfrak{p}$, and denote its isomorphic image x'. Form $\mathfrak{h} = k_0[x, x']$ and the principal order $\mathfrak{H}(z)$ of $KK'(z)$ with respect to $\mathfrak{h}(z)$, with the indeterminate z. It is easy to see that, when K and K' are identified, $\mathfrak{H}(z)$ becomes the entire field $K(z)$, and this is the reason why $\mathfrak{H}(z)$ cannot serve our purposes. By $\mathfrak{H}_1(z)$ we denote the set of elements of $\mathfrak{H}(z)$ which are mapped onto elements of $\mathfrak{I}(z)$ under $K = K'$, where now $\mathfrak{I}(z)$ is the principal order of $K(z)$ with respect to $\mathfrak{i}(z) = k_0[x](z)$. That $\mathfrak{H}_1(z)$ is an integral domain is obvious. Moreover, we maintain it to be a principal ideal domain in the sense that any finite set of elements $\alpha_i(z) \in \mathfrak{H}_1(z)$ not all divisible by the diagonal have a g.c.d. in $\mathfrak{H}_1(z)[\alpha_1(z), \alpha_2(z), \ldots]$. This is proved by an easy adaptation of the argument of II, §1,3. The elements $\alpha_i(z)$ given, like any elements of $\mathfrak{H}_1(z)$, can be uniquely decomposed to

$$\alpha_i(z) = \alpha_{i0}(z) + \delta_i(z)$$

with $\alpha_{i0}(z) \in K(z)$ and $\delta_i(z)$ divisible by D. The g.c.d. is

$$\alpha(z) = \alpha_0(z) + \delta(z) = \sum_i \alpha_{i0}(z) z^{h_i} + \sum_i \delta_i(z) z^{h_i},$$

with suitable h_i. It must still be shown that the quotients

$$\frac{a_i(z)}{\alpha(z)} = \frac{\alpha_{i0}(z) + \delta_i(z)}{\alpha_0(z) + \delta(z)}$$

are mapped onto elements of $\mathfrak{I}(z)$ by $K' = K$. But this is also obvious, as the $\alpha_i(z)$ were assumed not all divisible by D, so that the quotients $\alpha_{i0}(z)\alpha_0(z)^{-1}$ are defined and contained in $\mathfrak{I}(z)$.

For simplicity's sake we again write $\mathfrak{H}(z)$ in place of $\mathfrak{H}_1(z)$. The divisor associated with a correspondence $\mathfrak{A}$ not divisible by D is given by a finitely generated ideal and therefore by a Kronecker divisor with respect to $\mathfrak{H}(z)$. As such it is represented by an element $\alpha(z)$, uniquely determined by $\mathfrak{A}$ up to the factor of a unit $\varepsilon(z)$ of $\mathfrak{H}(z)$. Now identify the fields K and K' in the sense of the given isomorphism. The divisor D does not occur in the prime decomposition of $\mathfrak{A}$, so that $\alpha(z)$ is mapped onto a nonzero element $\bar{\alpha}(z)$ of $K(z)$. Moreover, elements of $\mathfrak{H}(z)$ are mapped into $\mathfrak{J}(z)$, and the property of being a unit is retained by definition. This means that the coset $\alpha(z)\mathfrak{E}_{\mathfrak{i}}(z)$ is a Kronecker divisor of K with respect to $\mathfrak{i}$. Its $\mathfrak{p}$-component is defined to be the $\mathfrak{p}$-component of $\mathfrak{f}(\mathfrak{A})$.

It is implicit that the $\mathfrak{f}_{\mathfrak{p}}(\mathfrak{A})$ really define a divisor $\mathfrak{f}(\mathfrak{A})$, that is, that they are almost all (1).

It is not yet clear, though, that the definition is independent of the choice of x. But, say x_1 is some other element whose demoninator is not divisible by $\mathfrak{p}$. Then so is $x_2 = x_1 x$. The principal order $\mathfrak{J}_2$ of K with respect to $k_0[x_2]$ consists of all functions with denominators composed of the prime divisors of the denominator of x_2. The analogous statement holds for the principal order $\mathfrak{H}_2$ of KK' with respect to $\mathfrak{h}_2 = k_0[x_2, x_2']$. Thus, $\mathfrak{J}_2 \supseteq \mathfrak{J}$, $\mathfrak{H}_2 \supseteq \mathfrak{H}$, and for the groups of units $\mathfrak{E}_{\mathfrak{i}_2}(z) \supseteq \mathfrak{E}_{\mathfrak{i}}(z)$, $\mathfrak{E}_{\mathfrak{h}_2}(z) \supseteq \mathfrak{E}_{\mathfrak{h}}(z)$. The denominator of x_2 is not divisible by $\mathfrak{p}$, so that identification of K and K' makes the units of $\mathfrak{E}_{\mathfrak{h}_2}(z)$ to Kronecker divisors with respect to $\mathfrak{i}$ not divisible by $\mathfrak{p}$. Hence $\mathfrak{f}_{\mathfrak{p}}(\mathfrak{A})$ comes out the same, if x is replaced by x_2. The same argument can be applied for x_1 and x_2.

The components of divisors not containing D satisfy

$$\mathfrak{f}_{\mathfrak{p}}(\mathfrak{A} + \mathfrak{B}) = \mathfrak{f}_{\mathfrak{p}}(\mathfrak{A}) + \mathfrak{f}_{\mathfrak{p}}(\mathfrak{B}).$$

For if $\alpha(z)$ and $\beta(z)$ are elements representing A and B, then $\alpha(z)\beta(z)$ represents the divisor $\mathfrak{A} + \mathfrak{B}$. The identification $K = K'$ maps the product $\alpha(z)\beta(z)$ onto the product of the images, which is what must be shown.

If D occurs in the prime decomposition of $\mathfrak{A}$ exactly a times, then set

$$\mathfrak{f}(\mathfrak{A}) = \mathfrak{f}(\mathfrak{A} - aD) - a\mathfrak{d}. \tag{8}$$

This convention then yields

$$\mathfrak{f}(\mathfrak{A} + \mathfrak{B}) = \mathfrak{f}(\mathfrak{A}) + \mathfrak{f}(\mathfrak{B}). \tag{9}$$

The following remark will be needed later: if $\mathfrak{A}$ is an integral divisor not containing D, then $\mathfrak{f}(\mathfrak{A})$ is integral. This is an immediate result of our construction.

A correspondence (A) *of the principal class in the strict sense satisfies*

$$\mathfrak{f}((\mathrm{A})) \sim 0. \tag{10}$$

Thus, $\mathfrak{A} \to \mathfrak{f}(\mathfrak{A})$ *is a homomorphism of the classes of correspondences onto the classes of divisors.*

Proof. First consider only (A) free of factors of the diagonal D. The element A can then be taken as representative of the Kronecker divisor with respect to $\mathfrak{h}$ for (A). This means

$$\mathfrak{k}((\mathrm{A})) = (\mathrm{A}_{K'=K}). \tag{11}$$

Now, let D divide (A) exactly a times. By III, §5,5 we see that the product $\mathrm{A}(x - x')^{-a}$ is free of D-components if x is a separating element of K. By assuming the validity of

$$\mathfrak{k}((x - x') - D) \sim \mathfrak{d}, \tag{12}$$

Eq. (8) implies (10), as we already know that assertion holds for $\mathrm{A}(x - x')^{-a}$.

It remains to prove (12). The ideal $D_{K\times K'}$ defining D is generated by differences $y - y'$ with $y \in K$, a system of $y = y_0, y_1, \ldots$ generating K over k_0 sufficing. Now, let $(y_i - y_i')$ generate $D_{K\times K'}$. Then

$$\delta(z) = (y_0 - y_0') + (y_1 - y_1')z + \cdots$$

is associated with a Kronecker divisor having only constant prime divisors aside from D, with exactly one factor D. As any y_i may be replaced by y_i^{-1}, we may also assume all the y_i to be integral at some place $\mathfrak{p}$. Finally, the difference $p - p'$ for some prime element p of $\mathfrak{p}$ could be chosen as one of the generators of $D_{K\times K'}$ allowing us to choose $y_0 = p$.

Thus prepared, we use the corollary of III, §5,5 to compute

$$\left(\frac{\delta(z)}{p - p'}\right)_{K'=K} = 1 + (dy_1/dp)z + \cdots. \tag{13}$$

The y being integral at the place $\mathfrak{p}$, so are the differential quotients dy/dp. Thus, (13) gives us

$$\mathfrak{k}_{\mathfrak{p}}\left(\frac{D}{p - p'}\right) = (1)$$

(where we have inconsistently used the multiplicative notation to emphasize the relation between the quotient and differential quotient). By again using the corollary of III, §5,5 (and once more in multiplicative notation) we have

$$\mathfrak{k}_{\mathfrak{p}}\left(\frac{x - x'}{D}\right) = \mathfrak{k}_{\mathfrak{p}}\left(\frac{x - x'}{p - p'}\right)\mathfrak{k}_{\mathfrak{p}}\left(\frac{p - p'}{D}\right) = \mathfrak{k}_{\mathfrak{p}}\left(\frac{x - x'}{p - p'}\right) = \left(\frac{dx}{dp}\right)_{\mathfrak{p}}.$$

Compiling these equations for all $\mathfrak{p}$, we have found that

$$\mathfrak{k}\left(\frac{x - x'}{D}\right) = (dx), \tag{14}$$

and (12) is proved. ▼

We can now define the *number of fixed points* of A as

$$f(\mathfrak{A}) = g(\mathfrak{f}(\mathfrak{A})). \tag{15}$$

Because of (10), $f(\mathfrak{A})$ *is a function of the correspondence class in the strict sense, and because of* (9),

$$f(\mathfrak{A} + \mathfrak{B}) = f(\mathfrak{A}) + f(\mathfrak{B}). \tag{16}$$

Another means of calculating the divisor $\mathfrak{f}(\mathfrak{A})$ is easier to handle in applications, but would not have sufficed to prove (9) and (16), as we do not know whether the $\mathfrak{J}_\mathfrak{p} \times \mathfrak{J}_\mathfrak{p}'$-ideals form a group.

The $\mathfrak{p}$*-component* $\mathfrak{f}_\mathfrak{p}(\mathfrak{A})$ *for an integral divisor* $\mathfrak{A}$ *can also be found as follows: decompose* $\mathfrak{A}$ *as in* (4) *and set*

$$\mathfrak{f}_\mathfrak{p}(\mathfrak{A}) = \mathfrak{f}_\mathfrak{p}(\mathfrak{a} + \mathfrak{b}' + A) = \mathfrak{a}_\mathfrak{p} + \mathfrak{b}_\mathfrak{p} + \mathfrak{f}_\mathfrak{p}(A),$$

where the $\mathfrak{J}_\mathfrak{p}$*-ideal associated with* $\mathfrak{f}_\mathfrak{p}(A)$ *and with the same notations is found as*

$$\mathfrak{f}_\mathfrak{p}(A) = \text{ideal of residue classes of } A_{K\times K'} \cap \mathfrak{J}_\mathfrak{p} \times \mathfrak{J}'_{\mathfrak{p}'} \bmod D_{K\times K'} \text{ in } \mathfrak{J}_\mathfrak{p}.$$

($\mathfrak{J}_\mathfrak{p}$ is the integral domain of elements of K integral at $\mathfrak{p}$.)

Proof. Let A_i be a system of generators of the ideal $A_{K\times K'}$. As the A_i could be multiplied with elements $a_i \in K$, $b_i' \in K'$ so that the products lie in $\mathfrak{J}_\mathfrak{p} \times \mathfrak{J}'_{\mathfrak{p}'}$, and are divisible neither by $\mathfrak{p}$ nor $\mathfrak{p}'$, we may assume them already to be such. The polynomial $\alpha(z) = \Sigma \mathrm{A}_i z^i$ thus represents a Kronecker divisor $\mathfrak{B}$ differing from A by a constant factor and, in particular, by a factor not divisible by $\mathfrak{p}$ or $\mathfrak{p}'$. Thus $\mathfrak{f}_\mathfrak{p}(A) = \mathfrak{f}_\mathfrak{p}(\mathfrak{B})$. But $\mathfrak{f}_\mathfrak{p}(\mathfrak{B})$ is the $\mathfrak{p}$-component of the divisor of $\bar{\alpha}(z) = \Sigma(\mathrm{A}_i)_{K'=K}\, z^i$, while the $(\mathrm{A}_i)_{K'=K}$ are the residues of the A_i in K with respect to the module $D_{K\times K'}$. The assertion is thus proved. ▼

We must finally get to the equation

$$f(A^1 A^{*1}) = [K_1 : K] f(AA^*), \tag{17}$$

which we used in §4,2 for the extension A^1 of the correspondence A between K and K' to a correspondence between K_1 and K'. It is a consequence of

$$f(B^{11'}) = [K_1 : K] f(B),$$

where the isomorphic extensions K_1/K and K_1'/K' lead from B to $B^{11'}$. The latter equation immediately follows from the fact that the degree of a divisor is multiplied by the degree of the extension when it is brought up to an extension field.

5. The Connection with §2

We interrupt our line of thought to show that the number of fixed points defined by (15) coincides with the number

$$f(A) = \sum f_{\bar{\mathfrak{p}}} \tag{18}$$

given in §2, where the $f_{\bar{\mathfrak{p}}}$ are those of §2,(23). We again make the assumptions of §2, that k_0 is algebraically closed and that A is a prime correspondence which defines a common separable extension $\bar{K} = (K \times K')/A_{K \times K'}$ of the two fields K and K'. Formula (18) was, in particular, the foundation of §4,2, where it was used for the classical case.

For the proof, we first select another field K'' independent of, but isomorphic with, K and K'. Set

$$\mathfrak{F}_{\mathfrak{p}}(A) = \text{ideal of residue classes of } (A_{K \times K'} \times K'') \cap (\mathfrak{J}_{\mathfrak{p}} \times \mathfrak{J}'_{\mathfrak{p}'} \times \mathfrak{J}''_{\mathfrak{p}''}) \quad \text{mod } K \times D_{K' \times K''} \quad \text{in} \quad \mathfrak{J}_{\mathfrak{p}} \times \mathfrak{J}''_{\mathfrak{p}''}. \tag{19}$$

The algorithm at the end of the last section shows that $\mathfrak{f}_{\mathfrak{p}}(A)$ is gotten from this by identifying $K = K''$, the distinction between the divisor $\mathfrak{f}_{\mathfrak{p}}(A)$ and the associated $\mathfrak{J}_{\mathfrak{p}}$-ideal being dropped.

Now extend K to K_1 until the extended correspondence is a sum $A^1 = \Sigma A_\varrho$ of prime correspondences of degree $g'(A_\varrho) = 1$. We give a construction of such an extension. By §1,6, A defines a relation between K and K' such that $\bar{K} = (K \times K')/A_{K \times K'}$ is a finite separable extension of both K and K'. Every $x' \in K'$ satisfies a polynomial equation $f(x') = 0$ over K, and $A_{K \times K'}$ is generated by all these polynomials. Each of these polynomials has a zero in $\bar{K}$, so that a summand A_1 of degree $g'(A_1) = 1$ can be split off A over $\bar{K}$. Then, if $\bar{K}_1 = \bar{K}, \bar{K}_2, \ldots$ are all the conjugate fields of $\bar{K}$ over K and K_1 is their composite field, the desired decomposition is found in K_1.

Now set this extension into (19), which is done by simply multiplying both sides by the principal order $\mathfrak{J}_{1\mathfrak{p}}$ of K_1 with respect to $\mathfrak{J}_{\mathfrak{p}}$. For the extended ideal on the left we find

$$\mathfrak{F}_{1\mathfrak{p}}(A^1) = \text{ideal of residue classes of } (A^1_{K_1 \times K'} \times K'') \cap (\mathfrak{J}_{1\mathfrak{p}} \times \mathfrak{J}'_{\mathfrak{p}'} \times \mathfrak{J}''_{\mathfrak{p}''}) \quad \text{mod } K_1 \times D_{K' \times K''} \quad \text{in} \quad \mathfrak{J}_{1\mathfrak{p}} \times \mathfrak{J}''_{\mathfrak{p}''}.$$

The separability of $\bar{K}/K$ assures us that the prime correspondences A_ϱ are distinct, and thus the ideals $(A_\varrho)_{K_1 \times K'}$ mutually prime. Hence,

$$A^1_{K_1 \times K'} = \Big(\sum_\varrho A_\varrho\Big)_{K_1 \times K'} = \bigcap_\varrho (A_\varrho)_{K_1 \times K'}.$$

The validity of this equation is not impaired by multiplication on both sides with K'' and intersection with $\mathfrak{J}_{1\mathfrak{p}} \times \mathfrak{J}_{\mathfrak{p}}' \times \mathfrak{J}''_{\mathfrak{p}''}$. Thus

$$\mathfrak{F}_{1\mathfrak{p}}(A^1) = \bigcap_\varrho \text{ideals of residue classes of } ((A_\varrho)_{K_1 \times K'} \times K'') \cap (\mathfrak{J}_{1\mathfrak{p}} \times \mathfrak{J}'_{\mathfrak{p}'} \times \mathfrak{J}''_{\mathfrak{p}''}) \quad \text{mod } K_1 \times D_{K' \times K''} \quad \text{in} \quad \mathfrak{J}_{1\mathfrak{p}} \times \mathfrak{J}''_{\mathfrak{p}''}.$$

Because the A_ϱ are mutually prime we have, for $\varrho \neq \sigma$,

$$((A_\varrho)_{K_1 \times K'} \times K'' \cup (A_\sigma)_{K_1 \times K'} \times K'') \cap (\mathfrak{J}_{1\mathfrak{p}} \times \mathfrak{J}'_{\mathfrak{p}'} \times \mathfrak{J}''_{\mathfrak{p}''}) = \mathfrak{J}_{1\mathfrak{p}} \times \mathfrak{J}'_{\mathfrak{p}'} \times \mathfrak{J}''_{\mathfrak{p}''}$$

so that, moreover,

$$\mathfrak{F}_{1\mathfrak{p}}(A^1) = \prod_{\varrho} \text{ideals of residue classes of } ((A_\varrho)_{K_1 \times K'} \times K'') \cap (\mathfrak{J}_{1\mathfrak{p}} \times \mathfrak{J}'_{\mathfrak{p}'} \times \mathfrak{J}''_{\mathfrak{p}''})$$
$$\text{mod } K_1 \times D_{K' \times K''} \quad \text{in} \quad \mathfrak{J}_{1\mathfrak{p}} \times \mathfrak{J}''_{\mathfrak{p}''}.$$

As $g'(A_\varrho) = 1$, the symmetry law for multiplication of §1,3 can be applied to the individual factors on the right, yielding

$$\mathfrak{F}_{1\mathfrak{p}}(A^1) = \prod_{\varrho} \text{ideals of residue classes of } (K_1 \times D_{K' \times K''}) \cap (\mathfrak{J}_{1\mathfrak{p}} \times \mathfrak{J}'_{\mathfrak{p}'} \times \mathfrak{J}''_{\mathfrak{p}''})$$
$$\text{mod } (A_\varrho)_{K_1 \times K'} \times K'' \quad \text{in} \quad \mathfrak{J}_{1\mathfrak{p}} \times \mathfrak{J}''_{\mathfrak{p}''}.$$

We saw that this gives $\mathfrak{f}_\mathfrak{p}(\mathfrak{A})$ if K and K'' are identified; actually it gives the extended ideal of $\mathfrak{f}_\mathfrak{p}(\mathfrak{A})$ in K_1, but a distinction between them need not be made here. Thus

$$\mathfrak{f}_\mathfrak{p}(A) = \prod_{\varrho} \text{ideals of residue classes of } K_1 D_{K' \times K} \cap (\mathfrak{J}_{1\mathfrak{p}} \times \mathfrak{J}'_{\mathfrak{p}'})$$
$$\text{mod } (A_\varrho)_{K_1 \times K'} \quad \text{in} \quad \mathfrak{J}_{1\mathfrak{p}}. \tag{20}$$

According to III, §5,5, if x is a prime element for $\mathfrak{p}$, then $x - x'$ is a prime element for D. The residue classes of $K_1 D_{K' \times K} \cap \mathfrak{J}_{1\mathfrak{p}} \times \mathfrak{J}'_{\mathfrak{p}'}$ modulo the A_ϱ are the principal $\mathfrak{J}_{1\mathfrak{p}}$-ideals $(x - x_\varrho')$, the x_ϱ' being the images of the x in the fields K_ϱ. Hence (20) can be written as

$$\mathfrak{f}_\mathfrak{p}(A) = \prod_{\varrho} (x - x_\varrho')_\mathfrak{p} = (n_{\bar{K}/K}(x - x'))_\mathfrak{p},$$

the subscript $\mathfrak{p}$ indicating that only the components at the place $\mathfrak{p}$, or the places lying over $\mathfrak{p}$, are to be taken. The subscript ϱ can also be eliminated, then choosing $x' \in \bar{K} = (K \times K')/A_{K \times K'}$.

This is exactly the configuration of §2,5, but with K and K' interchanged. The power series §2,(21) are set up for all fixed points $\bar{\mathfrak{p}}$ of $\bar{K}$ lying over $\mathfrak{p}$, and it is seen that $\mathfrak{f}_\mathfrak{p}(\mathfrak{A})$ is always divisible by the same power of $\mathfrak{p}$ as $n_{\bar{K}/K}(\Pi \mathfrak{p}^{f_{\bar{\mathfrak{p}}}})$. Here we have, of course, taken the fixed points in the sense of §2. Noting that, because k_0 is algebraically closed by hypothesis, all prime divisors of $\bar{K}$ are of the first degree and $n_{\bar{K}/K}(\bar{\mathfrak{p}}) = \mathfrak{p}$, we see that $g(\mathfrak{f}_\mathfrak{p}(A)) = \Sigma f_{\bar{\mathfrak{p}}}$, which verifies the assertion.

6. The Trace

For a correspondence in the extended sense,

$$s(\mathfrak{A}) = g(\mathfrak{A}) + g'(\mathfrak{A}) - f(\mathfrak{A}) \tag{21}$$

defines the trace. With (16) it satisfies

$$s(\mathfrak{A} + \mathfrak{B}) = s(\mathfrak{A}) + s(\mathfrak{B}).$$

For constant correspondences it is obvious that

$$s(\mathfrak{a}) = s(\mathfrak{b}') = 0.$$

Also for a correspondence (A) of the extended principal class the trace is

$$s((\mathrm{A})) = 0.$$

Hence, the trace is a linear function of the class of a correspondence in the extended sense.

We also mention the immediate result,

$$s(\mathfrak{A}) = s(\mathfrak{A}^*). \tag{22}$$

The *principal theorem*, which we now want to prove, is stronger than inequality (3). *In a conservative function field K,*

$$s(\mathfrak{A}\mathfrak{A}^*) > 0 \tag{23}$$

for all $\mathfrak{A}$ *not in the extended principal class. K and K′ need not be assumed isomorphic.* The excepted case clearly has $s(\mathfrak{A}\mathfrak{A}^*) = 0$. The assumption that the field be conservative assures the coincidence of differential and canonical classes.

Clearly (23) is equivalent to

$$s(AA^*) > 0 \quad \text{if} \quad A \nsim 0 \tag{24}$$

for the nonconstant component A of $\mathfrak{A}$ in the sense of (4). Correspondences in the extended sense were first introduced here to define the number of fixed points, and will also play a part in this proof.

We first take care of an easily demonstrated special case in a lemma: for a nonconstant prime correspondence $\mathfrak{A}$ of degree $g'(\mathfrak{A}) = 1$, we have

$$s(\mathfrak{A}\mathfrak{A}^*) = 2g'g(\mathfrak{A}), \tag{25}$$

with the genus g' of K'.

Proof. From (4) we see that $\mathfrak{A}$ coincides with its nonconstant component A. By §1,1 A is determined by an isomorphic mapping $x' \to x_0$ of K' onto a subfield $K_0 \subseteq K$, and $g(A) = [K : K_0]$. The ideal $A_{K \times K'}$ is generated by the differences $x_0 - x'$, a system of generators of K' being taken for x'. Thus the ideal $(AA^*)_{K \times K''}$ is generated by the differences $x_0 - x_0''$, so that AA^* is an extension of the unit correspondence D_0 between K_0 and K_0'' to a correspondence between K and K''.

For the unit correspondence we saw in (7) that

$$s(D_0) = 2 + g(\mathfrak{d}_0) = 2g', \tag{26}$$

with a divisor $\mathfrak{d}_0$ of the canonical class of K_0. Now choose a correspondence B_0 between K_0 and K_0'', strictly equivalent but prime to D_0. Extensions of B_0 and D_0 to correspondences between K and K'' are also equivalent. We saw that the extension of D_0 is AA^*, and will denote the extension of B_0 by B. The definition of divisors of fixed points shows that

$$\mathfrak{f}(B) = \mathfrak{f}_0(B_0).$$

Thus the degrees $g(B_0)$, $g''(B_0)$, $f(B_0)$ are all multiplied by $[K : K_0] = g(A)$ when B_0 is extended to B. The same is then true of $s(B_0)$. As B is equivalent to AA^*, (25) is a consequence of (26). ▼

The principal theorem itself is easily derived from this lemma for the case of an elliptic function field. The algebraic closure of the field k_0 of constants is first taken; all the degrees and therefore the trace remain unchanged. There exist constant divisors of any given degree. Thus any given class can be represented by a correspondence $\mathfrak{B}$ of degree $g'(\mathfrak{B}) = 1$. Apply the Riemann–Roch theorem to KK'/K; there exists an integral divisor $\mathfrak{A}$ equivalent to $\mathfrak{B}$, and of degree $g'(\mathfrak{A})=1$. So long as the extended class is not principal, $\mathfrak{A}$ is a nonconstant prime correspondence. But the trace is a function of the extended class, so that (25) implies (24).

Thus further consideration is necessary for fields of genus $g > 1$ only.

7. Second Proof of the Principal Theorem: Preparations

Extension of the field k_0 of constants leaves the divisor degrees, and therefore the trace, invariant, because K and K' were assumed conservative and separably generated (III, §3,4). It is thus no loss of generality to assume k_0 to be algebraically closed.

By a *nonspecial divisor system* we mean g distinct prime divisors $\mathfrak{p}_i$ of K such that $\dim(W(\mathfrak{p}_1 \cdots \mathfrak{p}_g)^{-1}) = 0$, where g is the genus of K and W its canonical class. Such a system can be constructed as follows. Let $\mathfrak{v}_i$ be a basis of the module of integral divisors in W. Choose some prime divisor $\mathfrak{p}_1$ by which not all $\mathfrak{v}_i$ are divisible, and determine constants $\varrho_i \in K_0$ such that $v_i - \varrho_i$ is divisible by $\mathfrak{p}_1$ for all i, where $v_i = \mathfrak{v}_i\mathfrak{v}_1^{-1}$. Then, $\mathfrak{v}_i' = (\mathfrak{v}_i - \varrho_i\mathfrak{v}_1)\mathfrak{p}_1^{-1}$ $(i = 2, \ldots)$ are linearly independent integral divisors of the class $W\mathfrak{p}_1^{-1}$. No further linearly independent integral divisors can exist in this class, for, with such an $\mathfrak{a}$, the $\mathfrak{v}_i'\mathfrak{p}_1$ and $\mathfrak{a}\mathfrak{p}_1$ would be linearly independent integral divisors of W, so that every integral divisor in W could be expressed as $\mathfrak{w}_1 = \sigma\mathfrak{a}\mathfrak{p}_1 + \sigma_2'\mathfrak{v}_2'\mathfrak{p}_1 + \cdots$. This would contradict our choice of $\mathfrak{p}_1$. Thus $\dim(W\mathfrak{p}_1^{-1}) = \dim(W) - 1 = g - 1$. Continuation of this process finally leads to the system desired.

The Riemann–Roch theorem shows that any nonspecial divisor system has dimension $\dim(\mathfrak{p}_1 \cdots \mathfrak{p}_g) = 1$.

Lemma. Let $\mathfrak{p}_1, \ldots, \mathfrak{p}_m$ $(m \leqq g)$ be contained in a nonspecial divisor system. There then exist m functions $w_i \in K$ with common denominator $\mathfrak{h}$ of degree

$$g(\mathfrak{h}) \leqq g + m - 2, \tag{27}$$

prime to an arbitrarily given integral divisor, and satisfying the congruences

$$w_i \equiv \begin{Bmatrix} 1 & \text{for} \quad i = j \\ 0 & \text{for} \quad i \neq j \end{Bmatrix} \bmod \mathfrak{p}_j. \tag{28}$$

Proof. For each $i \leqq m$ find an integral divisor $\mathfrak{w}_i$ of W divisible by $\mathfrak{p}_1 \cdots \mathfrak{p}_g \mathfrak{p}_i^{-1} = \mathfrak{r}_i$. It will be of the form $\mathfrak{w}_i = \mathfrak{r}_i \mathfrak{g}_i$ with an integral $\mathfrak{g}_i$ in the class $W\mathfrak{r}_i^{-1}$. The Riemann–Roch theorem shows that $\dim(W\mathfrak{r}_i^{-1}) = \dim(\mathfrak{r}_i) \leqq \dim(\mathfrak{p}_1 \cdots \mathfrak{p}_g)$, which is 1 by hypothesis. But, as trivially $\dim(\mathfrak{r}_i) \geqq 1$, necessarily $\dim(W\mathfrak{r}_i^{-1}) = 1$, so that up to constant multiples only a single integral divisor $\mathfrak{g}_i$ can lie in the class of $W\mathfrak{r}_i^{-1}$. As $\dim(W\mathfrak{r}_i^{-1}\mathfrak{p}_i^{-1}) = 0$, then that $\mathfrak{g}_i$ cannot be divisible by $\mathfrak{p}_i$.

Now set

$$w_i = \frac{\mathfrak{w}_i}{\sum \varrho_j \mathfrak{w}_j}$$

with constants $\varrho_j \neq 0$ to be determined. Multiplication of the w_i by suitable constants leads to (28). A common divisor cancels out of the $\mathfrak{w}_i$, and contains at least $\mathfrak{p}_{m+1} \cdots \mathfrak{p}_g$. Then suitable choice of the constants ϱ_i assures that the common denominator $\mathfrak{h}$ is prime to any divisor given. $\mathfrak{h}$ itself divides the divisor $\Sigma \varrho_i \mathfrak{w}_i (\mathfrak{p}_{m+1} \cdots \mathfrak{p}_g)^{-1}$, whose degree is on the right in (27). ▼

We have tacitly reverted to the multiplicative notation for the divisor group, but now return to the additive notation introduced in this chapter.

Choice of a suitable class representative. In §4,3 we showed multiplication of correspondences to be an operation between classes in the extended sense and, in the last section, the trace to be a function of these classes. $\mathfrak{A}$ may be replaced by a divisor equivalent in the extended sense, in the proof of (24). For $\mathfrak{A}$ not in the principal class,

$$s(\mathfrak{A}\mathfrak{A}^*) \geqq 2g(\mathfrak{A}) \tag{29}$$

can also be proved, which is more than (23). An extension K_1/K multiplies both sides of this equation by $[K_1 : K]$ [cf. (17)], so that we may extend K finitely whenever convenient in the proof.

Consider $\mathfrak{A}$ as a divisor of KK'/K, and let $\mathfrak{b}'$ be some divisor of K' such that $g'(\mathfrak{A} + \mathfrak{b}') = g'$, the genus of K'. The Riemann–Roch theorem then states that in the same class there exists an integral divisor $\mathfrak{B}$. This $\mathfrak{B} = \mathfrak{A} + \mathfrak{b}' + \mathfrak{c} + (\Gamma)$ with a constant divisor $\mathfrak{c}$ yet to be determined, and $g'(\mathfrak{B}) = g'$. A $K \times K'$-ideal is associated with $\mathfrak{B}$, that is, an integral nonconstant component in the sense of (4). Determine $\mathfrak{c}$ so that $\mathfrak{B}$ has no constant component in K. Then $\mathfrak{B}$ is integral taken as a divisor of KK'/k_0. Replace $\mathfrak{A}$ by $\mathfrak{B}$, that is, assume $\mathfrak{A}$ to be integral and $g'(\mathfrak{A}) = g'$.

Now choose an extension K_1/K for which $\mathfrak{A} = \mathfrak{P}_1 + \cdots + \mathfrak{P}_g$, with $g'(\mathfrak{P}_i) = 1$, but still write K. Also, choose some arbitrary divisor $\mathfrak{p}$ of K, but hold it fixed, and form

$$\mathfrak{p}\mathfrak{A} = \mathfrak{a}' = \mathfrak{q}_1' + \cdots + \mathfrak{q}_{g'}', \qquad \mathfrak{q}_i' = \mathfrak{p}\mathfrak{P}_i,$$

setting $\mathfrak{p}\mathfrak{P}_i = \mathfrak{P}_i = \mathfrak{q}_i'$ for any $\mathfrak{P}_i$ constant in K'. By §1,(18) and because $g'(\mathfrak{P}_i) = 1$ the $\mathfrak{q}_i'$ are prime divisors of K', so that $g'(\mathfrak{a}') = g'$.

After this preparation, we take some nonspecial divisor system $\mathfrak{p}_i'$ of K' and form an integral divisor $\mathfrak{B}$ of KK'/K equivalent to $\mathfrak{A} - \mathfrak{a}' + \mathfrak{p}_1' + \cdots + \mathfrak{p}_{g'}'$; that is,

$$\mathfrak{B} = \mathfrak{A} - \mathfrak{a}' + \mathfrak{p}_1' + \cdots + \mathfrak{p}_{g'}' + \mathfrak{c} + (\Gamma).$$

As before, $\mathfrak{c}$ is determined as a constant divisor in K making $\mathfrak{B}$ integral. We are interested in the product

$$\mathfrak{p}\mathfrak{B} = \mathfrak{p}_1' + \cdots + \mathfrak{p}_{g'}' + (\gamma'), \qquad \gamma' \neq 0 \in K',$$

which we compute by the second rule of §1,5. Residue class formation mod $\mathfrak{p}$ reduces the function field KK'/K modulo the prime divisor $\mathfrak{p}$ of the constant field, a summand like $\mathfrak{c}$ having no effect. (Incidentally, this coincides with the definition (6) of the product of generalized correspondences.) As $\mathfrak{B}$ is integral, so is $\mathfrak{p}\mathfrak{B}$, which is equivalent to $\mathfrak{p}_1' + \cdots$. But, the $\mathfrak{p}_1'$ are a nonspecial system of divisors, so that $\dim(\mathfrak{p}_1' + \cdots) = 1$. This implies that $\mathfrak{p}\mathfrak{B} = \mathfrak{p}_1' + \cdots + \mathfrak{p}_{g'}'$ (in other words, the principal divisor (γ') above is the unit divisor).

Again write $\mathfrak{A}$ in place of $\mathfrak{B}$ and extend K again, so that $\mathfrak{A}$ still decomposes to prime divisors $\mathfrak{P}_i$ of degree $g'(\mathfrak{P}_i) = 1$. Suitable indexing gives $\mathfrak{p}\mathfrak{P}_i = \mathfrak{p}_i'$. Some constant prime divisors may occur among the $\mathfrak{P}_i$. They coincide with the $\mathfrak{p}_i'$ and can be dropped, as they contribute nothing to $s(\mathfrak{A}\mathfrak{A}^*)$. The remaining nonconstant divisors will be denoted by roman letters. The configuration at hand has finally been reduced to the following.

Every class in the extended sense contains a representative of the form

$$A = P_1 + \cdots + P_m, \qquad g'(P_i) = 1, \qquad m \leqq g'.$$

We have $\mathfrak{p}\mathfrak{P}_i = \mathfrak{p}_i'$ with some prime divisor $\mathfrak{p}$ of K, the $\mathfrak{p}_i'$ being contained in a nonspecial divisor system of K'. Clearly no two $\mathfrak{P}_i$ can coincide, as this is excluded for the $\mathfrak{p}_i'$. In case $m > 0$, we will now prove (29).

8. Second Proof of the Principal Theorem: Conclusion

Using the class representatives determined we have

$$s(AA^*) = \sum_i s(P_iP_i^*) + 2\sum_{i<j} s(P_iP_j^*).$$

The first sum can be computed by (25) and is

$$\sum_i s(P_iP_i^*) = 2g' \sum_i g(P_i) = 2g'g(A). \tag{30}$$

Because of §1,(10) the second sum is

$$2\sum_{i<j} s(P_iP_j^*) = \sum_{i\neq j} (g(P_i) + g(P_j)) - 2\sum_{i<j} f(P_iP_j^*) = 2(m-1)g(A) - \cdots.$$

We shall demonstrate the inequality

$$\sum_{i<j} f(P_iP_j^*) \leqq (g' + m - 2)g(A), \tag{31}$$

which then implies

$$2\sum_{i<j} s(P_iP_j^*) \geqq 2(1-g')g(A).$$

Taken with (30), this inequality is equivalent to (29).

It is the proof of (31) in which the difficulty lies, and for which we needed the preparations of the last section. We start with m elements w_i', constructed to suit the $\mathfrak{p}_i'$ as in the lemma. The P_j define isomorphisms $K' \to K_j$ onto subfields of K of indices

$$[K\colon K_j] = g(P_j). \tag{32}$$

These isomorphisms map the w_i' onto elements $w_{ij} \in K_j$. The lemma showed that the w_i' could be chosen so that the w_{ij} are integral at a given place $\mathfrak{p}$ of K. We must now demonstrate three properties of the w_{ij}: (a) their determinant $|w_{ij}| \neq 0$; (b) its denominator is divisible by the divisor of fixed points $\mathfrak{f}(\sum_{i<j} P_iP_j^*)$; and (c) the denominator of $|w_{ij}|$ is of degree $\leqq (g' + m - 2)g(A)$. This suffices, for the numerator cannot have a larger degree than the denominator, nor can $\mathfrak{f}(\sum_{i<j} P_iP_j^*)$. The last statement is (31).

(a) By definition

$$w_i' = w_{ij} + \Pi_{ij}, \qquad \Pi_{ij} \equiv 0 \bmod P_j, \tag{33}$$

The w_{ij} are integral at the place $\mathfrak{p}$, as are also the w_i' and the Π_{ij}. Consider the w_{ij} mod $\mathfrak{p}$. Certainly there exist $\omega_{ij} \in k_0$ congruent to the w_{ij} mod $\mathfrak{p}$. Then

$$\omega_{ij} \equiv w_i' + \pi_{ij}, \qquad \pi_{ij} \equiv 0 \bmod \mathfrak{p}_j',$$

according to (33) and $\mathfrak{p}P_j = \mathfrak{p}_j'$, so that the ω_{ij} are the residue classes of the w_i' mod $\mathfrak{p}_j'$ in k_0. We can apply (28) to the primed symbols to see that $\omega_{ij} = 1$ or 0, depending upon whether $i = j$ or $i \neq j$. Hence, the determinants $|\omega_{ij}|$ and $|w_{ij}| \neq 0$.

(b) We show the denominator of $|w_{ij}|$ to be divisible by $\mathfrak{f}_\mathfrak{q}(\sum_{i<j} P_iP_j^*)$ for every place $\mathfrak{q}$ of K. By $\mathfrak{J}_\mathfrak{q}$ we denote the integral domain of all elements of K integral at $\mathfrak{q}$, we set $\mathfrak{J}_{\mathfrak{q},i} = K_i \cap \mathfrak{J}_\mathfrak{q}$, and write $\mathfrak{J}_\mathfrak{q}'$ for the isomorphic image of $\mathfrak{J}_{\mathfrak{q},i}$ in K'. The common denominator of the w_i' can still be adjusted so that the w_i' lie in each of the finitely many $\mathfrak{J}_\mathfrak{q}'$ for which $\mathfrak{f}_\mathfrak{q}(\sum_{i<j} P_iP_j^*) \neq 0$. (The 0 here is the additive notation for the unit divisor.)

Now let x_i traverse all elements of $\mathfrak{I}_{\mathfrak{q},i}$ and denote by x_j the corresponding elements of K_j under the isomorphisms $x_i \leftrightarrow x' \leftrightarrow x_j$. Also, let x''_j be the associated elements in a field K'' isomorphic to, but independent of, K. The differences $x_i - x''_j$ generate the ideal $(\mathfrak{P}_i\mathfrak{P}_j^*)_{K\times K''}$, so that computation in the manner of §4,4 using ideal theory shows $\mathfrak{f}_\mathfrak{q}(\mathfrak{P}_i\mathfrak{P}_j^*)$ to be generated by the differences $x_i - x_j$.

Let, in particular, a prime element for the prime divisor $\mathfrak{q}_i$ of K_i, divisible by $\mathfrak{q}$, be chosen as x_i. The degree of $\mathfrak{q}_i$ must be 1, as k_0 is algebraically closed, and by III, §1,2 there exist constants $\gamma_{ij\mu}$ in k_0 such that

$$w_{ij} = \sum_{\mu=0}^{r} \gamma_{ij\mu} x_j^{\mu} + \varphi_{ij,r}, \qquad \varphi_{ij,r} \equiv 0 \bmod \mathfrak{q}_j^{r+1}.$$

Remembering that the fields K_j are isomorphic, we see that the $\gamma_{ij\mu}$ do not depend upon j, so that

$$w_{ij} = \sum_{\mu=0}^{r} \gamma_{i\mu} x_j^{\mu} + \varphi_{ij,r}, \qquad \varphi_{ij,r} \equiv 0 \bmod \mathfrak{q}_j^{r+1}.$$

A choice of r greater than the power of $\mathfrak{q}$ contained in $\mathfrak{f}_\mathfrak{q}(\sum_{i<j} P_iP_j^*)$ assures the difference between the determinants $|w_{ij}|$ and $|\Sigma_0^r \gamma_{i\mu} x_j^{\mu}|$ to be divisible by $\mathfrak{q}$ more often than $\mathfrak{f}_\mathfrak{q}(\sum_{i<j} P_iP_j^*)$. But it is well known that the latter determinant is a polynomial in x_i and x_j divisible by $\prod_{i<j}(x_i - x_j)$, and thus also by $\mathfrak{f}_\mathfrak{q}(\sum_{i<j} P_iP_j^*)$. The same must be true of $|w_{ij}|$.

(c) For any fixed index j the w_{ij} have a common denominator $\mathfrak{h}_j$. Because of (27) its degree in K_j must be

$$g(\mathfrak{h}_j) \leqq g' + m - 2.$$

The degree of $\mathfrak{h}_j$ in K must be $g(P_j)$ times as much, because of (32). Thus the degree in K of the denominator divisor of the determinant $|w_{ij}|$ must be $\leqq \Sigma(g' + m - 2)g(P_j) = (g' + m - 2)g(A)$. The proof is complete. ▼

9*. Remarks Concerning the Ring of Correspondence Classes

By §1,4 the classes of correspondences form a ring. In §4,6 we saw that the trace is a function of these classes.

If $A \neq 0$ is a class and m a natural number then $mA = A + \cdots + A \neq 0$. For $s(AA^*) \neq 0$ means that $s((mA)(mA)^*) = m^2 s(AA^*) > 0$.

This makes it possible to write formal sums $\Sigma q_\varrho A_\varrho$ of correspondence classes A_ϱ with arbitrary rational q_ϱ, and to define addition and multiplication naturally. The result is the hypercomplex system $\mathfrak{C}$ over the rational number field Q.

$\mathfrak{C}$ *is semisimple*. It must be proved that no nilpotent elements are contained in the center of $\mathfrak{C}$. Say $N^{h-1} \neq 0$, $N^h = 0$, and $h \geqq 2$. It is certainly no limitation of generality to assume N to be a correspondence class in the original sense. The principal theorem says that $s(N^{h-1}N^{*h-1}) > 0$, so that $N^{h-1}N^{*h-1} \neq 0$. Also, $s(N^{h-1}N^{*h-1}(N^{h-1}N^{*h-1})^*) > 0$, so that $N^{h-1}N^{*h-1}(N^{h-1}N^{*h-1})^* \neq 0$. Certainly $(N^{h-1}N^{*h-1})^* = N^{h-1}N^{*h-1}$. By the center property of N we see that then $N^{2h-2}N^{*2h-2} \neq 0$. But, as $h \geqq 2$ we have $2h-2 \geqq h$, which gives the contradiction desired. ▼

The ring of correspondence classes can be represented by matrices of $2g$ rows if the characteristic is 0, as shown for the classical case in §4,1. (The representation decomposes to the sum of two representations of g rows.) $s(A)$ is the trace of the representation. For any characteristic A. Weil† could prove the following.

Let l be any prime not the characteristic and consider the totality of divisor classes of K which give the principal class when multiplied by l^h with some $h > 0$ (or, in multiplicative notation, whose l^hth power is that class). It is a group, $\mathfrak{G}_l$, isomorphic to the additive group of vectors with $2g$ components, which are residue classes of l-adic numbers modulo l-adic integers. The correspondence classes A define homomorphisms of $\mathfrak{G}_l$ into itself, and are thus representable as matrices $G(A)$ with $2g$ rows of l-adic integers. The trace of this representation $G(A)$ coincides with $s(A)$, which in turn implies that the representation is equivalent to one with rational coefficients.

The analogous group $\mathfrak{G}_l$ where l is the characteristic involves a more difficult configuration, which was investigated by J.-P. Serre.‡

Notes

The first general proof of the principal theorem is due to Weil [4], but we followed the argument of Roquette [3]. Another proof is found by close examination of the representation $G(A)$ mentioned in §4,9. The principal theorem then turns out to be a special case of a more general statement which is valid for all abelian varieties (cf. Lang's book, IV, §3,6 [1]). A comparatively short proof was given by Mattuck and Tate [2]. It is based on the Riemann–Roch theorem for two-dimensional algebraic varieties, which cannot enter the scope of this book. From this point of view the inequality $s(AA^*) > 0$ is related to other facts of algebraic geometry (cf. Grothendieck [1]).

---

† *Variétés abéliennes et courbes algébriques*, Actualités Sci. Ind. No. 1064 (1948).

‡ *Quelques propriétés des variétés abéliennes en charactéristique p*. Amer. J. Math. **80**, 715–739 (1958).

A common feature of all these proofs is their resort to algebraic geometry of several dimensions and, in particular, to the intersection theory for subvarieties. Our proof limited the use of intersection theory. Actually, though, the number of fixed points of a correspondence A is the number $s(A, D)$ of intersection points between the curve $A = 0$ on the "variety" $K \times K'$ and the diagonal D which is defined by identification of K and K':

$$f(A) = s(A, D).$$

More generally, the intersection number $s(A, B)$ of two correspondences A, B can be defined by the method of §4,4, first choosing a prime correspondence B with $g(B) = 1$ but not occurring among the prime factors of A. It defines a mapping of K onto a subfield $K_0{}' \subseteq K'$, and one can set

$$s(A, B) = g'((A_{\mathfrak{J} \times \mathfrak{J}'})_{K \to K_0'}).$$

As in §1,3 (multiplicative symmetry), it can then be shown that

$$s(A, B) = s(B, A).$$

Another easy result is

$$s(A, B) = s(AB, D).$$

This implies that $f(AB) = f(BA)$ and thus the important law of symmetry for the trace,

$$s(AB) = s(BA).$$

References

[1] A. Grothendieck, *Sur une note de Mattuck-Tate*. J. Reine Angew. Math. **200**, 208–215 (1958).

[2] A. Mattuck und J. Tate, *On the inequality of Castelnuovo-Severi*, Abh. Hamburger Math. Sem. **22**, 295–299 (1958).

[3] P. Roquette, *Arithmetischer Beweis der Reimannschen Vermutung in Kongruenzzetafunktionenkörpern beliebigen Geschlechts*, J. Reine Angew. Math. **191**. 199–252 (1953).

[4] A. Weil, *Sur les courbes algébriques et les variétés qui s'en déduisent*, Actualités Sci. Ind. No. 1041 (1948).

§5. Applications in Number Theory

1. The Zeta Function of a Field of Functions

Let K be a field of functions over a finite exact constant field k_0 with $q = p^r$ elements; p is the characteristic. These fields have much in common with finite algebraic number fields. This relationship is particularly emphasized by the zeta function, which we now define and investigate. As for the zeta function of an algebraic number field, we must refer to the texts on algebraic number theory, but no such information will be presupposed here.

In extreme contrast to the arithmetic case we can prove an analog of the Riemann conjecture for fields of functions, which turns out to be a result of Castelnuovo's inequality. In this regard the theory of functions proves to be easier than number theory, in the end.

To a divisor $\mathfrak{a}$ is associated its *absolute norm*

$$N(\mathfrak{a}) = q^{g(\mathfrak{a})}. \tag{1}$$

For a prime divisor $\mathfrak{a} = \mathfrak{p}$, the residue class field of the integral domain $\mathfrak{J}_\mathfrak{p}$ of functions integral at $\mathfrak{p}$, modulo $\mathfrak{p}$, is an extension of k_0 of degree $g(\mathfrak{p})$. All residue classes modulo $\mathfrak{p}$ are of the form $\Sigma \varrho_i \omega_i$, $\varrho_i \in k_0$, where ω_i is a basis of the field. Thus, there are exactly $N(\mathfrak{p})$ residue classes. The relationship mentioned between algebraic number fields and the function fields being considered is primarily rooted in the finiteness of this residue class number; indeed, the residue classes of a number field modulo a prime ideal are finite in number.

Obviously

$$N(\mathfrak{a}\mathfrak{b}) = N(\mathfrak{a})N(\mathfrak{b}). \tag{2}$$

The *zeta function* is defined by the sum

$$\zeta(s) = \sum_{\mathfrak{a}} N(\mathfrak{a})^{-s} \tag{3}$$

taken over all integral divisors $\mathfrak{a}$ of K. We drop the problem of convergence for the moment, as it will solve itself in the quite explicit form in which we can write $\zeta(s)$ below. The uniqueness of prime decomposition of divisors together with (2) yields the *Euler product decomposition*

$$\zeta(s) = \prod_{\mathfrak{p}} (1 - N(\mathfrak{p})^{-s})^{-1} \tag{4}$$

taken over all prime divisors $\mathfrak{p}$. The prime divisors arise from the numerator divisors of the prime polynomials $p(x)$ in some variable x and the denominator divisor of x by means of the extension $K = k_0(x)$. Now, the number of prime polynomials of some fixed degree, indeed, the number of arbitrary polynomials of any degree, is finite. This means that there can be but a finite number of prime divisors in K of any degree. Hence the product (4) and with it the sum (3) is meaningful, the problem of convergence being postponed.

It becomes opportune, at times, to use the variables

$$u = q^{-s} \tag{5}$$

in place of s, and to write

$$Z(u) = \zeta(s). \tag{6}$$

The function

$$u \frac{d \log Z(u)}{du} = \sum_{n=1}^{\infty} N_n u^n \tag{7}$$

can be interpreted as follows: *for every n the coefficient N_n in the expansion gives the number of first degree prime divisors of the extended field Kk_{0n}, where k_{0n} is the familiar unique extension of k_0 of degree n.* In fact the definition of $Z(u)$ placed into the Euler products yields

$$u \frac{d \log Z(u)}{du} = \sum_{\mathfrak{p}} \sum_{\nu=1}^{\infty} g(\mathfrak{p}) u^{\nu g(\mathfrak{p})}, \tag{8}$$

so that

$$N_n = \sum_{g(\mathfrak{p})/n} g(\mathfrak{p}).$$

All prime divisors $\mathfrak{p}$ of K whose residue class fields are isomorphically represented in k_{0n}/k_0 decompose to prime divisors of the first degree in that field. But these are exactly the prime divisors whose degrees over k_0 divide n. And then, these $\mathfrak{p}$ decompose into $g(\mathfrak{p})$ first degree divisors.

2. The Functional Equation

The functional equation which we now want to derive for $\zeta(s)$ will turn out to be a result of the Riemann–Roch theorem.

The number of divisor classes of a given degree n is the same for all n. For, if $A_1, A_2,\ldots$ are differing classes of degree n, then $A_1 A_1^{-1} = (1)$, $A_2 A_1^{-1},\ldots$ are differing classes of degree 0, and if $B_1, B_2,\ldots$ are differing classes of degree 0, then $B_1 A_1, B_2 A_1,\ldots$ are differing classes of degree n. This number h is usually called the *class number* of K, and we will show it to be finite.

Evidently all natural integers occurring as degrees of integral divisors are multiples of an integer r. In III,§3,5 we saw that $r = 1$ or 2 in function fields of genus 0. Under the condition that the constant field k_0 is finite $r = 1$, since Eq. (14) in that paragraph is always solvable in such a k_0. We shall prove here that, with a finite k_0, $r = 1$ regardless of the genus. The result will be derived from properties of the zeta function, so we have to investigate the zeta function without this information. Yet one point is clear since the canonical class contains integral divisors and has degree $2g - 2$; r must be a divisor of $2g - 2$.

If the genus g of K is 0, then the class number $h = 1$ and there are $n + 1$ linearly independent integral divisors $\mathfrak{a}_1,\ldots, \mathfrak{a}_{n+1}$ in the class of degree n. All integral divisors of this class are of the form $\mathfrak{a} = \Sigma \varrho_i \mathfrak{a}_i$ with $\varrho_i \in k_0$, not all 0. As $\mathfrak{a}$ depends only upon the ratios of the ϱ_i, there are exactly $(q^{n+1} - 1)(q - 1)^{-1}$ such $\mathfrak{a}$. Thus,

$$\zeta(s) = \sum_{n=0}^{\infty} \frac{q^{n+1} - 1}{q - 1} q^{-ns} = \frac{1}{(1 - q^{-s})(1 - q^{1-s})}. \tag{9}$$

Now assume $g > 0$. We group the $\mathfrak{a}$, over which (3) sums, into classes A, and these into degrees. We separate the A of degree $g(A) \leqq 2g - 2$ from the rest, for which $\dim(A) = n - g + 1$, by the Riemann–Roch theorem. Such a class thus contains $(q^{n-g+1} - 1)(q - 1)^{-1}$ integral divisors, and there are h classes per degree. Thus the second sum is

$$\zeta_2(s) = \sum_{n > 2g-2} \sum_{g(A)=n} \frac{q^{n-g+1} - 1}{q - 1} q^{-ns}$$

$$= \frac{h}{q-1} \frac{q^{1-g+(2g-2+r)(1-s)}}{1 - q^{r(1-s)}} - \frac{h}{q-1} \sum_{n > 2g-2} q^{-ns}.$$

In these sums n runs through all integers $> 2g - 2$ which occur as degrees of integral divisors or, as we have seen, over all multiples of r which are $> 2g - 2$. This equation assures the finiteness of the class number.

The first partial sum is

$$\zeta_1(s) = \sum_{n=0}^{2g-2} \sum_{g(A)=n} \frac{q^{\dim(A)} - 1}{q - 1} q^{-ns}$$

$$= \frac{1}{q-1} \sum_{n=0}^{2g-2} \sum_{g(A)=n} q^{\dim(A)-ns} - \frac{h}{q-1} \sum_{n \leq 2g-2} q^{-ns},$$

where n runs through all multiples of r between 0 and $2g - 2$. We combine the last sums of $\zeta_1(s)$ and $\zeta_2(s)$ and see that

$$\zeta(s) = \frac{1}{q-1} \sum_{n=0}^{2g-2} \sum_{g(A)=n} q^{\dim(A)-ns} + \frac{h}{q-1} \left(\frac{q^{1-g+(2g-2+r)(1-s)}}{1 - q^{r(1-s)}} + \frac{q^{rs}}{1 - q^{rs}} \right),$$

which again has the form of two sums $\zeta_3(s)$ and $\zeta_4(s)$.

Obviously $\zeta_4(s) q^{(g-1)s}$ remains invariant when s and $1 - s$ are interchanged. We see the same for $\zeta_3(s) q^{(g-1)s}$:

$$\zeta_3(s) q^{(g-1)s} = \frac{1}{q-1} \sum_{n=0}^{2g-2} \sum_{g(A)=n} q^{\dim A - (n-g+1)s}.$$

Write $A' = WA^{-1}$ for the canonical conjugate class of A, whose degree $g(A') = n' = 2g - 2 - n$ must also lie between 0 and $2g - 2$. Its dimension is related to that of A by $\dim(A) = n - g + 1 + \dim(A')$. Hence

$$\zeta_3(s) q^{(g-1)s} = \frac{1}{q-1} \sum_{n'=0}^{2g-2} \sum_{g(A')=n'} q^{\dim(A') - (n'-g+1)(1-s)}.$$

Because A' traverse all these classes along with A the asserted invariance holds.

We finally also see that

$$q^{(g-1)s}\zeta(s) = q^{(g-1)(1-s)}\zeta(1-s), \tag{10}$$

where the case $g = 0$ need not be excluded due to (9).

This relationship leads further. There exists a polynomial $L(q^{-s}) = L(u)$ of degree $2g$ such that if $r = 1$ (see next section), then

$$\zeta(s) = \frac{L(q^{-s})}{(1-q^{-s})(1-q^{1-s})}. \tag{11}$$

$L(u)$ has rational integers as coefficients, and

$$L(u) = 1 + (N_1 - q + 1)u + \cdots + q^g u^{2g}, \tag{12}$$

where N_1 is the number of prime divisors of first degree of K. We also have the functional equation

$$q^{gs}L(q^{-s}) = q^{g(1-s)}L(q^{s-1}). \tag{13}$$

Proofs. Equation (13) is a simple consequence of (10) and (11). If $r = 1$, the above computations show that $L(u)$ is a polynomial $L(u) = \Sigma a_n u^n$ with rational integers a_n. Because of (13),

$$\sum a_n u^n = \sum a_n q^{g-n} u^{2g-n}.$$

Thus the highest power possible is u^{2g}. As clearly $a_0 = 1$, (13) yields $a_{2g} = q^g$. The coefficient of u is found by comparing (3) and (11). ▼

As long as we do not yet know that $r = 1$ we can only conclude, however, that, instead of (11),

$$\zeta(s) = \frac{L(q^{-s})}{(1-q^{-rs})(1-q^{r(1-s)})} \tag{11a}$$

with a polynomial $L(q^{-s})$.

$\zeta(s)$ has first-order poles at $s = 1$ and $s = 0$. The residue at $s = 0$ coincides with that of the partial sum $\zeta_2(s)$, and is thus

$$\frac{-1}{\log q^r}\frac{L(1)}{1-q^r} = -\frac{h}{1-q^r}\frac{1}{\log q^r},$$

so that

$$h = L(1). \tag{14}$$

3. Extension of the Field of Constants

If k_0 is replaced by the unique extension k_{0n} of degree n, a new field $K_n = Kk_{0n}$ of functions is found. A simple relation holds between its zeta function $\zeta_n(s)$ and $\zeta(s)$. Let some prime divisor $\mathfrak{p}$ of K be given. The residue class field modulo $\mathfrak{p}$ of the integral domain $\mathfrak{J}_\mathfrak{p}$ of functions of K integral at $\mathfrak{p}$ is an extension of k_0 of degree $g(\mathfrak{p})$. Choose a $\vartheta \in \mathfrak{J}_\mathfrak{p}$ whose residue class generates this field. There then exists an irreducible polynomial $f(\vartheta)$ of degree $g(\mathfrak{p})$ over k_0 such that

$$f(\vartheta) \equiv 0 \bmod \mathfrak{p}.$$

Now, $f(\vartheta)$ is separable because k_0 is perfect, so that suitable choice of ϑ within its residue class modulo $\mathfrak{p}$ assures that the numerator of $f(\vartheta)$ is divisible by $\mathfrak{p}$ only once. In the extension of k_0 to k_{0n} we know, from the theory of Galois fields, that $f(\vartheta)$ decomposes to $t = \text{g.c.d.}(n, g(\mathfrak{p}))$ distinct irreducible factors $f_\tau(\vartheta)$. This extension of constants therefore also decomposes $\mathfrak{p}$ to

$$\mathfrak{p} = \mathfrak{p}_1 \cdots \mathfrak{p}_t, \qquad t = \text{g.c.d.}(n, g(\mathfrak{p})).$$

The irreducibility of $f_\tau(\vartheta)$ makes the $\mathfrak{p}_\tau$ prime divisors. Divisor degrees remain invariant under extensions of constants, so that

$$g(\mathfrak{p}_\tau) = g(\mathfrak{p})t^{-1}, \qquad \tau = 1, \ldots, t,$$

There are q^n elements in k_{0n}, so that q must be replaced by q^n to form the zeta function of K_n.

The individual prime divisors of K_n arise from those of K as described. Thus $\zeta_n(s)$ is found by replacing each factor of the Euler product (4) by

$$(1 - q^{-g(\mathfrak{p})nt^{-1}s})^{-t} = \prod_{\nu=0}^{n-1} (1 - q^{-g(\mathfrak{p})(s - 2\pi i\nu/n\log q)})^{-1}.$$

This means that the zeta function of K_n is

$$\zeta_n(s) = \prod_{\nu=0}^{n-1} \zeta\left(s - \frac{2\pi i\nu}{n\log q}\right). \tag{15}$$

Writing this equation for the function (6) we have

$$Z_n(u^n) = \prod_{\nu=0}^{n-1} Z(u\, e^{2\pi i\nu/n}). \tag{16}$$

Applying (15) with $n = r$ to (11a) we find

$$\zeta_r(s) = \frac{L_r(q^{-rs})}{(1 - q^{-rs})^r(1 - q^{r(1-s)})^r},$$

which exhibits poles of order r; so $r = 1$ because of §5,2. ▼

Certainly the denominator of (11) satisfies the functional equation (15). This means that (15) also holds for the numerator $L(q^{-s})$ and (16) for the polynomial $L(u)$. Decomposing it to

$$L(u) = \prod_{\nu=1}^{2g} (1 - \omega_\nu u) \tag{17}$$

we have, for the corresponding polynomial of K_n,

$$L_n(u) = \prod_{\nu=1}^{2g} (1 - \omega_\nu{}^n u). \tag{18}$$

In particular this yields the equation

$$\sum_{\nu=1}^{2g} \omega_\nu{}^n = q^n + 1 - N_n, \tag{19}$$

where N_n is the number of first degree prime divisors of K_n. This equation takes on importance later.

In the decomposition (17) we should note that the ω_ν, which are the reciprocals of the zeros of $L(u)$, can be indexed so as to make

$$\omega_\nu \omega_{\nu+g} = q, \qquad \nu = 1, \ldots, g, \tag{20}$$

valid, because of the functional equation (13).

4. Riemann's Conjecture

Riemann's conjecture is literally identical to the conjecture for the Riemann zeta function. It states that the real component of any zero of the function (3) is equal to $\frac{1}{2}$. Here no "trivial zeros" occur, as for the Riemann zeta function. Combined with (11) and the decomposition (17) of the polynomial $L(u)$ this conjecture requires that all the ω_ν have the absolute value

$$|\omega_\nu| = \sqrt{q}. \tag{21}$$

To prove this we make use of the Frobenius correspondence F between K and an independent isomorphic field K', as considered in §1,8. First, the trace of F must be found. The automorphism of k_0 denoted κ in §1,8 is the identity under our present hypotheses, and F maps any $a \in K$ onto $a'^q \in K'$. As a traverses an integral domain $\mathfrak{J}$ of K, like that of §4,4, the $\mathfrak{J}$-ideal found as a representation of $\mathfrak{f}(F)$ contains all the differences $a - a^q$. (Of course, this ideal only represents those local components $\mathfrak{f}_\mathfrak{p}(F)$ of the divisors $\mathfrak{f}(F)$ for which $\mathfrak{J} \subset \mathfrak{J}_\mathfrak{p}$.)

Let $\mathfrak{p}$ be a first degree prime ideal not contained in the denominator of any function of $\mathfrak{J}$ and $\alpha \in k_0$ be the residue class of a mod $\mathfrak{p}$. Then $a - a^q = a - \alpha - (a - \alpha)^q$ is divisible by $\mathfrak{p}$. Choose a divisible by $\mathfrak{p}$ to exactly the first power; the same is then true of $a - a^q$, and then $\mathfrak{p}$ also goes into $\mathfrak{f}(F)$ exactly once. Suitable choice of $\mathfrak{J}$ then assures that all fixed points are included.

Now let $\mathfrak{p}$ be a prime divisor of higher degree that does not occur in the denominator of a function of $\mathfrak{J}$. Choose some $a \in \mathfrak{J}$ not congruent to any $\alpha \in k_0$ modulo $\mathfrak{p}$. Then $a - a^q \not\equiv 0$ mod $\mathfrak{p}$, so that $\mathfrak{p}$ does not divide $\mathfrak{f}(F)$.

We have shown that $\mathfrak{f}(F)$ is the product of prime divisors of the first degree of K, and therefore $\mathfrak{f}(F) = N_1$. Thus

$$s(F) = s(F^*) = q + 1 - N_1. \tag{22}$$

By Castelnuovo's inequality we know that

$$s((mD + F)(mD + F)^*) \geqq 0$$

for all rational integers m. Further, §1,(22) states that $FF^* = qD$ and Eqs. (2) and (7) in §4 state that $s(D) = 2g$. Altogether

$$2g(m^2 + q) + 2m(q + 1 - N_1) \geqq 0$$

must always hold. But this can only happen if

$$|q + 1 - N_1| \leqq 2g\sqrt{q}. \tag{23}$$

The same argument applied to the extension K_n yields

$$|q^n + 1 - N_n| \leqq 2g\sqrt{q^n}$$

which, because of (19), means that

$$\left|\sum_{\nu=1}^{2g} \omega_\nu^n\right| \leqq 2g\sqrt{q^n}. \tag{24}$$

We can now extract the result. Use (17) to form

$$\log L(u) = -\sum_{n=1}^{\infty} n\left(\sum_{\nu=1}^{2g} \omega_\nu^n\right)u^n.$$

This series converges absolutely for $|u| < q^{-1/2}$ because of (24). This means that all the $\omega_\nu \leqq (q)^{1/2}$. But, (20) shows that $q\omega_\nu^{-1}$ occur among the ω_ν, so that (21) must hold. ▼

Another result is also contained in our considerations and it serves to place Riemann's conjecture into the right perspective:

The class of the Frobenius correspondence F satisfies the equation $L'(F) = 0$, where $L'(u) = u^{2g}L(u^{-1})$. (Set $F^0 = D$.)

The proof requires Castelnuovo's inequality in the strong form of the principal theorem of §4. If F is taken as a correspondence of the extended field K_n, then its nth power

$$F^n = F_n$$

is the Frobenius correspondence of K_n. By replacing F with F_n in (22) we find that

$$s(F^n) = q^n + 1 - N_n.$$

Because of (22) we thus have

$$s(F^n) = \sum_{\nu=1}^{2g} \omega_\nu^n \tag{25}$$

for $n = 1, \ldots$. This equation is trivial for $n = 0$.

The F^{*n} also have the same traces as the F^n, so that §1,(22) implies that

$$s(F^{-n}) = s(q^{-n}F^{*n}) = (q^n + 1 - N_n)q^{-n} = \sum_{\nu=1}^{2g} \omega_\nu{}^n q^{-n}.$$

The fact that the $q\omega_\nu^{-1}$ also take on the values of all ω_ν [which is a consequence of the functional equation (13) and of (17)] shows that (25) holds for negative n as well.

We know that $L'(\omega_\nu) = 0$, so that (17) and (25) imply that

$$s(L'(F)F^n) = \sum_\nu \prod_\mu (\omega_\nu - \omega_\mu)\omega_\nu{}^n = 0$$

for all n. Renewed application of §1,(22) gives

$$s(L'(F)L'(F)^*) = s(L'(F)L'(qF^{-1})) = 0$$

so that $L'(F)$ is the zero class by the principal theorem of §4. ▼

The inequality (23), which is essentially equivalent to the Riemann conjecture, can be interpreted elementarily. Generate the field in question as $K = k_0(x, y)$ with $f(x, y) = 0$. Prime divisors of the first degrees not contained in the denominators of x and y map x and y onto elements ξ and η of k_0, and these satisfy the equation $f(\xi, \eta) = 0$. Conversely, let ξ, η be a pair with this property. Then $x \to \xi$, $y \to \eta$ defines a homomorphism of the integral domain generated by x and y over k_0, whose kernel is a prime ideal corresponding to a prime divisor of the first degree.

This shows that N_1 is essentially the number of solutions of $f(x, y) = 0$ in k_0, and only "infinitely distant" solutions would have to be defined and adjoined. Then (23) states that this number of solutions differs from $q + 1$ by at most $2gq^{1/2}$.

For the congruence $f(x, y) = x^2y^2 + x^2 + y^2 - 1 \equiv 0 \bmod p$ this was conjectured by Gauss.†

5. Modular Functions

The generalization of Ramanujan's conjecture mentioned in §3,6 can be derived from the Riemann conjecture and thus proved by the last section, insofar as it applies to modular forms $u(\tau)$ of dimension -2, that is more correctly to modular forms corresponding to first kind differentials $u(\tau)\, d\tau$. Certain preparation is necessary.

First we remark on the model of the field $K_{\Gamma(N)}$ of modular forms of level N. $K_{\Gamma(N)}$ is a Galois extension of the field $K = C(J(\tau))$ of modular functions of level 1, and its group is defined in IV, §4,(10) as $\mathfrak{M}(N)$. It was shown there

† G. Herglotz first proved this conjecture in: *Zur letzten Eintragung im Gaußschen Tagebuch*, Ber. Verh. Sächs. Akad. Wiss. Leipzig Math.-Phys. Kl. **73**, 271–276 (1921).

that $\mathfrak{M}(N)$ is faithfully represented by the permutations suffered by the classes $\Gamma(N)a_1/a_2$ of cusps. The following section showed the same permutations to act on the Eisenstein series, $G_k(\tau; a_1, a_2, N)$. Thus the second degree differentials $G_4(\tau; a_1, a_2, N)d\tau^2$ of $K_{\Gamma(N)}$, when transformed by substitutions of Γ, undergo a group of permutations which faithfully represents the Galois group. Therefore the quotients

$$w_a(\tau) = \frac{N^4}{360} \frac{G_4(\tau; a_1, a_2, N)}{G_4(\tau; 0, 0, 1)}$$

generate the field $K_{\Gamma(N)}$ over K_Γ.

From IV, §4,(23) and (24) we see that the $w_a(\tau)$ can be expanded into Fourier series $\Sigma c_{an} \exp(2\pi i n N^{-1}\tau)$ whose coefficients c_{an} are numbers in the Nth cyclotomic field $Q_N = Q(\exp[2\pi i N^{-1}])$.† Such a Fourier series, with coefficients even in Q, exists for the modular function $j(\tau)$ which generates K_Γ. Thus $K_{\Gamma(N)}$ is generated by functions having Fourier series expansions with coefficients in Q_N. These functions generate a subfield with constant field Q_N which shall again be denoted $K_{\Gamma(N)}$ in the sequel and be the object of our study.

Any function $\varphi(\tau)$ with Fourier coefficients in Q_N and invariant under $\Gamma(N)$ belongs to this field $K_{\Gamma(N)}$. For, let $\varphi(\tau) = fg^{-1}$ with polynomials f, g in the $w_a(\tau)$. Let the coefficients of f and g belong to the field L over Q_N, let ω_ν be a basis of L/Q_N and $f = \Sigma\omega_\nu f_\nu$, $g = \Sigma\omega_\nu g_\nu$, where f_ν, g_ν are polynomials in the $w_a(\tau)$ with coefficients in Q_N. Substitute the Fourier series for the $w_a(\tau)$ into

$$\varphi(\tau) \sum \omega_\nu g_\nu = \sum \omega_\nu f_\nu$$

and then compare coefficients, to find that

$$\varphi(\tau) g_\nu = f_\nu$$

which proves the assertion.

Now let $x = j(\tau)$, y_ν be a system of generators of $K_{\Gamma(N)}$; a system of polynomial equations $f_\mu(x, y_\nu) = 0$ must hold. By comparing coefficients after substitution of the Fourier expansions a system of linear equations with coefficients in Q_N is found, and the coefficients f_μ can be computed from these.

† Clearly, we must also show that the constant terms of the expansion, $(2\pi i)^{-4} \sum'_{m \equiv a_2(N)} m^{-4}$, belong to the field Q_N. But this follows from the Fourier expansion

$$x^2(1-x)^2 = \frac{1}{30} + \frac{24}{(2\pi i)^4} \sum_{n=1}^{\infty} \frac{e^{2\pi i n x} + e^{-2\pi i n x}}{n^4}, \qquad \text{for } 0 \leqq x \leqq 1,$$

after setting

$$(2\pi i)^{-4} \sum_{m \equiv a_2(N)} m^{-4} = \frac{1}{24N} \sum_{\nu=0}^{N-1} \left(\frac{\nu}{N}\right)^2 \left(1 - \frac{\nu}{N}\right)^2 \exp[-2\pi i a_2 \nu N^{-1}] \qquad (a_2 \not\equiv 0 \bmod N).$$

These coefficients may thus be assumed to be elements of Q_N. The equations $f_\mu(x, y_\nu) = 0$ define a finite extension of $Q_N(x)$ which becomes the field of all modular functions of $\Gamma(N)$ under the extension of constants $Q \to C$. The exact constant field of $K_{\Gamma(N)}$ is Q_N.

Second we note that the elements of Γ permute the functions $w_a(\tau)$, according to IV, §4,4. They thus define automorphisms of $K_{\Gamma(N)}$. The form §3,(5) of the modular correspondences $T(n)$ shows that they are correspondences of $K_{\Gamma(N)}$ with itself (or, in other words, that they are defined over Q_N). For the symmetric functions of $\varphi(U_a^{-1}\left(\begin{smallmatrix} a & b \\ 0 & d \end{smallmatrix}\right)(\tau))$ for any $\varphi(\tau) \in K_{\Gamma(N)}$ have Fourier coefficients in Q_N and are invariant under $\Gamma(N)$.

For a number r prime to N, IV, §4,(23) and the footnote on page 308 show

$$w_{ra}(\tau) = \frac{N^4}{360} \frac{G_4(\tau; a_1, ra_2, N)}{G_4(\tau; 0, 0, 1)}$$

to have Fourier coefficients derived from those of $w_a(\tau)$ by the automorphism σ_r; $\exp[2\pi i N^{-1}] \to \exp[2\pi i r N^{-1}]$ of Q_N. Now let λ be an element of Q_N such that the λ^{σ_r} (r varying mod N) form what is called a normal basis of Q_N/Q. By setting

$$w_a(\tau) = \sum_{(s,N)=1} \lambda^{\sigma_s} \varphi_{a,s}(\tau), \tag{26}$$

with Fourier series $\varphi_{a,s}(\tau)$ having rational coefficients,

$$w_{ra}(\tau) = \sum_s \lambda^{\sigma_s \sigma_r} \varphi_{a,s}(\tau). \tag{27}$$

As the determinant $|\lambda^{\sigma_s \sigma_r}| \neq 0$, the $\varphi_{a,s}(\tau)$ can be expressed linearly in the $w_{ra}(\tau)$, making them modular functions of level N.

Thus $K_{\Gamma(N)}$ can already be generated by modular functions with rational Fourier coefficients. The previous argument shows that these form a function field $K_{\Gamma(N)}^Q = Q(x, y_1, \ldots)$ with generating equations $f_1(x, y_1) = \cdots = 0$ having coefficients in Q. Adjoining the Nth roots of unity we get $K_{\Gamma(N)}$:

$$K_{\Gamma(N)} = K_{\Gamma(N)}^Q Q_N.$$

We consider this field as an extension of $Q(j(\tau))$. The elements of the Galois group are generated by two types of automorphisms: the automorphisms σ_r of Q_N/Q, operating on the Fourier coefficients of the functions, and the substitutions of Γ on τ. The effect of the former on the functions $w_a(\tau)$ is seen by comparison of (26) and (27):

$$w_a(\tau)^{\sigma_r} = w_{ra}(\tau).$$

From this equation we deduce that the automorphisms U_s defined by §3,(2) commute with the σ_r. *Therefore the U_s are automorphisms of the field $K_{\Gamma(N)}^Q$.*

This latter fact is quite important in what follows. The first consequence we derive from it is: *The T_n, T_n^*, as defined by* §3,(5) *and* (8), *are correspondences of the field $K_{\Gamma(N)}^Q$.* For the proof we have to form the ideal §1,(27) and to show that its Fourier expansions with respect to τ have rational coefficients. In the case of T_p, p a prime, we have (writing τ' instead of τ'/p does not matter in this connection)

$$\left(ux(\tau') + vy(\tau') - ux(U_p^{-1}(p\tau)) - vy(U_p^{-1}(p\tau))\right)$$
$$\times \prod_{b=0}^{p-1}\left(ux(\tau') + vy(\tau') - ux\left(\frac{\tau+b}{p}\right) - vy\left(\frac{\tau+b}{p}\right)\right),$$

where $x(\tau)$ and $y(\tau)$ generate the field $K_{\Gamma(N)}^{Q}$. Because U_p is an automorphism, $x(U_p^{-1}(\tau))$ and $y(U_p^{-1}(\tau))$ have rational Fourier expansions in τ, as does the rest of our product.

Because the T_n are products of the T_p they are also correspondences of $K_{\Gamma(N)}^{Q}$, and the T_n^* are derived from them by the Rosati antiautomorphism.

Incidentally, it is easy to show (cf. G. Shimura, §3,7,[9]) that $K_{\Gamma(N)}$ is normal over $Q(j(\tau))$ and that the Galois group is isomorphic to $GL(2, Z/N)/\{\pm\left(\begin{smallmatrix}1 & 0\\ 0 & 1\end{smallmatrix}\right)\}$. The subgroup belonging to $K_{\Gamma(N)}^{Q}$ is that generated by $\sigma_r = \left(\begin{smallmatrix}r & 0\\ 0 & 1\end{smallmatrix}\right)$.

Third we reduce the $K_{\Gamma(N)}^{Q}$ modulo a regular prime number p_0 to a field $\bar{K}_{\Gamma(N)}^{Q}$ of functions over the finite field $\bar{k}_0 = Q/p_0$ of constants. The theory developed in III, §6 must be used.

A homomorphism $\mathfrak{a} \to \bar{\mathfrak{a}}$ of the divisor group of $K_{\Gamma(N)}^{Q}$ into that of $\bar{K}_{\Gamma(N)}^{Q}$ was defined there for each regular p_0. It maps classes into classes of the same degree, and also maps correspondences and their classes. The divisor of fixed points of A is mapped onto the divisor of fixed points of $\bar{A}$. Finally, for the trace defined in terms of divisors degrees,

$$s(\bar{A}) = s(A). \tag{28}$$

We can now assert, in full generality: *a field K of functions over a finite algebraic number field k_0 reduces modulo a regular $\mathfrak{p}_0$ in such a manner that its ring of correspondence classes is mapped isomorphically onto a subring of the ring of correspondence classes of $\bar{K}$ (in the extended sense).*

Proof. The mapping is certainly a homomorphism. Let $A_1,\ldots$ be correspondence classes of K and $\bar{A}_1,\ldots$ their images. Let a polynomial relation $f(\bar{A}) = 0$, with rational integers as coefficients, hold. Then, $s(f(\bar{A})f(\bar{A})^*) = 0$.

By III, §6,6 residue formation modulo (a prime divisor of) $\mathfrak{p}_0$ and extension of constants commute. Multiplying correspondences amounts to the latter. Thus,

$$\overline{AB} = \bar{A}\bar{B} \tag{29}$$

for two correspondences A, B. By (28), then,

$$s(f(\bar{A})f(\bar{A})^*) = s(\overline{f(A)f(A)^*}) = s(f(A)f(A)^*) = 0. \tag{30}$$

This means $f(A)$ is the principal class, as was to be proved. ▼

6. The Eigenvalues of Modular Correspondences

Our further considerations depend upon the important equation

$$\overline{T_p^*} = F^* + \overline{U}_p F \tag{31}$$

which is valid for the reduction of $K^{\mathbf{Q}}_{\Gamma(N)}$ *modulo the same prime number* p, *with the sole assumption that* p *is regular in the sense of* III,§6. F *is the Frobenius correspondence and* U_p *the automorphism of* $K^{\mathbf{Q}}_{\Gamma(N)}$ *defined in* §3,1. (We now write p instead of p_0.)

Proof. Let $K = K^{\mathbf{Q}}_{\Gamma(N)}$ be the field of modular functions in τ, and let K' be a duplicate but independent field of modular functions in another variable τ'. By §3,1 and Eq. §1,(27) the ideal $(T_p{}^*)_{K \times K'}$ is generated as the product of the ideals which contain all the differences

$$\varphi(\tau') - \varphi\left(U_d\begin{pmatrix} a & b \\ 0 & d \end{pmatrix}(\tau)\right),$$

where a, b, d take on the values of §3,(6); that is, the matrices

$$\begin{pmatrix} a & b \\ c & d \end{pmatrix} = \begin{pmatrix} p & 0 \\ 0 & 1 \end{pmatrix}, \qquad U_p\begin{pmatrix} 1 & Nb \\ 0 & p \end{pmatrix}, \qquad b = 0, 1, \ldots, p-1$$

occur. Only such functions φ need be considered whose Fourier coefficients do not contain p in their denominator.

The Fourier series immediately show that

$$\varphi(p\tau) \equiv \varphi(\tau)^p \bmod p$$

or

$$\varphi(\tau') - \varphi(p\tau) \equiv \varphi(\tau') - \varphi(\tau)^p \bmod p. \tag{32}$$

We can then set $\varphi(U_d(\tau)) = \psi(\tau) = \Sigma c_n \exp[2\pi i n N^{-1}\tau]$ and form

$$\prod_b \left(\varphi(\tau') - \psi\left(\frac{\tau + Nb}{p}\right)\right) = \prod_b \left(\varphi(\tau') - \sum_n \zeta^{nb} c_n \exp[2\pi i n N^{-1} p^{-1}\tau]\right),$$

with a primitive pth root of unity ζ. The residue class of this expression modulo p is the same as that modulo the prime $\pi = 1 - \zeta$ of $Q(\zeta)$. But, $\zeta^{nb} \equiv 1 \bmod \pi$. Thus

$$\begin{aligned}\prod_b \left(\varphi(\tau') - \psi\left(\frac{\tau + Nb}{p}\right)\right) &\equiv \varphi(\tau')^p - \sum_n c_n \exp[2\pi i n N^{-1}\tau] \\ &\equiv \varphi(\tau')^p - \psi(\tau) \qquad \bmod \pi.\end{aligned}$$

As both sides of this congruence have only rational coefficients, it must also hold mod p, so that

$$\prod_b \left(\varphi(\tau') - \varphi\left(U_p\begin{pmatrix}1 & Nb\\ 0 & p\end{pmatrix}(\tau)\right)\right) \equiv \varphi(\tau')^p - \varphi(U_p(\tau)) \bmod p. \tag{33}$$

Because of (32) and (33) the residue class $\overline{T}_p{}^*$ is the sum of F^* and U_pF as asserted in (31). ▼

From (31) we can yet derive

$$\overline{T}_p = F + \overline{U}_p^{-1}F^* \tag{34}$$

by using §3,(5) and (8).

We saw in §4,9 that the correspondence classes of $\overline{K}^{Q}_{\Gamma(N)}$ form a semisimple ring. The elements $\overline{U}_p$, F, and F^* commute, so there exists a faithful representation of this ring in which those elements have diagonal form:

$$F = \begin{pmatrix}\varphi_1 & & \\ & \cdot & \\ & & \cdot \\ & & & \cdot\end{pmatrix}, \quad F^* = \begin{pmatrix}\varphi_1{}^* & & \\ & \cdot & \\ & & \cdot \\ & & & \cdot\end{pmatrix}, \quad \overline{U}_p = \begin{pmatrix}\zeta_1 & & \\ & \cdot & \\ & & \cdot \\ & & & \cdot\end{pmatrix}.$$

Moreover, by §1,(22), $\varphi_\nu\varphi_\nu{}^* = p$. The ζ_ν are roots of unity of an order $\varphi_{(N)}$ dependent upon N and p. The φ_ν are from among the ω_ν of §5,4. By (34) then, $\overline{T}_p$ satisfies the equation

$$f(\overline{T}_p) = \prod\left(\overline{T}_p - \left(\omega_\nu + \zeta\frac{p}{\omega_\nu}\right)\right) = 0,$$

where the product is taken over all hth roots of unity and all zeros of the polynomial $L'(u)$ of §5,4. The coefficients of the polynomial $f(\overline{T}_p)$ are rational integers. The last theorem of the last section shows that T_p also satisfies this equation. Hence the eigenvalues $\tau_\nu(p)$ of T_p must occur among the $\omega_\nu + \zeta p/\omega_\nu$ in every representation. By (21) these eigenvalues can be majorized by

$$|\tau_\nu(p)| \leqq 2\sqrt{p}. \tag{35}$$

Comparison of these results with those of §3,6 leads to the conjecture that the roots of unity ζ_ν that occur here in the eigenvalues $\tau(p)$ of T_p for some fixed character $\chi(n)$ are equal to the $\chi(p)$, and that the roots $\omega_\nu = \omega_\nu(p)$ have the same meaning here as in §3,6. This is, in fact, a consequence of §3,(34), as the $\omega_\nu(p)$ here are uniquely determined by the left side up to the possibility of interchanging $\omega_\nu(p)$ and $\chi(p)p\omega_\nu(p)^{-1}$. The question as to how the roots ω_ν of $L'(u) = 0$ are each associated with an eigenvalue $\tau_\nu(p)$ for modular forms of a given character is left open here, though. We must refer to the literature for the answer.†

7. Modular Functions to the Principal Character

A simplification of the problem is found by limiting it to the subfield $K^Q_{\Gamma(N),0}$ of functions invariant under all automorphisms U_p. Its differentials $du_i = u_i(\tau)\, d\tau$ yield modular forms $u_i(\tau)$ of the principal character, using the language of §3. In place of (34) we then have

$$\overline{T}_p = F + F^*, \tag{36}$$

and the eigenvalues of any representation of T_p, and in particular of the representation $(t_{ij}(p))$ of §3, are

$$\tau_\nu(p) = \omega_\nu + p/\omega_\nu, \tag{37}$$

where the ω_ν are certain zeros of the polynomial $L'(u)$. This is precisely the conjecture of §3,6.

We now assert, moreover, that $L'(u)$ is the g-row determinant

$$|(u^2 + p)E - u(t_{ij}(p))| = L'(u), \tag{38}$$

Because of (37) the zeros on the left occur on the right, as well, and the degrees and constant terms on both sides coincide. We must know, though, that even the multiplicities of the zeros are correct. This is found by considering the trace.

The trace of the matrix $(t_{ij}(p))$ is that of the representation §2,(8) by first kind differentials. It is real, as (37) and $p/\omega_\nu = \overline{\omega}_\nu$ show the eigenvalues to be real; thus it is one-half the trace $s(T_p)$ in §4,(2). The same holds for traces of powers of $(t_{ij}(p))$. By (26) and (36) we thus have

$$\operatorname{trace}(t_{ij}(p)^n) = \tfrac{1}{2}s((F + F^*)^n), \qquad n = 0, 1, \ldots. \tag{39}$$

By numbering the ω_ν as in (20), and using (25) and $FF^* = pD$, we see that

$$\tfrac{1}{2}s((F + F_\nu^*)^n) = \tfrac{1}{2}\sum_{\nu=1}^{2g} (\omega_\nu + p/\omega_\nu)^n = \sum_{\nu=1}^{g} (\omega_\nu + p/\omega)_\nu^n.$$

Then (39) implies that the $\omega_\nu + p/\omega_\nu$ are the eigenvalues of the $(t_{ij}(p))$ with the right multiplicities, proving (38).

We conclude with a remarkable consequence of (38). By §5,5, $K^Q_{\Gamma(N),0}$ is a field of functions over the exact constant field Q. The field $K^Q_{\Gamma(N),0}$ can be reduced modulo all regular prime numbers p, the zeta functions $\zeta_p(s)$ of all

---

† G. Shimura, *Correspondences modulaires et les fonctions ζ des courbes algébriques*, J. Math. Soc. Japan **10**, 1–28 (1958).

these residue class fields $\bar{K}^{Q}_{\Gamma(N),0}$ are formed, and their product is as follows:

$$\prod \zeta_p(s) = \prod_p (1 - p^{-s})^{-1} \prod_p (1 - p^{1-s})^{-1} \prod_p L(p^{-s}). \tag{40}$$

The absolute convergence of this infinite product for $\mathrm{Re}(s) > 3/2$ results from (21). The factors of the first and second kinds essentially give the Riemann zeta functions $\zeta(s)$ and $\zeta(s-1)$ if the fact that finitely many irregular prime numbers have been omitted from the product is ignored. The product of the $L(p^{-s})$ is, according to (38), the reciprocal of the determinant of the matrix zeta functions §3,(28), a finite number of factors again being left out. This reduces the product (40) to known functions satisfying functional equations of the Riemann type. But the Riemann conjecture proved in §5,4 for $\zeta_p(s)$ is not related to the classical Riemann conjecture for $\zeta(s)$.

NOTES

As already mentioned in §5,4, the history of the problems delved into here goes back to Gauss. After the proof by Herglotz of Gauss's conjecture, Artin [1] took up the study of the zeta function of quadratically generated fields. Schmidt [9] generalized the theory to arbitrary fields of functions. Our presentation in §5,1–3 follows Hasse [5]. Hasse [6] was first able to prove the Riemann conjecture for elliptic fields. In addition, Davenport and Hasse [2] could prove the conjecture in certain fields permitting explicit computation. The zeros of $\zeta(s)$ could be determined as special types of Gauss sums.

The turning point came when Weil (cf. the paper cited on page 298) recognized the connection of this problem with Castelnuovo's inequality. The papers which followed all use this idea, differing but in their proof of that inequality. This literature was discussed in §4,10.

The product (40) of zeta functions of fields $\bar{K}$ arising from reduction of a field K of functions over a finite algebraic number field modulo the regular $\mathfrak{p}_0$ was first brought up by Hasse. Weil [11] and Hasse [7] investigated these products for fields $K = k_0$ with $\alpha x^m + \beta y^n + \gamma = 0$, showing them to be meromorphic functions and to satisfy functional equations of familiar types. Deuring [3] achieved similar results for certain (so-called singular) elliptic fields. The work of Hasse and Deuring pays particular attention to the finitely many irregular $\mathfrak{p}$. Their contributions $\zeta_{\mathfrak{p}}(s)$ can be defined meaningfully, and they make the product (40) into a function of particularly simple type.

A third class of function fields for which the product (40) is essentially known is that of the modular functions considered in §5,5–7. The present author [4] attained the first of these results. They were soon generalized by Shimura (cf. footnote, page 313). Igusa (cf. III, §6,7,[3]) could show that only the prime divisors of the level N are exceptional irregular primes. Rangachari [8] studied the products of certain L series corresponding to (40) in the same manner for these fields.

A zeta function can also be defined for an algebraic variety of several dimensions (Weil [12]). In spite of the expected complications its construction is similar to (11), in that it is a quotient of certain polynomials in q^{-s} which are associated with the dimensions 0, 1,..., n, where n is the dimension of the variety. Weil presents a conjecture similar to Riemann's for the zeros of these polynomials, but to date only special cases have been verified. Due to Taniyama [10] certain classes of abelian varieties have been mastered, and their products of type (40) are known.

References

[1] E. Artin, *Quadratische Körper im Gebiet der höheren Kongruenzen II*, Math. Z. **19**, 153–246 (1924).

[2] H. Davenport und H. Hasse, *Die Nullstellen der Kongruenzzetafunktion in gewissen zyklischen Fällen*, J. Reine Angew. Math. **172**, 151–182 (1934)

[3] M. Deuring, *Die Zetafunktion einer algebraischen Kurve vom Geschlecht Eins*, Nachr. Akad. Wiss. Göttingen, Math.-Phys. Kl. *I*, 84–94 (1953); *II*, 13–42 (1955); *III*, 37–76 (1956); *IV*, 55–80 (1957).

[4] M. Eichler, *Quarternäre quadratische Formen und die Riemannsche Vermutung für die Kongruenzzetafunktion*, Arch, Math. **5**, 355–366 (1954).

[5] H. Hasse, *Über die Kongruenzzetafunktion*, S.-B. Preuss. Akad. Wiss. Berlin, Math.-Phys. Kl. XVII, (*1934*).

[6] H. Hasse, *Zur Theorie der abstrakten elliptischen Funktionenkörper*, J. Reine Angew. Math. **175** II: 69–88, III: 193–208 (1936).

[7] H. Hasse, *Zetafunktion und L-Funktionen zu einem arithmetischen Funktionenkorper vom Fermatschen Typus*, Abh. Deutsch. Akad. Wiss. Berlin Kl. Math. Allg. Naturwiss. (1954).

[8] S. S. Rangachari, *Modulare Korrespondenzen und L-Reihen*, J. Reine Angew. Math. **205**, 119–155 (1961).

[9] F. K. Schmidt, *Analytische Zahlentheorie in Körpern der Charakteristik p*, Math. Z. **33**, 1–32 (1931).

[10] Y. Taniyama, *L-functions of number fields and zeta-functions of abelian varieties*, J. Math. Soc. Japan **9**, 330–366 (1957).

[11] A. Weil, *On Jacobi sums as «Grössencharaktere»*, Trans. Amer. Math. Soc. **73**, 487–495 (1952).

[12] A. Weil, *Numbers of solutions of equations in finite fields*, Bull. Amer. Math. Soc. **55**, 497–508 (1949).

§6. Elliptic Function Fields

We conclude with a report on the correspondences of elliptic function fields. Although we are dealing with the oldest and most established part of the theory we cannot carry out proofs here. A complete presentation would require the presupposition of a broader basis of algebraic number theory than has been possible to include in the scope of this book. Another part of the theory, and in particular the more recent results, build upon the study of the group $\mathfrak{G}_l$ mentioned in §4,9. A decisive reason for waiving a complete presentation is seen in the following. The special importance attached to elliptic function fields within algebraic function theory is due to the fact that they are one-dimensional abelian varieties. Their special treatment not including a general study of abelian varieties would hardly seem appropriate today.

1. The Ring of Correspondence Classes

In the sequel K and K' will be two isomorphic but algebraically independent function fields of genus $g = 1$ over an algebraically closed field k_0 of constants. Every class of correspondences in the extended sense between K and K' other than the principal class contains a correspondence M of degree $g'(M) = 1$ (cf. end of §4,5). M maps K' isomorphically onto a subfield $K_0 \subseteq K$ of index $[K : K_0] = g(M)$.

Now, K and K' and thus K and K_0 are isomorphic. Thus an isomorphism of K onto its subfield K_0 is given by M. Such a mapping is called a *meromorphism*. Meromorphisms can exist for rational and elliptic function fields only, incidentally. For, by III, §3,2, the genera g and g_0 of K and K_0 and the different $\mathfrak{d}_{K/K_0}$ satisfy the equation

$$g - 1 = [K: K_0](g_0 - 1) + \tfrac{1}{2}g(\mathfrak{d}_{K/K_0}).$$

If K and K_0 were isomorphic and $g > 1$, then $g = g_0$ would follow, so that the degree of the different would be negative, impossible for an integral divisor. We also see that in the elliptic case $g(\mathfrak{d}_{K/K_0}) = 0$, so that K is an unramified extension of K_0.

We must take note of another important detail. The addition theorem of IV, §2,2 showed that the prime divisors of K of the first degree form an abelian group, and a meromorphism M must map this group onto a subgroup of the corresponding group of K_0. The factor group turns out to be of order $g(M)$, and also to be the Galois group of K/K_0, which is, as is seen, a Galois extension.

Composition can be used in the natural manner to define multiplication of meromorphisms. It corresponds to the multiplication of the correspondence classes involved. The addition of meromorphisms is defined by the addition of their correspondence classes, which in turn is nothing but the addition of the first degree prime divisors of KK'/K in the sense of the addition theorem of IV, §2,2. By excluding the "constant" prime divisors of KK'/K, that is, those of K', a ring of meromorphisms coinciding with the ring of correspondence classes is attained.

The theory of meromorphisms can be carried over to arbitrary abelian varieties, no distinguished position being held by the elliptic, that is, one-dimensional varieties. A very comprehensible presentation of the theory of meromorphisms for the elliptic case is given by Hasse in §5,8,[6].

The structure of the ring of meromorphisms or of correspondence classes is closely linked with the formula

$$MM^* \sim g(M)D. \tag{1}$$

In the proof of the lemma of §4,5 we saw that MM^* is the extension of the unit correspondence D_0 between the fields K_0 and K_0'' to a correspondence between K and K''. Thus, only

$$D_0 \sim g(M)D$$

need be proved. D_0, as a prime divisor of K_0K_0'', decomposes to

$$D_0 = D_1 + \cdots$$

in KK''. We know that K/K_0 is unramified and Galois, and that its group is the additive group of the prime divisors of K. Thus this group consists of translation automorphisms as considered in IV, §2,3. This means that the D_i are all equivalent in the extended sense, proving (1).

If only correspondence classes are to be considered,

$$MM^* = g(M)D \tag{2}$$

may be written in place of (1). MM^* is called the *norm* of M:

$$MM^* = n(M)$$

Clearly

$$n(M_1M_2) = n(M_1)n(M_2). \tag{3}$$

It is not difficult to see that no divisors of zero can occur. For, were $M_1M_2 = 0$, then either $n(M_1)$ or $n(M_2)$ would vanish, which would imply M_1 or $M_2 = 0$ by the principal theorem of §4.

A ring with a norm such as (3), associating a natural number with each element must belong to one of the following types:

(a) it is isomorphic to Z;
(b) it is isomorphic to an integral domain of integers of an imaginary quadratic number field (singular case); or
(c) it is isomorphic to an integral domain of integers of a definite quaternion algebra (supersingular case).

In fact, all these three types do occur among the rings of correspondence classes. The third case is excluded, though, if the characteristic is 0. For then the correspondence classes are faithfully represented by the only differential of the first kind:

$$du\, M = \mu\, du.$$

By constructing elliptic function fields to given periods it is easy to generate fields whose correspondence classes are given integral domains of imaginary quadratic number fields.

The third type of ring occurs only for fields of characteristic $p > 0$. In particular, Deuring† showed that only maximal integral domains of those quaternion algebras over Q ramified nowhere except ∞ and the prime p can occur. Conversely, any maximal integral domain of such an algebra really occurs as a ring of correspondence classes. The supersingular case is characterized by the first kind differential being mapped onto 0 by the Cartier operator (cf. III, §4,4).

2. Complex Multiplication

Let the constant field be $k_0 = C$ and the ring of correspondence classes be the principal order of an imaginary quadratic number field Λ. If $\mathfrak{k}$ is an ideal of Λ with the basis κ_1, κ_2 over Z, then the elliptic functions with base periods κ_1, κ_2 generate a field K with the given ring of correspondence classes. K depends only upon the class of $\mathfrak{k}$. The invariant $j(K)$ which uniquely characterized the field K in IV, §2,6 can thus be taken to be a function of the class of $\mathfrak{k}$, and written $j(K) = j(\mathfrak{k})$. To the h different classes of ideals of Λ there are h different fields K and invariants $j(\mathfrak{k})$.

It turns out that these $j(\mathfrak{k})$ are algebraically conjugate algebraic integers with the following remarkable properties: (a) they generate a Galois extension Λ_1 of degree h over Λ, whose group is isomorphic to that of the ideal classes of Λ; (b) a prime ideal $\mathfrak{p}$ of Λ decomposes in Λ_1 to prime ideals of residue class degree f, where f is the smallest exponent such that $\mathfrak{p}^f$ lies in the principal class; (c) Λ_1/Λ is unramified; (d) every ideal of Λ becomes a principal ideal in Λ_1; and (e) any prime ideal $\mathfrak{p}$ of Λ satisfies the congruences

$$j(\mathfrak{p}\,\mathfrak{k}) \equiv j(\mathfrak{k})^{n(\mathfrak{p})} \mod \mathfrak{p} \tag{4}$$

in Λ_1. These congruences are the key to the theory and lead rather quickly to properties (a) through (c), property (d) lying deeper. Equation (4) has the same roots, incidentally, as does §5,(34).

Along with the invariant $j(\mathfrak{k})$, the so-called *partial values* of elliptic functions play an equally outstanding role. They are the constants onto which

---

† *Die Typen der Multiplikatorenringe elliptischer Funktionenkörper*, Abh. Math. Sem. Univ. Hamburg **14**, 197–272 (1941). "Multiplikatoren" are correspondence classes.

the functions x, y are mapped by a first degree prime divisor $\mathfrak{p}$ of K, and between which the Weierstrass normal equation IV, §2,(21) is valid, insofar as $\mathfrak{o}$ can be represented by an integral multiple of $\mathfrak{p}$, in the sense of the addition theorem. These partial values also generate Galois extensions of Λ with abelian groups permitting individual description. It can even be shown (Kronecker's "Jugendtraum") that any extension of Λ with an abelian Galois group can be generated by partial values.

A presentation of this theory is found in M. Deuring, *Die Klassenkörper der komplexen Multiplikation*, Enzyclopädie d. math. Wiss., new ed., Vol. I(2), Part 10_{II}(23). Stuttgart, 1958.

As mentioned above, this theory permits an extensive analog for abelian varieties, which can be found in the monograph G. Shimura und Y. Taniyama, *Complex multiplication of abelian varieties*, Publ. Math. Soc. Japan **6** (1961).

Author Index

Numbers in parentheses are reference numbers and indicate that an author's work is referred to although his name is not cited in the text. Numbers in italics show the page on which the complete reference is listed.

Subject Index